Biotechnological Applications of Microorganisms

A Techno-Commercial Approach

Biotechnological Applications of Microorganisms

A Techno-Commercial Approach

Editors
D.K. Maheshwari
R.C. Dubey
S.C. Kang

I.K. International Publishing House Pvt. Ltd.
NEW DELHI • BANGALORE

Published by

I.K. International Publishing House Pvt. Ltd.
S-25, Green Park Extension, Uphaar Cinema Market
New Delhi - 110 016 (India)
E-mail : ik_in@vsnl.net

ISBN 81-88237-97-3

10 9 8 7 6 5 4 3 2

Published by Krishan Makhijani for I.K. International Publishing House Pvt. Ltd., S-25, Green Park Extension, Uphaar Cinema Market, New Delhi - 110 016 and Printed by Rekha Printers Pvt. Ltd., Okhla Industrial Area, Phase II, New Delhi - 110 020.

Preface

Recent concern over the commercial exploitation of potential microorganisms has necessitated to provide a holistic view due to their pivotal role in the biotechnology industry. Microorganisms thrive in diverse habitats in various ecosystems. Now-a-days microorganisms have been developed in different viable forms both *in situ* as well as indirectly in some form or the other. This facilitated the effective development of the bioinoculants industries. They are as effective as pure chemicals in addition having market-friendly and above all true to work assigned to them. Psychrophiles persist where others perish due to their extreme habits after comparing with standard models such as *E. coli* or *Bacillus subtilis.* Having secrets to microbial successes proved potential gold mine for industries. Lactic acids as microbial metabolites have been extensively discussed. Enzymes are machines to biotechnological innovations. Microbially-mediated enzymatic reactions have been found useful due to their environment friendly processes. Production of secondary metabolites such as ergot alkaloids proved as one of the emerging molecules in medicinal biotechnology. Similarly, fatty acids are used in a wide range of industrial products but their value enhanced if converted microbially. Actinomycetes and their potential products in the form of antibiotics still captured about 60% of market in the biotechnology industries. Beside, they have proved successful in control of harmful plants/ insect pathogens also being ecofriendly and cheap. Prevention and suppression of diseases in insects by use of entomopathogenic fungi renders safety for humans and non-target organisms. Their impact has necessitated the development of suitable bioformulation for field applications. Modern biotechnology has given some expectations from *Fusarium* as never before. Diversity of fusaria, their recent applications and new developments have diversified the directions for the commercial applications. Microbial production of 1-3, propanediol and its applications have been highlighted and industrial methods discussed for the production of a wide variety of aldehydes, organic acids and their derivatives.

On the other hand, the role of GM in crop production focusing on the input of transgenes from the microbial world may be the only solution to feed more world population. In spite of tremendous properties, the other side of coin of these microbes i.e. in disease development, cannot be ignored but microbes and their products in the form of novel compounds are the solution to act against soil-borne pathogens. Genomics or the organisms total DNA and its complete sequencing is a recent development to resolve various constraints in biotechnology-related industries including medicine, agriculture, food and environmental biotechnology-related business. Indeed, the largest number of commercial applications of microorganisms predominantly are applied in bioaugmentation, bioremediation of heavy metals, agrochemicals including pesticides. Even, fungal biomass suitable system act as bioabsorbents for removal and recovery of heavy metals from contaminated sites. Conventional therapeutic methods are now being well recognized throughout the world. Biological molecules in coffee beverages acted as antimicrobial agents to combat the ever- increasing menace of bacterial resistance due to their continuous applications. All the articles of this volume reflect the application of the microorganisms to industries. The processes and products may lead to industrial sectors so that the society may be truly benefited.

D.K. Maheshwari
R.C. Dubey
S.C. Kang

Contents

Editors

D.K. Maheshwari started his teaching carrier as a lecturer in the Department of Botany, D.A.V. (PG) College, Muzaffarnagar and thereafter served as Reader in the Department of Microbiology, Barkatullah University, Bhopal. He joined Gurukul Kangri University, Haridwar as Professor in the year 1990 and served as Dean, Faculty of Life-Sciences (2004-2006). Prof. Maheshwari is an active member of several scientific bodies of international repute and in the board of panels of various academic and administrative bodies including UPSC and NAAC of Govt. of India. He has been elected as Editor of the *Journal of Indian Botanical Society* (1999-2002). He was an awardee of young scientist "Prof. Y.S. Murthy Medal" for his outstanding contribution.

As a young scientist, he was selected under UNESCO programmes and worked in Biological Research Centre, Szeged (Hungary) in the year 1983-84 and also visited Czechoslovakia, Belgium, Holland, Germany, Japan and S. Korea. He was visiting Professor in Science University of Tokyo, in 1993 and 1998, and Guest Professor, in the University of Ulm (Germany). Prof. Maheshwari visited S. Korea in the year 2000, 2003 and 2006 under bilateral scientific exchange programmes sponsored by Indian National Science Academy (INSA), New Delhi.

Prof. Maheshwari has many research papers in leading peer-reviewed journals both Indian as well as foreign. More than 25 candidates have been awarded Doctorate (Ph. D.) degree under him. He has completed 10 major research projects sponsored by various funding agencies. Prof. Maheshwari is the editor of *Microbes: Agriculture, Industry and Environment* (2000) and *Innovative Approaches in Microbiology* (2002). He is co-author of *Practical Microbiology* (2001) and *A Text Book of Microbiology* (2nd ed, 2005) and member Editorial board, *Korean J. Agric. Chem. and Biotechnology*.

He is a Fellow of the Indian Botanical Society and Indian Phytopathological Society and life-member of Indian Science Congress, Phytopathological Society of India, Indian Botanical Society and Association of Microbiologists in India.

R.C. Dubey, Professor, Deptt of Botany & Microbiology, earned M.Sc. and Ph.D. degrees from the Banaras Hindu University in 1981 and 1986, respectively. He served Kumaun University (Nainital) as lecturer from 1987 to 1996; thereafter, he joined Gurukul Kangri University (Haridwar) in 1996.

His field of interest is Soil Microbiology having expertise in biological control of soil-borne plant pathogens, mycorrhizae, rhizosphere microbiology and botanical pesticides, besides enzymology and heavy metal toxicity. He has published more than 70 research and review papers in the national and international journals of repute. He has produced 10 Ph.Ds and is supervising several research students on varied aspects of microbiology. He has handled Research Projects sponsored by UGC and CSIR.

He has authored three books viz., *Practical Microbiology* (2001), *A Text Book of Biotechnology* (4th edn., 2006) and *A Text Book of Microbiology* (2nd edn., 2005). He is also co-editor of *Himalayan Microbial Diversity* (1997); *Microbes: Agriculture, Industry and Environment* (2000) and *Innovative Approaches in Microbiology* (2002).

He was the Organising Secretary of the national seminar on Bioinoculants for Holistic Sustainable Rural Development jointly organized by the Department of Botany & Microbiology and DDU State Institute for Rural Development (U.P. Govt.) (1998).

He is the Life Member and Fellow of the Indian Botanical Society; Indian Phytopathological Society and the International Society for Conservation of Natural Resources. He was Associate Editor of the *Journal of the Indian Botanical Society* (2000-2004), and at present he is in the Editorial Board of the *Journal of Environmental Biology and Conservation*.

Sun Chul Kang earned his Doctorate degree in 1988 and thereafter pursued as Research Scientist at National Institute of Health (U.S.A). He joined as Professor in the Department of Biotechnology, Daegu University, Korea in 1988 and at present serving as Chairman.

He is recipient of a number of awards including best 30 scientists of Korea (2002) and outstanding Research Scientist award from Daegu University (2005). He has several papers in leading peer reviewed journals and granted US and Korean patents on different technologies involved in the area of Microbiology and Biotechnology.

Contributors

A. Mehta *Microbial Technology Laboratory, Department of Botany, Dr HS Gour University, Sagar-470 003, (India)*

A.Y. Al-Mahdi *Department of Microbiology, University of Pune, Pune-411007, (India)*

B.J. Naveena *Department of Microbiology, Osmania University Hyderabad-500 007, (India)*

B.P. Kapadnis *Department of Microbiology, University of Pune, Pune-411007, (India)*

Bhavesh Kumar *Department of Botany & Microbiology, Gurukul Kangri University, Haridwar-249404, Uttaranchal, (India)*

Byung-Hyuk Kim *Division of Civil and Environmental Systems Engineering, Korea Maritime University, Busan 606-791, (Korea)*

Chun Keun Lim *Division of Biological Environment, College of Agriculture and Life Sciences, Kangwon National University, Chuncheon 200-701K (Korea)*

D.C. Sharma *Department of Microbiology, University of Delhi South Campus, Benito Juarez Road New Delhi-110 0211, (India)*

D.K. Maheshwari *Department of Botany & Microbiology, Gurukul Kangri University, Haridwar-249404, Uttaranchal, (India)*

Daljit S Arora *Microbial Technology Lab., Department of Microbiology, Guru Nanak Dev University, Amritsar-143 005, (India)*

Doman Kim *Laboratory of Functional Carbohydrate Enzymes and Microbial Genomics, Faculty of Applied Chemical Engineering, Chonnam National University, Gwang-Ju, 500-757 (Korea)*

Duck Hwan Park *Division of Biological Environment, College of Agriculture and Life Sciences, Kangwon National University, Chuncheon-200-701K, (Korea)*

Eun-Seong Seo *Laboratory of Functional Carbohydrate Enzymes and Microbial Genomics, Department of Materials and Biochemical Engineering, Chonnam National University, Gwang-Ju, 500-757 (Korea)*

Gopal Reddy *Department of Microbiology, Osmania University Hyderabad-500 007, (India)*

Gurinder Jeet Kaur *Microbial Technology Lab., Department of Microbiology, Guru Nanak Dev University, Amritsar-143 005, (India)*

Hakryul Kim *Department of Animal Science and Biotechnology, Kyungpook National University, Daegu-702-701, (Korea)*

Hee-Kyoung Kang *Laboratory of Functional Carbohydrate Enzymes and Microbial Genomics, Engineering Research Institute, Chonnam National University, Gwang-Ju, 500-757 (Korea)*

Hwa-Ja Ryu *Department of Biochemistry, Biophysics and Molecular Biology, Iowa State University, Ames, IA, 50011 (USA)*

Jang Hyun Hur *Division of Biological Environment, College of Agriculture and Life Sciences, Kangwon National University, Chuncheon 200-701K (Korea)*

Jasmine Isar *Department of Microbiology, University of Delhi South Campus Benito Juarez Road, New Delhi-110021, (India)*

Javed Musarrat *Department of Agricultural Microbiology, Faculty of Agricultural Sciences, Aligarh Muslim University, Aligarh-202002, (India)*

Jeom Soon Kim *Alpine Agricultural Experiment Station, Pyeongchang, Gangwon-232-950, Korea*

Jin-Ha Lee *Laboratory of Functional Carbohydrate Enzymes and Microbial Genomics, Engineering Research Institute, Chonnam National University, Gwang-Ju, 500-757 (Korea)*

Jun Mo Cho *Division of Biological Environment, College of Agriculture and Life Sciences, Kangwon National University, Chuncheon 200-701K (Korea)*

Jung-Hye Choi *Division of Civil and Environmental Systems Engineering, Korea Maritime University, Busan 606-791, (Korea)*

K. Annapurna *Division of Microbiology, Indian Agricultural Research Institute, New Delhi-110012, (India)*

K.K. Joshi *Department of Botany & Microbiology, Gurukul Kangri University, Haridwar-249404, Uttaranchal, (India)*

Kimiyasu Isobe *Department of Agro-bioscience, Iwate University, 3-18-8 Ueda, Morioka-020-8550, (Japan)*

L. Agarwal *Department of Microbiology, University of Delhi South Campus Benito Juarez Road, New Delhi-110021, (India)*

M. Senthil Kumar *Division of Microbiology, Indian Agricultural Research Institute, New Delhi-110012, (India)*

M.S. Shejul *Department of Microbiology, University of Pune, Pune-411007, (India)*

Mahejibin Khan *Department of Microbiology, College of Basic Sciences Humanities, G.B.P.U.A.&T., Pantnagar, (India)*

Manisha Kapur *Institute of Microbial Technology Sector 39-A, Chandigarh-160 036, (India)*

Md. Altaf *Department of Microbiology, Osmania University Hyderabad-500 007, (India)*

Meenu Saraf *Department of Microbiology, Gujarat University, Ahmedabad-380009, (India)*

Myung-Hyun Hong *Division of Civil and Environmental Systems Engineering, Korea Maritime University,* Busan 606-791, (Korea)

Myung-Suk Ha *Engineering Research Institute, Chonnam National University, Gwang-Ju, 500-757 (Korea)*

N.N. Nawani *Department of Microbiology, University of Pune, Pune-411007, (India)*

P. Dureja *Division of Agricultural Chemicals, Indian Agricultural Research Institute, New Delhi-110012, (India)*

P. Mehta *Microbial Technology Laboratory, Department of Botany, Dr HS Gour University, Sagar-470 003, (India)*

Pallapotu Vikrant *Fungal Biotechnology and Invertebrate Pathology Laboratory, Department of Biological Sciences, R.D. University, Jabalpur- 482001, (India)*

Piyush Pandey *S.B.S.P.G. Institute of Biomedical Sciences and Research, Balawala, Dehradun- 248161, (India)*

Purnima Anand *Department of Microbiology, University of Delhi South Campus, Benito Juarez Road, New Delhi-110 021, (India)*

R.C. Dubey *Department of Botany & Microbiology, Gurukul Kangri University, Haridwar-249404, Uttaranchal, (India)*

R.K. Jain *Institute of Microbial Technology Sector 39-A, Chandigarh-160 036, (India)*

R.K. Saxena *Department of Microbiology, University of Delhi South Campus Benito Juarez Road, New Delhi-110021, (India)*

Ram Charan Rajak *Director (Academics), Seedling Academy of Design, Technology and Management, Seedling Campus, Jagatpur Jaipur- 302 025, (India)*

Reeta Goel *Department of Microbiology, College of Basic Sciences Humanities, G.B.P.U.A.&T., Pantnagar, (India)*

Sabina Zaidi *Department of Agricultural Microbiology, Faculty of Agricultural Sciences, Aligarh Muslim University, Aligarh-202002, (India)*

Sardul Singh Sandhu *Fungal Biotechnology and Invertebrate Pathology Laboratory, Department of Biological Sciences, R.D. University, Jabalpur- 482001, (India)*

Saurabh Saran *Department of Microbiology, University of Delhi South Campus Benito Juarez Road, New Delhi-110021, (India)*

Se-Eun Lee *Division of Civil and Environmental Systems Engineering, Korea Maritime University, Busan 606-791, (Korea)*

Soon Wo Kwon *Division of Molecular Genetics, National Inst. of Agri. Science and Technology, Suwon 441-707 (Korea)*

Sumeet Labana *Institute of Microbial Technology Sector 39-A, Chandigarh-160 036, (India)*

Sun Chul Kang *Department of Chemical Engineering and Biotechnology, Daegu University, Kyungsang, Daegu-702-701, (Korea)*

Sung-Cheol Koh *Division of Civil and Environmental Systems Engineering, Korea Maritime University, Busan 606-791, (Korea)*

Sung-Kyu Shin *Division of Civil and Environmental Systems Engineering, Korea Maritime University, Busan 606-791, (Korea)*

Sung-Woo Shin *Division of Civil and Environmental Systems Engineering, Korea Maritime University, Busan 606-791, (Korea)*

T. Satyanarayana *Department of Microbiology, University of Delhi South Campus, Benito Juarez Road, New Delhi-110 0211, (India)*

1

Constraints of Bacterial Bio-Inoculant Production and its Quality Assurance

Bhavesh Kumar, K.K. Joshi, R.C. Dubey and D.K. Maheshwari
Department of Botany & Microbiology, Gurukul Kangri University, Haridwar-249404 Uttaranchal, (India)

Farming based on chemical fertilizers is known to disturb environment, subvert ecology, deteriorate soil-productivity and mismanage water resources. Nowadays, attempts are being made to blend ecology and economy with bio-inoculants in a cost-effective framework. More than a century has passed since the development of the first bio-inoculant, but it lags still far behind as compared to a general development of mankind. It seems clear that world-wide production of inoculants is static or in a declining trend. This is a paradoxical situation where the improved strategies for augmenting biological nitrogen fixation have the potential to contribute enormous returns in the quality and quantity of food for humans and domestic animals.

This review assesses from the history to the future of the bio-inoculant development apart from reasons for the slow and not so effective development in the bio-inoculant industry. It has also been emphasized to evaluate what is available and what should be explored in the future to uplift the stature of the bio-inoculants? For being competent to chemicals, the inoculants apart from being eco-friendly, should also be market-friendly, and above all true to work assigned to them.

INTRODUCTION

Inoculation of plants with beneficial bacteria can be traced back to centuries. Farmers knew when they grew a crop after legume, its yield often increased. Hence, from experience they could know the value of crop rotation. By the end of 19th century, the practice of mixing 'naturally inoculated' soil with seeds became a recommended method of legume inoculation in the U.S.A. (Smith, 1992). Since 600 B.C., practice for increasing microbial population in cultivated field has been done by green manuring in India and China without the known causes of enhanced soil fertility and crop yield. However, inoculation of specific microbial preparations for improved crop yield is a common practice in many regions of the world including India. In 1895, for the first time, an inoculant 'nitragin' containing rhizobial cells was marketed as a commercial product (Nobbe and Hiltner, 1896).

At present rhizobial inoculants are well established as agricultural products in the U.S.A.,

Europe, Australia and Asia. In 1937, an *Azotobacter* preparation, under the trade name 'Azotogen' and later named 'Azotobaktrin' was commercialized in the Soviet Union (Russia) for soil and seed treatments. Increased crop yield was reported for vegetables and cereals. In 1959 Soviet microbiologists introduced another product 'phosphobakterin' consisting of *Bacillus megaterium* var. *phosphaticum,* with a view to enhance phosphorus availability. Application of *Azospirillum brasilense* in Brazil led to its use as an inoculum for cereals (Dobereiner *et al.,* 1976). More recently, *Pseudomonas fluorescens* and *P. syriangae* under the name Blight Ban A506 and Bio-Save10, respectively have been commercialized due to their plant growth-promoting activities (Glick *et al.,* 1999; Chet and Chertnin, 2002). During 1996 the first microbial combination became available. In recent years, various wild and recombinant microorganisms providing a wide range of nutrients as well as protection to plants are being used as inoculants. Besides, numerous biocontrol agents were explored, developed and produced as microbial weapons to be applied in crop fields for better growth and yield.

PLANT GROWTH-PROMOTING RHIZOBACTERIA (PGPR)

The bacteria useful to plants are characterized in two general types: bacteria forming a symbiotic relationship with the plant, and the free-living ones found in the soil but are often found near, on or even within the plant tissues (Kloepper *et al.,* 1988; Frommel *et al.,* 1991). Beneficial free-living soil bacteria that enhance plant growth are usually referred to as 'plant growth promoting rhizobacteria' (PGPR) (Kloepper *et al.,* 1989) or yield increasing bacteria (YIB) (Piao *et al.,* 1992). PGPR originally defined (Kloepper and Schrath, 1978) as root colonizing bacteria (rhizobacteria) that cause either growth promotion or biological control of plant diseases. Bashan and Holguin (1998) proposed for the division of the plant growth promoting rhizobacteria into two categories, biocontrol-PGPB (plant growth promoting bacteria) and PGPB. A number of different types of bacteria have been considered PGPR that are associated with plant growth promotion such as *Azotobacter, Azospirillum,* bacilli, *Burkholderia, Enterobacter, Flavobacterium,* pseudomonads, *Gluconacetobacter, Serratia* etc. (Brown, 1974; Elmerich, 1984; Kloepper *et al.,* 1989; Bashan and Levanony, 1990; Okon and Labondera-Gonzalez, 1994; Tang and Yang, 1997).

PLANT GROWTH-PROMOTING ACTIVITIES

The mechanisms of plant growth-promotion by non-pathogenic, plant-associated bacteria have not been completely elucidated, but the important mechanisms are categorized into the direct and indirect plant growth-promoting activities. There are several mechanisms by which different PGPRs have been reported to facilitate directly and indirectly the proliferation of plant host and antagonist to phytopathogens.

Nitrogen Fixation

More than 100 million tons of fixed nitrogen are needed annually to sustain global food production. Chemical fertilizers account for about half of nitrogen supply and most of the remainder is derived from nitrogen-fixing bacteria called diazotrophs. These bacteria belong to the genera: *Allorhizobium, Azorhizobium, Bradyrhizobium, Mesorhizobium, Rhizobium, Sinorhizobium, Azotobacter, Azospirillum, Gluconacetobacter, Herbaspirillum, Burkholderia.* All these bacteria form a host-specific symbiosis with leguminous plants. The symbiosis is initiated by the formation of root or stem nodules in response to the presence of the bacterium. Molecular signal molecules (lipo-oligosaccharide) secreted by these bacteria play a critical role in this process (Lang, 2000; Spaink, 2000; Perrot *et al.,* 2000).

Association of rhizobia with legumes is well known. But it was Trinic (1988) who highlighted the existence of rhizobia symbiosis beyond leguminaceae. Rhizobia, isolated from legumes can be used as potential PGPR with non-legumes (Antaun *et al.,* 1998). Several rhizobia have also proven their abilities by nodulating the non-legumes (Trinik and Hadobas, 1995).

Phosphate Solubilization

The biogeochemical cycle of phosphorus is influenced by various microorganisms which affect phosphorus metabolization in soil. The precipitated inorganic phosphate is solubilized by the action of minerals and organic acids produced by soil bacteria (Seshadri *et al.,* 2000; Deshwal *et al.,* 2003) and other organisms. Organic compounds containing phosphorus are decomposed and mineralized by bacteria and other group of microorganisms. Several bacteria such as *Azotobacter, Bacillus, Clostridium,* pseudomonads and rhizobia are reported to solubilize inorganic phosphorus (Gupta *et al.,* 1999; 2001; 2002; Arora *et al.,* 2001; Deshwal *et al.,* 2003).

Plant Growth Hormones

A large number of studies done so far indicate the involvement of plant hormones produced by a bacterial inoculant in altering plant growth and development. There are different types of plant hormones that are secreted from microorganisms such as indole acetic acid (Mordukhova *et al.,* 1991; Gupta *et al.,* 1999), gibberrelic acid (Mahmoud *et al.,* 1984), cytokinins (Tien *et al.,* 1979) and ethylene (Arshad and Frankenberger, 1992; Glick *et al.,* 1995).

Some plant growth-promoting rhizobacteria, namely, *Enterobacter cloacae* (Li and Glick, 2001), *Pseudomonas* spp. (Honma and Shimomura, 1978; Glick *et al.,* 1998); *Rhizobium* spp. encode the enzyme 1-aminocyclopropane-1-carboxylate (ACC) deaminase (Jacobson *et al.,* 1994; Glick *et al.,* 1994a,b). This enzyme hydrolyzes ACC, which is the immediate precursor of ethylene synthesized in plant tissues (Yang and Hoffmann, 1984). Paradoxically in bacteria, a pathway that probably synthesizes ethylene does not include ACC (Aeshad and Frankenberger, 1992). This enzyme might act to stimulate plant growth via root elongation by sequestering and then hydrolyzing ACC from germinating seeds, thereby lowering the level of ACC, and hence lowering the level of ethylene in seeds (Fig.1. 1) (Glick *et al.,* 1994).

Siderophore Production

Siderophores are the low molecular weight (~400–1000 Da) iron-chelating compounds that bind Fe^{3+} and transport back to the microbial cell. Iron is not available to microbial cells in Fe^{2+} form (Briat, 1992). The secreted siderophore molecules have a very high affinity (kd = 10^{-20} to 10^{-50}) for iron-binding. Siderophore production has been reported by a number of workers in different groups of plant growth-promoting bacteria such as rhizobia (Arora *et al.,* 2001; Saraf and Sood, 2002; Deshwal *et al.,* 2003), pseudamonads (Gupta *et al.,* 2002; Bhatia *et*

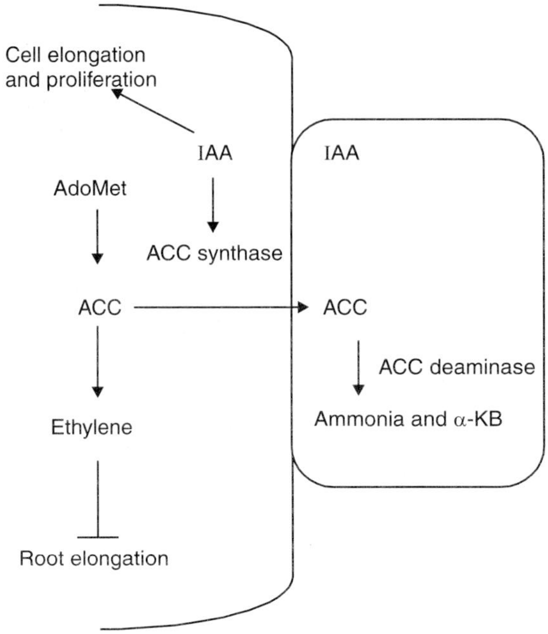

Fig. 1. Schematic representation of how a plant growth-promoting bacterium binds to either a seed or plant root lowers ethylene concentration, and thereby, prevents ethylene inhibition of root elongation (Adopted from Glick *et al.*, 1998)

al., 2003), *Azotobacter* (Budzikiewicz *et al.*, 1992) and *Azospirillum* (Shah *et al.*, 1992). There are different type of siderophores such as hydroxamate, catechol, pyoverdin, etc, reported in different groups of microorganisms.

Antibiotics

One of the most effective mechanisms, PGPR employ to prevent proliferation of phytopathogens, is the synthesis of antibiotics. A large number of antibiotics are reported to be produced by pseudomonads including agrocin-84, agrocin-434, 2,4-diacetyl phloroglucinol, herbicolin, oomycin, phenazine, pyoluteorin, pyrrolnitrin, pyrroles, etc. Growth inhibition of pathogens by antibiotics producd by bacteria has also been reported (Colyer and Mount, 1984; Gutterson *et al.*, 1986; James and Gutterson 1986; Kerr, 1980). Recently, Kang *et al.* (2004) isolated 2-Hydroxy-methyl-chromo-4-one produced by *Burkholderia* sp. MSSP1 showing antifungal activities. Antibiotics from rhizobia are not new as several antibiotics have been reported such as rhizobitoxin from *Bradyrhizobium* spp. (Chakraborty and Purkayastha, 1984; Deshwal *et al.*, 2003), anthranilic acid from *R. leguminosarum* (Rioux *et al.*, 1983), 4,3-Hydroxy-2,3,9-trimethoxy pterocarpan phytoalexin from *R. leguminosarum* bv *viciae* and and trifolitoxin (TFX) from *R. leguminosarum* bv *trifoli* T24 (Chakraborty and Chakraborty, 1988).

Lytic Enzymes

Microorganisms capable of lysing the other organisms are widespread in natural ecosystems.

Hyperparasitism and lysis of propagules in soil is a logically satisfying method of biological control of soil-borne plant pathogens by microbial antagonists. This mechanism operates through the production of lytic enzymes such as chitinase (EC 3.2.1.14), β-1,3-glucanase (EC 3.2.1.39), proteases (EC 3.4.21.14) and lipase (EC 3.1.1.3) (Fridlender *et al.*, 1993; Lim and Kim, 1995; Dunne *et al.*, 1997) that degrade fungal cell walls, and thereby, lyse the cells.

Cyanogens

It has been suggested that the ability of some PGPR to synthesize hydrogen cyanide, to which they are resistant, may be linked to the ability of those strains to inhibit some pathogenic fungi. Cyanides play a role in the suppression of root rot in tobacco caused by *Thilaviopsis basicola* (Ahl *et al.*, 1986). Cyanide produced by rhizosphere pseudomonads has detrimental effect on plant establishment in several crops but it also suppresses root pathogens (Defago *et al.*, 1990). However, rhizobia do contain low HCN in comparison to that of pseudomonads and only 12.5% and 3% strains were found by Beauchamp *et al.* (1991) and Antoun *et al.* (1998), respectively to be HCN producers.

Induced Systemic Resistance

Introduction of plant defense genes by prior application of inducing agents is called induced resistance (Hammerschmidt and Kus, 1995). The defense gene products include peroxidase, polyphenol oxidase, etc. which catalyze the formation of lignin and phenylalanine ammonia-lyase, involved in phytoalexin and phenolic synthesis. Other defense enzymes include pathogenesis-related proteins (PRP) such as β-1, 3 glucanase, chitinases, etc. that degrade the fungal cell wall and lyse the cells (Bloemberg and Lugtenberg, 2001).

Hydrolysis of Toxins

Several phytopathogenic fungi, namely, *Fusarium* spp. produce toxins (fusaric acid) harmful to the host plants and some bacteria, such as strains of *Cladosporium werneekii, Burkholderia capacia* and *Pseudomonas salanacearum*, have been reported to hydrolyse the fusaric acid (Toyoda and Utsumi, 1991). As a consequence of the ability to hydrolyse fusaric acid, these bacterial strains can prevent wilt diseases in plants. It is the rhizosphere competent PGPR that is required to establish successfully in the rhizosphere region for any effective disease suppressive mechanism.

NEED FOR INOCULATION

Population density of plant growth-promoting rhizobacteria in the rhizosphere zone is the determinant factor for the need of the inoculum. Various abiotic and biotic factors influence the population of the PGPR. Therefore, continuous assessment of these beneficial bacteria is necessary. In case of rhizobia, soil may be devoid of effective strains or contain low number of effective strains or may contain high number of ineffective or partially effective strains. When should the soil be inoculated is a critical question and it has pondered at length. Although it should be stated that there are far less problems with inoculating the plants when not needed (i.e. over–inoculating) than not using inoculants and producing N-deficient crops (Herridge,

2002). Allen and Allen (1961) listed four indicators that, if positive, would necessitate inoculation: (*i*) the absence of same or symbiotically-related legume in the immediate past history of the land, (*ii*) poor nodulation when the same crop was grown on the land previously, (*iii*) when the legume followed a non-legume in the rotation, (*iv*) when the land was undergoing reclamation, and (*v*) the perfect indicator for the inoculum requirement is the number of strains in the soil. Population of rhizobia >1000 cfu/g soil is required for optimum nodulation and N_2 fixation. Thies *et al.* (1991) developed the simple functions to predict the need for inoculation based on the numbers of rhizobia in soil and soil nitrate levels. Time to time techniques for evaluation of the protocol to assess the need of inoculum have been developed by many workers (Bonish, 1979; Brockwell *et al.*, 1988; Ballard and Charman, 2000).

COMMERCIAL PRODUCTION

Although, rhizobial inoculants have been commercialized for a century, it is only recently that the first commercial preparation of PGPR has been introduced in the market. In India, use of *Rhizobium* inoculant was initiated in 1920, but the systematic production began after 1960 with gradual introduction of other biofertilizers. During 2001-2002, production of biofertilizer in India was roughly 10,000 metric tons whereas the production of *Rhizobium, Azotobacter, Azospirillum, Acetobacter*, PSB was 1603, 1553, 1258, 166 and 4088 metric tons, respectively.

Commercial production of any inoculant is decided by two factors viz., quality in terms of acceptable bacterial population surviving for definite time period (acceptable to standard) and cost of production. Quality and cost of the inoculant production depend on the characteristics of the selected carriers as well as the incubation period and the sterilization procedures (Khavazi and Rejali, 2002).

Inoculant-manufacturer must accept this reality and assure that the probability of successful inoculation is maximized, even though the end user may not understand the biological principle on which the product is based. The crop response to inoculation in most instances is due to the result of appropriate strain-selection and production of target organism (Muldoon *et al.*, 1980; Amarger. 1981; Brockwell *et al.*, 1988; Hume and Blair, 1992; Fesenko *et al.*, 1995). The final assessment of the strain should be made only after evaluation in the field soils with large population of indigenous bacteria. Another concern with manipulated strain is their genetic stability under field conditions over a period of time. Although in India, genetically modified strains are not allowed in field applications.

Singleton *et al.* (1997) suggested the appropriate roles of both private and public sectors in the production of inoculants. Such partitioning of roles is applicable to the industry. The involvement of private sector institutions in manufacturing and marketing and public sector in quality control and R & D, appear to be the most successful model for the production and use of inoculant. The future of the inoculant industry and its potential benefits for world-agriculture depends on improving inoculant quality, both, numerically and in terms of strain effectiveness, establishment of R & D and adaptation of new technology by the industrialists. The search for a new inoculant strain is an ongoing process. Therefore, new strains are required which should be more adaptable to soil and environmental constraints (Herridge, 2002).

General requirements for strains to be used as inoculants are: broad host-range effectiveness within an inoculant group, survival on seed surface, ability to be cultured on cheap nutrients and their genetic stability for performance purposes. The other desired characteristics include that the bio-inoculant must be able to tolerate physiological and ecological stresses besides their competitive ability against the indigenous strains and satisfactory growth and survival (Howieson *et al.*, 2000). Despite more than 100 years of bio-inoculant production, today the efforts made on the quality of inoculants produced all over the world are still improving (Olsen *et al.*, 1995).

A microbe may function optimally under laboratory conditions but formulating it into an end user affordable product, that is consistently able to bring about equivalent results under field conditions, is a difficult task. Top performing strains for a given market area must be compatible with the commonly grown crop cultivars in that area and should display the effective and efficient adaptability over a range of geographic and environmental conditions. Lynch (1983) stated that each microbial species have an optimum for each physical or chemical factors and its growth or activity will decline either side of that optimum governing its contribution to the total population. Microbial activities and performance often change according to soil texture (Osa-Afiana and Alexander, 1982), aeration, pH (Cassmane *et al.* 1981), temperature and available moisture (Sprent *et al.*, 1986), nutrient status, organic matter content (Semu and Hume, 1979), and weather. Sattar *et al.* (1995) reported that the most effective strain for a given crop in a given geographic region must always be isolated from that region. Hence, strategy of testing and evolving the indigenous strains should not be sidelined.

The selected strains for inoculation purposes in the form of inoculants are prepared in sterile carriers because the sterile carriers contain high number of desired organism as compared to non-sterile carriers. Although, due to high cost of sterilization (α-irradiation and autoclaving) process most of the inoculants produced in Asian countries are made by using non-sterile carriers. Brockwell *et al.* (1995) were pessimistic about the prospects for the inoculant industry and its capacity for large-scale production of high quality inoculants. Industrialists should adopt the newly developed R & D technology and a lot of skill is needed for development of effective inoculants complied with the quality standards. Unfortunately, in many countries thirty to forty years old technology is still executed or mastered.

Most of the bio-inoculants are prepared in powdered form and organic support material such as carriers. These act as the favoured base, particularly when sterilized by α-irradiation (Roughley and Vincent, 1967). Jawson *et al.* (1989) used several cellulose gels as inoculant carrier and reported excellent survival of *B. japonicum* and good nodulation of soyabean in sand culture and in the field where there was resident population of competing strains. On the other hand, Kremer and Peterson (1982) prepared inoculants by re-suspending lyophilized cultures of rhizobia in vegetable oil that survived on seeds which performed well in the field trials.

An Australian company has recently experimented the production of broth (liquid) inoculant. Sterile fermentation liquor was used as nutrient broth for cultivation of rhizobia to get a population of $>1 \times 10^9$ cells ml^{-1}. Lie *et al.* (1992) have recently described an analogous system. Preliminary experiments indicate that this form of broth inoculant has good characteristics during storage or on the seed, as the microorganisms adhere tenaciously to the seed coat without the need of adhesive. Byproducts of processed food such as cheese, whey, corn steep, liquor, malt sprouts, malt baggase have been reported to be acceptable carbon and nitrogen sources (Balatti *et al.*, 1991). Major texts on the principle and practice of inoculant preparation and use have been published by a large number of workers (Vincent, 1970; Burton, 1976; Date and Roughley, 977; Brockwell, 1977 and Thompson, 1980). Demonstrations of *Rhizobium* biofertilizers were compiled by Tilak *et al.* (1992) (Table 1). When the number of rhizobia in the broth has attained the required standard (10^8 - 10^9 cells ml^{-1}), the broth should be added to the carrier for the preparation of carrier-based inoculum.

Free-living PGPR Inoculants

Although *Rhizobium* inoculants have been in use for a century, yet several free-living rhizobacteria and other microorganisms have also been commercialized as bio-fertilizers and bio-pesticides. Free-living PGPRs are most intensively studied for agriculture applications. Researchers in the former Soviet Union and India conducted the widespread tests in the early middle of the 20th century studying the effects of PGPRs on different crops. Plant growth promotion by PGPR includes an increase in seed germination, root growth, crop yield, leaf area, chlorophyll content, nitrogen content, protein content, hydraulatic activity, tolerance to drought, root and shoot and delayed leaf senescence. Several studies showed that growth promotion effects are seen early in plant development, and subsequently translate into higher yield (Glick *et al.*, 1997; Polyanskaya *et al.*, 2000). Evidence of late season grain weight increase has also been reported in rice (Tranvan *et al.*, 2000). Commercial preparations using free living PGPRs are also available to administered crops (Table 2).

Inoculants for Bio-control

Two major breakthroughs in plant inoculation technology occurred in the late 1970s by involvement of the use of biocontrol agents, mainly of *Pseudomonas fluorescens* and *P. putida*

Table1: Optimum fermentation conditions for mass multiplication of *Rhizobium* strains (Tilak *et al.*, 1992)

Parameters		Specification
Type of reactor	:	Stirred tank
Type of operation	:	Batch
Carbon source	:	Sucrose and molasses (3-5 g L^{-1})
Nitrogen source	:	Corn steep liquor or yeast extract
pH	:	7.8 (controlled)
Temperature	:	28°C
Inoculum rate	:	10% v/v
Inoculum count	:	10^9 cells ml^{-1}
Antifoam	:	PPG
Aeration	:	0.5VVM
Agitation	:	Depends on fermentor size
Fermentation time	:	Variable (18-40 h)
Cell processing	:	Centrifugation/ membrane filtration followed by lyophilization
ISI specification	:	10^8 g^{-1} carrier at the time of manufacture
Shelf-life	:	6 months from the date of manufacture

groups (Defago *et al.*, 1992; Kloepper and Schroth 1981; Glick *et al.*, 1995). In recent years, various other bacterial genera such as *Bacillus, Flavobacterium, Acetobacter, Azospirillum* related microorganisms and some fungi such as *Trichoderma, Gliocladium* etc. (Table 3) have also been evaluated (Kleopper 1994a; Tang and Yang 1997; Maheshwari *et al.*, 2001).

Table 2: Commercially available inoculants using free-living plant growth-promoting rhizobacteria (Chet and Chertnin, 2002; Glick *et al.*, 1999)

Bacterial strains	Product	Intended crop
Agrobacterium radiobacter	Diegall Galltrol-A Nogall Norbac 84 C	Fruit, nut, ornamental nursery stock and trees
Azospirillum brasilense	Azo-Green	Turf and forage crops
Azospirillum brasilense Cd	Zea-Nit	Maize
Azospirillum lipoferum Br-17		
Azospirillum lipoferum CRT1	AZOGREEN-m	Maize
Bacillus amyloliquefaciens GB99	Quantum 4000 Bio Yield™	Broccoli, cabbage, canthaloupe, cauliflower, celery, cucumber, lettuce, ornamentals
Bacillus subtilis CG 122		
Bacillus subtilis	Epic, HiStick N/T, Kodiak, Rhizo-Plus, Serenade, Subtilix, System 3	Barley, beans, cotton, legumes peanut, pea, rice, and soya bean
Burkholderia cepacia	Blue Circle, deny, Intercept	Alfalfa, barley, beans, peas, sorghum, cotton, maize, vegetables and wheat
Pseudomonas fluorescens	BlightBan A506, Conquer, Victus	Almond, apple, cherry, mushroom, peach, pear, potato, strawberry and tomato
Pseudomonas syringae	Bio-save10	Citrus and poam fruit
Streptomyces griseovirdis K61	Mycostop	Field, ornamental and vegetable crops

Table 3: Commercially available bacterial inoculants for control of economic agricultural crops.

Product	Antagonist	Target pathogen	Intended crop	Source country
Blue circle	*Pseudomonas cepacia*	*Rhizoctonia solani, Pythium* sp.	Vegetable crops	Stine Microbials, Madison WI (USA)
Dagger G	*Pseudomonas fluorescens*	*Rhizoctonia solani, Pythium ultimum*	Cotton	Ecogen Inc, Langhorne PA (USA)
Epic	*Bacillus subtilis*	*Rhizoctonia solani, Fusarium* sp.	Cotton, legumes	Gustafson Inc, Dallas (USA)
Galltrol, Norbac 84-C	*Agrobacterium radiobacter*	*Agrobacterium tumefaciens*	Trees, shrubs	Ag Biochem, Inc. Urinda CA (USA)
Kodiak	*Bacillus subtilis*	*Pythium ultimum, Rhizoctonia solani*	Cotton, legumes	Gustafson, Inc. Dallas TX (USA)
Myostop	*Streptomyces griseovirides*	*Alternaria* sp., *Fusarium* sp.	Vegetable, ornamentals	Kermira Biotech, Helsinki, (Finland)
Victus	*Pseudomonas fluorescens* bv 5	*Phytopithora tolaasii*	Mushrooms	Sylvan, Kattanning, PA (USA), Millar, FC; Sylvan, PC)
Bakuterohzu	*Agrobacterium radiobacter*	Crown gall disease	Rose	Japan
Bio Coat	*Pseudomonas fluorescens*	*Fusarium* sp.	Radish	Australia

The use of PGPR to increase crop yield has been limited due to the variability and inconsistency of results between laboratory, greenhouse and field studies (Misustin and Naumora, 1962). In addition, response to soil inoculation with associative, non-symbiotic PGPRs varies considerably depending on the bacteria, plant species, soil types, inoculant density and environmental conditions. Soil is an unpredictable environment and an intended result is some times difficult to obtain (Bashan, 1998). Shortly after the introduction of bacteria into the soil, the bacterial population of undesirable bacterial strains declines progressively (van Elsas *et al.*, 1986; Bashan, 1998). This may prevent the build-up of sufficiently large number of PGPRs.

Bacterial Strains, Maintenance and Storage

Search for new inoculant strains is an ongoing process so as to maximize the overall productivity. New strains are required with the capability to adapt in a new soil environment besides having effective and versatile plant growth promoting activities. Keyser *et al.* (1992) outlined the desirable characteristics for rhizobial strains. According to them rhizobia should (*i*) form nodules and fix N_2 with the target legume, (*ii*) compete in nodule formation with earlier population of rhizobia already present in the soil, (*iii*) fix N_2 with a wide range of host genotypes and across different environments, (*iv*) form nodules and fix N_2 in presence of soil nitrate, (*v*) grow in artificial media, inoculant carrier and in the soil, (*vi*) persist in the soil, particularly for annually regenerating legumes, (*vii*) migrate from the initial site of inoculation, (*viii*) colonize the soil in the absence of the legume host, (*ix*) maintain genetic stability, and (*x*) be compatible with agro-chemicals.

Mother cultures of bacterial strains are stored at low temperatures under lyophilized condition. Sub-culturing may cause loss of some properties of culture; therefore, time-to-time comparison between mother and working cultures is essential. Bacterial cultures are maintained on their growth-supporting medium eg. rhizobia on yeast extract mannitol agar and pseudomonads on King's B agar medium, *Azotobacter* on nitrogen free medium, and other general PGPR on nutrient agar medium.

Carriers

Inoculation of soil by microbial inoculants is required for a number of applications such as plant growth-promotion, inhibition of plant pathogens, biodegradation of toxic compounds, soil structure improvement and microbial leaching of metals. To facilitate the introduction of high cell numbers and increased survival of microorganism in soil, inoculum formulations using carrier material have widely been used. The term 'carrier' is generally used for a medium which carries live microorganisms. According to Khavazi and Rejali (2000) carrier refers to the solid, semi-solid or liquid substances that can sustain a given number of the particular bacteria for a given period of time. Carrier materials may act to enhance the survivability of inoculants by providing microorganism with a protective environment in order to escape from unfavourable conditions in soil. Carrier materials must enhance survival of inoculants during storage and after their introduction in soil.

Properties of suitable carrier: Smith (1992) has noticed that carriers must display two fundamental properties: (*i*) they must support the life of target organism and (*ii*) they should maintain the desired populations of inoculant strains over an acceptable recommended period.

To achieve these goals a carrier must also display chemical and physical uniformity. They should be non-toxic to inoculant strains and environmentally safe and eco-friendly.

On the other hand, Keyser *et al.* (1992) stated that, a good inoculant carrier should: (*i*) have high water-holding capacity, (*ii*) be non-toxic to inoculants, (*iii*) be easily sterilizable to avoid contamination, and (*iv*) readily available in ample amount, (*v*) sufficiently act as adhesive for effective application to seed, (*vi*) have pH buffering capacity, (*vii*) contain anion-exchange capacities, (*viii*) be free from lump forming materials, (*ix*) be easily processed, (*x*) be non-toxic and biodegradable, (*xi*) provide nutrient for the microorganisms and favour their survival, both, during storage and inoculation, (*x*) be cost effective. During our observations, most of the carrier-based inoculant preparations did not show their desirable characteristics during field applications.

The available literature reveals that a single carrier cannot have all these properties. Till now very little work has been done on their economics. However, more and more raw and waste materials must be evaluated.

Type of carriers: Carrier materials being used worldwide can be categorized as following:

(i) Different soil materials including peat, peat mixed with other materials, coal or mixture of coal and other materials, mineral soils and mixture of mineral soils and other materials,

(ii) Plant materials including baggase, rice husk, plant composts, wheat straw, sawmill waste,

(iii) Neutral ingredients including vermiculite, mixture of vermiculite and other materials, perlite rock phosphate, calcium sulphate, charcoal, lignite and polyacrylic gel synthetics

(iv) Talc based carriers, and

(v) Liquid carriers.

Perlite (Daza *et al.*, 2000) and peat (Roughly, 1970) are widely accepted carriers. Khavazi and Rejali (2002) showed that perlite by itself or combined with baggase, coal, malt residue, rice husk, charcoal and calcium carbonate is able to maintain a bacterial population of more than 10^9 cells g^{-1} for a period of six months or more. Both, liquid and the solid based carriers are recommended but solid carrier based inoculants in contrast to a liquid carrier may increase strain survivality by reducing desiccation or heat stress to the cells—a major factor involved in the establishment of strains of rhizobia in soil (Hansen, 1994).

Although, liquid products are easier to handle, (Hynes *et al.*, 1995) liquid inoculant formulation is an easy alternative to small producer. It overcomes many problems associated with processing of solid carrier. It possesses a good field performance characteristic that uses low cost materials. Liquid inoculation equipment is inexpensive to fabricate. Australian operators prepare liquid inoculants as a suspension of peat culture in water, while frozen concentrated broth cultures have been used in the U.S.A. (Gault and Brockwell, 1980). On the other hand, liquid inoculants have some limitations such as limited shelf-life and high end user cost which is increased after cold temperature storage. The poor results in the liquid inoculant treatments could be due to decline in the number of bacteria resulting from desiccation during and after planting when it was generally dry and windy at planting sites.

Preparation of carrier: As per BIS specifications, the powdered form of carrier should pass through 150-212 micron (72-100 mesh) ISI sieve. Powdered form of inoculant supports better seed adhesion but granular form of inoculant is also quite popular. However, selection of inoculant type to be used is based on the need of farmers, purpose of inoculation, and type of equipment available with the farmers and the cost-benefit ratio.

Carrier processing: The processes that are needful and cost-effective aspects of inoculant production are drilling, drying, cleansing and milling (Somasegran and Hoben, 1994). Sun-dried carriers should bear a moisture level of 5%. It is mixed with finally powdered calcium carbonate to bring the pH 6.5 -7.0. The amount of calcium carbonate to be added depends on the acidity of the carrier. To be a good carrier, it is necessary to go through minimum processing to make it ready for inoculant production.

Sterilization of carriers: Now-a-days preference is given to sterile carrier material for the preparation of inoculum. Roughley and Vincent (1967) found that sterile carriers generally support higher populations and display much longer shelf-life. They demonstrated a 10-fold improvement in numerical quality by the use of peat carrier sterilized by irradiation. Bacterial survival is enhanced by incubating cells in sterile, autoclaved soil prior to introduction into non-sterile agricultural soil (van Dyke and Prosser, 1998). Date and Roughley (1977) stated that inoculants prepared with non-sterile peat may contain 100-fold fewer rhizobial cells than those made with sterilized peat, because mortality of rhizobia increases in non-sterilized peat during storage. Further, the use of sterile carrier significantly reduces threats to bio-inoculant quality due to the presence of contaminants.

The pre-treated carrier is filled upto two-thirds of the capacity of the containers (steel trays, 2×1-1/2×1/2 feet) and is sterilized at 121°C (15 lbs pressure inch^2) for 3-4 hrs. continuously or more each on three successive days (Somasegran, 1985). Although the process is laborious, costly and time-consuming, and in commercial context results in bio-inoculant quality compromised by the production of toxins during sterilization. However, sterilization of some carrier material like soil by autoclaving is known to release nutrients that can be utilized for bacterial growth. Other methods like nuclear and α-irradiation (Date and Roughly, 1977) and electron acceleration have been adopted to sterilize the carrier material but these methods are expensive and slow.

Electron acceleration is the most recent technology being used for the sterilization of carriers. This process is non-nuclear and depends on the exploitation of a series of acceleration cavities that generate 10×10^6 eV with which the carrier is excited and hence sterilized. The greatest benefit of electron beam sterilization is time-saving, because carrier material has to be exposed for a few minutes to electron beams (Stephenes and Rask, 2000).

Mixing and curing: The inoculants based on the sterile carriers do not require any sterilization. However, non-sterile carriers need to be sterilized before mixing the culture broth. The broth culture, which possesses a minimum population of 10^9 cells/ml, is blended with the finely powdered and sterilized carrier. Blending is carried out in ribbon or paddle type mixer or a concrete mixer (Smith, 1992). The broth is added to one-third the water-holding capacity of the carrier. The ideal moisture content (30% -60%) must support the growth of the micro-organisms, its survival and anticipated moisture loss from the packets during storage. After proper mixing of carrier with the broth culture, the blended carrier is kept for 24 hrs. for curing i.e. acclimatization of bacteria in the new environment of carrier. Curing is important to gain maximum number of cells or spores. The curing period depends on the growth rate of the microorganisms to be inoculated. During curing the inoculant is to be spread on a shallow tray at 20 to 30°C for 3 hrs. and covered with a plastic film so as to protect desiccation.

Packing and labelling: After curing and immobilization, inoculant is readily packed in polythene bags in the form of packets. Polythene bags are made up of 50-75 m thick low density and flexible pouch-type sheets 6×10 inches in size. The packing material should possess some properties such as: (*i*) stability towards (*i*) irradiation, (*ii*) autoclavablity, (*iii*) high gas exchange capacity, and (*iv*) high rates of moisture loss.

Inoculant bags are sealed leaving about 2/3 vacant spaces in the bag to give proper aeration to the inoculant. Bags should also be pinned on few places so that air can pass through the holes to avoid oxygen tense conditions. The packets of the carrier-based inoculants should bear the labels marking some important information such as brand name of the inoculant, manufacturer's name with complete address, quantity of the inoculant, manner of application, safety measures during use, expiry date, information about handling and storage and a tentative list of field crops on which the inoculant is more effective.

Incubation and storage: The inoculant must be incubated at 25-30°C for a week in a room condition. When population of the organisms has reached the required standard, the packet may then be stored in a cold room (4-15°C) till its use.

Availability, Transportation and Marketing of Bio-inoculants

In India, there are more than 120 biofertilizer production units supplying different types of bio-inoculants. Out of these 120 units, 73 units have been supported by the Ministry of Agriculture and another 10 units have been strengthened by Ministry of Fertilizer, Government of India. Poor transportation and storage facility in remote areas of bio-inoculants is a matter of concern. Mostly, bio-inoculant available in remote areas contain less amount of viable population of organisms due to improper storage conditions during transportation and also in the stores where it is kept for months. Methods to improve bacteria under such situations need immediate attention.

The marketing of bio-inoculants is one of its constraints. At present, the production capacity of biofertilizer industry alone in the country is 18,000 metric tons, but the production has not adequate momentum because of lack of marketing network. All units are still looking for Government indent, seems impractical. It is the responsibility of manufacturers to invite the consumers in the open market on competitive basis. It is possible if the supply of quality bio-inoculants along with proper price structure is insured. The production units may be categorised into: (*i*) government unit, (*ii*) government undertaking unit, (*iii*) co-operative unit, (*iv*) production/seed production units, (*v*) NGO's, and (*vi*) private enterprises.

Till date, there is no mechanism for registration system of entrepreneurs which is urgently required. Singleton *et al.* (1997) determined the role of public and private sectors in production and marketing of inoculants (Table 4).

METHODS OF INOCULATION

Often the success of inoculation is limited by several factors including environmental conditions, the number of infective cells applied, the presence of competing indigenous rhizobacteria, and the inoculation method (Brockwell *et al.*, 1995). Method of inoculation depends on the types of bacterial carrier and preservation methods (Hegde and Brahmaprakash,

1992). The inoculants are applied generally by two methods (*i*) soil inoculation, and (*ii*) seed inoculation.

Table 4: Role of public and private sectors in bio-fertilizer manufacturing and marketing.

Public Sector	Private Sector
Rhizobium strain evaluation	Product development
Rhizobium culture maintenance	*Rhizobium* culture maintenance
Applied research	Manufacturing
Technical assistance	Market development
Training: extension, industry, farmers	Marketing/distribution
Quality standards	Internal quality control (QC)
External QC	

Soil Inoculation

The microorganisms are released into the soil by soil drench method. Inoculants are commonly delivered into the furrows where seeds have to be sown. Another method of inoculation is broadcast application in the field prior to sowing. Preparation and procedures for inoculant application directly into the seed-beds are now in trend viz., solid inoculant (Hegde and Brahmaprakash, 1992) which may be introduced into the seed-bed through an insecticide attachment to the seed drill. Liquid inoculants are applied through irrigation or specially modified sprayers. Liquid inoculation requires an inoculant tank, a pump, a manifold of capillary tubes to deliver the liquid into the seed-bed and beneath seed (Brockwell, 1982).

In case of rhizobia Brockwell *et al.* (1980) found that when conditions were unfavourable for the survival, seed germination was delayed due to environmental conditions, soil inoculation resulted in better nodulation, growth and yield than seed inoculation. Inoculants are applied directly to the soil to avoid direct contact with chemical treated seeds, and to withstand adverse environmental conditions. Scudder (1975) by using granular inoculant in the seed furrow obtained 38% increase in yield over seed inoculation in soyabean (*Glycine max*). Greater seed yield for inoculant applied to the soil (seed row) compared with seed-applied inoculant was reported for soyabean (Bezdicek *et al.*, 1978; Muldoon *et al.*, 1980) and faba bean (Dean and Clark, 1977). Previous reports on soyabean indicate that soil inoculation produced higher N_2 fixation (Hardarson *et al.*, 1989) and seed-protein contents (Muldoon *et al.*, 1980) than the seed inoculation. Major advantage of this method is avoiding of seed mixing, which may disrupt the delicate seed coat and higher ability to withstand low moisture conditions. The depth of inoculum placement is an important factor that can influence the benefits of soil inoculation. It is well established that movement of flagellated bacteria in the soil is limited (Madsen and Alexander, 1982). Therefore, much attention should be paid on depth of inoculum placement during soil inoculation. Deeper placement of rhizobial inoculants in the soil results in substantial nodule formation on the lower part of the root system (Hardarson *et al.*, 1989; Wadisirisuk *et al.*, 1989).

On the other hand, there are some limitations with direct soil inoculation such as: (*i*) the inoculated microorganism is subjected to competition with native soil microorganisms, which are better adapted to their own ecological niche, (*ii*) inoculated microorganism also has to pass through the hostile soil environment from the inoculation site to colonize the root system.

The high rate of mortality during the travelling of inoculated organism through hostile soil environment results in poor root colonization, (*iii*) delivery into the furrows requires more labour and time, thereby increasing the cost, and (*iv*) in broadcast application mortality of inoculated microbe is quite high and requirement of more amount of inoculant makes this method expensive.

Seed Inoculation

Seed inoculation is a more widely used method which ensures the desired number of inoculated microorganisms on the seed surface. Seed coating (Fig. 2) can be done at the seed processing plant months before the sowing or on the site in the fields, just before seed-sowing (Bashan, 1986a). Application of inoculum to the seed surface prior to sowing is the traditional, most commonly used and easiest way of inoculation, although the bacteria are subjected to the hazards of drying (Saleema *et al.*, 1982), fertilizer and pesticide contacts (Kremer and Peterson, 1982; Saraf and Sood, 2002) seed coat toxicity (Materon and Weaver 1984), and mineral additives (Skipper *et al.*, 1980; Gault and Brockwell 1980). There are numerous adhesive materials suitable for attaching the inoculant to seed surface (Elegba and Rennie, 1984; Hoben *et al.*, 1991). The adhesive must be free from any chemical preservative that might diminish the viability of microorganisms and natural adhesive, such as gum-arabic, molasses, sugar syrups, starch should be applied. Although, additional coating of seeds with calcium carbonate, rock phosphate or other pelleting materials enhances the efficiency of inoculation by protecting the coated microorganism from adverse weather conditions, insecticides, pesticides, and dry hot or acidic environment (Somasegaran and Hoben, 1994).

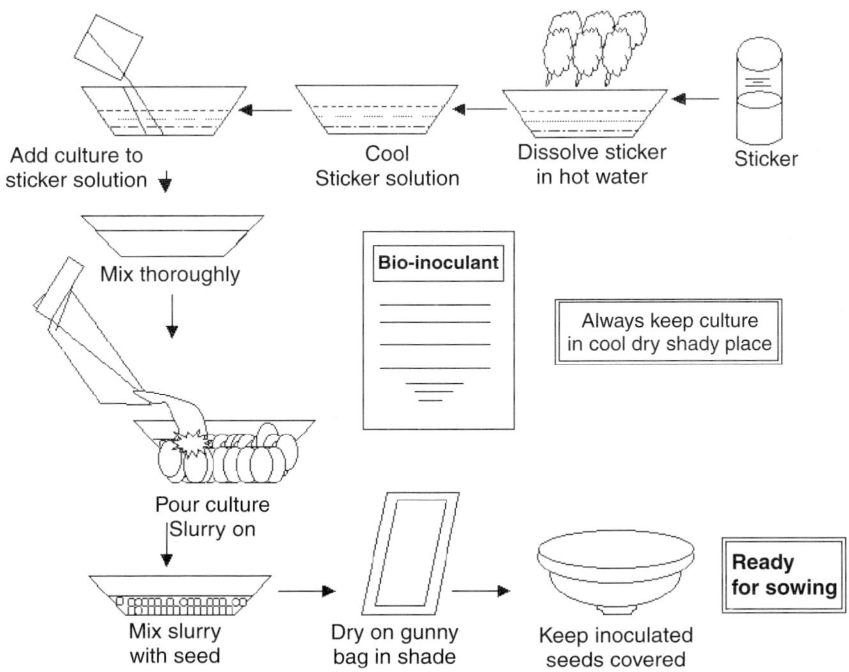

Fig. 2: Method of rhizobial application for farmers.

Due to cost effective nature, it is a widely acceptable method of inoculum application. But it has some disadvantages, such as (*i*) lack of proper training regarding seed bacterization to the farmers increases the chances of improper mixing of inoculant with the seed, (*ii*) probability of seed damaging is very high during the improper mixing of inoculant on seed surface, (*iii*) poor adherence of inoculant on the seed surface, (*iv*) seed dressed with fungicide, insecticides and other chemicals can kill or inhibit the microbial population coated on the seeds, and (*v*) poor mobility of rhizobia in the soil when placed on the seed (Madsen and Alexander, 1982).

In case of biofertilizers, when comparison was made between seed inoculation treatments (liquid or peat-based powder) and soil inoculation (granular inoculant) either placed in the seed furrow or side-banded (i.e. 2.5 cm to the side) at the depths of either 2.5 or 8 cm below the seed, it was observed that the method of inoculation and the depth of inoculant placement influenced the distribution of the nodules on the root system. Inoculating the seed resulted predominantly in crown nodulation due to the restriction of rhizobia to the vicinity of the seed and its inability to contact the younger roots at the lower part of the root system. In contrast, granular inoculant applied to the soil enhanced lateral root nodulation, and this phenomenon increased with the depth of inoculant placement. Economic status of the farmers, availability of the trained personnel, type of soil and the weather of the region play a significant role in adaptation of method of application. However, during our field trials best result of plant inoculation was obtained by the use of both, seed as well as soil application procedures.

QUALITY ASSURANCE

Quality assurance approach for inoculum production includes integration of equipment performance, raw material quality and other variables required for a quality product. Inoculant quality and products and their success in field are prerequisite necessity for quality control system (Roughley *et al.*, 1983).

Many factors affect the quality or potency of inoculants. As a matter of fact, evaluation of inoculant quality by enumerating the viable microorganisms present is usually an accurate index of inoculant potential. High quality inoculants contain effective strains in large number in a formulation that is stable and easy to use (Lupwayi *et al.*, 2000). The aim of effective inoculation is to have as many bacteria as possible alive on the seed when it germinates and forms a root. The bacteria must be able to colonize successfully on the emerging root. Numeric data regarding the number of bacteria alive when a seed germinate can be monitored and the effective inoculation can be assessed by: (*i*) enumerating the number of bacteria present in a bio-inoculant when produced, (*ii*) survival of bacteria in bio-inoculant i.e. shelf-life, (*iii*) amount of the inoculant that remains attached to seed during application, and (*iv*) survival of the inoculant on seed during application.

Certain aspects on quality control relating to bio-inoculant products have been included by a number of workers (Thompson, 1980; Lupwayi *et al.*, 2000; Bihari *et al.*, 2004; Durai *et al.*, 2004; Dwivedi *et al.*, 2004). One of the greatest barriers to predict and consistently achieve responses to inoculation is the lack of knowledge relating to target organism, fate over time and under commercial crop production conditions. A successful inoculant can only be formulated through a process of addressing the known product substrate variable that affects the viability of target organism. Formulation of any product and final product quality must be assured through a process of strict quality assurance.

Countries like Canada, Uruguay, France, etc. have some norms for quality control which is supported by legislation (Table 5). But in Australia, Thailand, New Zealand and South Africa, it may be voluntary on the part of inoculant manufacturer. In the U.S.A. regulatory control has not been considered necessary since 1940s. However, results of independent tests published from time to time indicate that substantial proportions of inoculants produced in the countries are unsatisfactory (Hiltbold *et al.*, 1980). In India, quality standards are governed by BIS (Bureau of Indian Standards) earlier ISI (Institute of Indian Standards). This bureau has established some standards related to carriers, shelf-life, CFUs., and growth-promoting activities of bio-inoculant.

Table 5: Total production, cost and quality standards of bio-fertilizers in different countries

Name of country	Total production	Cost/Pkt	Quality standard (cfus/g carrier)
Australia (M.B. Poples)	5 lakh pkts	2$/100-200gm	10^9
USA (P.W. Singleton)	5 million Kg	5$/Kgm	10^9
France (G. Catroux)	60-70 tons	2$/200-250g	10^9
UK (Lovell)	50 tons	2$/200-250gm	10^9
Bangladesh (M.A. Sattar)	5 tons	7 Rs./100gm (20 Taka/ pkt)	—
Mayanmar (M.M. Thein)	200000 packets	1.2$/150 gm (10 Kyata/pkt)	$10^8 - 10^9$
Nepal (S. Bhattarai)	10000 packets	25 Rs. N/200 gm	—
Pakistan (M. Aslam)	10 tons	10 Rs./pkt (500gm/acre)	10^8
Thailand (S. Kongugoon)	211 tons	0.4$/100-200 gm	10^8
Vietnam (P.V. Thoam)	5 tons	0.006$/40-45 gm	—
Nigeria (N. Singinga)	1 ton	0.05$/100 gm	—
Philippines (J.K. Ladha)	5 tons	1$/100 gm	10^9
Srilanka (K. Theduchanamurthy)	4.5 tons	15 Rs./100 gm	10^8
India (T. Singh)	5000 tons	4-10 Rs./200 gm	10^8

Source: BNF International Workshop on 20-24 August 1996 at ICRISET. The current information in r/o India is: Production: 10000 metric tons (roughly); cost: Rs. 7-12/200 g (roughly); Quality standard (CFUs/g carrier): 10^7.

Thereafter, standards of inoculants and their regulations vary with countries. In Canada, standards for inoculum are covered in the Fertilizers Act (Rennie, 1991). If any company wants to sell inoculants in Canada, it must submit registration to the Agriculture Canada with all information regarding product efficacy. France also has legislation covering inoculants although standard and procedures are slightly different (Wadoux, 1991). The inoculant products must be registered for sales in France and should not be harmful to non-target crops, animals, humans and the environment, whereas Thailand has no legislation for inoculant but the national standards are existing. However, Thai department also conducts independent testing of inoculants (Boonkerd, 1991). Inoculant quality standards for selling in Australia are not covered by any legislation. Therefore, there is neither any compulsion to meet effective number of bio-inoculant population nor the number of contaminants. Countries, including France, Rwanda and Australia also have a requirement on the level of contamination (Olsen *et al.*, 1994) and/

or require a serological test to demonstrate the presence of recommended strain. In Australia, these standards have been enforcing for more than twenty years. Hiltbold *et al.* (1980) examined the commercial inoculant for quality and efficacy and observed a number of bacteria widely varying in the commercial inoculants. Despite nearly 100 years of experience, it is unfortunately true that most of the inoculants produced in the world today are of relatively poor quality (Olsen *et al.,* 1995).

At present, BIS has raised the earlier norms and issued fresh guidlines for preparation of bio-inoculants (Tables 6 and 7). It insists the manufacturers to get ISI mark so that farmers can be assured of good quality biofertilizer. State government should bring about Quality Control Act like Fertilizer Control Act so as to implement licence requirement. There is a need to have a regulatory act concerning biofertilizers and for the revision of the BIS specifications. The BIS should have a network to test samples, and once ISI mark is given to a product of an organization the sample should be routinely checked thereafter too. At present there is no quality-testing laboratory in the country to evaluate the biofertilizers produced despite the existing BSI rules

Table 6: Indian standard specifications for *Rhizobium, Azotobacter* and *Azospirillum* inoculants.

Parameters	Rhizobium	Azotobacter
Base	Carrier based	
Cell number at the time of manufacture	10^7 g^{-1} carrier	10^7 g^{-1} carrier
Cell number at the time of expiry	10^7 g^{-1} carrier	10^7 g^{-1} carrier
Expiry period	6 months	6 months
Permissible contamination	Nil at 10^5 dilution	Nil at 10^5 dilution
pH	6.5 – 7.5	6.5 – 7.5
Moisture (%)	30-40	30-40
Nodulation test	Should be positive	
Nitrogen fixation	50% increase in dry mass of plants over un-inoculated	Not less than 10 mg/g of source consumed
Strain	Should be checked serologically	—
Carrier	Should pass through 150-212 micron, (IS -72-100 mesh)	Should pass through 106 micron IS sieve

Table 7: Indian standard specifications for *Azospirillum* and phosphate solubilizing microorganisms (PSM) inoculants.

Parameters	Azospirillum	PSM
Base		
Cell number at the time of manufacture	10^7 g^{-1} carrier	10^7 g^{-1} carrier
Cell number at the time of expiry	10^7 g^{-1} carrier	10^7 g^{-1} carrier
Expiry period	6 months	6 months
Permissible contamination	Nil at 10^5 dilution	Nil at 10^5 dilution
pH	6.5 – 7.5	6.5 – 7.5
Moisture (%)	30-40	30-40
Nitrogen fixation	—	15 mg/g malic acid
Growth in Pikovaskaya's medium	Positive	—
Phosphate solubilization zone	Min, 10 mm	
Growth in S.S. malate medium	—	Positive
Pellicle formation	—	Positive

for the biofertilizer production. It has been observed that manufacturers are not following minimum protocols of biofertilizer production. The competent authority of the state governments can handle the issues with the assistance of Package Commodity Rules 1977 framed under the standards of Weight and Measure Act 1997 or Consumer Protection Act 1986 for unfair trade practice, if any. So far nowhere the quality assurance of PGPR-based bio-inoculant preparations has been mentioned.

RESIDUAL EFFECT OF BIO-INOCULATIONS

Inoculant applied once in crops irrespective of any means should be effective for subsequent following crop. Introduced strains re-isolated from the applied field by using suitable markers viz. intrinsic resistance to high level of antibiotic, serological features, *gfp* proteins, *gus* marker, transposons, etc. may provide an evidence of their influence on next crop raised in such field. Enumeration of applied bacterial strain tagged with any one of the markers mentioned above from the field is done by using serial dilution technique or standard plate count (Dubey and Maheshwari, 2002). Effective inoculation of bio-inoculants not only provides a sufficient number of bacteria but also protects the introduced bacteria from the hostile environment in soil. Therefore, initially carrier material helps the bacteria in enhancement of their population and simultaneously support in the acclimatization of bacteria in soil. Once bacteria adapt the environment, it grow luxuriantly and also stabilize its appropriate population density in soil. Study of residual bio-inoculants especially in the context of microbial survival and its effect on plant growth promoting activities has great concern with the inoculant type and method of inoculation. An inoculant *Rhizobium* strain introduced in soil once is unable to become dominant in a soil already harbouring the same species of *Rhizobium* (Brockwell *et al.*, 1982; van Renbeurg and Strijdom, 1985). Revellin *et al.* (1996) demonstrated that bacterial strains introduced through seed or soil bacterization usually survived for many years. Obaton *et al.* (2002) reported that introduced bradyrhizobial strain predominant in 70% of the nodules of soyabean and after four to five years these were localized in about 50% of the nodules. It survives for a period of sixteen to twenty years of introduction in soil.

FUTURE PROSPECTS

Selection of suitable rhizobacterial strain for different crops grown under hostile soil conditions such as low moisture, high temperature, salinity and other stress conditions is the first thrust area to be explored. Strain improvement based on physiological and genetic aspects must be an effective step towards the better goal of bio-inoculants production strategies. Simultaneously, shelf-life of bacteria in different carriers and standardization quality control of inoculants are the interesting and challenging research direction for future. Mode of inoculation is also important equally that affects the efficacy of inoculants. At the time of sowing, non-availability of the product (biofertilizer) in the remote areas in a viable state exists because of poor transportation and storage facilities. Methods to improve bacteria under such situations need immediate attention.

Beneficial improvement in the inoculants development required to be increased the survival of the bacteria during inoculation on the seed surface and into the soil, using development of new formulations, granular inoculants and seed-coating technique that may protect bacteria from stress and physically keep them apart from the toxic chemicals.

Appropriate use of molecular techniques to transfer the useful genes coding for direct product will benefit the inoculated plant and monitoring of bacterial survival and performance. Development of rhizobacterial strains may be done by rDNA technology. Such modified strains are capable to withstand certain adverse conditions such as high temperature and competition with indigenous strains in soil. Development of single bacterial strain for a wide range of legumes and non-legumes as well, is a major area of interest for researchers and should be possible to undertake genetic manipulations of the bacterial strains towards the expected results.

Moreover, development of an effective microbial consortium/biofilm of plant growth-promoting rhizobacteria is still an unexplored area of research in bio-inoculant technology. Introduction of new beneficial microbes to soil and crops individually or in consortium (for example PGPR and microbial agent for biological control of pest disease and weeds) is widely accepted. The inoculum technology has been successfully exploited in many developed countries, whereas in developing countries it has been hampered by several factors. A major problem, especially relevant to bio-inoculants development programme is lack of coordination between institutions and industry. An active coordination and exchange of knowledge between the research institutions and industry are essential.

Establishment of R & D is essential with the motive to reduce the cost of inoculants without hampering the plant growth potential of the inoculants. There are few R & D programmes on quality improvement of bio-inoculant in developed countries. Although, R & D is expensive and companies involving inoculant business have limited capacity to invest, R & D programmes are also difficult to launch in public research, probably due to the competition between basic and applied research. The deadline for the development of new technology based on research programme should often be set that lasts for five to ten years, because of the shared need of both funding agencies and scientists to suggest that they are conducting research of immediate importance in terms of economic impact of their research.

We conclude that the major immediate issues to be stressed upon in inoculant technology are: (*i*) to set up appropriate guideline or act on quality aspect, (*ii*) to establish appropriate marketing structure and pricing mechanism, (*iii*) to develop agro-chemically/rhizosphere-competent microbial strains, (*iv*) to develop technology on longer shelf-life and storage, (*v*) to arrange adequate training, demonstration and publicity, and (*vi*) to develop cost-effective certification/inspection process.

REFERENCES

Ahl, P., Volsard, C. & Defago, G. (1986) *J. Phytopathol.* **116**: 121-134.

Allen, E.K. & Allen, O.N. (1961) In *Recent Advances in Botany*, Vol 1, University of Toronto Press, Toronto, pp. 585-588.

Amarger, N. (1981) *Soil. Biol. Biochem.* **13**: 475-480.

Antoun, H., Beauchamp, C.J., Goussard, N., Chabot, R. & Lalande, R. (1998) *Plant Soil.* **204**: 57-67.

Appoma, U.P., Szczapan, E.W. & Vishwanath, T. (1987) *Curr. Microbiol.* **15**: 11-15.

Arora, N.K., Kang S.C. & Maheshwari, D.K. (2001) *Curr. Sci.* **81**: 673-677.

Arshad, M. & Frankenberger, W.T. Jr. (1992) *Plant Soil.* **133**: 1-8.

Bakker, A.W. & Shippers, B. (1987) *Soil Biol. Biochem.* **19**: 249-156.

Balatti, A.P., Alcaaraz, M., Ronchi, A.I. & Grassano, A. (1991) *Rev. Lat. Amer. Microbiol.* **33**: 219-223.
Ballard, R.A. & Charman, N. (2000) *Aus. J. Exp. Agric.* **40**: 939-948.
Bashan, Y. (1986) *Appl. Enviorn. Microbiol.* **51**: 1089-1098.
Bashan, Y. (1998) *Biotechnol. Adv.* **4**: 37-40.
Bashan, Y. & Holguin G.C. (1998) *Soil Biol. Biochem.* **30**: 1225-1228.
Bashan, Y. & Levanony, H. (1990) *Can. J. Microbiol.* **36**: 591-608.
Beauchamp, C.J., Dion, P., Kloepper, J.W. & Antoun, H. (1991) *Plant Soil.* **132**: 273.
Bezdicek, D.K., Evans, D.W., Abede, B. & Witters, R.E. (1978) *Agron. J.* **70**: 865–868.
Bhatia, S., Bhatia, S., Dubey, R. C. & Maheshwari, D. K. (2003) *Ind. J. Exp. Biol.* **41**: 1442-1446.
Bihari, K., Dwivedi, V. & Bhattachraya, P. (2004) In *National conference on quality control of biofertilizers,* (Eds. Bhattachraya, P. & Dwivedi, V.). National Biofertilizer Development Centre, Ghaziabad (India).
Bloemberg, G.B. & Lugtenberg, B.J.J. (2001) *Curr. Open. Plant. Biol.* **4**: 343-350.
Bonish, P.M. (1979) *New Zealand J. Agri. Res.* 22: 89-93.
Boonkered, N. (1991) In *Inoculant quality control and standards in Thailand.* Report on the Expert Consultation on Legume Inoculant Production and Quality Control. FAO, Rome, pp. 121-130.
Briat, J.F. (1992) *J. Gen. Microbiol.* **138**: 2475-2483.
Briskot, G., Taraz, K. & Budzikiewicz, H. (1986) *FEMS Microbiol. Rev.* **104**: 200-220.
Brockwell, J. (1977) In *A Treatise on nitrogen fixation.* Section IV. *Agronomy and Ecology* (Eds. R. W. F. Hardy & A. H. Gibson), Whiley. Newyork, pp. 277-309.
Brockwell, J. (1982) In *Nitrogen Fixation in Legumes* (Ed. Vincent, J.M.), Academic Press, Sydney, pp. 211-227.
Brockwell, J., Bottomley, P.J. & Thies, J.E. (1995) *Plant Soil* **174**: 143-180.
Brockwell, J., Diatloff, A., Roughley, R. J. & Date, R. A. (1995) In *Nitrogen fixation in legumes,* (Ed. Vincent, J.M.) Academic Press, Sydney, pp. 173-191.
Brockwell, J., Gautk, R.R., Chase, D.L., Hely, F.W., Zorin, M. & Corbin, E.J. (1980) *Aus. J. Agric. Res.* **31**: 47-60.
Brockwell, J., Holliday, R.A & Pilka, A. (1988) *Plant Soil* **108**: 163-170.
Brown, M.E. (1974) *Ann. Rev Phytopathol.* **12**: 181-197.
Budzikiewicz, H., Schroder, H. & Taraz, K. (1992) *Naturforsch.* **47**: 26-32.
Burton, J.C. (1976). In *Symbiotic nitrogen fixation in plants,* (Ed. Nutman, P.S.). University Press, Cambridge, U.K., pp. 175-189.
Carson, K.C., Meyer, J.M. & Dilworth, M.J. (2001) *Soil Biol. Biochem.* **32**: 11-21.
Cassmane, K.G., Munns, D.N. & Beck, D.P. (1981) *Soil Sci. Soc. Amer. J.* **45**: 520-523.
Chakraborty, U. & Chakraborty, B.N. (1988) *Can. J. Bot.* **68**: 1698-174.
Chakraborty, U. & Purkayastha, R.P. (1984) *Can. J. Microbiol.* **30**: 285-289.
Chet, I. & Chertnin, L. (2002) In *Encyclopedia of Environmental Microbiology* (Ed. Bilton, G.), John Villy and Sons, Inc., New York, USA, pp. 455-465.
Colyer, P.D. & Mount, M.S. (1984) *Plant Dis.* **68**: 703-706.
Date, R.A. & Roughley, R.J. (1977) In *A Trreatise on Dinitrogen Fixation.* (Eds. Hardy, R.W.F. & Gibson, A.H.) John Wiley & Sons. Chichester, U.K., pp. 243-276.
Daza, A., Sontamaria, D.N., Rodriguez-Navarro, C.M., Olive, R. & Temprano, F. (2000) *Soil Biol. Biochem.* **32**: 567-572.
Dean, J.R., & Clark, K.W. (1977) *Can. J. Plant Sci.* **57**: 1055–1061.
Defago, G., Berling, C.H., Burger, U., Hass, D., Hahr, G., Keel, C., Voisard, C., Wirthner, P.H. & Wutrich, B. (1990) In *Biological control of soil-borne plant pathogens,*(Ed. Hornby, D.), CAB international, pp. 93-108.
Defago, G., Berling, C.H., Burger, U., Hass, D., Kahr, G., Keel, C., Voisard, C., Wirthner, P. & Wuthrich, B. (1992) In *Biological control of soil borne plant pathogens.* (Ed. Horny, D.), CAB International, Walington, Oxon UK, pp. 93-108.
Deshwal, V., Dubey, R. C. & Maheshwari D. K. (2003) *Curr. Sci.* **84**: 443-448,
Dilworth, M.J., Carson, K.C., Giles, R.G.F., Byrne, L.T. & Glan, A.R. (1998) *Microbiol.* **144**: 781-791.

Dobereiner, J., Marrel, I.E. & Nery, M. (1976) *Can. J. Microbiol.* **22**: 1464-1473.

Dubey, R.C. & Maheshwari, D.K. (2002) In *Practical Microbiology.* S Chand & Co. New Delhi. India.

Dunne, C., Crowely, J.J., Moenne-Loccoz, Y., Dowiling, D.N., de Bruijn, F.J. & O'Gara, F. (1997) *Microbiol.* **143**: 3921-3931.

Dura, P., Ponnusami, N., Janakiraman, C. & Alok kumar, G.V. (2004) In *National conference on quality control of biofertilizers.* (Eds. Bhattachraya, P. & Dwivedi, V.). National Biofertilizer Development Centre, Ghaziabad (India).

Dwivedi, V., Ghosh, T.K., Singh, J. & Rajput, A.S. (2004) In *National conference on quality control of biofertilizers.* (Eds. Bhattachraya, P. & Dwivedi, V.). National Biofertilizer Development Centre, Ghaziabad (India).

Elegba, M.S. & Rennie, R.J. (1984) *Can. J. Soil Sci.* **64**: 631-636.

Elmerich, C. (1984) *Biol. Technol.* **2**: 967-978.

Fesenko, A.N., Provorov, N.A., Orlowa, I.F., Orlov, V.P. & Simarov, B.V. (1995) *Plant Soil* **172**: 189-198.

Fridlender, M., Inbar, J. & Chet, I. (1993) *Siol Biol. Biochem.* **25**: 1211-1221.

Frommel, M.I., Nowak, J. & Lazarovits, G. (1991) *Plant Physiol.* **96**: 928-936.

Gault, R.R. & Brockwell, J. (1980) *Aust. J. Agric. Anim. Husb.* **20**: 63-76.

Glick, B.R., Jabcobson, C.B., Schwarze, M.M.K. & Pasternak, J.J. (1994b) *Can. J. Microbiol.* **40**: 911-915.

Glick, B.R., Jabcobson, C.B., Schwarze, M.M.K. & Pasternak, J.J. (1994a) In *Improving plant productivity with rhizosphere bacteria.* (Eds. Ryder, P.M. & Stephens, G.D.) Bowen. Commonwealth Scientific and Industrial Organization, Adelaide, Austrelia, pp. 150-152.

Glick, B.R., Karaturobic, D.M. & Newell, P.C. (1995) *Can. J. Microbiol.* **41**: 535-536.

Glick, B.R., Liu, C., Ghosh, S. & Dumbrofr, E.B. (1997) *Soil Biol. Biochem.* **29**: 1233-1239.

Glick, B.R., Patten, C.L., Holguin, G. & Penrose, D.M. (1999) In *Biochemical and genetic mechanisms used by plant growth-promoting bacteria.* Imperial College Press, London, UK.

Glick, B.R., Penrose, D.M. & Li, J. (1998) *J. Theor. Biol.* **190**: 63-68.

Gupta, C.P., Dubey, R.C. & Maheshwari, D.K. (2002) *Biol. Fertil. Soil.* **35**: 399-405.

Gupta, C.P., Sharma, A., Dubey, R.C. & Maheshwari, D.K. (1999) *Cytobios* **99**: 183 – 189.

Gupta, C.P., Sharma, A., Dubey, R.C. & Maheshwari, D.K. (2001) *Ind. J. Exp. Biol.* **39**: 8-1321.

Gutterson, N.I., Layton, T.J., Ziegle, J.S. & Warren, G.J. (1986) *J. Bacteriol.* **165**: 696-703.

Hammerschmidt, R. & Kuc, J. (1995) *Physiol. Plant Pathol.* **20**: 73-82.

Hansen, A.P. (1994) In *Symbiotic N_2 fixation of crop legumes: Achievements and perspectives.* Margraf Verlag, Weikersheim, Germany.

Hardarson, G., Golbs, M. & Danso, S.K.A. (1989) *Soil Biol. Biochem.* **21**: 783–787.

Hedge, S. & Brahmaprakash, G.P. (1992) *Plant Soil.* **144**: 309-311.

Herridge, D. (2002) *ACIAR Proce.* **109**: 52-66.

Hiltbold, A.E., Thurlow, D.L. & Skipper, H.D. (1980) *Agron. J.* **72**: 675-681.

Hoben, H.J., Aung, N.N., Somasegaran, P. & Kang, U.G. (1991) *W. J. Microbiol. Biotechnol.* **7**: 324-330.

Hohlneicher, U., Hartmann, R., Taraz, K. & Budzikiewicz, H. (1995) *Naturforsch.* **50**: 337-344.

Homna, M. & Shimomura, T. (1978) *Agric. Biol. Biochem.* **42**: 1825-1831.

Howieson, J.E., O' Hara, G.W. & Carr, S.S. (2000) *Crop Res.* **65**: 107-122.

Hume, D.J. & Blair, D.H. (1992) *Can. J. Microbiol.* **38**: 588-593.

Hynes, R.K., Craig, K.A., Covert, D., Smith, R.S. & Rennie, R.J. (1995) *J. Prod. Agric.* **8**: 547-552.

Jacobson, C.B., Pastemak, J.J. & Glick, B.R. (1994) *Can. J. Mirobiol.* **40**: 1019-1025.

Jacques, P., Ongena, M., Gwose, I., Scinsche, H., Delfosse, P., Thonart, P., Taraz, K. & Budzikiewicz, H. (1995) *Naturforesch. Soc.* **50**: 622-629.

Jadhav, R.S. & Desai, A.J. (1992) *Curr. Microbiol.* **24**:1371-1376.

James, D.W. Jr. & Guttorson, N. (1986) *App. Environ. Microbiol.* 52: 1183-1189.

Jawson, M.D., Franzleubbers, A. J. & Berg, R. K. (1989) *Appl. Environ. Microbiol.* **49**: 127-131.

Kang, J.G., Shin, S.Y., Kim, M.J., Bajpai, V., Maheshwari, D.K. & Kang, S.C. (2004) *J. Antibiot.* **57**: 726-731.

Kerr, A. (1980) *Plant Dis.* **64**: 25-30.

Keyser, H.H., Somasegaran, P. & Bohlool, B.B. (1992) In *Soil Microbial Ecology. Application in Aricultural and Environmental Management.* (Ed. Metting, F.B.) Marcel Decker, New York, pp. 205-256.

Khavazi, K. & Rejali, F. (2002) In *17th WCS Symposium No.47,* Thailand, p. 1009.
Khavazi, K. & Rejali, F. (2000) In *Xth International Colloquium for the optimization of the plant nutrition.* Cairo, Egypt.
Kloepper, J.W. & Schroth, M.N. (1978) In *Proceedings of the 4th International Congress on plant pathogen bacteria.* Vol (II). Stat. de Pathologie, vegetable, Et Phytobacter INA Angeris, France, pp. 879-882.
Kloepper, J.W. & Schroth, M.N. (1981) *Phytpathol,* **71**: 642-644.
Kloepper, J.W., Lifshitz, R. & Schroth, M.N. (1988) *Plant Dis.* **72**: 42-43.
Kloepper, J.W., Lifshitz, R. & Zablotowich, R.K. (1989) *Trends Biotechnol.* **7**: 39-43.
Kremer, R.J. & Peterson, H.L. (1982) *Soil Sci.***134**: 117-125.
Lang, S.R. (2000) *Plant Physiol.* **125**: 69-72.
Li, J. & Glick, B.R. (2001) *Can. J. Microbiol.* **47**: 359-367.
Lie, T.A., Muilenburg, M., Hiep, N.H. & Ayhan, K. (1992) *Can. J. Microbiol.* **38**: 569-572.
Lim, H.S. & Kim, S.D. (1995) *J. Microbiol.* **33**: 295-301.
Lumsden, R.D. & Locke, J.C. (1989). *Phytopathol.* **79**: 361-366.
Lupwayi, N.Z., Olsen, P.E., Sande, E.S., Kayser, H.H., Collins, M.M., Singleton, P.W. & Rice, W.A. (2000) *Field Crops. Res.* **65**: 259-270.
Lynch, J.M. (1983). In *Soil Biotechnology: Microbiological Factors in Crop Porductivity.* Blackwell Scientific, Oxford, Lundon, pp. 5-24.
Madsen, E.L. & Alexander, M. (1982) *Soil Sci. Soc. Am. J.* **46**: 557–560.
Maheshwari, D.K., Dubey, R.C. & V.K. Sharma. (2001) *Ind. J. Microbiol.* **41**: 409–413.
Mahmoud, S.A.Z., Ramadan, E.M., Thabet, F.M. & Khater, T. (1984) *Mikrobiol.* **139**: 227-232.
Materon, L.A. & Weaver, R.W. (1984) *Agron. J.* **76**: 471- 473.
Mishustin, E.N. & Naumora, A.N. (1962) *Mikrobiol.* **31**: 543-555.
Modi, M., Shah, K.S. & Modi, V.V. (1995) *Arch. Microbiol.* **144**: 156-158.
Mordukhova, E.A., Skvortsova, N.P., Kochaetko, V.V., Dubeikovskii, A.N. & Boronin, A.M. (1991) *Mikrobiol.* **60**: 494-500.
Mossialos, D., Meyer, I.M., Budzikiewicz, H., Wolff, U., Koedam, N., Baysse, C., Anjaiah, V. & Cornelis, P. (2002) *Appl. Environ. Microbiol.* **66**: 487-492.
Muldoon, J.F., D.J. Hume & Beversdorf, W.D. (1980) *Can. J. Plant Sci.* **60**: 399–409.
Nobbe, F. & Hiltner, L. (1896) *US Patents,* 570-813.
Obaton, M., Bounols, A., Piva, G. & Vadez, V. (2002) *Plant Soil.* **245**: 315-326.
Okon, Y. & Labandera-Gonzalez, C.A. (1994) In *Improving plant productivity with rhizospheric bacteria,* (Eds. Ryder, M.H., Stephens, P.M. & Bowen, G.D.). Commonwealth Scientific and Industrial Research Organisation, Adelaide, Austeralia, pp. 274-278.
Olsen, P.E., Rice, W.A. & Collins, M.M. (1995) *Soil. Biol. Biochem.* **27**: 699-702.
Olsen, P.E., Rice, W.A., Bordeleau, L.M. & Diederbeck, V.O. (1994) *Plant Soil.* **161**: 127-134.
Ordentlich, A., Elad, Y. & Chet, I. (1988) *Phytopathol.* **78**: 84-88.
Osa-Afiana, L.O. & Alexander, M., (1982) *Soil Sci. Soc. Amer. J.* **46**: 285-288.
Park, S.K., Lee, H.Y. & Kim, K.C. (1995) *Kor. J. Plant Pathol.* **11**: 304-311.
Paudyal, S.P., Deshwal, V.K. & Maheswari, D.K. (2002) *J. Ind. Bot. Soc.* **82**: 113-118.
Perrot, X., Stachelin, C. & Broughton, W.J. (2000) *Microbiol. Mol. Biol. Rev.* **64**: 180-201.
Persmark, M., Pittman, P., Buyer, J.S., Schwyn, B., Gill, P.R. & Neilands, J.B. (1993) *J. Amer. Chem. Soc.* **115**: 3950-3956.
Piao, C.G., Tang, W.H. & Chen, Y.X. (1992) *Can. J. Microbiol.* **4**: 55-62.
Plessener, O., Klapathch, T. & Guerionot, M.L. (1993) *Appl. Environ. Microbiol.* **59**: 1688-1690.
Polyanskay, L.M., Vedina, O.T., Lysak, L.V. & Zvyagintev, D.G. (2000) *Microbiol.* **71**: 109-115.
Poole, K., Young, L. & Neshat, S. (1990) *Nature* (London) **287**: 833-834.
Rennie, R.J. (1991) In *Canadian legume inoculants: Evolution of an industry.* Report on the Expert Consultation on Legume Inoculant Production and Quality Control. FAO, Rome, pp. 51-60.
Revellin, C., Pinochet, X., Beauclair, P. & Catrroux, G. (1996) *Eur. J. Soil Sci.* **47**: 505-510.
Rioux C.R., Jordan, D.C. & Rattray, J.B.M. (1983) *Can. J. Soil Sci.* **65**: 555-562.

Roughley, R.J (1970) *Plant Soil.* **32**: 675-701.
Roughley, R.J. & Vincent J.M. (1967) *J. Appl. Bacteriol.* **30**: 362-376.
Roughley, R.J., Bullard, G.K. & Gemell, L. (1983) In *The Australian Inoculants Research and Control Services*, New South Wales. Department of Agriculture, Gosford.
Saleema, M.P., Parker, C.A. & Kidby, D. K. (1982) *Soil Biol. Biochem.* **14**: 13-14.
Saraf, M. & Sood, N. (2002) *J. Ind. Bot. Soc.* **81**: 313-316.
Sattar, M.A., Quader, M.A. & Danso, S.K.A. (1995) *Soil Biol. Biochem.* **27**: 725-727.
Scudder, W.T. (1975) *Soil Crop Sci. Soc. Fla. Proc.* **34**: 79–82.
Semu, E. & Hume, D.J. (1979) *Can. J. Microbiol.* **25**: 739-745.
Seshadri, S., Muthukumarswami, R., Laksminarasimhan, C. & Ignacimuthu, S. (2000) *Curr. Sci.* **79**: 565-567.
Shah, H., Karkhanis, V. & Desai, A. (1992) *Curr. Microbiol.* **25**: 347-351.
Singleton, P.W., Boonkered, N., Carr, T.J. & Thompson, J.A. (1997) In *Extending Nitrogen Fixation Research to Farmers' Fields.* (Eds. Rupela, O.P., Johnsen, C. & Herridge, D.F.) ICRISAT, Patancheru, A.P., India, pp. 17-38.
Skipper, H.D., Palmer, J.H., Giddens, J.E. & Woodruff, J.F. (1980) *Agron. J.* **72**: 673-674.
Smith, M.J. & Neilands, J.B. (1984) *J. Plant Nutr.* **7**: 449-485.
Smith, M.J., Shoolery, J.N., Schwin. B., Holden, I. & Nelands, J.R. (1985) *J. Amer. Chem. Soc.* **107**: 1739-1746.
Smith, R.S. (1992) *Can. J. Microbiol.* **38**: 485-492.
Somasegaran, P. & Hoben, H.J. (1994) In *Hand book of rhizobia: Methods in legume-Rhizobium technology.* Springer- Verlag, New York.
Somasegran, P. (1985) *Appl. Environ. Microbiol.* **44**: 330-332.
Spaink, H.P. (2000) *Ann. Rev. Microbiol.* **54**: 277-288.
Sprent, J.I., Stephens, J.H. &, Rupela, O.P. (1986) In *Abstracts of the International Food Legume Research Conference on Pea, Lentil, Fababean and Chickpea*, (Ed. Summer, F.) Matinus Nijhoff, Dordrecht, p. 47.
Stephens, J.H.G. & Rask, H.M (2000) *Field Crop Res.* **65**: 249-258.
Tang, W.H. (1994) In *Improving plant productivity with rhizospheric bacteria.* (Eds. Ryder, M.H., Stephens, P.M. & Bowen, G.D.), Commonwealth Scientific and Industrial Research Organisation, Adelaide, Austeralia, pp. 267-278.
Thies, J.E., Singleton, P.W. & Bohlool, B.B. (1991) *Appl. Environ. Microbiol*, **57**: 29-37.
Thompson, J.A. (1980) In *Methods for Evaluating Biological Nitrogen Fixation*, (Ed. Bergresen, F.), John Wiely, Chicheste, pp. 489-533.
Thompson, J.A. (1992) Consultant Report to UNFAO. IND/86/003. FAO, Rome, Italy, p. 48.
Tien, T.M., Gaskins, M.H. & Hubbell, D.H. (1979) *Appl. Environ. Microbiol.* **37**: 1016-1024.
Tilak, K.V.B.R., Jauhri, K.S. & Saxena, A. K. (1992) In *Rhizobium Biofertilizer Technology for Legumes,* IARI, New Delhi. p. 32.
Toyoda, H. & Utsumi, R. (1991) *U.S. Patent* No. **4**: 988-586.
Tranvan, V., Berge, O., Ngoke, S., Balandreau, J. & Heulim, T. (2000) *Plant Soil.* **281**: 273-284.
Trinic, M.J. (1988) *Plant Soil* **110**: 177-185.
Trinic, M.J. & Hadobas, P.A. (1995) *Plant Soil* **172**: 207-219.
Tweddell, R.J., Jeubaji-Hare, S.H. & Charest, P.M. (1994) *Appl. Environ. Microbiol.* **60**: 489-495.
Vaidya, R.J., Shah, I.M., Vyas, P.R. & Chhatpar, H.S. (2001) *W. J. Microbiol. Biotechnol.* **17**: 691-696.
van Dyke M.I. & Prosser, J.I. (1988) *Soil. Biol. Biochem.* **32**: 1377-1382.
van Elsas, J.D., Kijkstra, A.F., Govaert, J.M. & van Veen, J.A. (1986) *FEMS Microbiol. Ecol.* **38**: 151-160.
van Rensburg, H.J & Strijdom, B.W. (1985) *Appl. Environ. Microbiol.* **49**: 127-131.
Vincent, J.M. (1970) In *Manual for the Practical Study of Root –Nodule Bacteria,* IBP Handbook No. 15. Blackwell. Oxford.
Wadisirisuk, P., Danso, S.K.A., Hardarson, G. & Bowen, G.D. (1989) *Appl. Environ. Microbiol.* **55**: 1711-1716.
Wadoux, P. (1991) In *Inoculant production in industry using sterile carriers.* Report on the Expert Consultation on Legume Inoculant Production and Quality Control. FAO, Rome, pp. 33-42.
Yang, S.F. & Hoffmann, B.E. (1984) *Ann. Rev. Plant. Physiol.* **35**: 155-189.

2

Psychrophiles: Their Diversity and Industrial Applications

Mahejibin Khan and Reeta Goel
Department of Microbiology,
College of Basic Sciences Humanities, G.B.P.U.A.&T., Pantnagar (India)

Psychrophiles persist where others perish—thriving in deep ocean trenches, abysmal plains, polar oceanic regions and continental margin cold seeps. These are of interest to both basic and applied biology. On basic front these organisms hold many interesting biological secrets, such as the biochemical limits to macromolecular stability and the blueprint for constructing macromolecules stable to one or another extreme. In an applied field, these organisms have yielded an amazing array of enzymes capable of catalyzing specific biochemical reactions under extreme conditions. Such enzymes have found their way into the grist of industry in applications as diverse as laundry detergent additives (proteases, lipases).

Membranes of psychrophiles may contain surfactants bearing unique stability that can be used in pharmaceutical formulations. Other important innovative products are cyclodextrins, compatible solutes and polyunsaturated fatty acids and anti-freeze proteins that are produced in psychrotrophs and help in protection freeze labile enzymes. Microbes also exist in nature that feed on toxic materials such as methylene chloride, detergents, phenol, sulphur and polychlorinated biphenyls (PCBs). Microbes metabolize the toxic compounds to produce harmless end-products such as carbon dioxide, water and salt. These secrets to microbial successes are potential gold mines for industry.

INTRODUCTION

Microorganisms represent the largest reservoir of undescribed biodiversity, and hence possess the greatest potential for the discovery of new natural products. It is estimated that the earth currently supports between three and thirty million species of organisms. Of these, approximately 1.4 million have been described by science. This includes virtually all the species of birds and mammals (~ 13,500) too. In contrast only around 200,000 of the estimated 1.0 to 1.5 million species of fungi have been characterized (i.e. 13-20%). For the bacteria this percentage is even lower with estimates ranging from only 1-10% of probable species being described in culture (Bull, 1992). Several of the divisions (e.g. Proteobacteria, Actinobacteria) are well represented by cultivated strains. However, the majority are poorly represented by cultured organisms. In fact 13 of the 36 divisions are defined by environmental sequences only (Hugenhaoltz *et al.*, 1998). The proportion of cultivated versus uncultivated sequences obtained from environmental molecular analyses may be used as an indication of the level of described biodiversity for a particular division.

The search for new or better byproducts or technological processes from biological sources may be defined as bioprospecting. In particular, bioprospecting relies on provision of a bioresource, a supply of novel biodiversity. The isolation and characterization of novel psychrotolerant microbes has the potential for discovery of new enzymes of value for biotechnological applications in the food/feed, biopharmaceutical and environmental industries. The use of microorganisms for remediation of contaminated sites is a promising environmental technology. In cold environments, however, the slow rate of biodegradation of organic contaminants is a matter of concern. Psychrophilic and psychrotrophic microbes have been characterized, with a focus on petroleum hydrocarbons and fuel additives degradation. A detailed understanding of the microbial community structure and ecological function of microorganisms functioning at low temperatures is essential for development of environmental technologies for clean-up of contaminated soil, sediment and water in the Arctic.

Cold-adapted microorganisms are generally understood to achieve their physiological and ecological successes in cold environments as a result of unique features in their proteins and membranes and their genetic responses to thermal shifts. Relative to mesophilic counterparts, a cold active enzyme tends to have reduced activation energy, leading to high catalytic efficiency, which may possibly be attributed to an enhanced local or overall flexibility of the structure of the protein. Membranes appear to incorporate specific lipid constituents to maintain fluidity and the critical ability to transport substrates and nutrients under very cold, otherwise rigidifying conditions (Allen and Deming, 2002).

Cold-shock proteins (CSPs) enabling specific continued activities as the temperature drops are induced in psychrophiles, as in other thermal classes of microorganisms. Khan *et al.* (2003) studied the role of cold shock proteins in cold tolerant mutants of pseudomonas fluorescens and suggested that CSPs are cytosolic in nature and expression of these proteins increases with the decrease in temperature. Maximum expression was at 4°C but at 37°C it was negligible. With limited information from a small number of 'model' organisms, however, these potentially universal themes for defining cold adaptation have been embraced only cautiously in the literature. This situation is poised to change, as the focus on microorganisms in very cold habitats sharpens and the research enterprise on the subject of cold adaptation broadens (Deming and Baross, 2000; Sheridon *et al.*, 2000). In the latter case, the main focus is on very cold, typically subzero, aquatic environments, including the deep sea (–1°C to 4°C), Arctic and Antarctic marine habitats (sea-water and sediments near –1°C and sea ice, where internal fluids remain liquid to –35°C in wintertime), and glacial and lake ice (down to –5°C). The polar regions of the earth, especially the Arctic, are undergoing relatively rapid environmental change on a global perspective (Brambilla *et al.*, 2001), such that inhabitants dependent on the frozen environment of sea-ice appear threatened with extinction in this century. Motivations to document biodiversity at high altitudes, especially in ice, are strong, just as the new field of astrobiology demands research in the earth's frozen polar regions as habitat analogues for possible life elsewhere (Deming and Barross, 2001; DesMarais and Walter, 1999). The most likely candidates for harbouring microbial life in our solar system, now or in the past, are Mars, with its present polar ice-caps and indications of past water (Baker, 2001), and Europa, Jupiter's ice-covered moon with a possibly vast ocean below the ice (Chyta and Philips, 2002). Both bodies present us with frozen surfaces for evaluation and sampling.

MICROBIAL DIVERSITY IN POLAR AND COLD DEEP-SEA ENVIRONMENTS

With the advent of molecular phylogeny and methods for assessing unculturable micro-organisms in the environment came the initiation of biodiversity surveys based on the16S rRNA gene sequence. Although cold regions were not the first environmental targets of interest, they are now receiving increased attention (Table 1). When both, bacteria and archaea have been surveyed across a range of polar and other cold environments (Thomas and Dieckmann, 2002; Raymond and Fritsen, 2001), the bacteria are observed to dominate and be present in greater diversity than the archaea. The important exceptions are the cold interior and deep waters of the ocean, well below the upper mixed layer. There, the archaea appear in numbers elevated above their counterparts in surface waters (Wells and Deming, 2001; Loper-Garcia *et al.*, 2001) or actually higher than the bacteria at the same depths (Junge *et al.*, 2002). In the latter case in the deep Pacific Ocean the populations of crenarchaeota (when added to the euryarchaeota that were also present) largely accounted for the archaeal numbers matching or exceeding the bacterial numbers. Certain archaeota are also found in deep Antarctic waters (Beja *et al.*. 2002) though only euryarchaeota groups were detected among the archaea in a recent study (Lonhienne *et al.*, 2001).

Table 1: Microbial diversity in polar environment.

Environment	Approach	Selected findings
Pacific Ocean waters Including cold deep sea	Quantitative culture independent survey of bacteria and archea	Archea especially crenarcheota, dominant in deep sea
Arctic sea water and nepheloid (particle rich) layers	Quantitative culture independent survey of bacteria and archea	Bacteria especially CFB,dominant in surface water,archea elevated the deeper nepheloid layer
Arctic marine surface sediments	Quantitative culture independent survey of bacteria and archea	High fraction especially sulphate reducersand CFB of diverse bacteria, low fraction of archea, limited diversity
Deep Antarctic sea-water	Culture independent survey of bacteria and archea	Diverse bacteria, γ proteobacteria dominant, archea limited to euryarcheota
Antarctic lake microbial mat	Culture based and culture independent survey of bacteria and archea	Diverse bacteria, limited diversity of archea
Antarctic and Arctic sea ice	Culture based and culture independent survey of bacteria and archea	Diverse bacteria, with some common to both poles, no archea
Deep cold Pacific Ocean	Quantitative culture based survey of salt tolerant organisms	Cold adapted holotolerant *Halomonas* in high numbers
Arctic sea ice	Quantitative culture based survey of cold adapted facultative oligotrophs	Diverse bacteria including novel organisms, high culturable percentage and low diversity in upper ice horizons
Arctic sea water	Culture independent survey of bacteria	Diverse bacteria, depth related and seasonal patterns
Arctic and Antarctic sea water	Culture based survey of cold adapted oligotrophs	Diverse bacteria among the 173 isolated strains
Antarctic sea water (and deep water elsewhere)	Culture independent survey of archea	Euryarcheota groups II,III and IV present

The deep Arctic Ocean remains entirely unexplored from this perspective. Because the study of archaeal diversity in polar and other cold environments is still in its infancy, use of the phrase 'limited diversity' in (Table 2) is best taken with a grain of salt. Archaea can be tracked in time and space not only by 16SrRNA gene surveys but also via their unique lipid signatures. These features have made possible the dating of the expansion of the Crenarchaeota from high-temperature environments (where they were first detected and thrive today) to the cold deep sea (Junge *et al.*, 2002) (and elsewhere) from analyses of deep-ocean drilling cores (Kuypess *et al.*, 2001). The expansion appears to have occurred during the mid-Cretaceous period, over 100 million years ago and, on the basis of stable carbon isotope analyses of the same core samples, to have been dominated by a chemolithotrophic life-style.

Given this information and the chemolithotrophic nature of culturable nitrifiers, the documentation of bacterial nitrifying genes in cold oceans (Hollibaugh *et al.*, 2002) the evidence for lateral gene transfer between bacteria and contemporary cold-ocean crenarchaeota and various correlations between physical–chemical hydrographic and archaeal (and other microbial) parameters in the Northwest Passage of the Arctic Ocean (Thomas and Dieckmann, 2002), the currently unknown metabolic functions of the cold archaea are suggested to include nitrification the cold crenarchaeota will enter centrestage, not simply for their elevated numbers in cold waters, but for their participation in the nitrogen cycling process considered responsible for continued primary production throughout the ocean. Continued studies of the various species, genera and phyla of the bacteria that are present in a range of polar environments (Table 2) indicate that even this better-explored phylogenetic domain holds surprises in store.

Conventional quantitative cultivation methods newly applied to salt-tolerant organisms in the cold deep sea (Krembs and Engel, 2001) and to psychrophilic bacteria in sea ice (32), have yielded many isolates with novel 16S rRNA sequences that grow at –1°C. They have also revealed that culturable organisms can comprise surprisingly large fractions of the total populations, especially in sea-ice (Junge *et al.*, 2001). Although environments exposed to the coldest of temperatures (such as the upper horizons of Arctic sea ice, which can experience temperatures to well below –20°C in winter (Junge *et al.*, 2001) do appear to support the lowest observed diversity of bacteria (32), the rarity of studies of natural environments at any temperature below that of polar sea-water (–1.7°C) makes any conclusion about temperature-limited diversity at the low end of the temperature scale premature.

MICROBIAL LIFE IN FROZEN ENVIRONMENTS

Ice, whether in the form of tundra, glacier, snow, lake or sea ice, presents a special environment for microbial life (Deming and Barass, 2000; Deming and Huston, 2000; Thomas and Dieckmann, 2002). It arguably provides the more selective environment for cold adaptation than milder unfrozen bodies of water or sediment. The seasonal process of encasement of marine microbial communities into sea ice, exposure to severe winter conditions and release again in summer time is postulated to seed the global ocean with psychrophiles annually (Junge *et al.*, 2002). Continued activity during winter time would ensure a positive selection process for cold adaptation rather than only a destructive one against freeze-intolerant organisms. Indeed, the lower temperature limit for metabolic activity (as opposed to preserved life) is sought in various forms of ice, particularly freshwater forms, because of astrobiology

Table 2: Cold adaptation at molecular genetic level.

Organism source	Targeted cellular constituent	Selected findings
Antarctia Bacterium, *Cytopphaga* sp KUC1	Valine dehydrogenase	Amino acid residue implicated in conformational flexibility for cold activity
Icelandic Bacterium strain PA43	Alanine malate dehydrogenase Leucine	Typical cold active enzyme traits
Arctic bacterium, *Colwellia psych rery-therea* strain 34 H	amino peptidase	Lowest temperature for optimal activity of an extra cellular protease
Fish pathogenic bacterium, *Flavobacterium psychrophillum*	Metaloproteases	Environmental host factors governing bacterial enzyme production
Antarctic bacterium *Pseudoalteromonas haloplankits*	β galactosidase	Typical cold active enzyme traits expressed in *E.coli*
Marine bacterium *Moritella marina mp1*	β lactamase	Heat labiality of enzyme retained when expressed in *E.coli*
Antarctic bacterium	α amylase	Many evolutionary steps to cold activity evaluated using mutants
Pseudomonas haloplankits	α amylase	First recombinant protein production in
Arthrobacter spTAD20	Chitinase	Antarctic host
Arthrobacter spTAD20	Chitibiases	Cold activity on both dissolved and particulate substrate
Vibrio sp strain F1: 7	Chitinase	Key cold adaptation located in domain
Methanococcoides burtonii	Elongation factor 2 protein	of active site
Bacterium *Listeria monocytogenes* Lo 28	Cold shock proteins	Cold activity expressed in *E.coli* Cold active trait of overproduction, greater affinity for GTP
Arthrobacter agilis	Carotenoids	Enhanced survival at low temperature
Arctic bacterium, *C.psychrrerythraea*	Whole genome	Carotenoid contributing to membrane stability at low temperature Amino acid substitution important to cold adaptation

defined interests in habitat analogues for soil-covered Mars (Arctic tundra) and ice-covered Europa (Antarctica's LakeVostok buried deep below glacial ice) (Price, 2000; Siegert *et al.*, 2001). Theoretical considerations of deep glacial and lake-ice matrices imply adequate energy resources for bacteria and room for their movement in the veins between ice crystals that remain liquid-filled at the *in situ* temperatures relevant to these environments (down to –5°C) (Price, 2002). Incubation experiments using tritiated thymidine and leucine with Antarctic snow-ice samples, which experience colder atmospheric temperatures, have indicated metabolic activity to –17°C (Carpenter, 2000).

Other research with frozen samples of Siberian tundra, first slurried (slightly warmed) to amend with 14C-acetate, has pushed the lower limit for detectable activity to –20°C (Rivkina *et al.*, 2000). Until recently, interest in sea-ice as a natural medium for exploring the lower temperature limits for microbial activity has been limited by comparison (Junge *et al.*, 2002; Thomas and Dieckmann, 2002). This delayed interest is perhaps surprising, given the extreme temperatures to which the upper horizons of sea ice are exposed during winter time, the capacity for brine networks in Arctic sea ice to remain liquid to –35°C, and the presence of ample organic resources in that network (Hollibaugh *et al.*, 2002). Advances are coming quickly, however, as researchers develop novel microscopic means to examine bacteria directly within the matrix of (unmelted) winter sea ice for the first time (Karner *et al.*, 2001), document respiratory activity and presumably competent protein-synthesizing machinery (as reflected by the fluorescent *in situ* hybridization [FISH] method for quantifying microbial diversity) to -20°C (Junge *et al.*, 2002) uncover novel means of cryoprotection in the form of high concentrations of exopolymeric substances (EPS) (Krembs and Engel, 2001). EPS as cryoprotectants add to the discovery of microbially produced ice-active substances that serve a similar role in Antarctic photosynthetic mats (Raymond and Fritson, 2001). Also clear, however, is the absence of a need for specific antifreeze compounds when the inhabitable space within ice becomes limited to the size of the bacterium itself. At that point, the known ordering behaviour, surface charge and polarity of water molecules in confined space alone can account for the persistence of the liquid phase (Mindock *et al.*, 2001). The inevitable presence of some liquid even within the coldest of natural ice formations promises more surprises regarding the lower temperature limit for microbial life and the mechanisms supporting it.

ADVANCES AT THE MOLECULAR GENETIC LEVEL

Microbial Adaptations to Antarctica

Microbial adaptation to a permanently cold environment includes the optimization of basic cell processes necessary for growth and survival. These can be summarized into three categories: enzyme function, nutrient transport and cell membrane function. The adaptations of cellular processes in each of these areas represent potential biotechnology products for exploitation. Two examples are the production of polyunsaturated fatty acids (PUFA) and the production of cold-active enzymes by bacteria from Antarctic sea ice. Investigations of sea ice microbial communities have demonstrated the dominance of bacterial populations highly adapted to life at constant low temperature. These bacteria are known as 'psychrophiles'. The term 'psychrophile' (*psychro*—cold, *phile*—loving) was first coined in 1902 to describe bacteria capable of growth and survival at 0°C (Russell, 1992). However, it soon became apparent that a more stringent definition was required to separate those organisms requiring cold temperatures for growth (true psychrophiles) from others that were capable of growth at or near 0°C, but possessed quite high optimal growth temperatures (around 20-30°C). The latter organisms became known as psychrotolerant. The term psychrophile is therefore an operational definition, to describe the temperature-growth relationship of microorganisms, while there is not necessarily a phylogenetic relationship between psychrophilic organisms, such relationships may exist through the evolution of adaptive strategies for low temperature growth. An example is the genus *Shewanella* which includes a number of psychrophilic species (*S. pealeana, S. hanedai, S. benthica*) including two isolated from Antarctic sea ice (*S. gelidimarina, S. frigidimarina*), in addition to species which are non-cold-adapted, or mesophilic (Russell and Nichols, 1999).

Most research at the molecular genetic level in cold-adapted microorganisms continues to focus on enzymes, although other cellular constituents are also receiving attention now (Table 2). Joining the list of enzymes already known and at least partially characterized from cold-adapted bacteria are the first xylanases and laminarases (Huphrey *et al.*, 2001), valine dehydrogenase (Oikawa *et al.*, 2001), chitobiase and chitinases (Lonhienne *et al.*, 2001; Bendt *et al.*, 2001), and pectate lyase (Truong *et al.*, 2001). Increasingly sophisticated work with some of these and with previously characterized enzymes have reinforced the general themes of cold activity especially that key adaptations in the form of specific amino-acid residues are located in the domain of the active site or, if distant, nevertheless influence the conformational flexibility required of the site for reactivity in the cold (Kuypers *et al.,* 2001). The expression of genes for enzymes from cold-adapted microorganisms has been advanced through the use of conventional *E. coli* (Tanaka *et al.,* 2001) and novel vectors, including the first recombinant protein produced in an Antarctic bacterial host (Tutino *et al.,* 2001).

The traditional approach of using mutants to understand gene expression has been put to good advantage in evaluating the number of evolutionary steps that may have been required to achieve cold activity in an alpha-amylase (D'Amico *et al.,* 2001). More research has also been directed towards understanding how an extra cellular enzyme behaves in a very cold environment, once released by the producing organism, and thus the importance of enzyme lifetime in the cold (Demin and Baross, 2002). The possibility of co-evolutionary processes contributing to enzyme behaviour in a cold environment has also been recognized, in the form of vertebrate (fish) host control of bacterial enzyme production (Secades *et al.,* 2001). Advances with proteins other than enzymes include new work with elongation factor 2 proteins from an Antarctic methanogen (Thomas *et al.,* 2001), addressing some of the cold-active features of the protein synthesizing machinery.

The most recent discovery of bacterial genes for cold shock proteins in some archaea, the uncultured cold crenarchaeota from Antarctic waters (Beja *et al.,* 2002) contributes to the discussion of environmentally driven gene transfers across phylogenetic domains. Adaptations to other forms of environmental challenges experienced at low temperatures, such as salt stress, via the production of carotenoids suggest that research on carotenoids (and perhaps pigments in general) in bacteria from the briny pockets of sea ice may reveal new links between psychrophily and halophily or halotolerance.

To enhance the growth of agriculturally important bacterial strains at low temperature, cold resistant mutants of Pseudomonas fluorescens were developed using N-methy l-Nnitro N-nitrosoguanidine (Lonhienne *et al.,* 2002). These mutants show maximum 'P'solubilization at 10°C and thus contribute to growth promotion of plants at low temperature (Deming and Baross, 2000).

APPLICATIONS

Cold-adapted organisms and their products have potential applications in a broad range of industrial, agricultural and medical processes. In order for growth to occur in low-temperature environments, all cellular components must adapt to the cold. This fact, in combination with the diversity of archaea, bacteria and eucarya isolated from cold environments, highlights the breadth and type of biological products and processes that might be exploited for biotechnology. Relative to this undisputed potential, psychrophiles and their products are under-utilized in biotechnology; however, recent advances, particularly with cold-active enzymes, herald rapid growth for this burgeoning field (Feller and Gerday, 1997).

Polyunsaturated Fatty Acids (PUFAs)

PUFAs have remained as a natural product due to the synthetic difficulty of reproducing the methylene interrupted double bond sequence (Fig. 1) by industrial chemistry. Their significance to humans lies in their biological activity, as precursors for groups of nutritionally important compounds and as essential cellular components. PUFAs act as precursors to groups of compounds known as eicosanoid hormones encompassing the prostaglandins, leukotrienes and thromboxanes. These hormones are short-lived, but have a diversity of effects in mediating cardiovascular disease conditions (Torssell, 1983). They are also essential in animals for the proper development of retinal and nervous tissue (Takahata *et al.*, 1998). Of importance is the fact that higher animals lack the ability to synthesize ω3 PUFA de novo, hence these components or their immediate C18 precursors, must be provided in the diet (Meyer *et al.*, 1999).

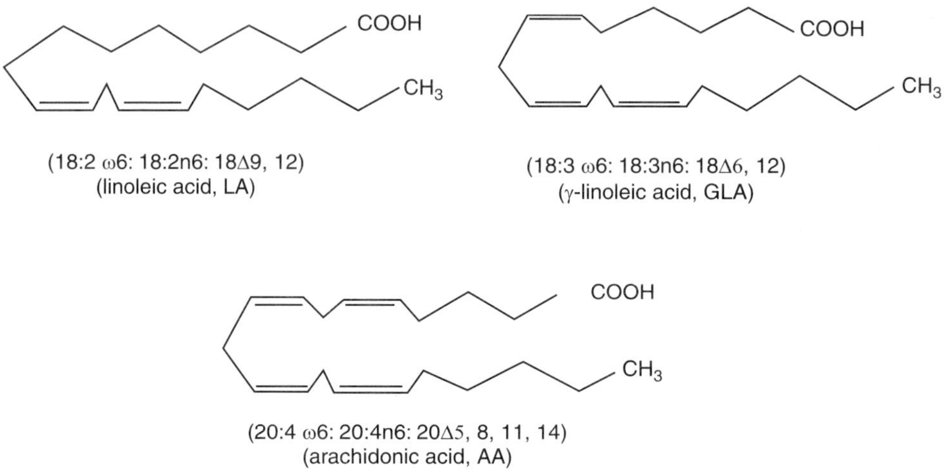

Fig. 1. Chemical structures of the common polyunsaturated fatty acids (PUFA) from the ω6 family.

In bacteria PUFAs are component fatty acids of certain phospholipids that occur in the cell membrane. Generically, long-chain PUFA consists of a long chain of carbon atoms (usually C20 or C22) containing a number (usually 4-6) methylene interrupted double bonds. Studies of psychrophilic bacteria from the Antarctic have shown that high proportion of cold-adapted bacteria from sea ice possesses the ability to produce PUFA, such as eicosapentaenoic acid (20:5ω3; EPA) or docosahexaenoic acid (22:6ω3; DHA) (Nichols et al., 1999; Russell and Nichols, 1999). It is considered that the low melting temperature of these highly unsaturated membrane components, combined with their unique molecular geometry in the membrane, yields a particular advantage at low temperature in balancing the competing homeoviscous and homeophasic forces in the cell membrane (Russell and Nichols, 1999). It can be inferred that PUFA production represents a particular physiological strategy for cold-adaptation. This is supported by empirical observations such as in the genus *Shewanella.* Here, there is a good correlation between those species, which are cold adapted and produce PUFA (highlighted in bold, Fig. 2), in contrast to those which do not produce PUFA and grow at higher temperatures (mesophiles) (highlighted in grey, Fig. 2).

The current commercial sources of EPA and DHA are restricted to fish and algal-derived oils. Problems exist with both of these sources. Algal-derived oils require a relatively high investment of technology and expense compared to the prospect of bacterial fermentation,

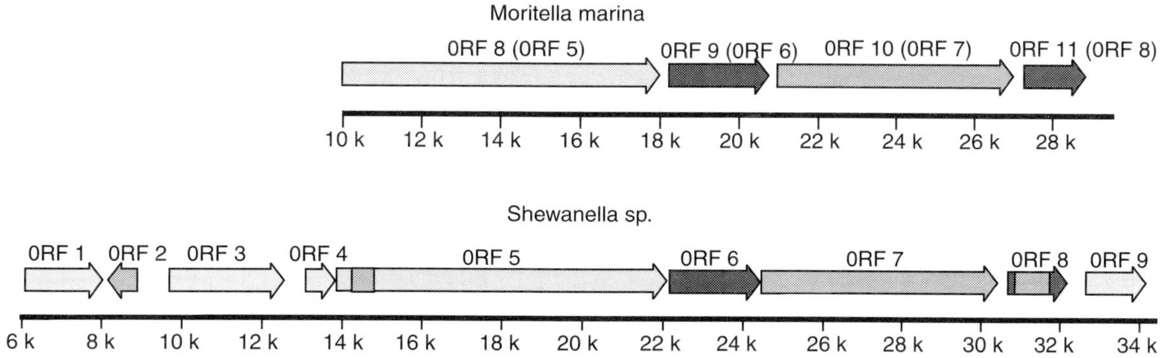

Fig. 2. Schematic of the phylogenetic relationship of the genus *Shewanella* based on 16S rDNA sequence. Species known to produce polyunsaturated fatty acids (PUFA) and their lines of descent are highlighted in bold. Species and lines of descent known not to produce PUFA are highlighted in grey. Arrows indicate sites of divergence where the expression of PUFA synthesis has been lost (adapted from Russell and Nichols (1999).

although bacteria contain a lower proportion of lipid (Nichols *et al.*, 1996). A key advantage of bacterial PUFA production is that only a single PUFA is produced, rather than the complex mixture yielded from fish or algal oils (Russell and Nichols, 1999). Thus bacterial sources of PUFA remove the expense of preparative purification in the production of high-purity PUFA oils. In addition to their potential use as 'cell factories' bacteria in particular offer the biotechnological opportunity to investigate the genes and enzymes responsible for PUFA production.

Interest in the production of PUFA from alternative sources for use in aquaculture feeds and human nutraceuticals (health supplements) has fuelled recent research into the molecular biology of PUFA production in prokaryotes (Lewis *et al.*, 1999). Insight has also been gained from related research into the mechanisms of gene expression in deep-sea microorganisms (Nakasone *et al.*, 1998) many of which belong to PUFA-producing genera (Katao *et al.*, 1998). The transfer of a gene cluster from *Shewanella putrefaciens* SCRC-2738 into *Escherichia coli* and a *Synechococcus* sp. resulted in the successful expression of EPA in these organisms (Takeyama *et al.*, 1997; Yazawa, 1995). However, the level of expression achieved was low. The gene cluster used in both cases consisted of a 38 kb fragment containing eight open reading frames with three of these possessing homology with genes that encode for enzymes involved in fatty acid biosynthesis (Fig. 3). These studies were extended to other PUFA-producing strains (*Moritella (Vibrio) marinus, Photobacterium* (sic) and *Shewanella* sp.). Further characterisation using these organisms has identified five genes responsible for PUFA biosynthesis, four of which may have been grouped as a possible operon within the DNA fragment. Strong homologies of both DNA composition and organisation were noted between the strains, with PUFA genes from either EPA or DHA producing bacteria able to replace and/or complement each other in *E. coli*. Furthermore, a single gene (*orf 8*) controlled the synthesis of either EPA or DHA in the strains examined (Fig. 3). In future years it will be interesting to observe whether the transgenic potential of bacterial PUFA genes is realized.

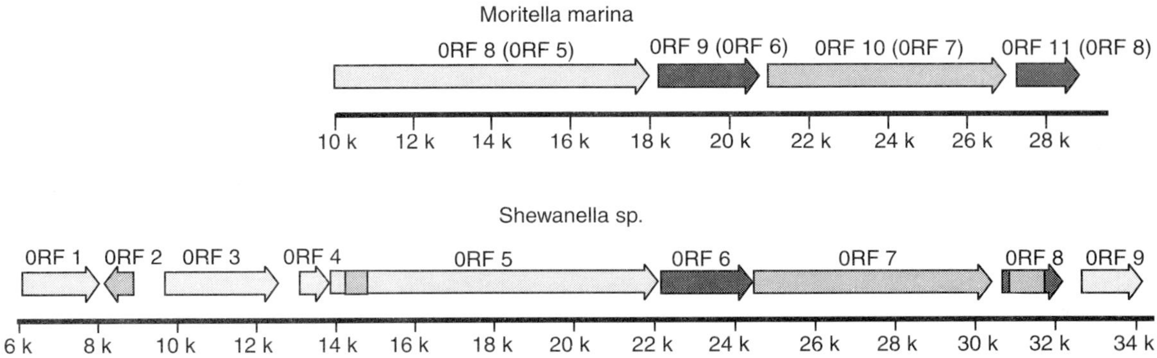

Fig. 3: Analysis of open reading frames encoding enzymes involved in polyunsaturated fatty acid biosynthesis in the genera Shewanella and Moritella. Based on the sequences (adapted from Nicholas *et al.*, 1999).

The strong homology of PUFA genes matrix with the correlation of the current 16S rDNA phylogenetic tree for the genus *Shewanella* (Bowman, 2001) and with the known ability of different species to produce PUFA (Russell and Nichols, 1999). This correlation implies that the progenitor of the genus possessed this ability. It also appears that while the ability to synthesize PUFA has been retained by all the psychrophilic and several of the psychrotrophic species, the ability (or the expression of PUFA synthesis) has disappeared on two distinct occasions during the divergence of psychrotrophic and/or mesophilic species. The apparent conservation of PUFA genes in this genus may point to the potential utilisation of these 'functional genes' as phylogenetic markers in culture-independente ecological studies.

Cold-active Enzymes

More than three-quarters of the earth's surface is occupied by cold ecosystems, including the ocean depths, and polar and alpine regions. These permanently cold environments have been successfully colonized by a class of extremophilic microorganisms. The ability to thrive at temperatures that are close to, or below, the freezing point of water requires a vast array of adaptations to maintain the metabolic rates and sustained growth compatible with life in these severe environmental conditions (Feller and Gerday, 2003).

Cold-active or cold adapted enzymes may be produced by organisms existing in permanently cold habitats located in polar zones, at high altitudes or in the deep sea. These enzymes provide opportunities to study the adaptation of life to low temperature and the potential for biotechnological exploitation (Aguilar, 1996; Morita *et al.*, 1997). Applications exist in a range of industries for cold-active enzyme applications, eg. cleaning agents, leather processing, degradation of xenobiotic compounds in cold climes, food processing (fermentation, cheese manufacture, bakery, confectionery, meat tenderization) and molecular biology (heterologous gene expression) (Margesin and Schinner, 2001). Cold-active enzymes typically have maximal catalytic activity at temperatures below 40°C and usually display some degree of thermolability.

Recent research has focused on determining the structural characteristics, which confer cold adaptation in enzymes. The tertiary and quaternary structures of cold-active enzymes have more open and flexible arrangements, thus providing better access of substrates to the active site at lower temperatures. Rigid secondary structures and disulphide bridges are practically absent, thus accounting for increased thermolability (Feller *et al.*, 1999). Individual enzyme types possess different structural strategies to gain overall increased flexibility. Certain structural features thought to be indicative of cold adaptation have also been found in similar non-cold adapted enzymes (Schroder *et al.*, 1998).

In addition, the same enzyme from different organisms, containing an identical amino acid sequence, has been found to possess different thermal properties (Love and Maeshall, 1998). This suggests that protein folding has a critical role in conferring activity at low temperature. With more knowledge of the mechanisms of cold adaptation our fundamental understanding of protein function and structure will be advanced. Preliminary data have been obtained for a variety of psychrophilic and psychrotolerant (cold-tolerant) enzymes from Antarctic bacterial isolates (Buia, 1999). The isolates were obtained from sea ice and lake habitats in the Vestfold Hills. Whole cell and cell-free assays indicated the presence of protease, β-galactosidase, phosphatase, and amylase enzyme(s) exhibiting strong cold adaptation in several strains (Table 3).

Table 3: Optimal temperature and relative activity of cold-active enzymes from a variety of Antarctic sea ice bacteria. Adapted from Buia (1997)(9) and Nichols *et al.* (1996).

Bacterial Strains	Enzymes	Optimum temperature	% Activity (10°C)	% Activity (25°C)
Colwellia demingae	Protease(azocasine)	28	75	25
	Protease(azoalbumin)	30	39	30
	Trypsin	14	90	29
			90	85
Colwellia-like strain	β galactosidase	26	100	53
	phosphatase	17		
Cytophaga-like strain	phosphatase	19	85	85
	Protease(azocasine)	20	75	70
Cytophaga-like strain	Protease(azoalbumin)	27	68	65
	Trypsin	30	70	55
	β galactosidase	15	72	60
	A amylase			
	Protease			
	trypsin			
	β galactosidase			

The *Cytophaga*-like strain IC166 showed particular promise as it elaborates several cold adapted enzymes. From this research it is apparent Antarctic bacteria, especially those isolated from ice, are good sources of cold-active enzymes (Buia, 1997).

In summary, Antarctic bacteria offer a unique biological resource. To date a high proportion of isolates represent novel biodiversity and biotechnological opportunities have been identified in the areas of the production of PUFA and cold-active enzymes. Some of known applications of cold active enzymes and their sources are given below in Table (4).

Table 4: Enzymes produced by Psychrophillic microorganisms and their industrial applications.

Enzymes	Micro-organisms	Applications
Proteases	*Serratia liquefaciens Bacillus* spp *Pseudomonas* spp *Xanthomonas maltophila Candida humicola*	Detergent additives dairy Industry
Lipase	*Pseudomonas tolasii Bacillus* sp	Detergent additive,in pharmaceuticals
α Amylases	*Micrococcus* spp *Alteromonas haloplanctis*	Detergent additive, food industry
β Galactosidases	*Arthrobacter* sp	Food Industry
Amylases	*Aromonas hydrophila micrococcus*	Detergent additive, food industry
Alkaline proteases	*Pseudomonas aeruginosa*	Textile industry
	Aeromonas hydrophilla	Detergent additive
β Lactamase	*Psychrophile*	Food industry

DETOXIFICATION OF ENVIRONMENT

Biodegradation is the natural degradation of matter in the environment in the absence of any human intervention. Bioremediation, in contrast, is characterized by human intervention and is the technology of pollution treatment, using biological systems to transform and convert various pollutant species in the environment to less toxic or non-toxic forms. An effective bioremediation will produce harmless water and carbon dioxide as the end-products, which are then able to re-enter natural ecosystems.

In recent years there has been an increased interest in the use of microorganisms for environmental restoration. The usefulness of psychrophillic microorganisms with a diversity of metabolic activities in wide ranging applications coupled with advances in the technology have lead to successful demonstrations in the ever-expanding bioremediation in polar and Arctic regions. The application of environmental biotechnology as a successful remediation tool depends on the ability to stimulate or enhance specific activity of indigenous or introduced microorganisms. Many of cold adapted bacteria have a ubiquitous distribution in the environment and stimulate bioremediation activities through enzymes such as mono-oxygenase. It allows for a relatively efficient, inexpensive, and safe means to manipulate the environment to accelerate bioremediation.

In temperate regions due to seasonal variations in temperature, effectivenss of micro-organisms reduces in degrading organic pollutants such as oils, lipids and xenobiotics ie phenolic and nonphenolic aromatic compounds. However bioagumentation and inoculation of contaminated environment with specific cold adapted microorganism in mixed cultures can help the biodegradation of recalcitrant. As a result of high catalytic efficiency of their enzymes and their unique specificity at low temperature, cold adapted microorganism should be ideal for bioremediation purpose (Margesin and Schinner, 1999).

Many hydrocarbon-contaminated environments are characterized by low temperatures. Hydrocarbon-degrading microorganisms, adapted to grow and thrive in these environments, play an important role in the biological treatment of polluted extreme habitats. The biodegradation (transformation or mineralization) of a wide range of hydrocarbons, including aliphatic, aromatic, halogenated and nitrated compounds, has been shown to occur in various extreme habitats. The biodegradation of many components of petroleum hydrocarbons has been reported in a variety of terrestrial and marine cold ecosystems. Cold-adapted hydrocarbon degraders are also useful for wastewater treatment (Margesin and Schinner, 2001).

The potential of eighty-nine culturable cold-adapted isolates from uncontaminated habitats, including 61 bacterial and twenty eight yeast strains, to utilize representative fractions of petroleum hydrocarbons (*n*-alkanes, monoaromatic and polycyclic aromatic hydrocarbons) for growth and to produce various enzymes at 10°C was investigated. The efficiency of bacterial and yeast strains was compared. The growth temperature range of the yeast strains was significantly smaller than that of the bacterial strains. Sixty percent of the yeasts but only 8% of the bacteria could be classified as true psychrophiles, showing no growth above 20°C. A high percentage (89%) of the yeast strains showed lipase activity. More than one-third of the 61 bacterial strains produced amylase, beta-lactamase, beta-galactosidase or lipase; more than two-thirds were protease producers. Only 6% of the bacterial strains but 79% of the yeast strains utilized n-hexadecane for growth; 13% of the bacterial strains and 21-32% of the yeast strains utilized phenol, phenanthrene or anthracene for growth. Only four yeast strains but none of the bacterial strains could grow with all hydrocarbons tested.

Polycarbonate degradation using psychrotolerant strains isolated from low temperature habitat was investigated and was observed that out of 30 bacterial strains 4 showed high potential for the polycarbonate degradation (Goel *et al.*, 2004)

The biodegradation of phenol was investigated in fed-batch cultures at 10°C. Three yeast strains degraded phenol concentrations as high as 10 mM (one strain) or 12.5 mM (two strains). Of eight bacterial strains, two strains degraded up to 10 mM phenol. The optimum temperature for phenol degradation was 20°C for all eight bacterial strains and for two yeast strains. Biodegradation by five yeast strains was optimal at 10°C and faster at 1°C than at 20°C. All phenol-degrading strains produced catechol 1, 2 dioxygenase activity (Margesin and Schinner, 2001).

Three strains capable of mineralizing nonylphenol as sole carbon source were isolated from a sample of contaminated soil and characterized as two *Pseudomonas* spp. and a *Stenotrophomonas* sp. The two *Pseudomonas* spp. expressed characteristics typical of psychrophiles growing optimally at 10°C and capable of growing at 0°C. The *Stenotrophomonas* sp. was more likely psychrotrophic because it had an optimal temperature between 14 and 22°C although it was not capable of growing at 4°C. At 14°C, one of the *Pseudomonas* spp. exhibited the highest rate of degradation of nonphenol (4.4 mg l-1 d-1), when compared with axenic or mixed cultures of the isolates (Soares *et al.*, 2003).

ANTIFREEZE PROTEINS

There are a number of substances described in the published literature which inhibit ice nucleation. Certain bacterial strains, mostly found among the non-fluorescent pseudomonade species, release material into the growth medium which reduces the nucleation temperature of water droplets to below that of distilled water (Holt, 2003). These substances include sucrose, unsaturated fatty acids and phospholipids. But in psychrophillic bacteria, some specific proteins are produced which reduce freezing temperature and protect them form freeze injury. These are known as antifreeze proteins.

Antifreeze proteins (AFPs) are a structurally diverse group of proteins that have the ability to modify ice crystal structure (Nichols *et al.*, 1999) and inhibit recrystallization of ice (Knight *et al.*, 1988). They have been extensively described in polar fish, some insects, plants and fungi. AFPs and antifreeze lipoproteins (AFLPs) have been also found in some bacteria. A novel AFP assay, designed for high-throughput analysis in Antarctica, demonstrated putative activity in 187 of the cultures. Subsequent analysis of the putative positive isolates showed 19 isolates with significant recrystallization inhibition (RI) activity. The 19 RI active isolates were characterized using ARDRA (amplified rDNA restriction analysis) and 16S rDNA sequencing. They belong to genera from the – γ Proteobacteria, with genera from the - α subdivision being predominant. The 19 AFP-active isolates were isolated from four physico-chemically diverse lakes (Jacket *et al.*, 2004).

Antifreeze proteins have been isolated and characterized by Katiyar (Katiyar 2004) from cold resistant mutants of *Pseudomonas fluorescens* and observed that these AFPs, were able to protect freeze labile alkaline phosphatase against freeze denaturation upto approximately 84%.

The structural and functional features of antifreeze proteins (AFPs) enable them to protect living organisms by depressing freezing temperatures, modifying or suppressing ice crystal growth, inhibiting ice recrystallization, and protecting cell-membranes from cold-induced damage. The versatility of the AFPs suggests that their production and commercialization would be a potentially profitable venture. The current commercial sources of antifreeze proteins are restricted to fish. Problems exist with source because it requires a relatively high investment of technology and expense compared to the prospect of bacterial production. Therefore use of Antarctic bacteria for commercial production of AFPs is preferable.AFPs and their genes can be used in fish and plants to enhance resistance to freezing. AFPs can be used in medicine to improve the cold protection of blood platelets (to extend their shelf life prior to transfusion); paradoxically, when used in conjunction with cryosurgery, they can help destroy malignant tumors. Their ability to inhibit recrystallization can improve the quality of frozen foods. In addition, antifreeze gene promoters are uniquely suited to drive the expression of functional genes, such as growth hormone that results in enhanced growth rates of salmonids (e.g. salmon, trout) and other fish species valuable to aquaculture.

Therefore, we propose commercial applications of AFPs in the following areas: (*i*) cold protection of mammalian cells, tissues, and organs, *(ii)* enhanced tumor cell destruction during cryosurgery, *(iii)* longer shelf-life for and better quality of frozen foods; and improved growth characteristics in transgenic fish by using AFP gene promoters.

CROP PROTECTION WITH AFPs

The damaging effects of low temperatures and freezing conditions on plant material consist of mechanical injury (cell and tissue disruption), which results from ice formation and dehydration injury caused by water loss associated with ice formation. Under deep-freezing conditions, intracellular bulk water and water oriented on the surface of macromolecules and on the polar heads of lipids in cellular membranes are effectively removed, causing severe dehydration and structural and functional damage to plasma membranes. Although low and freezing temperatures kill many plants, some species and varieties have developed physiological processes to help them survive the winter. With the onset of cold conditions, some plants produce colligative cryoprotectants such as sucrose and proline; in others, changes in membrane lipids and proteins that render membranes more cold stable have been reported.

The plant growth promoting rhizobacterium *Pseudomonas putida* GR12-2 was isolated from the rhizosphere of plants growing in the Canadian High Arctic. This bacterium was able to grow and promote root elongation of spring and winter canola at 5°C, a temperature at which only a relatively small number of bacteria are not able to proliferate and function. In addition, the bacterium survived exposure to freezing temperatures, i.e. –20 and –50°C it was discovered that following growth at 5°C, *P. putida* GR12-2 synthesized and secreted to the growth medium a protein with antifreeze activity (Schroder *et al.*, 1998).

PLANT GROWTH PROMOTION AT LOW TEMPERATURE

Temperature is below 20°C in Arctic and Alpine region. Therefore, the growth of plants is highly affected and results in freeze injury in plants and loss of yield. Use of cold adapted PGPR strains in such climates can be a better option for plant growth promotion. The impact of rhizobacteria (those members of the total rhizosphere bacteria that are able to colonize roots) on plant growth may be neutral, deleterious or beneficial. The term plant growth-promoting rhizobacteria (PGPR) was given to beneficial rhizobacteria in 1978. Most PGPRs are fluorescent *Pseudomonas* or strains of *Bacillus subtilis*. Plant growth promoting rhizobacteria can be inoculated on to some crops and can subsequently improve growth. The beneficial effects of PGPRs fall into two categories: growth promotion and plant disease suppression. Growth-promotion is evidenced by increase in seedling emergence, vigour, seedling weight, root system development, and yield. Various mechanisms of PGPR, which enhance growth, are not yet well understood. Some PGPR strains are supposed to increase plant growth by positively interacting with various plant-symbiotic microorganisms, such as *Rhizobium, Bradyrhizobium, Frankia* and mycorrhizal fungi. For example, several *Pseudomonas* and *Bacillus* sp. are capable of enhancing modulation.

Some PGPR strains produce siderophores, which is a low molecular weight protein and has high affinity for iron. At low temperature cold resistant mutant of *Pseudomonas* CRPF8 and CRPf9 overproduces siderophores and show better growth promotion in wheat (29%) and mung bean (51.1%) at 10°C(35). There are also PGPRs, which produce plant growth hormones in culture, but there is no actual proof of such activity under normal field conditions. Some PGPR strains do not promote growth per se, but the inoculated bacteria are controlling minor root pathogens, which are not producing enough obvious symptoms to be noticed, but which hinder the complete expression of the plant's potential.

Katiyar and Goel (2004) reported siderophores mediated plant growth promotion at low temperature both under *in situ* and *in vitro* conditions by cold tolerant mutant of *Pseudomonas fluorescens*. It was observed that cold tolerant mutants induces a significant increase in root (35%) and (28%)of mung bean plants in unsterlized soil system.

COLD-ADAPTED RHIZOBIA

Cold-adapted rhizobia were selected with the aim of improving productivity of legumes that are subjected to cool temperatures during the growing season. One approach was to use rhizobia associated with legume species indigenous to Arctic and subarctic regions: (*i*) *Mesorhizobium* sp. isolated from *Astragalus* and *Oxytropis* spp. and (*ii*) *Rhizobium leguminosarum* from *Lathryrus* spp. The majority of these rhizobia were psychrotrophs because they can grow at 0°C. The advantages of cold adaptation of Arctic *Mesorhizobium* to improve legume symbiosis were demonstrated with the temperate forage legume sainfoin (*Onobrychis viciifolia*). In laboratory and field studies, Arctic rhizobia were more efficient than temperate (commercial) rhizobia in improving growth of sainfoin and were more competitive in forming nodules. Biochemical studies on cold adaptation showed higher synthesis of cold shock proteins in cold adapted than in non-adapted Arctic rhizobia. Since Arctic *Mesorhizobium* cannot nodulate agronomically important legumes, the nodulation genes and the bacterial signals (Nod factors) were characterized as a first step to modifying the host specificity of nodulation. Another valuable approach was to screen for cold adaptation i.e. rhizobia naturally associated with agronomic legumes cultivated in temperate areas. A superior strain of *Sinorhizobium meliloti* adapted for nodulation of alfalfa at low temperatures was selected and was the most efficient for improving growth of alfalfa in laboratory and field studies. This strain also performed well in improving regrowth of alfalfa after over-wintering under cold and anaerobic (ice encasement) stresses, indicating a possible cross-adaptation of selected rhizobia for various abiotic stresses inherent to temperate climates.

Co-inoculation of plant growth-promoting rhizobacteria (PGPR) with *Bradyrhizobium* has been shown to increase legume nodulation and nitrogen fixation at optimal soil temperatures. Feng *et al.* (1996) tested nine rhizobacteria co-inoculated with *Bradyrhizobium japonicum* 532C for their ability to reduce the negative effects of low root zone temperature (RZT) on soyabean [*Glycine max*(L.) Merr.] nodulation and nitrogen fixation at three temperatures: 25°C (optimal), 17.5°C (somewhat inhibitory), and 15°C (very inhibitory). At each temperature some PGPR strains increased the number of nodules formed and the amount of fixed nitrogen when co-inoculated with *B. japonicum,* but the stimulatory strains varied with temperatures, at 15°C, *Serratia proteamaculans* 1-102; 17.5°C, *S. proteamaculans* 1-102 and *Aeromonas hydrophila*P73; at 25°C *Serratia liquefaciens* 2-68 showed maximum stimulatory effects.

CONCLUSION AND PERSPECTIVES

In summary, it is clear that psychrophiles have a great potential in many industrial processes including agriculture, chemicals, and pharmaceutical applications; many consumer products will increasingly benefit from addition or further exploitation of psychrophiles; the tool-box to select and make such products is steadily expanding.

It has been suggested that less than 10% of organisms in a defined environment will be cultivated and so further improvement of gene expression technologies will accelerate the exploration of microbial diversity. Concurrent development of protein engineering and directed evolution technologies will result in further tailoring and improving the industrial traits which will increase the application of psychrophiles, by-products in industries.

REFERENCES

Aguilar, A. (1996) *FEMS Microbiol. Rev.* **18:** 89-92

Allen, D.M. & Deming, J.W. (2002) American Society for Microbiology 102nd Annual Meeting (Abstract I-52), 19–23 May, Salt Lake City.

Baker, V.R. (2001) *Nature.* **412**: 228-236.

Beja, O., Koonin, E.V., Aravind, L., Taylor, L.T., Seitz, H., Stein, J.L., Bensen, D.C., Feldman, R.A., Swanson, R.V. & DeLong, E.F. (2002) *Appl. Environ. Microbiol.* **68:** 335-345.

Bendt, A., Huller, H. Kammel, U. Helmke, E. & Schweder, T. (2001) *Extremophiles* **5:** 119-126.

Bowman, J.P. (2001) In *Bergey's Manual of Systematic Bacteriology,* (Eds. Krieg, N.R. & Holt, J.G.) Williams & Wilkins, Baltimore.

Brambilla, E., Hippe, H., Hagelstein, A., Tindall, B.J.& Stackebrandt, E. (2001) *Extremophiles* **5**: 23-33.

Brown, M.W. & Bowman, J.P. (2001) *FEMS Microbiol. Ecol.* **35:** 267-275

Buia, A. (1997) *BSc (Hons) Thesis*, University of Tasmania.

Bull, A.T. (1992) *Annu. Rev. Microbiol.* **46:** 219-252.

Carpenter, E.J., Kin, S. & Capone, D.G. (2000) *Appl. Environ. Microbiol.* **66:** 4514-4517.

Chyta, C.F. & Phillips, C.B. (2002) *Orig. Life Evol. Biosph.* **32:** 47-68.

D'Amico, S., Gerday, C. & Feller, G. (2001) *J. Biol. Chem.* **276:** 25791-25796.

Das, K., katiyar,V. & Goel,G. (2003) *Microbios. Res.***158:** 359-62.

Deming, J.W. & Baross, J.A. (2000) In *Non-Culturable Microorganisms in the Environment* (Eds. Colwell, R.R & Grimes, D.J.), American Society for Microbiology Press, Washington, DC, pp. 147-197.

Deming, J.W. & Baross, J.A. (2001) In *Enzymes in the Environment.* (Eds. Dick, R.P. & Burns, R.G.) Marcel Dekker Publishers, New York, pp. 327-362.

Deming, J.W. & Huston, A.L. (2000) In *Cellular Origins and Life in Extreme Habitats.* (Eds. Seckbach J.), Kluwer Publishers, Dordrecht, pp. 149-160.

Desmarais, D.J. & Walter, M.R. (1999) *Annu. Rev. Ecol. Syst.* **30:** 397-420.

Feller, G., Arpigny, J.L., Narinx, E. & Gerday, C. (1997) *Biochem. Physiol,* **118:**495-499.

Feller G, & Gerday, C. (2003) *Nat. Rev. Microbiol.* **1 (3):** 200-8.

Feller G, & Gerday, C. (1997) *Cell Mol. Life Sci.* **53**:830-841.

Feng, Zhang, Narjes Dashti, Hynes, R.K. & Donald L. Smith. (1996) *Annals Bot.* **77**: 453-460.

Goel,R., Kusum lata, Zaidi, M.G.H., Khan, M. & Sarita, (2004) In *Challenges and opportunity to harness the modern biology for socio-economic development in genomic era*, 73rdannual meeting , Nov 21-24. pp. 94

Herborg, L.M., Thomas, D.N., Kennedy, H., Haas, C. & Dieckmann, G.S. (2001) *Antarctic. Sci.* **13:**119-125.

Hollibaugh, J.T., Bano, N. & Ducklow, H.W. (2002) *Appl. Environ. Microbiol.* **68:** 1478-1484.

Holt, C.B. (2003) ***Error! Bookmark not defined.*, , 24 (6):** 269-274

Hugenholtz, P., Goebel, B.M. & Pace, N.R. (1998) *J. Bacteriol.* **180**: 4765-4774.

Humphrey, D.R., George, A., Black, G.W. & Cummings, S.P. (2001) *Int. J. Syst. Evol. Microbiol.* **51:** 1235-1243.

Irwin, J.A., Gudmundsson, H.M., Marteinsson, V.T. & Hreggvidsson (2002) *Appl. Environ. Microbiol.* **68:** 456-463.

Jack. A. Gilbert, Philip, J. H, Christine, E., Dodd, R. & Johanna Laybourn-Parry (2004) *Microbiology* **150**:171-180

Junge, K, Imhoff, J.F., Staley, J.T. & Deming, J.W. (2002) *Microb. Ecol.* **43**. 218-225.

Junge, K., Krembs, C., Deming, J., Stierle, A. & Eicken, H. (2001) *Ann. Glaciol.* **33:** 304-310.

Karner, M.B., DeLong, E.F. & Karl, D.M. (2001) *Nature,* **409: 507**-510.

Katiyar, V. & Goel, R. (2004) *Plant Growth regulation.***42(3):** 239-244.

Katiyar,V. & Goel, R. (2004) *J. Microbial. Biotechnol.* **14(4):** 653-657.

Katiyar,V., Mishra,D.P., Kumar, S. & Goel, R. (2004) *Ind. J. Biotech.***3:** 378-381.

Kato, C., Li, L., Nogi, Y., Nakamura, Y., Tamaoka, J. & Horikoshi, K. (1998) *Appl. Environ. Microbiol.* **64:** 1510-1513.

Kaye, J.Z., & Baross, J.A. (2000) *FEMS Microbiol. Ecol.* **32:** 249-260.

Khan,M., Bajpai,V.K., Ansari, S.A., Kumar, A. & Goel, R. (2003) *Microbiol . Immunol,* **47(12)**:895-901:

Knight, C. A., Hallett, J. & Devries, A. L. (1988) *Cryobiology* **25:** 55–60.

Krembs, C. & Engel, A. (2001) *Mar. Biol.* **138:** 173-185.

Kuypers, M.M., Blokker, P., Erbacher, J., Kinkel, H., Pancost, R.D., Schouten, S. & Sinninghe Damste, J.S. (2001) *Science,* **293**:92-95.

Lewis, T.E., Nichols, P.D. & McMeekin, T.A. (1999) *Mar. Biotechnol.* **1:** 580-587.

Lonhienne, T., Baise E., Feller, G., Bouriotis, V. & Gerday, C. (2001) *Biochim. Biophys. Acta Prot. Struct. Mol. Enzymol.* **1545:** 349-356.

Lonhienne. T., Zoidakis, J., Vorgias, C.E., Feller, G., Gerday, C., Bouriotis, V., Beja, O., Koonin, E.V., Aravind, L, Taylor, L.T., Seitz, H., Stein, J.L., Bensen, D.C., Feldman, R.A., Swanson, R.V. & DeLong, E.F. (2002) *Appl. Environ. Microbiol.* **68**: 335-345.

Lonhienne, T., Zoidakis, J., Vorgias, C.E., Feller, G., Gerday, C. & Bouriotis V. (2001) *J. Mol. Biol.* **310**: 291-297.

Lopez-Garcia, P., Lopez-Lopez, A., Moreira, D. & Rodriguez-Valera, F. (2001a) *FEMS Microbiol. Ecol.* **36:**193-202.

Lopez-Garcia, P., Moreira, D., Lopez-Lopez, A. & Rodriguez-Valera F (2001b) *Environ. Microbiol.* **3:** 72-78.

Love, C.A. & Marshall, C.J. (1998) *International Biology Symposium,* Christchurch, New Zealand **23:**108.61.

Margesin, A. I. & F. Schinner, A.I. (2001) *Applied Microbiol. Biotechnol.* **56**: 650 - 663

Margesin, R. & Schinner, F. (1999) *Biotechnological Applications of Cold-Adapted Organisms.* Springer-Verlag, Berlin. pp.338

Meyer,B.J, Tsivis, E., Howe., P.R.C., Tapsell, L. & Calvert, G.D.(1999) *Foods Australia* **51:**82-95

Mindock, C.A., Petrova, M.A. & Hollingsworth, R.I. (2001) *Biophys. Chem.* **89:** 13-24.

Mishra, M. & Goel, R. (1999*) J. Biotech.* 75:**71-74.**

Morita, Y, Nakamura, T., Hasan, Q., Murakami, Y., Yokoyama, K. & Tamiya, E. (1997) *J. Am. Oil Chem. Soc.* **74:** 441-444.

Nakasone, K., Ikegami, A., Kato, C., Usami ,R. & Horikoshi K. (1998) *Extremophiles* 2:149-154.

Nichols, D.S., Hart, P., Nichols, P.D. & McMeekin, T.A. (1996) *Aquaculture* **147:** 115-125.

Nichols, D.S., Sanderson, K., Bowman, J., Lewis, T., Mancuso, C.A., McMeekin, T.A. & Nichols, P.D. (1999) *Curr. Opin. Biotechnol.,* **10:** 240-246.

Oikawa, T., Yamanaka, K., Kazuoka, T. , Kanzawa, N. & Soda, K. (2001) *Eur. J. Biochem.* **268**: 4375-4383.

Pomeroy, L.R. & Wiebe, W.J. (2001) *Aquat. Microbiol. Ecol.* **23:** 187-204.

Price, P.B. (2000) *Proc. Natl. Acad. Sci. USA* **97:** 1247-1251.

Ravenschlag, K., Sahm, K. & Amann, R. (2001) *Appl. Environ. Microbiol.* **67**: 387-395.

Raymond, J.A. & Fritsen, C.H. (2001) *Cryobiol.* **43:** 63-70.

Raymond, J. A. & DeVries, A. L. (1977) *Proc. Natl. Acad. Sci. U S A* **74:** 2589–2593

Rivkina, E.M., Friedmann, E.I., McKay CP. & Gilichinsky, D.A. (2000) *Appl. Environ. Microbiol.* **66:** 3230-3233.

Russell, N.J. (1992) In: *Molecular Biology and Biotechnology of Extremophiles* (Eds. Herbert, R.A. & Sharpe, R.J., Blackie) Glasgow, pp. 203-244.

Russell, N.J. & Nichols, D.S. (1999) *Microbiology* **145:** 767-779

Schroeder, H.K., Willassen, N.P. & Smalas, A.O. (1998) *Acta Crystallogr D,* **54**:780-798.

Secades P, Alvarez B, Guijarro JA. 2001, *Appl. Environ. Microbiol.* **67**:2436-2444.

Sheridan, P.P., Lanasik, N., Coombs, J.M. & Brenchley, J.E. (2000) *Biochim. Biophys. Acta Prot. Struct. Molec. Enzymol.,* **1543**: 417-433.

Siegert, M.J., Ellis-Evans, J.C., Tranter, M , Mayer, C. , Petit, J.R. , Salamatin, A. & Priscu, J.C. (2001) *Nature.* **414:** 603-609.

Soares, A., Guieysse, B., Delgado, O. & Mattiasson, B. (2003) *Biotechnol. Lett.* **25(9):** 731-738.

Staley, J.T., Junge, K. & Deming, J.W. (2001) In *Biodiversity of Microbial Life: Foundation of Earth's Biosphere.* (Eds. Staley J.T. & Reysenbach, A.L.) New York: Wiley-Liss, Inc. pp. 423-438.

Sun, Xiuying, Griffith, M., Pasternak, J.J. & Glick, B.R. (1995) *Can. J. Microbiol.* **41:** 776-784

Takahata, K., Monobe, K., Tada, M. & Weber, P.C. (1998) *Biosci. Biotechnol. & Biochem.* **62:** 2079-2085.

Takeyama, H., Takeda, D., Yazawa, K., Yamada, A., & Matsunaga, T. (1997) *Microbiol.* **143:** 2725-2731.
Tanaka, M., Okuyama, H., & Morita, N. (2001) *Biosci. Biotechnol. Biochem.* **65:** 666-669.
Thomas, D.N. & Dieckmann, G.S. (2002) *Science.* **295:** 641-644.
Thomas, T., Kumar, N. & Cavicchioli R. (2001) *J. Bacteriol.* **183:** 1974-1982
Torssell, K.B.G. (1983) *Natural Product Chemistry*, pp. 401.
Truong, L.V., Tuyen, H., Helmke, E., Binh, L.T. & Schweder T (2001) *Extremophiles* **5:** 257-264.
Tutino, M.L., Duilio, A., Parrilli, E., Remaut, E., Sannia, G., Marino, G. (2001) *Extremophiles* **5:** 257-264.
Wells, L.E. & Deming JW (2001) Abstract N-206 of the American Society for Microbiology 101st Annual Meeting, May 20–24, Orlando, Florida.
Wells, L.E. & Deming, J.W. (2002) Viral abundance, diversity and processes in the cold, oligotrophic Northwest Passage of the Canadian Archipelago. Abstract N-181 of the American Society for Microbiology 102nd Annual Meeting, May 19–23, Salt Lake City.
Yager, P.L. (1996) *Ph.D Thesis,* Seattle: University of Washington.
Yazawa, K. (1996) *Lipids.* **31:** 297-300.

3

Lactic Acid: A Potential Microbial Metabolite

Gopal Reddy, B.J. Naveena and Md. Altaf
Department of Microbiology, Osmania University
Hyderabad-500 007 (India)

Lactic acid fermentation has received extensive attention for a long time. Holten, Lockwood and Budavari have given various physical and chemical properties of lactic acid. Lactic acid has wide applications in food, pharmaceutical, leather, textile industries and as a chemical feed stock. It has two enantiomers L(+) and D(–). L(+) lactic acid is in great demand due to its use as a starter for biodegradable polymer, polylactate, for medical, industrial and consumer products. Specialty chemical applications for optically active lactic acid and its derivatives are very high. Due to its significant role in almost all the sectors of industries, lactic acid production using cheaper raw materials to make fermentation process economical has accelerated to a greater extent. As per the available statistics, there is a large gap between the production and demand for lactic acid. From the beginning refined sugars are the choice as carbon source for lactic acid fermentation. Replacement of sugars with renewable inexpensive substrata (like starches) and also single step conversion of these substrates will lead to elimination of pretreatment steps which can make the fermentation process cost effective. Use of starchy agricultural wastes (wheat bran) can make the process further cheap. Mostly amylolytic *Lactobacillus* spp are preferred for industrial fermentations in comparison to fungal organisms for their higher yield efficiencies. Obligately homofermentative lactobacilli are one of the physiological groups which are frequently used for lactic acid fermentations. Laboratory studies were found to be successful for single step production of L(+) lactic acid from starchy substrates by *Lactobacillus amylophilus* GV6 in submerged, semi-solid state and solid-state fermentations. The overall view of lactic acid as a microbial metabolite, fermentation, its requirements, uses, economics and organisms involved are reviewed.

INTRODUCTION

Lactic acid (2-hydroxy propionic acid, α - hydroxypropionic acid, ethylideneactic acid or 1-hydroxyethanecarboxylic acid) is one of the earliest known fermentation products from microbial metabolism. In 1780, Carl Wilhem Scheele had discovered the 'milk acid' which was thought to be a milk component, later identified as lactic acid. Lactic acid [L(+)] is naturally present in animal and human tissues as well as in numerous food products (meat, milk products, pickles, beer, etc.). Lactic acid for centuries is in use as a natural preservative in many food products. Many years ago, people started using lactic acid bacteria to produce cultured foods (pickling of vegetables, baking, wine making, curing fish, meats and sausages) with improved preservation properties and with characteristic flavours and textures, without understanding the scientific basis. Later Louis Pastuer, Joseph Lister and Max Delbruck identified lactic acid as a microbial metabolite (Beninga, 1990; Gregor, 1999). Lactic acid bacteria are known for their role in the preparation of fermented dairy products by producing lactic acid as a product of the fermentation process.

Due to the importance of lactic acid in various sectors of industries, Charles E Avery at Littleton, Massachusetts, USA first commercially produced it in 1881 in the form of calcium lactate. Application of lactic acid in leather and textile industries began in 1894 (Garet, 1930). Lactic acid's use as a raw material for biodegradable plastics was identified in late 1940s and early 1950s (Vickroy, 1985).

Research towards production of higher grades of lactic acid (for polymer synthesis) and in reduction of costs with essential properties was unsuccessful. A decade later need for heat-stable lactic acid in baking industry has opened a way for synthetic lactic acid production (Anon, 1963). In 1942, around 50% of the lactic acid produced in the US was utilized in leather industries and 20% in food industries (Filachione, 1952). Slowly, food industries dominated the market for lactic acid consumption. In 1982, about 50% of the lactic acid produced was used as acidulant in food industry and 20% in baking industry as emulsifying agent. Overall, food industries surpassed the application of lactic acid in leather industries (Vickroy, 1985).

Although importance and multiple applications of lactic acid were well understood, it was not produced as a large volume chemical. By 1990 two major producers of lactic acid emerged, CCA Biochem b.v. (from Netherlands and its subsidiaries in Brazil and Spain) through fermentative process and Sterling Chemicals, Inc. (in Texas City, USA,) and Musashino (in Japan) through chemical technology (Datta *et al.*, 1995) in lower volumes. Major use of lactic acid (accounting to 85% of demand) is still in food and food related applications (Datta *et al.*, 1995). About 50% of lactic acid production is through fermentation technology (Vickroy, 1985). Presently, Purac in America, Galactic in Belgium, some small producers in Europe, the Far East and Cargill Dow in the US are the major producers of lactic acid through fermentation process except for Musashino Chemical Laboratory Ltd. (Mirasol, 1999). Purac is the world's largest and most experienced manufacturer of natural L(+) lactic acid. Existing manufactures in India are Lactochem Ltd. a Mahadi group company, Munirabad Chemical Company of Karnataka, Prathista Industries of Andhra Pradesh, Sangita Biotech Limited from Kolkata and Interspice Biotech of Chennai.

Both chemical process and fermentative processess are involved in commercial production of lactic acid. Chemical synthesis of lactic acid involves hydrolysis of lactonitrile formed by the reaction of acetaldehyde with hydrogen cyanide (Holten *et al.*, 1971; Benninga, 1990). Mostly lactic acid production is through fermentation technology using sugars obtained from variety of carbohydrate sources. The existing industries are producing L(+)/D(-) lactic acid using homolactic organisms (Litchfield, 1996).

PHYSICAL AND CHEMICAL PROPERTIES

Holten *et al.* (1971) gave extensive coverage to physical and chemical properties of lactic acid. Lactic acid is the smallest chemical molecule with an asymmetric carbon and therefore exists with two optically active stereo-isomers: the L (+) or S (+) and the D (-) or R(-). Racemic mixture (DL) results from chemical synthesis or through some lactic acid bacteria possessing both, D and L lactate dehydrogenases. The specific optical rotations of the zinc salts of lactic acid are opposite to those of free acids (Holten *et al.*, 1971). Lockwood *et al.*, (1965) stated that although L (+) form appears to be dextrorotatory, it may actually be leavorotatory as are its salts and esters. The apparent reversal in optical rotation may be due to the formation of an ethylene oxide bridge between carbon atoms 1 and 2 by tautomeric shift of the hydroxyl group on carbon atom 2 to the carboxyl radical. Salts and esters of L(+) lactic acid cannot form this epoxide ring and hence remain levorotatory only. Although both the L(+) and D(-) isomers are found in biological systems, only L(+) isomer is present in humans, possessing L lactate dehydrogenase (LDH) (Lockwood *et al.*, 1965). Some selected properties of general interest of lactic acid enantiomers are outlined in Table 1. Highly pure optically active lactic acid can form monoclinic crystals and it is hygroscopic (Lockwood *et al.*, 1965). Lactic acid is soluble in water with less volatility. Self-esterification of lactic acid is identified in solutions with 20% lactic acid and more due to presence of hydroxyl and carboxyl functional groups.

Table 1: Selected properties of lactic acid enantiomers (Holten *et al.*, 1971).

Property	L(+) Lactic acid	D(-) Lactic acid	DL Lactic acid
Molecular weight	90.08	90.08	90.08
Melting point ° C	52.8 – 53.6	52.8 – 53.6	16.8 – 33
Boiling point ° C	–	103	82 – 85
Optical rotation $[\alpha]_D^{20}$, degrees			
Acid	+ 2.5	– 2.5	–
Zinc salt	– 8.2	+ 8.18	–
Dissociation constant	1.90×10^{-4}	–	1.38×10^{-4}
pK (25° C)	3.79	3.83	3.73

Lactic acid forms the linear dimer lactoyl lactate, higher linear polymers in aqueous solutions and on prolonged heating at 140° C at low pressures (10 mm) it can also form lactide (a cyclic dimmer) and it is reversible by distillation. Lactic acid can participate in numerous types of chemical reactions because it can react both as an organic acid and organic alcohol (Holten *et al.*, 1971).

Lactic acid is analysed by both enzymatic and non-enzymatic methods. Drewes (1974) described enzymatic analysis of lactic acid through calorimetric assay. A non-enzymatic, non-stereo specific, colorimetric assay is given by Pryce (1969), Kimberly and Taylor (1996). Titration/gas chromatography is also used for lactic acid analysis (Van Ness, 1981). Identification of lactic acid enantiomer and esters is possible by variations of HPLC techniques (Pirkle *et al.*, 1981; Milind *et al.*, 2004).

COMMERCIAL USE OF LACTIC ACID

The lactic acid consumption market is dominated by the food and beverage sector since 1982. Even today, lactic acid market still exists for food and beverage industries. It has a wide range of beneficial uses in sectors related to food preservation, flavour enhancement etc. Lactic acid is non-volatile, odourless organic acid and is classified as GRAS (generally recognized as safe) for use as a general-purpose food additive by FDA in the US and other regulatory agencies (Datta *et al.*, 1995). Litchfield (1996) has summarized typical food applications for lactic acid and its salts (Table 2).

Main functions of lactic acid and its salts (88% pure) in foods are its antimicrobial nature, as calcium source for pH regulation and flavouring agent. Because of its natural occurrence and mild taste, it is used in formulations of mayonnaises, dressings, sauces and salads as preservative and pH controlling agent. It is used as pickling agent for sauerkraut, olives, pickled vegetables, canned vegetables and as gelling agent for jams and jellies. Usage of lactic acid and its salts can increase the shelf-life by 30 to 50% for products like sausages, hams, poultry, carcasses of fish, etc. by inhibiting the growth of spoilage organisms (Litchfield, 1996).

Table 2: Food related applications of lactic acid (Litchfield, 1996).

Compound (Chemical abstracts service Registry No.)	Food applications	Code for Federal Regulations Reference
L(+) lactic acid (79-35-4) D(-) lactic acid (10326-41-7) DL lactic acid (598-82-3)	Antimicrobial agent, curing and pickling, flavouring, enhancer, adjuvant, pH control, solvent and vehicle	21 CFR 184. 1061
Calcium lactate (814-80-2)	Flavouring enhancer, firming agent, leavening agent, nutrient supplement, stabilizer, thickener	21 CFR 184. 1207
Ferrous lactate (5905-52-2	Nutrient supplements and in infant formula	21 CFR 184. 1311
Potassium lactate (996-31-6)	Flavor enhancer, flavouring agent, humectant, pH control	21 CFR 184. 1639
Sodium lactate (72-17-3)	Flavor enhancer, flavouring agent, humectant, pH control, emulsifier	21 CFR 184. 1768
Calcium stearoyl-2-lactylate	Dough conditioner in bakery products, whipping agent in egg products, conditioning agent in dehydrated potatoes	21 CFR 172. 844
Sodium steroyl lactylates (25-383-997)	Dough conditioner, emulsifier, processing aid in baked products; emulsifier, stabilizer, processing aid in milk or cream substitutes., snack dips, imitation cheeses, dehydrated potatoes	21 CFR 172. 846
Lactylated esters of fatty acids	Emulsifiers, plasticizer, surface active agent in foods	21 CFR 172. 848
Lactylated fatty acid esters of Glycerol and propylene glycol	Emulsifiers, plasticizer, surface active agent in foods	21 CFR 172. 850
Glycero-lacto esters of fatty acids	Emulsifiers, plasticizer, surface active agent in foods	21 CFR 172. 852

In dairy, lactic acid is used for direct acidification of products such as cottage cheese, mozzarella, etc. to control production of unwanted by-products. Lactic acid is added to magarine, butter, yoghurts, etc. for pleasant taste. Calcium lactate is added to milk and other sports and diet drinks as mineral supplement. Lactic acid is used to adjust the pH of wine and cider beverages, as they are needed to maintain constant product quality. Beer fermentation needs lactic acid to adjust pH at various stages of the brewing process. Acidification of hard-boiled sweets to prevent sugar inversion and the resulting stickiness and opacity of the sweets is achieved by adding lactic acid (Litchfield, 1996).

A large fraction of (>50%) heat-stable fermentation grade lactic acid goes to produce emulsifying agents such as sodium and calcium stearoyl lactylates in bakery goods. Calcium salt (calcium lactate) is a good dough conditioner and sodium salt (sodium lactate) is both conditioner and emulsifier for yeast-leavened bakery products. In addition, esters of lactate salts (palmitates and stearates) are used in prepared cake mixes to improve texture, to increase cake value, to permit mixing tolerances and other bakery products in liquid shortenings. In addition to this, lactic acid is used for direct acidification of dough as a preserving agent and fermentation booster in products like rye bread (Litchfield, 1996; Archibald, 2003).

Lactic acid is used in pharmaceutical and cosmetic applications. Acyllactates are effective as emulsifiers, builders and stabilizers in cosmetics and personal care products (improving skin texture and appearance resulting from aging) (R.I.T.A. Corporation, 1985; Smith, 1993; Litchfield, 1996). Other applications of lactic acid include treatment of dermatological disorders, as intermediate in optically active drugs, in pH control for hemorrhoid preparations and as raw material for synthesis of polylactide polymer. Homopolymer of L-lactide is suitable for load bearing applications such as in orthopaedic fixation (pins, rods, tacks, staples for wound closure, dental applications) and sutures (Frost and Sullivan, 1995). Polylactide (DL-lactide) can be used in drug delivery system as their degradation rate is very rapid. Currently FDA has approved the devices made of homopolymers for marketing (John and Arthur, 1998). Polymers made of lactide have long history of safe and effective use. Lactic acid in the form of sodium salt acts as an electrolyte and sodium carrier in dialysis solutions, continuous ambulatory peritoneal dialysis solutions and parenteral solutions (Datta *et al.*, 1995; Malhotra *et al.*, 2000).

Mineral supplementation through calcium lactate is used in the treatment of osteoporosis, high blood pressure, colon cancer and as an anti-caries agent. Respective lactic acid salts are added in many food and health products to cure magnesium, iron and zinc nutrition deficiencies. Lactates are often superior to natural products and more effective than polyols as humectants in cosmetic applications (Van Ness, 1981). Ethyl lactate is the active ingredient in many anti-acne ointments (Datta, *et al* 1995).

Industrial applications of lactic acid are quite diverse. Technical grade lactic acid (88%) is extensively used in leather tanning industry, as an acidulant for deliming hides, in vegetable tanning, in acid dying of wool and in various textile-finishing operations (Datta *et al.*, 1995). Lactic acid is used in many types of technical products and processes as descaling agent, pH-regulator/pH buffer, neutralizer, chiral intermediate, solvent, cleaning agent, slow acid release agent, (metal) complexing agent and humectant (Datta, *et al.*, 1995).

Currently lactic acid is used in a wide variety of specialized industrial applications where the functional specialty of the molecule is desirable (Datta *et al.*, 1995). For instance, it is used for pH control in cellophane production and also as a plasticizer for cellulose. Lactic acid is also used in thermoplastic adhesives. In metal industry (electronics, aerospace and semiconductor), lactate esters are used for degreasing and precision cleaning since they provide excellent solvency power for oils, oligomeric stains with optimal environmental and toxicological properties (Ratindatta, 1998). Lactic acid with Ni forms unique complexing constant, which is exploited in Ni-plating. It acts as pH buffer and stabilizer for electropolishing baths. In detergents, lactic acid offers better descaling properties than conventional organic descalers. Therefore, it is used in many decalcification products such as cleaners for toilets, bathrooms and coffee machines (Datta *et al.*, 1995).

Lactic acid in the form of ethyl lactate offers an excellent safe solvent alternative for many mechanical cleanings and for cleaning top coats prior to the application in aerospace industry. Enantiomerically pure lactic acid (>98%) is suitable as a starter in the synthesis of chiral products in chemical industry eg. non-toxic herbicides and pesticides. A special high purity lactic acid is needed to meet the demands of semiconductor industry. It is used as a safe solvent in photoresist and in edgebead removal formulations in micro-electronics industry replacing toxic solvents. Paint and ink industry finds lactic acid application as pH regulator and complexing agent in various binder systems for water based coatings such as electrodeposition coatings. The optical activity of lactic acid opens the door to a wide range of stereospecific chemical reactions leading to optically active products (Van Ness, 1981).

POTENTIAL PRODUCTS OF LACTIC ACID AND ITS MARKET GROWTH

The future growth and demand for lactic acid is its use as a feedstock for potentially large-volume applications. These applications are in synthesis of biodegradable polymers, oxygenated chemicals, green chemicals/solvents, in bioremediation, and in plant-growth regulators (Lipinsky and Sinclair, 1986; Kinnersley *et al.*, 1990). There has been considerable research and development in biopolymers in recent years for polylactides. Use of lactic acid has been enormously successful in foods, but presently the companies are excited about lactic acid's use as a building block for polylactide. Cargill Dow, US based company is one of the first producers of biodegradable polymers for commercial applications. For polylactide preparation, pure enantiomeric form of lactic acid is required; impure form will lead to crystal formation that weakens and leads to rapid degradation. Usually L(+) form of lactic acid is preferred in the synthesis of biodegradable, biocompatible plastics and coatings (Rice, 1991; Litchfield, 1996; Gross and Kalra, 2002). These biodegradable polymers are transparent and they offer a good shelf-life because they are degraded slowly upon hydrolysis. This is achieved by adjusting the composition and molecular weight. These polymers have potential use in a wide variety of consumer products (paper coatings, films, moulded articles, foamed articles, fibres, etc.) (Ecochem, 1992). One of the exciting current areas for application of biodegradable polymers is in tissue engineering, where these materials are investigated as a matrix for living cells (John and Arthur, 1998).

With increasing consumer and political consciousness for environmentally sound products, large volume-low cost lactic acid could give rise to successful commercial development of cost competitive lactate esters (methyl, ethyl, butyl,etc.) as substitutes to those derived from petrochemical sources such as MEK, IPA, etc. (Chem systems report, 2002). Use of these green solvents are substantial because they are non-volatile, non-toxic and easily degradable compounds. The estimates of the volume of these are based on typical volumes and prices of several intermediate volume solvents, such as ethyl/butyl acetate or plasticizers (octyl phthalate).

Chlorinated solvents and other chemicals are often used in open systems and under controlled conditions in many industrial operations. In many cases, these solvents have leaked into ground where they pose a chronic danger to the immediate environment and ground water supply. Therefore, it is recognized world-wide that actions have to be taken to reduce the potential environmental and health threats. One of the most promising technologies to remove chlorinated solvents in the ground is by enhancing bioremediation by potential organisms. In this process various strains of bacteria obtained by enrichment culturing are used for breaking down of the contaminants. Substrate materials are often injected into the ground in order to enhance this process. Lactates are well known and widely applied as substrate materials in enhanced bioremediation. Lactates act as hydrogen donors by playing a crucial role in the food chain for the microorganisms responsible for dechlorination (Kastanek *et al.*, 1999). Mostly sodium salts and esters of L(+) lactic acid products are used in the bioremediation industry (Watkins and Karen, 2002).

Development of new technologies to reduce the cost of lactic acid would result in dramatic increase in existing markets, and evolution of a range of new applications. It has been estimated that a selling price of US$0.90 per pound of lactic acid, for solvent replacement would be US$4.5 billion per year, for biodegradable polymers US$0.9 billion per year and for oxygenated chemicals US$5.4 billion per year in world market. Lactic acid based biodegradable polymers are expected to remain more expensive than commodity polymers in the near future (Bezzari and Kishi, 2003). If polylactides and lactate esters are commercially successful, global demand will be around 14%-19% according to Stanford Research Institute (SRI) International (Chem systems report, 2002; Jarvis, 2003). However, if lactic acid could be produced from carbohydrates at lower costs and in larger quantities, the potential market for the acid and its derivatives could reach about $3.2-$4.4 billion/yr (Tsai, 1997).

L(+) -Lactic acid is used as plant-growth stimulator. Stimulation of growth in a variety of crops and fruits is discovered when L-lactic acid polymers (2-10) are applied at a low level (Datta *et al.*, 1995). These findings may lead to products and formulations that would incorporate L-polylactic acid into controlled release or degradable mulch films for large-scale agricultural applications. These advances in above fields could open specialty chemical opportunities for optically active lactic acid and its derivatives.

LACTIC ACID FERMENTATION

Raw Materials

Various carbohydrate raw materials are investigated for lactic acid production by bacteria and molds. Refined carbohydrates such as glucose, sucrose, starch, unrefined complex carbohydrates – (cellulose, cereal grains, cassava, sweet sorghum, corn, artichokes, potatoes, black strap molasses), waste materials – (cheese whey and permeate, municipal solid wastes) (Litchfield, 1996), agricultural by-products – (rice bran, wheat bran, corn fibre, rice kernels, cassava fibrous residue, pearled barley, pearled wheat), sulphite waste liquor and wood molasses are used (Pandey *et al.* 2001). Complex materials may require either, costly pretreatment, product recovery and purification processes or both. Table 3 shows selected raw materials for microbial production of lactic acid.

In the year 1994, in the United States there has been considerable interest in using cheese whey/permeate for lactic acid production by *Lactobacillus* spp. that ferment lactose, as permeate was largely produced by dairy industries. *L. delbruekii* subsp. *bulgaricus, L. helviticus, L. casei* are the organisms used for fermentation of cheese whey to lactic acid. Glucose or sucrose is generally preferred raw material for commercial-scale lactic acid fermentations. Presently in the US, glucose obtained from cornstarch is used whereas in Europe and Brazil sucrose from cane or beet sugar is used. However, international sugar prices may fluctuate widely and have a severe impact on the economics of lactic acid production processes based on these sugars.

Table 3. Selected raw materials for fermentative production of lactic acid.

Raw Material	Microorganisms	Reference
Refined carbohydrates		
Glucose	*L. delbrueckii* NRRL B-445 *R. oryzae, R. arrizus,* *R. stolonifer*	Inskeep *et al.* (1952), Snell and Lowery (1964), Soccol *et al.* (1994a,b)
Lactose	*L.lactis* subsp. *cremoris,* *L. delbrueckii* subsp. *bulgaricus*	Venkatesh *et al.* (1993), Veringa (1994)
Sucrose	*B. coagulans, L. delbrueckii*	Hariban *et al.* (1993) Beninga (1990)
Starch	*L. amylophilus, L.amylovorus,* *L.amylophilus* GV6, *L. lactis* and *A. awamori,* *R. arrhizus*	Mercier *et al.* (1992), Cheng *et al.* (1991), Vishnu *et al.* (1998, 2002) Kurusawa *et al.* (1988) Kristificova *et al.* (1991)
Complex carbohydrates and wastes		
Barley, Cassava, Corn, Oats, Rice	*R. oryzae*	Hang (1989), Yu and Hang (1989),
Wheat bran	*L. amylophilus* GV6	Hang (1990)
Cellulose	*L. delbruekii* and *Trichoderma reesei*	Naveena *et al.* (2003, 2004c)
Casein whey	*L. delbruekii* subsp. *bulgaricus*	Schmidt and Padukone (1997)
Casein whey	*L. delbruekii* subsp. *bulgaricus*	Prescott and Dunn (1959)
Cheese whey permeate	*L. casei* *L. delbruekii* subsp. *bulgaricus* *L. helveticus, L. lactis*	Stieber *et al.* (1977) Mehaia and Cheryan (1986)
Molasses Black strap Wood	*Lactobacillus* sp. Mixed lactobacilli and Yeast	Prescott and Dunn (1959) Griffithand Compere, (1977)
Municipal solid waste	*L. pentosus*	McCaskey *et al.* (1994)
Potatoes	*L. delbrueckii* NRRL B-445 *Lactyobacillus* sp., *L. pentosus*	Cordon (1950)
Sulphide waste liquor	*L. pentosus*	Leonard *et al.* (1948)

Starch produced from various plant products is a potentially interesting raw material based on cost and availability. Few laboratory-scale fermentations have been reported for lactic acid production from starch which is discussed in later sections. However, these processes are restricted to laboratory level.

PRODUCTION STRAINS

There is a wealth of information concerning the genetics of lactic acid bacteria and possible applications of genetic engineering in this field. Although all prerequisites for developing recombinant DNA system have been thoroughly investigated, industrial production of lactic acid apparently still relies on producer strains that are selected empirically (Litchfield, 1996; Danilo *et al.*, 1999). The choice of organisms to be applied and selected for their productivities largely depends on the raw materials. Prescott and Dunn (1959) state that molds in the genera *Rhizopus*, *Mucor* and *Monilia* produce lactic acid but not those in the genera *Aspergillus*, *Fusarium* or *Pencillium*. Various *Rhizopus* sp. are known to produce L(+) lactic acid aerobically utilizing glucose or sucrose. Also some *Rhizopus* spp., such as *R. arrhizus* and *R. oryzae,* have amylolytic

Table 4: Selected lactic acid producing bacteria and molds with specific ferm entation pattern and lactic acid enantiomer.

Organism	Fermentation type	Lactic acid enantiomer	Reference
Bacteria			
B. coagulans	Facultative, Heterofermentative	L (+)	Gibson and Gordan (1974)
B. laevolacticus	Facultative, Heterofermentative	D (–)	Gibson and Gordan (1974)
L. amylophilus	Homofermentative	L (+)	Kandler and Weiss (1986), Nakamura and Crowell (1979)
L. amylophilus GV6	Homofermentative	L (+)	Vishnu *et al.* (1998, 2002) Naveena *et al.* (2003) (2004a,b,c)
L. amylovorus	Homofermentative	DL	Kandler and Weiss (1986) Nakamura (1981)
L. casei subsp. *casei*	Facultative, Heterofermentative	L (+)	Kandler and Weiss (1986)
L. delbrueckii subsp. *bulgaricus*	Homofermentative	D (–)	Kandler and Weiss (1986)
L. helviticus	Homofermentative	DL	Kandler and Weiss 1986
L. rhamnosus	Facultative, Heterofermentative	L (+)	Collins *et al.* (1989) Kandler and Weiss, (1986)
Lactococcus lactis subsp. *lactis* subsp. *cremoris*	Homofermentative	L (+)	Holt *et al.* (1994)
Streptococcus feacalis	Homofermentative	L (+)	Holt *et al.* (1994)
S. thermophilus	Homofermentative	L (+)	Holt *et al.* (1994)
Sporolactobacillus Inulinis	Homofermentative	D (–)	Holt *et al.* (1994)
Molds			
R. arrhizus	Homofermentative	L (+)	Kristoficova *et al.* (1991) Soccol *et al.* (1994a)
R. delmar	Homofermentative	L (+)	Hang (1989), (1990) Soccol *et al.* (1994a)
R. oryzae	Homofermentative	L (+)	Hang, (1989, 1990) Soccol *et al.* (1994a,b)
R. stolonifer	Homofermentative	L (+)	Soccol *et al.* (1994a)

enzyme activity and convert starch or raw materials containing starch to L(+) lactic acid (Yu and Hang, 1989; Hang, 1989, 1990) (Table 4). The aerobic metabolism of glucose by *Rhizopus* sp. gives the theoretical yield of 1.5 mol of lactic acid per mole of glucose (Margulies and Visiniac, 1961). However, in industrial fermentations the use of various species of *Lactobacillus* is preferred because of their higher conversion efficiencies and increased rate of metabolism (Vickroy, 1985; Atkinson and Mavituna, 1991). These bacterial cultures are capable of producing lactic acid with a theoretical yield of more than 80% based on the weight of the substrata used.

Using glucose as raw material, any homofermentative member of genus *Lactobacillus* may be used. The preferred organism, however, is *Lactobacillus delbrukii,* now designated as *Lactobacillu delbrukii* subsp., *delbrukii* (Hammes *et al.*, 1991; Mattey, 1992; Teuber, 1993). With respect to starch, which cannot be directly fermented by lactic acid bacteria and has to be hydrolyzed prior to fermentation, the choice of suitable organism depends on the way of starch hydrolysis. If α and α amylases are used as in the traditional saccharifying procedures, e.g. in ethanol production, the resulting carbon source mainly will be maltose and glucose. In case of saccharification with α-amylases and glucoamylases, resulting in production of mainly glucose and oligosaccharides, suitable strains will be those described above. *L. amylophilus* JCM 1125 produced 53.4g/l of lactic acid using 100g/l liquefied starch as reported by Yumotto and Ikeda (1995). *L. plantarum* NCIM 2084 produced 72.9g/l of lactic acid when provided with 100g/l of liquefied starch (Krishnan *et al*, 1998). *L. amylophilus* (NRRL B4437) produced 29g/l lactic acid from 45g/l of corn starch and *L. amylovorous* was used in conversion of 120g/l liquefied starch to 92.5g/l lactic acid in submerged fermentation (Zhang and Cheryan, 1991; Mercier *et al.*, 1992). In the above reports enzymatic hydrolysis of starch was done prior to *Lactobacillus* fermentation which is expensive. *Lactobacillus amylophilus* (Nakamura and Crowell, 1979) *Lactobacillus amylovorus* (Nakamura, 1981) are exceptions that have been described to actively ferment starch to lactic acid to produce L(+) and (DL) lactic acid enantiomers respectively and this has lead to alternative process of industrial lactic acid production (Zhang and Cheryan, 1991; Cheng *et al.*, 1991; Mercier *et al.*, 1992; Zhang and Cheryan, 1994; Yumoto and Ikeda, 1995; Litchfield, 1996) (Table 4), whereas, the lactic acid production with these reported amylolytic strains was low at high starch concentrations. L(+) lactic acid as a specialty molecule is the most desired for its vast uses. Hence, production strains with high conversion efficiency at higher starch concentrations with maximum L(+) lactic acid yield are required. In view of this, a strain of *Lactobacillus amylophilus* GV6 isolated and maintained in our lab was found to be capable of degrading pure and crude starchy substrates (flours) and produce L(+) lactic acid in submerged fermentation initially, with high fermentation efficiency (Vishnu *et al*, 1998; 2000; 2002). Later it was studied successfully in solid-state fermentation (SSF) and semi-solid-state fermentations using wheat bran as substrate (Naveena *et al.*, 2003; 2004a; 2004b; 2004c).

Fermentation Processes

Batch, fed-batch or continuous types of fermentations include in lactic acid production processes. In addition, microorganism, strain, inoculum size, nutritional state, temperature, pH, substrate concentration and pretreatment, inhibition by product (lactic acid) with increasing concentrations and aeration in aerobic processes affect lactic acid productivities and yields.

On a commercial scale, lactic acid is manufactured by batch fermentation processes (Litchfield, 1996).

Among *Lactobacillus* spp, *L.delbrukii, L.bulgaricus, L.plantarum, L.helviticus* and in molds *Rhizopus* spp are generally used for commercial production of lactic acid for a specific stereo isomer (L (+), D(-) or LD) (Litchfield, 1996). The preferred isomer is L(+) lactic acid. The inoculum size in submerged fermentations is usually 5-10% volume of fermentation medium (Vickroy, 1985). The fermentation time is 1-2 days for lower concentration of sugars and 2-6 days with above 15% sugar concentrations. Temperature range for optimal growth of mesophilic lactic acid bacteria is 28 – 45° C and that of thermophilic lactic acid bacteria 45-62° C. In dairy industry, organisms operating at high temperatures are preferred. Thermophilic organisms belong to two genera: *Lactobacillus* and *Streptococcus. L. delbruekii* sp. *lactics* and *L. heiviticus* with *Str. thermophilus* are used as starters for cheese with high cooking temperatures (Annika and Bigret, 1998).

Lactic acid bacteria display complex nutrient requirement. These comprise many of the known vitamins, amino-acids and even small peptides. Based on the respective functional relationship between the requirement for one specific substance and the resulting growth or lactic acid production, specific and highly sensitive assays for vitamins and amino-acids were developed before other micro assay procedures became available (Barton-Wright, 1952). In the technical process, this indicates the necessity of including complex and sometimes expensive nutrients in the fermentation media. The presence of such complex nutrients, in turn, may contaminate the end-product and make purification steps even more expensive.

Most strains of the lactobacilli are micro-aerophilic and little oxygen is required for growth. However, they may vary in their sensitivity to oxygenation by aeration. *L.delbruckii* NRRL B-445 (now *L. rhamnosus*) is relatively insensitive to oxygen concentration in the medium. Facultatively heterophilic species may shift from homofermentation to heterofermentation with increasing dissolved oxygen concentrations (Thomas *et al.* 1979; Tseng and Montville, 1990). Therefore, in practice, low oxygen concentrations are maintained in the fermentation production processes.

Acidophilicity and acidostability of lactic acid bacteria depend on the membrane function and its permeability to undissociated weak acids. It is interesting to note that those lactobacilli (e.g. *L. casei*) which are able to metabolize a wider range of carbohydrates, survive longer than those with a restricted range of usable carbohydrates (e.g. *L. acidophilus*). The intra-cellular pH values of all organisms are close to neutrality. As lactic acid bacteria produce a single product/ mixture of acids, the undissociated lactic acid at high lactate concentrations stops the fermentation process by not maintaining the cell viability (by lowering the intracellular pH) (Lee and Wong, 1998). Thus, processes aiming at high consumption of carbon sources to produce high concentrations of lactic acid have to be investigated by neutralizing the acid formed to maintain the pH at optimum (pH 6-7.5).

In general, lactic acid bacteria utilize common carbon sources such as glucose, fructose, lactose, maltose and sucrose for growth and acid production. Starch (pure/raw) is normally not utilized except for very few reports stating that certain members of lactic acid bacteria can use liquefied starch for lactic acid production in submerged fermentations (Nakamura and Crowell, 1979; Nakamura, 1981; Mercier *et al.*, 1992). Even though the capacity of lactic acid bacteria to convert starch into lactic acid is known, it is not a frequent characteristic (Giraud *et al*, 1991). Amylolytic lactic acid bacterial fermentation has been receiving significant interest in recent past because of the cost-effective nature of the substrate (starch). In this direction, a few amylolytic lactic acid bacteria are reported such as *L. amylophilus, L. amylolyticus,* (Nakamura and Crowell, 1979, Bohak *et al.*, 1998), *L. amylovorus* (Nakamura, 1981; Zhang and Cheryan, 1991), *L. amylophilus* GV6 (Vishnu *et al.*, 1998, 2002), *L. plantarum* A6 (Giraud *et al.*, 1991) *L .amylophilus* (NRRL B4437) (Mercier *et al.*, 1992), *L. amylophilus* JCM (Yumotto and Ikeda, 1995), *Lactobacillus* strains LEM 220, 207, and 202 (Champ *et al.*, 1983), *Leuconostoc strains* (Lindgren and Refai, 1984), *L. acidophilus* (Kandler and Weiss, 1986), *L. cellobiosus* (Sen and Chakrabarthy, 1984), *L. plantarum* (Olympia *et al.*, 1995), *L. fermentum* Ogi E1 and Mw2 (Agati *et al.*, 1998), *L. manihotivorans* LMG 18010T (Morlon-Guyot *et al.*, 1998), *L. fermentum* K9, *L. plantarum* A6 (Sanni *et al.*, 2002), *Strep. bovis, Strep. macedonicus, Lactococus lactis* and *Enterococcus sulfurous* (Diaz *et al.*, 2003), *Bacillus* sp. (Milind *et al.*, 2004) are some reported amylolytic lactic acid bacteria.

Solid-state Fermentation

Solid-state fermentation (SSF) processes can be defined as 'the growth of microorganisms (mainly fungi) on moist solid materials in the absence of free-flowing water' (Moo-Young *et al.*, 1983; Pandey, 1992; Lonsane, 1994). These processes have been used for the production of food, animal feed, and both pharmaceutical and agricultural products. The production of organic acids by SSF is associated with the historical development of SSF. Koji's acid (citric acid) is the oldest SSF product which was produced in China during 2500 BC migrated to Japan by Buddhist priest in 7th century. During 18th century, for the first time, SSF was used to make vinegar from apple pomace and gallic acid (used for tanning, printing of leather). Solid waste treatment through SSF was seen in the 19th century and in the beginning of the 20th century. SSF was applied to produce enzymes and organic acids such as fumaric acid, oxalic acid and lactic acid only in the last few years using mold cultures (Pandey *et al.*, 2000a; 2000b). Apparently, much work has been done on the production of industrial enzymes using SSF and good commercial success has been achieved. Moreover, till date there has been no report of production of lactic acid in a single step through SSF using bacterial cultures. Substrata that have been traditionally fermented in solid-state include a variety of agricultural products such as rice, wheat, millet, barley grains, beans, corn and soyabeans. However, non-traditional substrata which may also be of interest in industrial process development include an abundant supply of agricultural, forest and food-processing wastes [such as wheat bran and soya grits (flakes remaining after extraction of oil)] (Pérez-Guerra *et al.*, 2003).

SSF is a more competitive process than is commonly thought. This technology has now gained renewed attention from industry, as it has become an attractive alternative to liquid fermentation for many products. SSF process is more relevant to developing countries because of their technological simplicity (Carrizales and Jaffe, 1986). The solid matrix (inert or noninert) contains available substrates such as carbohydrates (cellulose, starch, sugars), nitrogen sources (ammonia, urea, peptides) and mineral salts. The microorganisms adsorb and hydrolyze those substrates and convert into products. Critical analysis of the literature reveals a number of reasons for the resurgence of interest in SSF. Major reasons include high product titers, lower capital as well as recurring expenditure, a more stable product, with reduced energy requirements, simplicity, elimination of the need for solubilization of nutrients, absence of foam problem, lesser fermentation space, economical, less complex plant as well as machinery (Lonsane, 1994) and smaller volumes of polluting effluents (Pandey *et al.*, 2001).

Many bacteria, yeasts and fungi are capable of growth on solid substrates and find applications in SSF processes. Among these, fungi are well adapted and dominate in research on their use in SSF processes. Bacterial SSF processes are few in number, but are very important in nature and in the fermented food industry (Wang and Hesseltine, 1982). In non-traditional SSF processes, the involvement of bacteria is very scarce. Therefore, L(+) lactic acid production by SSF using a bacterial culture was reported for which wheat bran (a by-product of wheat milling industry) was selected by Naveena *et al.*, (2003). Certain non-inert materials, not only function as an attachment place but also supply nutrients to the microorganism (Couto *et al.*, 2001).

Lactic acid production from cheap renewable starchy resources is gaining importance. Some reports are available on amylolytic lactic acid producing bacteria. (Shamala and Sreekantaiah, 1988; Giraud *et al.*, 1991; 1994; Richter and Trager, 1994; Figueroa *et al.*, 1995; Morlon-Guyot *et al.*, 1998). However, presently low amylolytic activity and low L(+) lactic acid yields at high substrate concentration are major limitations in direct fermentation of starch to lactic acid. Agricultural by-products which are available abundantly, are a good source of starch. The starch present covering the outer layer of these materials is tightly bound. An efficient amylolytic organism can hydrolyse this starch to glucose for higher yields of lactic acid. Therefore, use of organism which can overcome these major hassles in direct fermentation of crude starch to L(+) lactic acid would be of immense importance. *L. amylophilus* GV6 is identified to produce high yields of L(+) lactic acid in SSF (90%) and semi-solid-state fermentation (98%) using wheat bran as substrate (Naveena *et al.*, 2003; 2004a, b, c).

A fungal strain of *R. oryzae* was evaluated in SSF for production of L(+) lactic acid using an inert solid support (sugarcane bagasse) impregnated with glucose. It could produce 13.7% lactic acid from 18% of glucose showing 77% fermentation efficiency (Pandey *et al.* 2001). Few non-amylolytic bacterial strains such as *L. paracasei, L. casei, L. helviticus* and *Streptococcus thermophilus* were used in SSF for lactic acid production. Pretreatment of wheat straw hemicellulose through enzymes/acid released 11-12g/l fermentable sugars, which were used for production of lactic acid in co-culture fermentation using *L. brevis* and *L. pentosus* (Grade *et al*, 2002). *L. casei* subsp. *casei* CFTRI 2022, *L. helviticus* CFTRI 2026 and *Strep. thermophilus* CFTRI 2034 were used in SSF with sugarcane pressmud as a substrate containing 14% total sugars on dry weight basis. Out of these *L. casei* subsp. *casei* CFTRI 2022 was reported to produce more lactic acid (6.14%) (Xavier and Lonsane, 1994). The above reports explain the use of inert supports with glucose as substrate or pretreatment of complex substrate to give sugars for lactic acid fermentation. Whereas, *L. amylophilus* GV6 was found to produce 36g of lactic acid from high concentration of raw starch (54.4g) present in 100g of wheat bran (Naveena *et al.*, 2004b). At high starch concentrations, lactic acid production is low with the known amylolytic organisms (Litchfield, 1996; Yumotto and Ikeda, 1995; Zhang and Cheryan, 1991; Mercier *et al.*, 1992). Comparable with other reported strains of *Lactobacillus, L. amylophilus* GV6 showed higher amylolytic efficiency which can successfully eliminate saccharification steps of raw starch with high L(+) lactic acid production efficiencies (Vishnu *et al.*, 2000; 2002; Naveena *et al.*, 2003; 2004a, b, c).

Recovery

For lactic acid to be used in a variety of applications, economical, efficient and environmentally sound manufacturing processes are needed for its production. In the past, efficient and economical technologies for the recovery and purification of lactic acid from crude fermentation broth and conversion of lactic acid to other chemical or polymer intermediates had been the key technology impediments and main process cost centres. The development and deployment of novel separation technologies such as electro dialysis (ED) with bipolar membranes, extractive distillations integrated with fermentation and chemical conversion can enable low cost production with continuous process in large-scale operations. The use of bipolar ED can virtually eliminate the salt or gypsum waste produced in the current lactic acid purification processes. Thus, emerging technologies can use environmentally sound process to produce environmentally useful products. Development of new lactic acid production plants by major chemical and agro processing companies may use new technologies for the efficient low cost manufacture of lactic acid and its derivatives for new applications (Datta *et al.*, 1995; Litchfield, 1996).

ECONOMICS OF LACTIC ACID

Lactic acid production in 1952 was around 5000 kg/year (Inskeep *et al.*, 1952). The world-wide production of lactic acid in 1982 was 24-28 × 10^6 kg/year. By 1990, world-wide lactic acid production has grown to 40 × 10^3 tons/year with the involment of industries like CCA Biochem b.v. and Sterling Chemicals from Netherlends and USA. (Datta *et al.*, 1995). The major use of L(+) lactic acid during this period was dominated by food related applications (nearly 85% of the demand). L(+) lactic acid polymer has proven to be immensely important in medical applications over the last three decades. US revenues over US $300 million in 1995 from commercial products developed from biodegradable polymers of which more than 95% revenues were from absorbable sutures (John and Arthur, 1998).

The scale of lactic acid production has been rising significantly in order to supply feedstock for the production of polylactide polymer since 1998. Demand for biodegradable polymers in the US, Western Europe and Japan has reached to 18,000 MT (95$ million) and by 2003 consumption was expected to increase to 91,000MT, with annual growth rate of 37%. Therefore, production should reach around 140, 000 MT per year to achieve price reductions. In 1998, market for biodegradable-polymers in the US alone was 50% of world consumption, in Western Europe 40% and in Japan, it was 10%. In 1998, Western Europe was producing 60% of world's proportion of lactic acid to satisfy the demands for lactic acid (Greg and Yoshida, 2000). As lactic acid raw material cost for polylactide is $2 to $3 per pound, which is high compared to polystyrene (Mirasol, 1999). Hence, development of technologies to bring down the cost of lactic acid are widely studied. In 1999, Hepner rated the demand to grow continually at 5-8% annual clip, estimating world-wide consumption of lactic acid at 130,000-150,000 MT per year (Mirasol, 1999). This emphasizes the importance of lactic acid in various applications. Lactic acid global demand was around 86,000 tons in 2001, with market dominated by food and beverage sector and the personal care industry (Chem systems report, 2002).

Between 2002 and 2007, demand of lactic acid and its derivatives in food and beverage sector is expected to grow around 3.5-4%, pharmaceutical and personal care sector about 5.2%, industrial applications 19%, polylactic acid alone is expected to grow 22% and others at 5% (Jarvis, 2003). It is observed from above statistics that industrial applications of lactic acid have surpassed food and beverages in 2002 as the awareness has grown towards use of environmentally sound products (Jarvis, 2003). The potential market for a lower priced lactic acid based on new technology for production would result in a dramatic increase in existing markets, and the evolution of a range of new applications. Conversion of lactic acid to polylactide polymer pioneered by Cargill Dow represents one of the first commercial applications of biodegradable polymers (Hepner, 2002). Purac and Cargill Dow LLC are the two major companies producing lactic acid and they are sorting out different technologies to meet the market demand in coming years.

In India, lactic acid demand was only 560 tons in the year 2001 (Ramesh, 2001). There was an increase in lactic acid global demand growth of upto 14% per year untill 2002 (Chem systems report, 2002). The annual production capacity of lactic acid presently is 6000 tons in India and there is estimated gap of 2300 tons in supply by 2015 AD (assuming no additional capacity creations) as given by Technology Information Forecasting and Assessment Council (TIFAC) in India (TIFAC, 2001). Plastic wastes generated by China are about 16 million tons, India 4.5 million tons and UK 1 million tons, of which more than 800,000 tons is waste polyethylene causing environmental pollution threat (AZOM. Com, 2002). The need for fully degradable plastic is pressing and demand for lactic acid is increasing due to its use as feedstock for production of environmentally friendly polylactate polymer (Gross and Kalra, 2002). L(+) lactic acid polymers can efficiently replace polyethylenes. If lactic acid could be produced from carbohydrates at lower costs and in larger quantities, the potential market for the lactic acid and its derivatives could reach about $3.2-$4.4 billion/yr on global terms. In this direction, the study was made to produce L(+) lactic acid by *Lactobacillus* fermentation using starch and wheat bran as cheap starchy substrate (Vishnu *et al.*, 2000; 2002; Naveena *et al.*, 2003; 2004a, b, c).

METABOLISM OF LACTIC ACID BACTERIA

Lactic acid bacteria (LAB) are generally very fastidious and require amino-acids, nucleotides or nucleotide precursors, and vitamins in addition to an energy source, generally a carbohydrate (Rogosa *et al.*, 1961; Chopin, 1993). For a rapid growth of these bacteria, an efficient transport system for the uptake of essential nutrients is required, which is highly connected to the bioenergetics of the cells. LAB metabolism is efficient in various carbohydrate (glucose, mannose, galactose, fructose, lactose, maltose etc.) fermentation. The metabolism of LAB is aimed at the generation of ATP. LAB do not possess an electron transport chain and hence ATP is generated by substrate level phosphorylation through oxidation of substrate. The oxidation results in the formation of NADH from NAD^+, which has to be regenerated for the cells to continue the fermentation. In major fermentation pathways used by LAB, pyruvate holds a key position in serving as an electron/hydrogen acceptor for this regeneration step. Under normal conditions ie., excess sugar and limited access to oxygen, pyruvate is reduced to lactic acid by an $NADH^+$-dependent lactate dehydrogenase (nLDH) to give NAD^+. A redox balance is thus obtained, lactic acid is virtually the only end-product, as seen in group I (obligately homofermentative) Lactobacilli (Axelsson, 1998).

In theory, homolactic fermentation of glucose (Embden-Meyerhof Pathway) results in 2 moles of lactic acid and a net gain of 2 ATP per mole of glucose consumed. Heterolactic fermentation of glucose (phosphoketolase pathway) gives 1 mole each of lactic acid, ethanol, and CO_2 and 1 ATP per mole of glucose consumed. In practice, these theoretical values are seldom obtained. A conversion factor of 0.9 from sugar to end-product carbon is common and reflects an incorporation of 0.1 sugar carbon into the biomass, even though most factors are supplied in excess in the rich media that is frequently used (Kandler, 1983). These complex media may also contribute to other fermentation balances and to the formation of other end-products, because compounds like organic acids, amino-acids and sugar residues can alter the fermentation (Kandler, 1983). Heterofermentative bacteria are involved in most of the typical fermentations leading to food or animal feed, whereas homofermentative bacteria are producers of industrial lactic acid. It is known, however, that the homofermentative behaviour of certain strains may change to heterofermentative, if the main carbon source is decreased to growth limiting levels (De Vries *et al.*, 1970). The presence of oxygen may also have a significant effect on the metabolism (Condon, 1987). During the fermentation of sugars, different species of LAB produce either exclusively L-lactic acid/ exclusively D-lactic acid or approximately equal amounts of both or predominantly one form (but measurable amounts of the other) (Garvie, 1980; Kandler and Weiss, 1986; Schleifer, 1986). This depends on the presence of specific $NADH^+$-dependent lactate dehydrogenases (nLDH) and their respective activities.

SYSTEMATICS OF LACTIC ACID BACTERIA

Many scientific evidences are available on taxonomy of lactic acid bacteria (LAB). The general description of these bacteria include Gram-positive, non-spore forming cocci, coccobacilli or rods, catalase negative, devoid of cytochromes, of non-aerobic habit but aerotolerant, fastidious, acid tolerant, and which strictly produce lactic acid as the major end-product during the fermentation of carbohydrates. The classification of LAB into different genera is largely based on morphology, mode of glucose fermentation, growth at different temperatures, configuration of the lactic acid produced, ability to grow at high salt concentrations and acid or alkaline tolerance. The present taxonomy relies on true phylogenetic relationship revealed by extensive work on determining ribosomal RNA (rRNA) sequences.

Two major groups of LAB based on fermentation (end-products) are homolactic fermenters and heterolactic fermenters. This is due to the presence of two major fermentation pathways. Embden-Meyerhof pathway is observed in homolactic fermentation where the end-product is exclusively lactic acid. The 6-phosphogluconate/phosphoketolase pathway results in significant amounts of other end-products such as ethanol, acetate, and CO_2 in addition to lactic acid, and the metabolism is referred to as heterolactic fermentation. Initially Orla-Jensen (1919) has worked on classification of LAB. His idea had a large impact on the systematics of LAB where he used few characters as classification basis: morphology, mode of glucose fermentation, growth at certain 'cardinal' temperatures (10°C and 45°C). During that period, only phenotypic characters were examined and evaluated as phylogenetic markers.

Today we have the means to examine in detail the phylogenetic positions with DNA-DNA homology studies (Jhonson, 1984). For determining phylogenetic positions of species and genera, rRNA is more suitable as it contains both well-conserved and less conserved regions. Initially the sequence of long stretches of rRNA was done with the reverse transcriptase technique (Lane *et al.,* 1985) but it is now being done more conveniently by sequencing the corresponding genes with methods involving the polymerase chain reaction (PCR) (Saiki *et al.,* 1988). Comparisons of these sequences are currently the most powerful and accurate techniques for determining phylogenetic relationships of microorganisms (Woese, 1987). With this technique, a clear picture of the phylogeny of LAB is emerging and the ideas of Orla-Jensen (1919) can be examined with some accuracy. In addition, rRNA sequencing is an important aid in the classification of LAB, as exemplified by the descriptions of new genera (Collins *et al.,* 1990; 1993, Wallbanks *et al.,* 1990). Another technique which has proven very useful in the classification of LAB is soluble protein patterns (Pot *et al.,* 1993).

The general basis for the classification of LAB into different genera has largely remained unchanged since the work of Orla-Jensen (1919). However, with the description of new genera and species, it is becoming increasingly difficult to use these classical tests for reliable genus identification. A summary of the differentiation of the LAB genera with classical phenotypic tests is shown in Table 5. It is noted that there are phenotypic overlaps between genera and exceptions to the general rules outlined in Table 5. Classification is becoming dependent on methods that are more sophisticated. *Bergey's Manual of Determinative Bacteriology,* 9th edition classifies the genera of lactic acid bacteria into the following groups, with the examples of representative species (Holt *et al.,* 1994).

Table 5: Differential Characteristics of lactic acid bacteria (LAB) (Axelsson, 1998).

	Rods			Cocci						
Character	**Carnob**	**Lactob**	**Aero**	**Enteroc.**	**Vagoc.**	**Oenoc.**	**Pedioc.**	**Streptoc.**	**Tetragenoc**	**Weissella[b]**
Tetrad formation	–	–	+	–	–	–	+	–	+	–
CO_2 from glucose	–[e]	–/+	–	–	–	+	–	–	–	+
Growth at 10° C	+	–/+	+	+	+	+	+	–	+	+
Growth at 45° C	–	– /+	–	+	–	–	–/+	–/+	–	–
Growth at 6.5% NaCl	ND[f]	– /+	+	+	–	–/+	–/+	–	+	–/+
Growth at 18% NaCl	–	–	–	–	–	–	–	–	–	+
Growth at pH 4.4	ND[f]	– /+	–	+	–/+	+	–	–	–	–/+
Growth at pH 9.6	–	–	+	+	–	–	–	–	+	–
Lactic acid[d]	L	D, L, DL[g]	L	L	L	D	L, DL[g]	L	L	D, DL[g]

[a]+, positive; – negative; –/+, response varies between species; ND, not determined.
[b]weisella strains may also be rod shaped.
[c]Test for homo- or heterofermentation of glucose; negative and positive denotes homofermentative and heterofermentative, respectively.
[d]Configuration of lactic acid produced from glucose.
[e]Small amounts of CO^2 can be produced from glucose.
[f]No growth in 8% of NaCl has been reported.
[g]Production of D-, L-, or DL- lactic acid varies among species

Group 17 – Gram-positive cocci: *Lactococcus* spp., *Enterococcus* spp., *Pediococcus* spp., *Saccharococcus* spp., *Streptococcus* spp.,

Group 18 – Endospore forming Gram-positive rods and cocci: *Bacillus* spp.,

Group 19 – regular, non-sporing Gram-positive rods: *Lactobacillus* spp.

Beijerinck proposed the genus *Lactobacillus* in 1901. The genus *Lactobacillus* represents one of the largest genera, comprising over 60 recognized species (Collins *et al.*, 1991, Natalia *et al.*, 2001). The presence of fructose-1-6-biphosphate aldolase resulted in the subdivision of these bacteria into three physiological groups: (*i*) obligately homofermentative lactobacilli, (*ii*) facultatively heterofermentative species, and (*iii*) obligate heterofermentative species which lack fructose-1-6-biphosphate aldolase.

These fermentation groups were consequently redefined by Kandler and Weiss (1986) and by Hames and Vogel, (1995). To match further the phenotypic division of the lactobacilli with the phylogenetic data obtained from rRNA sequencing, the subdivision of lactobacilli was proposed by Hames and Vogel, (1995). The different available genetic methods are applied in classifying LAB but it shows from the various studies that each method has advantages and disadvantages, and that one single method is not the solution for all applications but rather the methods are complementary.

CONCLUSION

The overall view of lactic acid as a microbial metabolite, fermentation, its requirements, uses, economics, and organisms involved etc., are summarized in this review. The last few years has brought an enormous increase of knowledge on varied applications of lactic acid. Current research programmes indicate the future potential of lactic acid and also the LAB as a product. The studies can be focused on the isolation and use of efficient strains with high production efficiencies and product tolerance to make the fermentation processes more economical. The industries which produce lactic acid by fermentation are dependent on sugar as a raw material whereas the use of amylolytic lactic acid producing organisms possibly will prove to be very beneficial. *L. amylophilus* GV6 isolated from starch processing industry is found to be efficient among all the amylolytic bacterial strains available in literature in utilizing cheap renewable raw materials such as flours, crude starches and brans at high substratum concentrations. Strain GV6 is studied successfully in all types of fermentations (submerged, solid-state and semi-solid state) possible at lab level and its progress industrially would be beneficial. A new and exciting era is opening with the possibility of using lactic acid and LAB in almost all sectors of commodities and industries.

REFERENCES

Agati, V., Guyot, J. P., Morlon – Guyot, J., Talamond, P. & Hounhouigan, D. J. (1998) *J. Appl. Microbiol.* **85**: 512 – 520.

Annika, M.M. & Bigret, M. (1998) In *Lactic Acid Bacteria, Microbiology and Functional Aspects* (Eds. S. Salminen and A. Wright, Marcel Dekker, Inc., New York. pp. 73-96.

Anon, (1963) *Chem. Eng. Progr.* **41**: 44 – 46.

Archibald, A. (2003) Acidifiers for flavor, nutrition. *Prepared foods*, Jan edition.

Atkinson, B. & Mavituna, F. (1991) In *Biochemical Engineering and Biotechnology* (Eds. Atkinson, B and Mavituna, F.), Handbook, Macmillan Publishers Ltd, NewYork, pp. 1181-1183.

Axelsson, L. (1998) In *Lactic Acid Bacteria, Microbiology and Functional Aspects* (Eds. S. Salminen and A. Wright), Marcel Dekker, Inc., New York, pp. 8-8.

Azom, Com (2002) The Premier On-line Materials Information Site, Supplier and Expert Directory, March, 27th.

Barton-Wright, E.C. (1952) *The microbiological assay of the vitamin–complex and amino-acids*. Publ. Pitman & Sons, London.

Beninga, H. (1990) *A History of lactic acid making*. Kluwer Academic Publ. Dodrecht/Norwell, MA.

Bizzari, S & Kishi, A. (2003) *Chemical Handbook Economics* (CHE) Report.

Bohak, I., Back W., Richter L., Ehrmann M., Ludwing W. &. Schleifer, K. H. (1998) *Syst. Appl. Microbiol.* **21**: 360-364.

Carrizales, V. & Jaffe, W. (1986) *Interscience* **11**: 9-15.

Champ, M. O.,. Szylit, P., Raimbault, M. & Abdelker, N. (1983) *J. Appl. Bacteriol.* **55**, 487-493.

Chem systems reports (2002) Biotech Routes to Lactic acid/Polylactic Acid (00/01S3), Process Evaluation/ Research Planning program (PERP) Program, June.

Cheng, P., Muller, R. E., Jaeger, S., Bajpai, R. &. Jannotti, E. L (1991) *J. Ind. Microbiol.* **7**: 27-34.

Chopin, A. (1993) *FEMS Microbiol. Rev.***12**: 21 – 38.

Collins, M. D., Phillips, B. A. & Zanoni, P. (1989) *Int. J. Syst. Bacteriol.* **39**: 105 – 108.

Collins, M. D., Williams, A. M. & Wallbanks, S. (1990) *FEMS Microbiol. Rev.* **46**: 269 – 280.

Collins, M. D., Rodrigues, U., Ash, C., Aguirre, M., Farrow, J. A. E., Martinezmurcia, A., Phillips, B. A., Wiiliams, A. M. & Wallbanks, S. (1991) *FEMS Microbiol. Lett.* **77:** 5-12.

Collins, M. D., Samelis, J., Metaxopoulos, J. & Wallbanks, S. (1993) *J. Appl. Bacteriol.* **75**: 595 – 603.

Condon. S. (1987) *FEMS Microbiol. Rev.* **46**: 269 – 280.

Cordon, T. C., Treadway, R. H., Walsh, M. D. & Osborne, M.F. (1950) *Ind. Eng. Chem.* **42**: 1833 – 1836.

Couto, R. S., Rivela, I. & Sanroman, A. (2001) *J. Chem. Technol. Biotechnol.* **76**, 78-82.

Danilo, P., Michele, M. B., Luca, B., Rossella, M.,. Davide, B, Jofferson, C., Chi-Li, L., Bianca, M. R., Laura, F. & Lilia, A. (1999) *Appl. Environ. Microbiol.* **65** (9): 4211-4215.

Datta, R.S., Sai, P.T., Patric, B., Moon, S.H. & Frank, J.R. (1995) *FEMS Microbiol. Rev.* **16**: 221-231.

De Vries, W., Kapteijn, W. M. C., Vander Beek, E. G. & Stouthamer, A. H. (1970) *J. Gen. Microbiol.* **63**: 333-345.

Diaz-Ruiz, G., Guyot, J. P., Ruiz-Teran, F., Morlon-Guyot, J. & Wacher, C. (2003) *Appl. Environ. Microbiol.* **69**: 4367-4374.

Drewes, P. A. (1974) In *Clinical Chemistry: Principles and Techniques* (Ed., R. J.

Henry, D., Cannona, C. & Winkelman, J. W. 2nd (Eds. Harper and Row), New York, pp. 1328 – 1334.

Ecochem, (1992) ECO=LMC[TM] Lactic acid, Publ. Adell. WI.

Condon, S. (1987) *FEMS Microbiol. Rev.* **46**: 269-280.

Filachione, E. M. (1952) In *Encyclopedia of Chemical Technology* (Eds. R. E. Kirk and D. F. Othmer), Vol 8, Wiley, New York, pp. 167 – 180.

Figuerao, C., Davila, A. M. & Pourquie, J. (1995) *Lett. Appl. Microbiol.* **21**:126-130.

Frost & Sullivan (1995) Absorbable and Erodible Biomaterials Products Markets, Frost and Sullivan Report, chap 10, Mountain View, CA, US.

Garret, J. F. (1930) *Ind. Eng. Chem.* **22**: 1153 – 1154.

Garvie, E. I. (1980) *Microbiol. Rev.* **44**: 106-139.

Gibson, T. & Gordon, R. E. (1974) In *Bergy's manual of determinative bacteriolog* (Eds. R. E. Buchanan and N. E. Gibbons). 8th Ed, Williams and Wilkins, Baltimore, pp. 529 – 550.

Giraud, E., Brauman, A., Keleke, S., Lelong, B. & Raimbault, M. (1991) Isolation and physiological study of an amylolytic strain of *Lactobacillus plantarum. Appl. Microbiol. Biotechnol.* **36**: 379-383.

Girau, E., Champailler, A. & Raimbault, M. (1994) *Appl. Environ. Microbiol.* **60** (12): 4319-4323.

Grade, A., Jonsson, G., Schmidt, A.S. & Ahring, B.K. (2002) *Bio. Technol.* Feb. **81** (3): 217-223.

Greg & Yoshida, Y. (2000) Biodegradable Polymers, Chemical Handbook Economics (CHE) Report, Feb.

Gregor, R. (1999) *Appl. Environ. Microbiol.* **60** (**9**): 3763 – 3766.

Griffith, W. H &. Compere, A. L. (1977) *Dev. Ind. Microbiol.* **18**: 135 – 143.

Gross, R.A & Kalra, B. (2002) *Science* **2**: 297 (5582), 803-807.

Hammes, W. P, Weiss, N. & Holzapfel, W. (1991) *The prokryotes,* 2nd ed., Vol 2 (Eds. Balow, A., H. G. Truper, M. Dworkin, W. Harder and K. H. Scheifer), Springer- Verlog, New York, pp. 1535 – 1594.

Hammes, W. P & Vogel, R. F. (1995) In *The lactic acid bacteria,* Vol. **2**. (Eds. B. J. B. Wood and W. H. Holzapfel), Glasgow, Chapman and Hall, pp. 19 – 54.

Hang, Y.D. (1989) *Biotechnol. Lett.* **11:** 299-300.
Hang, Y. D. (1990) *U S Patent* **No.** *4*, 963, 486.
Hariban, V., Sitkey, V., Sturdick, E. & Matus, P. (1993) *Acta. Biotechnol.* **13**: 283 – 288.
Hepner L. & Associates (2002). *Food Ingredients- Lactic acid current status and outlook by 2010,* Kluwer Academic Publ., pp. 450 – 450.
Holten, C. H., Muller, A. & Rehbinder, D. (1971) *Lactic acid-Properties and Chemistry of Lactic Acid and Derivatives*, VCH, Weinheim, Germany.
Holt, J. G., Krieg, N. R., Sneath, P. H. A., Staley, J. T. & Williams, S. T. (1994) In *Bergy's manual of determinative bacteriology,* 9th (Eds., Williams and Wilkins), Baltimore.
Inskeep, G. C., Taylor, G. G. & Breitzke, W. C. (1952) *Ind. Eng. Chem.* **44**: 1955-1966.
Jarvis, L. (2003) *Chemical Market Reporter*, Feb 10.
Jhonson, J. L. (1984) In *Bergy's Manual of systematic Bacteriology,* Vol. **1** (Eds. N. R. Krrieg and J. G. Holt), Willims and Wilkins, Baltimore, pp. 8-11.
John, C. M. & Arthur, J. T. (1998) *Synthetic Biodegradable Polymers as Medical Devices, Medical Plastics and Biomaterials*, Medical Devise & Diagnostic Industry, March Issue.
Kandler, O. (1983) *Antonei van Leeuwenhoek.* **49**: 209-224.
Kandler, O. & Weiss, N. (1986) In *Bergeys Manual of Systematic Bacteriology*, Vol. **2** (Eds. P. H. A., Sneath, N. C. Mair, M. . Sharpe & J. G. Holt), The Williams and Wilkins Co., Baltimore, pp. 1209 – 1234.
Kastanek, F., Demmerova, K., Pazlarova, J., Burkhard, J. & Maleterova, Y. (1999) *Int. J. Biodeterioration. Biodegradation.* **44**: 39-47.
Kimberley, A.C. & Taylor, C. (1996) *Appl. Biochem. Biotechnol.* **56:** 49-58
Kinnersley, A M., Scott III, T. C., Yopp, J. H. & Whitten, G. H. (1990) *Plant Growth Regul.* **9**: 137-146.
Krishnan, S., Bhattacharya, S. & Karanth, N. G. (1998) *Food Biotechnol.* **12:**105 – 121.
Kristoficova, L., Rosenberg, M., Vlnova, A., Sajbidor, J. & Certik, M. (1991) *Folia Microbiol.* **36**: 451 – 455.
Kurusava, H., Ishikawa, H. & Tanaka, H. (1988) *Biotechnol. Bioeng.* **31**: 183 – 187.
Lane, D. J., Pace, B., Olsen, G. J., Stahl, D. A., Sogin, M. L. & Pace, N. R. (1985). *Proc, Natl. Acad. Sci. USA*, **82**: 6955-6959.
Lee, Y.K. & Wong, S.F. (1998) In *Lactic Acid Bacteria, Microbiology and Functional Aspects,* (Eds. Seppo, S. and Attervon, W.), Marcel Dekker, Inc., pp. 103-115.
Leonard, R. H., Peterson, W. H. & Johnson, M. J. (1948) *Ind. Eng. Chem.* **40**: 57 – 67.
Lindgren, S. & Refai, O. (1984). *J. Appl. Bacteriol.* **57**: 221 – 228.
Lipinsky, E. S. & Sinclair (1986) *Chem.. Eng. Progr.* **82**: 26-32.
Litchfield, J.H. (1996) *Adv Appl Microbiol.* **42**: 48-95.
Lockwood, L. B., Yoder, D. E. & Zienty, M. (1965) *Ann. N. Y. Acad. Sci.* **119**: 854-867.
Lonsane, B. K. (1994) In *Solid State Fermentation* (Eds. Pandey A.), Weiley Eastern Ltd., New Delhi, pp.11-20.
Malhotra, V. P., Raina, R. K & Sanjay, R. (2000) *Techno-Economic News Magazene*. May, pp. 3-3.
Margulies, M., & Vishiniac (1961) *J. bacterial.* **81**: 1-9.
Mattey, M. (1992) *Crit. Rev. Biotechnol.* **12**: 87-132.
McCasky, T. A., Zhou, S. D., Britt, S. N. & Strickland, R. (1994) *Appl. Biochem. Biotechnol.* **45/46**: 555 – 568.
Mehaia, M.A. & Cheryan, M.(1986) *Enz. Microbiol. Technol.* 8: 289 – 292.
Mercier, P., Yerushalami, L., Rouleau, D. & Dochania, D. (1992) *J. Chem. Technol. Biotechnol.* **55**: 111-121.
Milind, P., Mark Ou., Ingram, L.O. & Shanmugam, K. T. (2004) *Biotechnol. Lett.* **26**: 865 – 868.
Mirasol, F. (1999) Lactic Acid Prices Falter As Competition Toughens. Chemical Market Reporter, March
Moo-Young, M., Moreira, A. R. & Tengerdy, R. P. (1983) In *The filamentous fungi*, (Eds. J. E. Smith, DE. R. Berry and B. Kristiansen), Vol. 4, Edward Arnold, London, pp. 117-144.
Morlon-Guyot, J., Guyot, J. P., Pot, B., Jacobe de Haunt, I. & Raimbault, M.(1998) *Int. J. Syst. Bacteriol.* **48**: 1101-1109.
Nakamura, L.K. & Crowell, C.D. (1979) *Dev. Ind. Microbiol.* **20**: 531-540.
Nakamura, L.K. (1981) *Int. J. Sys. Bacteriol.* **31** (1): 56-63.
Natalia, N., Carmen, W., Enevold, F., Bjorn, A., Matthew, D. C. & Paul, A. L. (2001) *Int. J. Syst. Bacteriol.* **51**: 2081-2085.

Naveena, B. J., Vishnu, C., Altaf, Md. & Reddy, G. (2003) *J. Sci. Ind. Res.* **62**: 453-456.
Naveena, B. J., Altaf, Md., Bhadrayya, K. & Reddy, G. (2004a) *Bioresource Technology* (In Press).
Naveena, B. J., Altaf, Md., Bhadrayya, K., Madhavendra, S. S. & Reddy, G. (2004b) *Proc. Biochem.* (In Press).
Naveena, B. J., Altaf, Md., Bhadrayya, K. & Reddy, Gopal (2004c) *Food Technol. Biotechnol.* (In Press).
Olympia, M., Fukuda, H., Ono, H., Kaneko, Y. & Takan, M. (1995) *J. Ferment. Bioeng.* **80**: 124-130.
Orla-Jensen, S. (1919) *The Lactic Acid Bacteria*, Host and Son Publ., Copenhagen.
Pandey, A. (1992) *Proc. Biochem.* **27**, 109-117.
Pandey, A., Soccol, C. R. & Mitchell, D. (2000a) *Proc. Biochem.* **35**, (10) 1153-1169.
Pandey, A., Soccol, C. R., Nigam, P., Soccol, V. T., Vandenberghe, L. P. S. & Mohan, R. (2000b) *Biores. Technol.* **74**, 81-87.
Pandey, A., Soccol, C. R., Rodriguez-Leon, Jose A & Nigam, P. (2001) In *Solid-State Fermentation in Biotechnology: Fundamentals and Applications*, Asiatech Publishers, New Delhi, pp. 11-127.
Pérez-Guerra, N., Torrado-Agrasar, A., López-Macias, C. & Pastrana, L. (2003) *Electronic. J. Environ. Agri. Food Chem.*, Jan **9**.
Pirkle, W. H., Finn, J. M., Schreiner, J. L. & Hamper, B. C. (1981) *J. Am. Chem. Soc.* **103**: 3964 – 3966.
Prescott, S. C. & Dunn, C. G. (1959) *Industrial Microbiology*, 3rd ed., McGraw – Hill, New York, pp. 304 – 331.
Pyrce, J. D. (1969) *Analyst.* **94**: 1151-1152.
Pot, B., Vandamme, P. & Kersters, K. (1993) In *Chemical Methods in Procaryotic Systematics* (Eds. M. Goodfellow and A. G. O' Donnell), John Wiley & Sons, Chichester, pp. 493 – 521.
R.I.T.A. Corporation (1985) RITA Corporation Publ., Crystal Lake, IL.
Ramesh, M. V. (2001) *India Markets Empowering Business,* April 2.
Ratindutt (1998) *Discover* **19**. 7 - 7.
Rice, J. (1991) *Food Processing*-USA, **52**: 34-42.
Richter, K. &. Trager, A. (1994) *Acta Biotechnol.* **14**: 367 – 378.
Rogosa, M. J., Franclin, G. & Perry, K. D. (1961) *J. Gen. Microbiol.* **25**: 473-482.
Saiki, R. K., Gelfand, D. H., Stoffel, S., Scharf, S. J., Higuchi, R., Horn, G. T., Mullis, K. B. & Erlich, H. A. (1988) *Science.* **239**: 487-494.
Sanni. A., Morlon – Guyot, J. & Guyot, J. P. (2002) *Int. J. food. Microbiol.* **72**: 53-62.
Schleifer, K. H. (1986) In *Bergy's Manual of systematic Bacteriology*, Vol. 2 (Eds. Sneath, P. H. A., N. S. Mair, M. E. Sharpe and J. G. Holt), Williams and Wilkins, Baltimore, pp. 999-1103.
Schmidt, S & Padukone, N. (1997) *J. Ind. Microbiol. Biotechnol.* **18**: 10 – 14.
Sen, S. &. Chakrabarthy, S. L. (1984) *J. Ferment. Technol.* **6**: 407-413.
Shamala, T. R. & Sreekantaiah, K. R. (1988) *J. Ind. Microbiol.* **3**: 175 – 175.
Smith, W.P. (1993) *Soap/Cosmetic Specialities,* **9**:54-76.
Snell, R. L & Lowery, C. E. (1964) *US Patent No.* 3, 125, 494.
Soccol , C., Marin, B., Raimbault, M. & Lebeault, J. M. (1994a) *Appl. Microbiol. Biotechnol.* **41**: 330-336.
Soccol, C.R., Marin, B., Raimbault, M. & Lebeault, J.M. (1994b) *Appl. Microbiol. Biotechnol.* **41**: 286-290.
Stieber, R. W., Coulman, G. A. & Gerhardt, P. (1977) *Appl. Environ. Microbiol.* 34: 733 – 739.
Teuber, M. (1993). In *Lactic acid Bacteria in Biotechnology*, Vol. 1 (Eds. Rehm, H. J and G Reed), Weinheim Publ., VC, pp. 325-366.
Thomas, T. D., Ellwood, D. C. & Longyyear, V. M. C. (1979) *J. Bacteriol.* **138**: 109-117.
TIFAC. (2001) Technology-linked Business Opportunity Publications. Code No. TMS157, **4.**
Tseng, C.P. & Montville, T. J. (1990) *J. Bacteriol.* **173**: 4411 – 4416.
Tsai, S.P. (1997). Argonne National Laboratory market analysis, May .
Van Ness, J. H. (1981) In *Kirk-Orthmer Encyclopedia of Chemical Technology,* Vol. 13, J. Wiley & Sons, New York, pp. 80-103.
Venkatesh, K. V., Okos, M. R. & Wankat, P. C. (1993) *Pro. Biochem.* **28**: 231 – 241.
Veringa, H. A. (1994). *US Patent No.* **5:** 322, 781.
Vickroy, T. B. (1985) In *Comprehensive Biotechnology*, Vol 3 (Ed. Moo–Young), Dic Pergamon Press, Toronto, pp. 761 – 776.

Vishnu, C., Sudha Rani, K., Reddy, G. & Seenayya, G. (1998) *J. Sci. Ind. Res.* **57**: 600-603.

Vishnu, C., Seenayya, G. & Reddy, G. (2000) *Bio. Eng.* **23:** 155-158.

Vishnu, C., Seenayya, G. & Reddy, G. (2002). *World. J. Microbiol. Biotechnol.* **18**: 429-433.

Wallbanks, S., Martinez-Murcia, A. J., Fryer, J. L., Phillips, B. A. & Collins, M. D. (1990) *Int. J. Syst. Bacteriol.* **40**: 224-230.

Wang, H. L & Hesseltine, C. W. (1982) In *Microbiol. Technology*, Vol. (Eds. Peppler H.L. and Perlman, D.) Academic Press, New York.

Watkins & Karen (2002) *Chemical & Engineering News*, **Jan.** 14, pp.15-16.

Woese, C. R. (1987) *Microbiol. Rev.* **51**: 221-271.

Xavier, S. & Lonsane, B.K (1994) *Appl. Microbiol. Biotechnol.* **41:** 291-295.

Yumoto, J. & Ikeda, K. (1995) *Biotechnol. Lett.* **17**: 543-546.

Yu, R. U. & Hang, Y. D. (1989) *Biotechnol. Lett.* **11**: 597- 600.

Zhang, D.X. & Cheryan, M. (1991) *Biotechnol. Lett.* **10**, 733-738.

Zhang, D.X. & Cheryan, M. (1994) *Proc. Biochem.* **29**: 145 – 150.

4

Production, Characteristics and Biotechnological Applications of Alkaline Pectinases

D. C. Sharma and T. Satyanarayana
Department of Microbiology, University of Delhi
South Campus, Benito Juarez Road
New Delhi 110 0211 (India)

Plant substances are complex colloidal acid polysaccharides that are found interlinked with other structural polysaccharides or proteins in the plant tissues. The enzymes that hydrolyze pectins are known as pectolytic/ pectinolytic enzymes or pectinases, and these include polygalacturonases, pectin esterases, pectin lysases and pectate lyases based on their mode of action. Pectinases that are active at alkaline pH are produced by the species of *Clostridium, Bacillus, Streptomyces, Erwinia, Thermoanaerobacter* and *Pseudomonas* in solid state and/or submerged fermentations. Alkaline pectinases have been found to be active in the pH and temperature ranges between 8 and 11, and 25 and 85°C, respectively. The molecular masses ranged between 20 to 115 kDa. The enzymes have been found to be useful in textile and paper pulp industries, degumming of plant fibres, treatment of pectic waste waters, coffee and tea processing and, in developing environment-friendly processes.

INTRODUCTION

Pectins, a major constituent of cereals, vegetables, fruits and fibres, are complex high molecular weight, heterogeneous and acidic structural polysaccharides. They are present in the middle lamella of plant cell-wall as a thin layer of adhesive extracellular material attached with cellulose microfibrils, surrounded by a matrix of hemicelluloses and proteins, acting as a 'cementing' agent. The pectin is synthesized in golgi apparatus as UDP-D-galactorunic acid during early stages of growth (Sakai *et al.*, 1993). The pectic polysaccharides received attention due to their role in ripening of fruits and widespread application as gelling agent in food industry. Texture of fruits and vegetables is greatly influenced by the type of constituting pectin present. One of the most characteristic changes during the ripening of fleshy fruits is softening or ripening. In unripe fruit, the pectin is bound to cellulose microfibrils in the cell-walls. Such pectin is insoluble, and therefore, confers rigidity with the cells. During ripening, these pectic substances are degraded and solubilized by naturally occurring enzymes in the fruits leading to softening of the fruits (Sakai *et al.*, 1993). During ripening, the neutral sugar composition of pectin, however, does not change (de Vries *et al.*, 1986).

Microbes produce a variety of pectin degrading enzymes, which play an important role in the degradation of pectic substances, and have extensive application in food processing, biological degradation of plant materials, fermentation and food spoilage. The major industrial applications of microbial acidic pectic enzymes include extraction and clarification of fruit juices and wine technology, maceration of vegetables and fruits and in the extraction of olive and citrus oils. A new emerging area is the utilization of alkaline pectinases in the fibre processing, and treatment of waste water from food processing industries for eliminating pectic substances. In this article, an attempt has been made to review the aspects such as production, characteristics and applications of microbial alkaline pectinolytic enzymes.

STRUCTURE OF PECTIN

The pectin molecules may consist of hundreds to thousands of D- galacturonic acid units that are joined together by α-(1,4)- linkages in axial arrangement, and form *trans*-1, 4- polysaccharide, hence the polymer has a tendency of coiling around the screw axis (Whistler *et al.*, 1953) (Fig. 1). The carboxyl group of D-galacturonic acid unit in pectin is partially esterified with methanol, or neutralized by mono- or divalent cations like K^+, Na^+, and Ca^{2+}. Degree of esterification is hundred per cent when the methoxyl content is 16.32%. Sometimes the hydroxyl groups at C_3 or C_4 positions are also acetylated; the degree of acetylation may vary from 0.18% - 2.5% (McComb *et al.*, 1957). The acetyl groups are important because they determine the gelling properties of the polymer. Additionally, rhamnose constitutes a minor component of pectin backbone, and it is present as α-L-rhamnopyranose residue. They are present in the segments

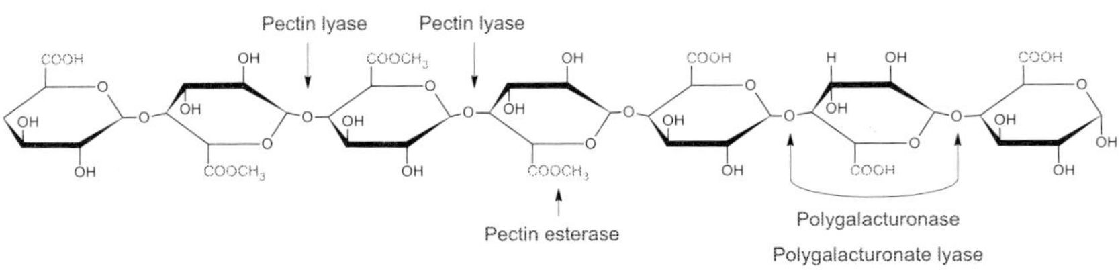

Fig. 1: Chemical structure of pectin showing sites of the action of enzymes.

of structure: α-D-galactopyranosyluronic acid-(1 → 2)-α-L-rhamanopyranosyl-(1 → 4) galactopyranosyluronic acid (Lau *et al.*, 1985; Aspinall *et al.*, 1970; Aspinall *et al.*, 1968a; Aspinall *et al.*, 1968b; Aspinall, 1981). Due to the presence of rhamnose, T-shaped 'kink' in the linear chain are formed. Other neutral sugars like L-arabinofuranose and D-galactopyranose occurs most frequently in the side chains, while D-xylopyranose, D-glucopyranose, and L-fucopyranose are less common units. D-Adipose, 2-O-methyl-D-xylose, and 2-O-methyl-L-fucose are usually very less, but widespread constituents of pectins (Darvill *et al.*, 1978; Aspinall 1981; Barrett *et al.*, 1965). These side chains are composed of neutral sugars that give the 'hairy' character to rhamnogalacturonan portion of the pectin (de Vries *et al.*, 1986). It is also observed that acidic and neutral pectins carry ferulic acid at non-reducing ends of the neutral arabinose and/or galactose containing domains. It was suggested that they play some important role in the regulation of cell expansion and lignification (Fry, 1982).

NOMENCLATURE

In 1944, the Committee for Revision of Nomenclature of Pectic Substances defined pectic substances as those complex colloidal carbohydrate derivatives, which occur in plants, and contain a large proportion of anhydrogalacturonic acid units that are thought to exist in a chain-like combination. The carboxyl groups may be partially esterified by methyl groups and partially or completely neutralized by one or more bases. The Committee of American Society defined complex pectic substances (Kertesz *et al.*, 1994) as described below:

Protopectin

Protopectin is water-insoluble parent pectic substance found in the plant cell-wall, in most plant tissues, and only this insoluble form of pectin occurs except in the ripe fruits. On hydrolysis it produces pectin and pectic acids. The insolubility of the protopectin depends on the polymer size and on the presence of divalent cations such as Ca^{2+}.

Pectinic acids

These are the colloidal polygalacturonic acids with various amounts of methyl ester groups. Pectinates are normal or acid salts of pectinic acids; pectinic acid alone has the unique property of forming a gel with sugar and acid or if suitably low in methyl content, with compounds such as calcium salt.

Pectin or Pectins

Pectin refers to the water-soluble polymeric material, which has various degrees of esterification with methanol, and is capable of forming gel with sugar and acid under suitable conditions.

Pectic acids

Pectic acid is a group designation applied to the pectic substances mostly composed of colloidal polygalacturonic acids and essentially free from methyl ester groups. Most of the pectic acids are formed after disintegration of the tissue, via the action of pectin methyl esterases. Normal or acid salts of pectic acid are called pectates.

PECTIN DEGRADING ENZYMES

The enzymes responsible for the degradation of pectin is widespread and have diverse modes of action, which is due to the variation in the structure of pectin and its universal occurrence. Pectinolytic enzymes include a group of enzymes that are able to catalyze the breakdown of pectin containing substrates (Loera *et al.*, 1999). On the basis of their optimum pH for activity, they can be categorized into acidic and alkaline pectinases

Acidic pectinases

The pectinases which have pH optima below 7 are called acidic pectinases and find wide application in the processes, where reaction conditions are acidic, like extraction and clarification of fruit juices, grape musts, and wine technology, maceration of vegetables and fruits, and in the extraction of olive and citrus oils.

Alkaline pectinases

Optimum pH for the activity of alkaline pectinases is above 7. Alkaline pectinases find application in fibre processing, and treatment of waste-water from food processing industries for eliminating pectic substances. Polygalacturonate lyase is considered to be an alkaline pectinase.

Esterases

Pectinesterases (PMGE, EC 3.1.1.11)/Pectin pectyl-hydrolase [SN]***/ Polymethylgalacturonate esterase*** [RN]**:** They de-esterify the pectin molecule to pectic acid, by attacking exclusively on methyl ester group of pectic substances. This nucleophilic substitution results in the formation of an intermediate acyl-enzyme complex with the release of methanol. This acyl-enzyme complex then undergoes de-acylation, leading to the regeneration of enzyme and pectic acid. The esterase preferentially acts only on the methyl ester groups which are adjacent to the free carboxyl groups, in a sequential manner along the length of the polymer in a block-wise manner (Sakai *et al.*, 1993). However, some pectin esterases attack pectin only at the reducing chain end, while others attack the non-reducing end (Miller and MacMillan, 1971).

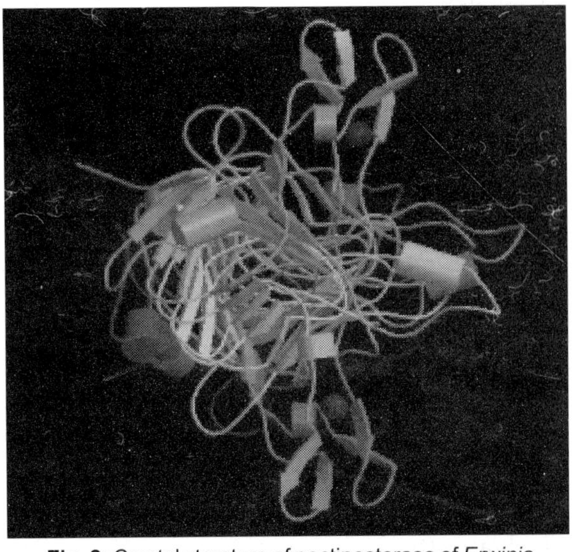

Fig. 2: Crystal structure of pectinesterase of *Erwinia chrysanthemi* (Jenkins *et al.*, 2001).

Jenkins *et al.*, (2001) studied the crystalline structure of pectin esterase of *Erwinia chrysanthemi*. The 36.9 kDa molecule contained 342 residues, and α and β contents of 6.43 and 34.80, respectively (Fig. 2). The molecule contained 4 sheets, two β hairpins and one disulphide linkage.

Pectin esterases of microbial origin are not always alkaline proteins. They are active at a pH range between 4.5-9.0. The optimum pH for pectin esterase of fungal origin is generally lower than that of bacteria (Sakai *et al.*, 1993). The esterases of *Aspergillus oryzae* and *Trichoderma reesei* were optimally active at pH 8.5 and 7.5, respectively (Lim *et al.*, 1983; Markovic *et al.*,

1985). *Erwinia chrysanthemi* secretes an esterase, which has a low pH optimum of 5.5 among bacterial esterases. The de-esterification patterns of alkaline and acidic pectin esterases are different. The de-esterified pectin formed by the action of alkaline pectin esterase forms weak gels with Ca^{+2}, whereas the product of acidic pectin esterase is capable of forming stronger gels with Ca^{2+} (Wong, 1995).

The temperature optima range between 25 and 55°C. The esterase of *Erwinia chrysanthemi* has the highest temperature optimum of 50°C among bacteria. Their molecular mass ranges between 34,000 (*Aspergillusoryzae*) and 400,000 (*Clostridium multifermentans*). The esterase of *E. chrysanthemi* is a lipoprotein that undergoes a side chain modification (Shevchik *et al.*, 1996). This is the most unstable esterase at room temperature since it loses activity in 5 minutes, but retains its activity for more than a week when stored at 4°C. Markovic *et al.*, (1985) reported an increase in the stability after immobilization of the enzyme. The main producers of this enzyme are fungi (Markovic *et al.*, 1985; Maldonado *et al.*, 1998; Kawano *et al.*, 1999) and bacteria (Rijssel *et al.*, 1993).

The most common and widely used method for the determination of pectinesterase activity is titrimetric method, in which the amount of carboxylic acid formed is measured in terms of alkali consumed to neutralize the reaction mixture. The enzyme activity can also be determined by measuring the amount of methanol formed by distillation followed by oxidation to formaldehyde as described by Holden (1945). Wood and Siddiqui (1971) proposed a simplified spectrophotometric method by removing the step of the distillation of methanol. Another accurate and reproducible method involves enzymatic digestion of pectin with enzyme in the presence of bicarbonate buffer in a Warburg respirometer. The most sensitive and accurate method is the conversion of methanol to methyl nitrite and its quantitation using HPLC (Bartolome and Hoff, 1972).

Polygalacturonases (PG, EC 3.2.1.15) / Poly (1, 4 - α - D - galactosiduronate) glycanohydrolase ^{SN} ***/ Endopolygalacturonase*** ^{RN}**:** They cleave the glycosidic bonds of pectin by adding water molecules. They are generally considered acidic pectinases, which selectively attack α-1, 4-glycosidic linkages in pectic acid (polygalacturonic acid). There are, however, a few reports on alkaline endopolygalacturonases. The optimum pH for activity lies in between 8.4 and 10.5. The enzymes of *Bacillus* sp. retained 30% activity at pH 8.5 and about 40% at pH 12. It also retained 30% and 70% activity at 50°C and 70°C, respectively (Horikoshi, 1972).

Lyases

The lyases (transeliminases) are a group of pectinolytic enzymes, which catalyze the breakdown of either pectic acid (polygalacturonate lyases) or pectin (polymethylgalacturonate lyases) by β-elimination reaction. Two kinds of lyases are known.

Endo polygalacturonate lyases (Endo-PGL, EC 4.2.2.2)/Poly (1, 4 - α - D - galactosiduronate) endolyase ^{SN} ***/Endopolygalacturonate lyase (endopectate lyase)*** ^{RN}**:** Polygalacturonate lyases are a group of hydrolytic enzymes, which randomly cleave α-1, 4-glycosidic linkages by β-elimination in pectates and pectic acid, resulting in rapid decrease in the viscosity, as compared to the number of bonds broken. The pectates lyase of *Bacillus* sp. contain 353 residues and one calcium ion with α-and β-contents of 8.5 and 28.9, respectively (Fig. 3). They can be easily distinguished from polygalacturonase by having pH optima of 8.0 (*Bacillus subtilis*) and 10.0

(*Bacillus* sp.), and requirement of calcium ions for the activity. The pH optima depends on the length of the substrate and the buffering medium (Ward and Fogarty, 1972).

Some species require other divalent cations like strontium, cobalt and magnesium and others instead of calcium (Hasegawa and Nagel, 1966; Nagel and Wilson, 1970.). The endopolygalacturonate lyase activity of *B. polymyxa* (Nagel and Wilson, 1970) increased proportionally when the ratio of calcium ion to carboxyl ion remained less than 0.5. It is thought that a calcium salt bridge of oligogalacturonides may be the true substrate for polygalacturonate lyase. The activity of endopolygalcturonate lyase decreases with decreasing chain length of substrate. Thus the Michaelis constant increases with decreasing chain length (Nagel and Wilson, 1970; Nagel and Hasegawa, 1967a; Atallah and Nagel, 1977), wheareas the maximum velocity remains constant (Nagel and Wilson, 1970). The endopolygalacturonate lyases are generally able to act on trigalacturonate and unsaturated trigalacturonate, but the enzyme of an unidentified *Bacillus* sp. (Hasegawa and Nagel, 1966; Nagel and Hasegawa, 1967a) and *Bacillus pumilus* (Dave and Vaughn, 1971) did not attack these substrates.

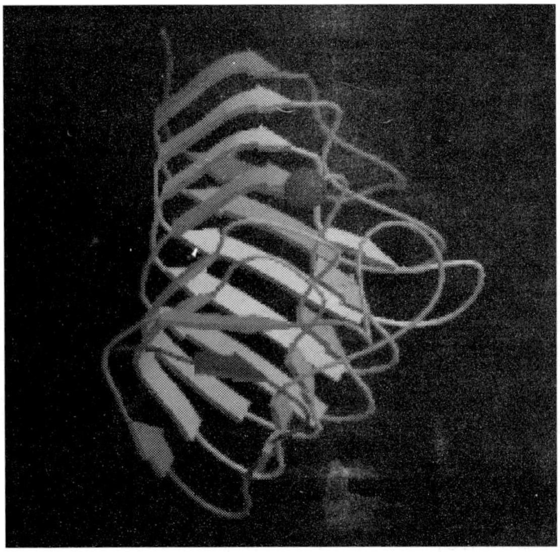

Fig. 3: Crystal structure of pectate lyase of *Bacillus* Sp. KSM-P15 (Akita *et al.*, 2000).

Although endopolygalacturonate lyases are resistant to denaturation, calcium chloride showed protective action against the heat denaturation of the endopolygalacturonate lyase of *Bacillus pumilus* (Dave and Vaughn, 1971). Endo polygalacturonate lyases are predominantly of microbial origin. Enzymes of the genera *Erwinia* and *Bacillus* are among the best known to produce soft rot symptoms in plants (Perombelon and Kelman, 1980; Collmer and Keen, 1986), although the enzyme is also found in other microorganisms, including *Aeromonas* (Hsu and Vaughn, 1969), *Pseudomonas* (Liao *et al.*, 1988; Liao, 1991), *Xanthomonas* (Nasuno and Starr, 1967), *Aspergillus* (Dean and Timberlake, 1989), *Fusarium* (Crawford and Kolattukudy, 1987).

Pectate lyase of *Bacillus* sp. was optimally active in the pH range between 8.5 and 12 (Horikoshi, 1972). The temperature optima was maximum for that of *Thermoanaerobacter italicus* (80°C). The pectate lyase of *Xanthomonas compestris* was optimally active at 25°C. All the reported polygalacturonate lyases are extracellular except that produced by *Bacteroides thetaiotaomicron,* which is present in the cell membrane (McCarthy *et al.,* 1985). The molecular weight ranges between 45,000 Da and 20,300 Da.

The activity of the enzyme could be assayed by measuring the increase in light absorption by reaction mixtures at 230 or 235 nm. The double bonds formed by trans-eliminative cleavage of the substrate absorb maximum light at this wavelength. In the second method, the property of thiobarbituric acid to produce red chromogen on reacting with unsaturated di- and oligo-uronides is used. This chromogen gave maximum absorption at 545-550 nm (Hasegawa and Nagel, 1962).

Exopolygalacturonate lyases (Exo-PGL, EC 4.2.2.9)/Poly (1, 4 - a - D - galactosiduronate) exolyase ^{SN}***/Exopolygalacturonate lyase (exopectate lyase)*** ^{RN}***:*** Exopolygalacturonate attacks preferentially on pectates than pectins, and polymethylgalacturontemethylglycoside is not attacked at all. They liberate unsaturated oligogalacturonic acid from the reducing end of the substrate chain and the smallest substrate that can be degraded is a trimer. They are optimally active at pH 8.0-9.5.

The polygalacturonate lyase and pectin esterase, as a complex, operate in a coordinate manner on pectin, with the pectin chain passing from the esterase site to the polygalacturonate lyase site without intermediate dissociation and rebinding (Sheiman *et al.*, 1976). On heating, the polygalacturonate lyase can be freed from pectin esterase (Miller and Macmillan 1971). Exopolygalacturonate lyases are generally stabilized by Ca^{+2} and Ma^{+2} but not Mg^{2+} or Zn^{2+}.

Polymethylgalacturonate lyases (PMGL)

They are all endo- acting enzymes, and therefore, cause rapid decrease in the viscosity by catalyzing the β-elimination between fourth and fifth carbon of pectin at the non-reducing end. They are mainly of fungal origin, and can degrade directly pectin although their preferential substrate is highly esterified pectin. They are optimally active in the pH range between 5.0 and 9.0. They do not require metal ions for activity, but few researchers have reported the requirement of Ca^{+2} and other cations depending on pH and degree of esertification of pectin (Edstrom and Phaff, 1964; Ishii and Vokotsuka, 1973).

The activity of polymethylgalacturonate lyases decreases rapidly as the chain length of the substrate decreases. The smallest substrate degraded is tetramethytetragalacturonate by the enzyme of *Aspergillus fonsecaeus* (Edstrom and Phaff, 1964), and trimethytrigalacturonate by that of *A. niger* (van Houdenhoven, 1974). On the basis of their cleavage pattern, they are further classified as endo-polymethylgalacturonate lyases and exo-polymethylgalacturonate lyases. Endo-polymethylgalacturonate lyases cause random cleavage in pectin molecule by a transelimination process. While exopolymethylgalacturonate lyases cause sequential cleavage in pectin by a trans-elimination process.

Protopectinases (EC 3.2.1.99)

The insoluble protopectin is converted to highly polymerized water-soluble pectin by the action of a group of enzymes known as protopectinase (PPase). They are classified as A- and B-type depending on their mode of action. A-type protopectinases react with the smooth regions in the protopectin, which are composed of partially methoxylated galacturonic acid residues. Protopectinases belonging to the B-type react with the hairy regions, which consists of rhamnogalacturonans and neutral sugar side chain, and thus, they are considered glycan hydrolases (arabinases, galactanases) [Esquivel *et al.*, 1999].

Sakai and Sakamoto (1990) reported a B-type PPase of *Bacillus subtilis* IFO 3134 that failed to degrade polygalacturonic acid. The enzyme called PPase-C has potent activity on the protopectin from sugar beet pulp but has much less on lemon peel protopectin. This PPase is optimally active at pH 8 and 60°C, and retains more than 80% activity for ten minutes at this temperature. Its molecular mass is 30, 000Da, and made up of a single polypeptide; the EDTA stabilizes the enzyme during purification (Sakai and Sakamoto, 1990). The protopectinase of

Cellvibrio cellulosa contains 315 residues and has a molecular mass of 35875 Da. The α and β content are 2.54 and 40.63, respectively. The enzyme contained 6 sheets with 40 β-turns and one disulphide bridge (Fig. 4). The enzyme contained one chlorine atom.

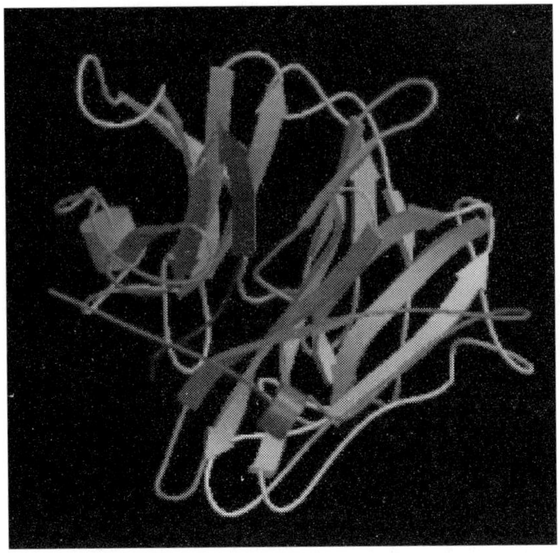

Fig. 4: Crystal structure of protopectinase C (Arabinan endo-1,5-alpha-L-arabinosidae) of Cellvibrio cellulose (Nurizzo *et al.*, 2002).

PRODUCTION OF ALKALINE PECTINOLYTIC ENZYMES BY MICROORGANISMS

Pectinases are produced by several microbes such as bacteria (Horikoshi, 1972; Karbassi and Vaughn; 1980; Sakellaris *et al.*, 1989; Rijssel *et al.*, 1993; Dosanjh and Hoondal, 1996), yeasts (Gainvores and Belarbi, 1995) fungi (Aguillar and Huitron, 1990; Borin *et al.*, 1996; Solis-Pereyra *et al.*, 1996) and actinomycetes (Bruhlmann *et al.*, 1994). The microbes that produce alkaline pectinases are listed in Table 1.

Optimization of Pectinase Production

The credit of recent advances in the production of pectinases goes to Aguilar and Huitron (1987), who optimized various aspects of pectinase production in submerged fermentation (SmF). Lonsane and Ghildyal (1992) reported production of pectinase in solid-state fermentation (SSF), while Solis-Pereira *et al.*, (1993) compared pectinase yield and productivity in SSF and SmF. Acuna-Arguelles *et al.*, (1995) showed differences in enzyme profiles and kinetics of pectinase produced in SSF and SmF.

Production in Submerged Fermentation

Effect of carbon sources: Carbon is the well-known major limiting factor for both cell growth and enzyme production. There are several reports, where pectin has been used as the sole source of carbon for pectinase production. Polygalacturonic acid or pectin initially induces

Table 1. Characteristics of bacterial alkaline pectinases.

Type of Enzyme	Organisms	Methods used for purification	Fold purification	Size Da	Temp. opt (°C)	Ph opt.	Km gL^{-1}	Inhibitors	Activator /Elicitors	References
Pectinesterases (PMGE, EC 3.1.1.11)	*Clostridium multifermentans*	—	—	400000	25	9	—	—	—	Versteeg (1979),
	Erwinia chrysanthemi	—	—	—	50	7.5	—	D-galacturonate SDS	—	Heikinheimo *et al.*, (1991), Pitkanen *et al.*, (1992), Shevchik *et al.*, (1996), Horikoshi (1972)
Polygalacturonases (PG, EC 3.2.1.15)	*Bacillus No. P-4-N*	Saphadex G-160, DEAE-cellulose followed by Sephadex G-200 Ammonium sulphate precipitaiton, calcium phosphate gel adsoporption and Sephadex G-100 DEAE-Cellulose	300							
	Bacillus pumilus	Sephadex G-120	16	20,000						Dave and Vaughn, (1971)
	Bacillus polymyxa		156							Nagel and vaughn, (1961)
	Clostridium multifermentans									Miller and MacMillan (1971)
	Bacillus sp. KSM-P576	Ultra filtration, Super Q-Tyopearl 650M, DEAE-Bio-Gel A	22	115,000	55	8.0	0.86	EDTA, bromosuccinimide, hydroxynitrobenzoate, Zn^{2+}, Cu^{2+}, Ni^{2+}	N- Mg^{2+}, Mn^{2+}, Pd2+ Ca^{2+}	Kobayashi et al., (2001).
Endo polygalacturonate lyases (Endo-PGL, EC 4.2.2.2)	*Erwinia carotovora*				60	10		Cd^{2+}, Cu^{2+}, Mg^{0}, Mn^{2+}, Ni^{2+}, Zn^{2+}	Ca2+	Sugiura *et al.* (1984)
	Bacillus sp. KSM-P15	Ammonium sulphate precipitation	66.7	26000	50	10.5		Zn^{2+}, Hg^{2+}, N–bromosuccinimide, 2, 4, 6-Tritrinitrobenzenesulfo	Ca^{2+}	Kobayashi *et al.*, (1999)
	Bacillus sp.			20300	60	10		2.3-butanedione, 2, 4, 6-trinitrobenzenesul-fonate, 2-hydroxy-5-nitrophenyl bromide, EDTA, Fe3+, Hg2+, N-bromosuccinimide, Zn2+	Ca^{2+} Mn^{2+}, Sr2+	Kelly and Fogarty (1978)
	Streptomyces nitrosporeus			39000	50	10	0.38	EDTA,	Ca^{2+}, Mg^{2+} Mn^{2+}	Kobayashi *et al.* (1999)
	Xanthomonas campestris				45	9.5	0.23 (pentagalacturonic acid), 0.19	Ba2+	Ca2+	Nasuno and Starr (1967)

							(tetragalacturonic 1.5 (trigalacturonic acid)			
***Erwinia** chrysanthemi*					60	9.3		Ba^{2+}, EDTA, C^{o+}, epicatechin, iodo-acetic acid, Mn^{2+}, polygalacturonic acid, salicylic acid, Sr^{2+}, Zn^{2+}	Ca^{2+}, Mn^{2+}	Tardy *et al.*, (1997).
	Streptomyces fradiae				30	9			Ca^{2+}	Sato and Kaji (1975).
	Erwinia aroideae			36,000	35	9		Ba^{2+}, Co^{2+}, EDTA, Hg^{2+}, Mn^{2+}, Ni^{2+}	Ca^{2+}	Kamimiya *et al.*, (1977)
	Thermoanaerobacter italicus			148000	85	9		EDTA	Ca^{2+}	Kozianowsky *et al.*, (1997)
	Bacillus macerans			45000	60	9		Ba^{2+}, EDTA, Zn^{2+}	Ca^{2+} Mn^{2+}	Miyazaki *et al.*, (1991)
	Bacteroides thetaiotaomicron					8, 7	0.040-0.070 mg/ml (poly-D- galacturonic acid)		Ca^{2+}	McCarthy *et al.*, (1985)
	Pseudomonas solanacearum					8.5				Ofuya *et al.*, (1984).
	Bacillus subtilis			32000	60	8.4	0.862 g/l (polygalacturonic acid)	$AgNo_3$, Ba^{2+}, $CdCl^2$, Co2+, Cu2+, diethyldicarbonate, EDTA, Fe2+, Hg2+, hydroxygmercuribenzoate, Mn^{2+}, NEM, Ni^{2+}, Zn^{2+}	Ca^{2+}	Nasser *et al.*, (1993)
Exoppolygalac-turonate lyases (Exo-PGL, EC 4.2.2.9)	*Streptomyces massasporeus*			54000	40	9.5		EDTA,	Ca^{2+}, Mn^{2+},	Sato *et al.*, (1980)
	Clostridium multifermentans					8.5		EDTA,	Sr^{2+}, Ba^{2+}, Ca^{2+}, Mg^{2+}, Mn^{2+},	Macmillan and Phate (1966)
	Erwinia chrysanthemi			79570	45	8.5		EDTA	Sr^{2+}, Co^{2+}, Cu^{2+}, Mn^{2+}, Ni^{2+},	
	Bacillus sp,	Ammomium sulphate precipitation, gel filteration, hydrophobic interaction and ion exchange chromatography	45.4	38,000	69	11			Ca^{2+},	Singh *et al.*, (1999)
Protopectinases (EC 3.2.1.99)	*Bacillus subtilis*	CM-cephadex C-50, Heat treatment, DEAE-Toyopearl.	17	30,000	37	8		2-mercaptoethanol, Ag^+, Ce^{+2}, Fe^{+3}, Hg^{+2}, Mn^{+2}, NaN_3, $NaNO_2$, 0-phenanthroline, p-chloromercuribenzoate	NaCl, Pb^{+2}, Sn^{+2}	Sakai *et al.*, (1989)

the pectinase production, and when acted upon by pectinases serve as carbon source. Hyperproduction of pectinase by *Aeromonas liquefaciens* was recorded when the carbon-limited culture was supplied with glucose and glycerol (Hsu and Vaughn, 1969). Polygalacturonic acid (2%) was preferred by *Erwinia* sp. for the production of transeliminase. *Pseudomonas marginalis* was shown to grow equally well in 0.5% glucose, polygalacturonic acid or pectin (Nasuno and Starr, 1966), Nagel and Vaughn (1961) used 3% pectin in the production medium for polygalacturonase production by *Bacillus polymyxa*. Alkalophilic *Bacillus* sp. RK9 grew well in the culture media supplemented with 1% sucrose/sodium pectate as carbon source (Kelly and Fogarty, 1978). Another strain of *Bacillus* sp. KSM-P15 utilized pectin for pectate lyase secretion (Kobayashi, 1999). Alkalophilic Bacillus GK-8 utilized 1% glucose for constitutive secretion of alkaline exopectinase (Dosanjh and Hoondal, 1996).

Effect of nitrogen sources : Commonly used nitrogen sources can be broadly categorized into organic and inorganic sources. Most widely used organic nitrogen source is the yeast extract. It has also been reported to serve as inducer (Minussi et al., 1996). Yeast extract along with organic/inorganic nitrogen sources has been for the production of pectinolytic enzymes. Yeast extract and ammonium sulphate supported polygalacturonic acid transeliminase production by Xanthomonas campestris (Nasuno and Starr, 1966). A combination of yeast extract and ammonium chloride was found suitable for pectinase production by Pseudomonas marginallis (Membre and Burlot 1994). Horikoshi (1972) used polypeptide along with yeast extract for alkaline pectinase production by Bacillus sp. P-4-N, while Kobayashi et al., (1999) reported maximal pectinase production by an alkaliphilic Bacillus sp. KSM-P15 with polypeptone.

Effect of initial pH of the medium: The optimum pH for pectinase production depends on the microbial strain used. The bacteria yielded maximum titer in the pH range between 7.0 and 10.0 (Horikoshi, 1972; Kelly and Fogarty, 1978; Karbassi and Vaughn, 1980; Magro *et al.*, 1994; Dosanjh and Hoondal, 1996; Kobayashi *et al.*, 1999). However, in a few cases like *Xanthomonas campestris*, slightly acidic pH (6.8) was favourable (Nasuno and Starr, 1967).

Effect of incubation temperature: Majority of bacteria and fungi produce pectinases at 30°C or less. Magro *et al.*, (1994) reported the pectinase production by *Pseudomonas suringae* pv. *glycinea* at 27°C, *Aeromonas liquefaciens* at 25°C (Hsu and Vaughn, 1969) and *Bacillus* sp. KSM-PI5 at 30°C (Kobayashi *et al.*, 1999); *Bacillus* No. P-4-N at 37°C (Horikoshi, 1972), and *B. stearothermophillus* at 65°C (Karbassi and Vaughn, 1980).

Effect of inducers on pectinolytic enzyme: Production of pectinase in some organisms is constitutive, whereas in others it is inducible. Constitutive polygalacturonic acid transeliminase production has been reported in *Aeromonas liquefaciens,* but catabolite repression was observed in the presence of excess substrate (Hsu and Vaughn, 1969). *Aspergillus* sp. CH-Y-1043 produced exopectinase constitutively on different carbon sources (Aguillar and Huitron, 1990). In organisms with inducible enzyme system, pectin was the best inducer followed by polygalacturonic acid (Ros *et al.*, 1991).

Effect of salts: Metal ions play very important role in the production and activity of various extracellular enzymes. Among all metal ions, Ca^{2+} plays the most important role. Kelly and Fogarty (1978) reported the production of polygalacturonase lyase in the absence of Ca^{2+}, but the production level increased considerably with increased levels of Ca^{2+}. The activation of enzyme was seen in the presence of Ca^{2+}(Horikoshi, 1972). Other monovalent and divalent cations have been shown to inhibit pectinase production alone or in the presence of Ca^{2+} (Kelly and Fogarty, 1978). EDTA inhibited pectinase production (Nagel and Vaughn, 1961, Horikoshi, 1972; Kelly and Fogarty, 1978).

Solid-state Fermentation (SSF)

Solid state fermentation (SSF) has received much attention in recent years for the production of bioactive molecules (Lonsane and Ghildyal, 1992; Solis- Pereyra *et al.*, 1996). The SSF has several advantages over the submerged fermentation (SmF) (Oriol *et al.*, 1988; Roussos *et al.*, 1991; Gessesse and Mamo, 1999), such as: (*i*) low water requirement, (*ii*) cheap media for the fermentation, (*iii*) less stringent aseptic conditions; due to inability of most contaminants to grow in the absence of free flowing water, (*iv*) use of high concentration of substrate, and (*v*) production of high concentration of products.

Agro-industrial residues such as wheat bran, rice bran, sugarcane bagasse, corncobs and apple pomace are generally considered the best substrates for the SSF (Pandey *et al.*, 2000). Pectinase production using wheat bran (Cavalitto *et al.*, 1996), citrus pulp-pellets (Fonseca *et al.*, 1991), coffee pulp (Antier *et al.*, 1993) and sugarcane bagasse (Acuna-Arguelles *et al.*, 1995; Solis-Pereyra (*et al.*, 1996) has been recorded. For a particular substrate, the particle size and moisture level/water activity (a_w) are the important factors, which affect the microbial growth and enzyme production (Pandey, 1992). The polygalacturonase production declined with decrease in water activity of the substrate (Acuna-Arguelles *et al.*, 1994). In a comparative study of pectinase production by *Aspergillus* sp. in SmF and SSF, a high titre of enzyme was reported in SSF (Solis-Pereyra *et al.*, 1993; Minjares-Carranco *et al.*, 1997). The pectinase produced by *A. niger* in SSF was more thermotolerant with a broader range of pH for activity than that produced in SmF (Solis-Pereyra *et al.*, 1996).

CLONING OF ALKALINE PECTINASES

There are very few reports on cloning and characterization of alakaline pectinases. Nasser *et al.* (1993) reported cloning and characterization of pectate lyase (*pel*) gene of *Bacillus subtilis,* a plant pathogen. The gene was expressed in *Escherichia coli* using the pUC18 vector. A single copy of the gene is present in the genome. The 1.6 kb fragment of gene was obtained using HindIII, which encoded pectate lyase in *E. coli*; it had 1, 260 bp open reading frame and encoded a 420 amino acid polypeptide that included a 21 amino-acid signal sequence. The cloned pectate lyase was located in the periplasm and the biochemical properties of the cloned enzyme were similar to those of extracellular pectate lyase of *B. subtilis*.

Guo *et al.* (1995) isolated pectate lyase gene,(*pelC*, 1.3-kb), from *F. solani pisi* using *pelA* cDNA as a probe. The coding region of *pelC* was amplified by reverse transcription-polymerase chain reaction using total RNA isolated from a pectin-induced *F. solani pisi* culture as template. The open reading frame of pelC was predicted to encode a 23.3-kDa protein of 219 amino acid residues, which shared 51% identity with PLA from *F. solani pisi*. The amplified pelC cDNA was expressed in *Pichia pastoris* that expressed a pectate lyase C (PLC) with a molecular mass of 26.0 kDa that contained carbohydrate. The purified PLC required Ca^{2+} for its activity and exhibited optimal lyase activity at pH 9.5 and 55°C. The western blot using antibodies raised against PLA and PLC, showed that PLC and PLA were immunologically related to each other.

Sakamoto *et al.* (1997) sequenced the gene that encoded an endo-1, 5-alpha-L-arabinase i.e. protopectinase C of *Bacillus subtilis.* The sequencing fragments were amplified by the cassette-ligation-mediated PCR (CLM-PCR). An approximately 1.5-kb amplicon was cloned into the vector pUC119. The ORF that encoded the arabinase composed of 324 amino-acids including a 33-amino-acid signal peptide. The deduced amino-acid sequence of the enzyme was compared with that of an *Aspergillus niger* endoarabinase; 37% identity with a 207-amino-acid overlap was recorded. The DH5a cells of *Escherichia coli* that harbour pPPC, did not show arabinase activity, either intracellular or extracellular, suggesting that the *B. subtilis* promotor is not functional in this transformant. The readily metabolizable carbohydrates completely suppressed the expression of pPPC in *B. subtilis* IFO 3134 strain. In contrast, productivity of the enzyme increased fourfold in the presence of 0.75 M potassium phosphate in the culture medium.

Purification of pectinases

Pectinases have been purified by using procedures such as precipitation technique (using ammonium sulphate organic solvent/organic polymers like polyethylene glycol), gel filtration, ion-exchange chromatography, and affinity chromatography. Some of the purified bacterial pectinases and their characteristics are listed in Table 1.

INDUSTRIAL APPLICATIONS OF ALKALINE PECTINASES

Alkaline pectinases have wider applications in degumming and retting of plant fibres, bioscouring of cotton fibres, recovery of plant viruses, tea industry, de-inking of recycled paper, pretreatment of pectic wastewater from fruit juice industries, paper making and as feed supplement.

Plant fibres are long, narrow, thick walled and lignified sclerenchymatous cells, which are dead and, therefore, serve as purely mechanical function of giving support and rigidity to the plant body (Dutta, 1980). The plant fibres are classified into three main types based on their origin and structure [soft or bast fibres, they are formed in groups outside xylem in the cortex, phloem or pericycle e.g. Ramie *(Boehmeria nivea),* sun hemp *(Crotolaria juncea)*; surface fibres, this types of fibres are produced on the surface of stems, leaves e.g. cotton *(Gossypium* spp); hard or structural fibres, these fibres are supportive and conducting fibrovascular bundles that are chiefly found in monocots, e.g. Manila hemp *(Musa textilis)*].

According to Bajpai (1999), more than 40% dry weight of the cambium cells in plant consists of pectin. Before industrial utilization of these fibres, they should be freed from this gummy material. Removal of this non-cellulosic gummy material from the plant fibre is called degumming and the process is called retting. The gummy material could be removed conventionally by mechanical processing and treatment of fibre with 12-20% NaOH (2% NaOH in case of sunn hemp and ramie) along with wetting and reducing agents. This is inefficient because gummy material of the fibre is not completely removed and non-specifically cellulose is removed leading to high weight loss of fibre. Due to the drawback of conventional methods, fibre treatment with alkaline solution containing pectinases is preferred (Deshpande *et al.,* 1985; Baracat-Pereira *et al.,* 1993).

Pectinolytic enzymes and microbes are also involved in the process of retting of jute, flax, hemp, ramie, kenaff *(Hibiscus sativa),* and coir from coconut husks (Bruhlmann *et al.,* 1994; Chesson, 1980). Certain bacteria (e.g. *Clostridium, Bacillus)* and fungi (e.g. *Aspergillus, Penicillium)* are also known to decompose the pectin of the bark and release fibre (Sharma and Robinson, 1983). Biological retting can be done in open field by spreading plant straw on the ground and left exposed to the action of the fungi and aerobic bacteria for 2-10 weeks. This is called dew retting (an aerobic process). The species of *Aspergillus, Penicillium, Rhodotorula* and *Cladosporium* have been isolated from dew-retted plants (Fogarty and Ward, 1972). The process does not lead to high quality of fibre due to the attack of cellulolytic microbes. In an alternate method the process of retting is carried out in closed water tanks or water in the presence of anaerobic microbes (Chesson, 1980). In due course the tank and water pits rapidly become depleted of dissolved oxygen encouraging the development of anaerobic microbial flora (Chesson, 1978). Species of *Clostridium,* especially *C. butyricum* and *C. felsineum,* are considered to be major retting agents (Hellinger, 1953). *Bacillus, Pseudomonas* and *Micrococcus* spp. have been isolated from retting flax (Rosemberg and de Franca, 1967), jute (Ahmed, 1963), sisal and coir (Jayasankar *et al.,* 1967), and have shown to macerate straw fibre in laboratory.

During retting of flax, fibre separation is brought about by pectinases produced by bacteria (Chesson, 1978). In a series of experiments conducted by Sharma (1987), a mixture of pectinase and ultrazyme (a combination of xylanase and traces of PG) and an enzyme extracted from *Ceraceomyces sublaevis* has been found to be the most suitable enzyme preparation for degumming at 45°C and pH 5.0 - 6.0. Due to low pH and moderate temperature, there is a high risk of contamination during the process, and hence, the enzyme preparations that are optimally active at alkaline pH and at higher temperature, are recommended (Hoondal *et al.,* 2002).

Ramie (*Boehmeria nivea*) Bast Fibres

The ramie fibres contain 20-35% ramie gum. After removal of the gum (pectin and hemicellulose), the fibre can be used in textile industry. The degumming of ramie fibre is carried out by chemical means as described above. In addition to the high consumption of energy, this process also results in serious environmental pollution. To solve this problem, a combined process was suggested, in which microbes and their enzymes carry out the process at low temperature without involving chemical agents (Paul and Bhattacharya, 1979; Gurucharanam and Deshpande, 1986). Cao *et al.,* (1992) reported degumming of fibres with neutrophilic bacteria, but due to difficulties in prevention of heterogenous contamination and some other disadvantages, these strains have not been applied on industrial scale. The pretreatment of decorticated ramie fibre with alkali accelerates the degumming process (Deshpande and Gurucharanam 1985). The enzyme of *Bacillus* sp. (strain NT-33) having a pH optima of 10 reported to remove more than 70% ramie gum after 24 hr fermentation under alkaline conditions (Cao *et al.,* 1992). The quality of degummed fibres produced by this technique satisfied the textile industry standards. Pectinolytic enzymes from Actinomycetes have also been shown to have good correlation between the pectate lyase activity and the degumming effects, resulting in good separation of the bast fibres (Bruhlmann *et al.,* 1994).

Buel (*Grewia optiva*) Bast Fibres

The bast fibres are removed from the stem by dumping the stems in stagnant water for months. The natural microflora present on the fibres helps in retting. The fibres are then separated by physical methods. The quality of the fibres obtained by this way was not good due to the attack of cellulolytic microbes in prolonged incubation. Villagers sell these fibres and their products in the rural markets.

Pretreatment of Pectic Wastewater

Wastewater coming from the citrus-processing industry contains high amount of pectinaceous materials. During the treatment of activated-sludge, pectic wastes are not degraded completely by the naturally occurring microbes present in the sludge (Tanabe *et al.*, 1986) and cause choking of filters. Various methods such as physical dewatering, spray irrigation, chemical coagulation, direct activated sludge treatment and chemical hydrolysis followed by methane fermentation are available for the treatment of wastewater from citrus processing industries (Tanabe *et al.*, 1986). All the processes have one or the other limitations, like lower efficiency due to chemical resistance of the pectic substances, high treatment cost, long treatment periods and complexity of the process. Tanabe *et al.*, (1987) developed a method for the pectic wastewater treatment using an alkaliphilic *Bacillus* sp. (GIR 621) (optimal pH for growth 10) that produces an endopectate lyase. *Erwinia carotovora* (FERM P- 7576) a well-known plant pathogen (cause soft-rot of carrot) that secretes endo- pectate lyase, was found to be useful in the pretreatment of pectinaceous wastewater (Tanabe *et al.*, 1986). Due to its high phytopathogenecity, it cannot be used directly. An indirect pretreatment method using pectolytic enzyme produced by bacterium, was reported to solubilize almost all the pectic substances present in the wastewater.

Production of Japanese Paper

Japanese paper is prepared from the retting of Mitsumata bast fibres (Horikoshi, 1990). Soda-ash cooking method is conventionally used for the processing of bast fibres. In an alternate method, Tanabe *et al.*, (1987), the fibres are treated with alkaline pectinase produced by *Bacillus* sp. and *Erwinia carotovora* due to its strong macerating activity. These retted bast fibres have been used for the preparation of Japanese paper. The strength of the pulp resulting from bacterial retting is as high as that obtained by conventional soda-ash cooking method. The paper sheets prepared from this pulp are very uniform and soft.

Paper Making

Application of enzymes in paper industry is increasing day by day as they transform the substrate of interest in the presence of other chemically similar compounds (Reid and Ricard 2000). Pectinase solves the paper making retention problems in mechanical pulps bleached with hydrogen peroxide. In papermaking the dilute suspension of fibres, fibre fragments (fines), and inorganic filler particles (clay or $CaCO_3$) are converted into sheets by continuous filtration process. The water must be drained continually. For this, a filter fabric is used with pores that allow the fibre fragment and filler particles to pass through it. Some drainage devices are used for retaining the 'fine' and 'filler' in the paper sheets, which help fast drainage of water. In retention aids, various cationic polymers with varying structures are commonly used (Horn and Linhart, 1996). Bleaching of pulps with alkaline peroxide, solubilizes the polysaccharides present in the pulp; the prominent among these polysaccharides are pectins or polygalacturonic acid and are troublesome substances (Holbom *et al.*, 1991). The polygalacturonic acids have a tendency to complex with cationic polymers (cationic demand) and form gels. The cationic demand depends strongly on their degree of polymerization. Monomers, dimers, and trimers do not have significant cationic demand, but hexamers and long chains have high cationic demand (Thornton *et al.*, 1994). This leads to choking of the retention devices. Hence, pectinase is used to degrade the polysaccharide into monomers, dimers and trimers to reduce the cationic demand of pectin solutions from peroxide bleaching (Reid and Ricard, 2000).

Recovery of Plant Viruses

Recovery of plant viruses found in the phloem is difficult due to the presence of cellulose and pectin. In such cases alkaline pectinase along with cellulase has been used for the recovery of viruses (Salazar and Jayasinghe, 1999).

Recovery of Plant DNA

The contaminating substances co-precipitate with the DNA or RNA. Steven *et al.* (2001) have suggested the use of alkaline pectinase in the last DNA-hydration step to overcome this problem.

CONCLUSIONS

Alkaline pectinases produced by alkali-tolerant/alkaliphilic bacteria have wide applications in various industries. Very few attempts have been made to clone and express alkaline pectinases. In view of application of alkaline pectinases in fibre processing, wastewater treatment and paper and pulp industry, further research efforts are needed on economical production of enzymes using efficient wild microbial strains and/or recombinants.

ACKNOWLEDGEMENTS

We are grateful to Ministry of Environment and Forests, Government of India for providing financial assistance during the course of writing this chapter.

REFERENCES

Acuna-Arguelles, M.E., Gutierrerz-Rozas, M., Viniegra-Gonzalez, G. & Favela-Torres, E. (1995) *Appl. Microbiol. Biotechnol.* **43**: 808-814.

Acuna-Arguelles, M.E., Gutierrerz-Rozas, M., Viniegra-Gonzalez, G. & Favela-Torres, E. (1994) *Biotechnol. Lett.* **16**: 808-814.

Aguillar, G. & Huitron, C. (1987) *Enz. Microb. Technol.* **9**: 690-696.

Aguillar, G. & Huitron, C. (1990) *Biotechnol. Lett.* **12**: 655-660.

Ahmed, M. (1963) *J. Appl. Bacteriol.* **26**: 117-126.

Akita, M., Suzuki, A., Kobayashi, T., Ito, S. & Yamane, T. (2000) *Acta Crystallogr.* **56**: 749-752.

Antier, P., Minjares, A., Roussoss, S., Raimbault, M. & Viniegra-Gonzales, G. (1993) *Enz. Microb. Technol.* **15**: 254-260.
Aspinall, G.O. & Cottrell, I.W. (1970) *Can. J. Chem.* **48**: 1283-1289.
Aspinall, G.O. (1981) *Plant Carbohydrates II* (Eds. Tanner, W. & Loewus, F. A.) Springer-Verlag, Berlin & New York, pp. 554.
Aspinall, G.O., Craig, J.W.T. & Whyte, J.L. (1968a) *Carbohydr. Res.* **7**: 442-452.
Aspinall, G.O., Gestener B., Molloy, J.A., & Uddin, M. (1968b) *J. Chem. Soc.* **23**: 2554-2559.
Bajpai, P. (1999) *Biotechnol. Prog.* **15**: 147-57.
Baracat-Pereira, M.C., Vanetti, M.C.D., Aruojo, E.F.D. & Silva, D.O. (1993) *J. Indian Microbiol.* **11**: 139-142.
Barrett, A.T. & Northcote, D.H. (1965) *Biochem. J.* **94**: 617-627.
Bartolome, L.G. & Hoff, J.E. (1972) *J. Agric. Food Chem.* **20**: 266–270.
Borin, M.D.F., Said, S. & Fonseca, M.J.V. (1996) *J. Agric. Food Chem.* **44**: 1616-1620.
Bruhlmann, F., Kim, K.S., Zimmerman, W. & Fiechter, A. (1994) *Appl. Environ. Microbiol.* **60**: 2107-2112.
Cao, J., Zheng, L. & Chen, S. (1992) *Enz. Microb. Technol.* **14**: 1013-1016.
Castaneda, G., Gunasekran, P & Vinegara-Gonzalez, G. (1996) *World J. Microbiol. Biotechnol.* **12**: 257-260.
Cavalitto, S.F., Arcas, J.A. & Hours, R.A. (1996) *Biotechnol. Lett.* **18**: 251-256.
Chesson, A. (1978) *J. Appl. Bacteriol.* **45**: 219-230.
Chesson, A. (1980) *J. Appl. Bacteriol.* **48**: 1-45.
Collmer, A. & Keen, N.T. (1986) *Ann. Rev. Phytopathol.* **24**: 383-409.
Crawford, M.S. & Kolattukudy, P.E. (1987) *Arch. Biochem. Biophys.* **258**: 196-205.
Darvill, A.G., McNeil, M. & Albersheim, P. (1978) *Plant Physiol.* **62**: 418-422.
Dave, B.A. & Vaughn, R.H. (1971) *J Bacteriol.* **108**: 166-174.
de Vries, J.A., Voragen, A.G.J., Rombouts, F.M. & Pilnik, W. (1986) *Chemistry and Function of Pectin* (Eds. Fishman, M.L. & Sen, J.J.) American Chemical Society: Washington, D.C. pp. 38-48.
Dean, R.A. & Timberlake, W.E. (1989) *The Plant Cell* **1**: 275-284.
Deshpande, K.S. & Gurucharanam, K. (1985) *Indian J. Bot.* **1**: 79-81.
Deuel, H. & Stutz, E. (1958) *Adv Enzymol.* **20**: 341- 382.
Dosanjh, N.S. & Hoondal, G.S. (1996) *Biotechnol. Lett.* **18**: 1435-1438.
Dutta, A.C. 1980, *A class book of Botany* (ed. Dutta, A.C.) Oxford University Press, London. pp.182-194.
Edstrom, R.D. & Phaff, H.J. (1964) *J. Biol. Chem.* **239**: 2409-15.
Esquivel, J.C.C., Hours, R.A., Voget, C.E. & Mignone, C.F. (1999) *J. Biosci. Bioeng.* **1**: 48- 52.
Fonseca, M.J.V., Spandaro, A.C.C. & Said, S. (1991) *Biotech. Lett.* **13**: 39-42.
Fry, S.C. 91982) *Biochem. J.* 203: 493-504.
Gainvors, A. & Belarbi, A. 1995, *Yeast* **10**: 1311-1319.
Gessesse, A. & Mamo, G. (1999) *Enz Microb Technol.* **25**: 68-72.
Guo, W., Gonzalez-Candelas, L & Kolattukudy, P., E. (1995) *Arch. Biochem. Biophys.* **323**: 352-60.
Gurucharanam, K. & Deshpande, K.S. (1986) *Indian J. Phytopathol.* **39**: 385-389.
Hasegawa, S. & Nagel, C.W. (1962) *J. Biol. Chem.* **237**: 619-621.
Hasegawa, S. & Nagel, C.W. (1966) *J. Food Sci.* **31**: 838-845.
Heikinheimo, R., Hemilä, H., Pakkanen, R. & Palva, I. (1991) *Appl. Microbiol. Biotechnol.* **35**: 51-55.
Hellinger, E. (1953) *Nature* 171: 1119-1123.
Holbom, B., Ekman, R., Sjoholm, R., Eckerman, C. & Thornton, J. (1991) *Das Papier,* 45 : V 16- V22.
Holden, M. (1945) *Biochem. J.* **39**: 172-178.
Hoondal, G.S., Tiwari, R.P., Tewari, R., Dahiya, N. & Beg, Q.K. (2002) *Appl. Microbiol. Biotechnol.* **59**: 409-18.
Horikoshi, K. (1972) *Agr. Biol. Chem.* **36**: 285-293.
Horikoshi, K. (1990) *Microbial Enzymes and Biotechnology,* 2 ed. (Eds. Fogarty, W.M. & Kelly, C.T.), Elsvier App1ied Science, London and New York, pp. 275-294.
Horn, D. & Linhart, F. (1996) *Paper Chemistry,* (Ed. Roberts, J.C.), Blackie Academic and Professional, London, pp. 64-82.
Hsu, E.J. & Vaughn, R.H. (1969) *J. Bacteriol.* **4**: 172-181.
Ishii, S. and Vokotsuka, T. (1973) *J. Agric Food Chem.* **21**: 269-272.

Jayasankar, N.P., Agate, A.D. & Batt. J.B. (1967) *J. Indian Inst Sci.* **49**: 10-17.
Jenkins, J., Mayans, O., Smith, D., Worboys, K. & Pickersgill, R. (2001) *J Mol. Biol.* **305**: 951-960.
Kamimiya, S., Itho, Y., Izaki, K. & Takahashi, H. (1977) *Agric. Biol. Chem.* **41**: 975-981.
Karbassi, A. & Vaughn, R.H. (1980) *Can. J. Microbiol.* **26**: 377-384.
Kawano, C.Y., Chellegatti, M.A.D.S.C., Said, S. & Fonseca, M.J.V. (1999) *Biotechnol. Appl. Biochem.* **29**: 133-140.
Kelly, C.T. & Fogarty, W.M. (1978) *Can. J. Microbiol.* **29**: 1164-1172.
Kertesz, T. & Clydesdale, S. (1994) *File of Neurology* **51**: 1226-1231.
Kobayashi, T., Koike, K., Yoshimatsu, T., Higaki, N. Suzumatsu, A., Ozawa, T., Hatada, Y. & Ito, S. (1999) *Biosc. Biotechnol. Biochem.* **63**: 65-72.
Kozianowsky, G., Canganella, F., Rainey, F.A., Hippe, H. & Antranikian, G. (1997) *Extremophiles* **1**: 171-182.
Lau J.M., McNeil, M., Darvill, A.G., & Albersheim, P. (1985) *Carbohydr Res.* **137**: 111-125.
Liao, C.H. (1991) *J. Bacteriol.* **173**: 4386-4393.
Liao, C.H., Hung, H.V. & Chatterjee, A.K. (1988) *Mol Plant- microbe Interact.* **1**: 199-206.
Lim, J.Y., Fujio, Y. & Ueda, S. (1983) *J. Appl. Biochem.* **5**: 91-98
Loera, O., Aguirre, J. & Viniegra-Gonzalez, G. (1999) *Enz. Microb. Technol.* **25**: 103-108.
Lonsane, B.K. & Ghidyal, N.P. (1992) *Exoenzymes* (Eds. Doelle, H.W., Mitchell, D.A.& Rolz, C. E.) Elsvier Science Publishers Ltd., London, pp. 191-209.
Macmillan, J.D. & Phaff, H.J. (1966) *Methods in Enzymology,* vol 8. (Eds. Neufeld, E.F. & Ginsburg, V.) Academic: New York, pp. 632–635.
Maldonado, M.C. & Strasser de Saad, A.M. (1998) *J. Ind. Microbiol. Biotechnol.* **20**: 34-38.
Maldonado, S.C., Navarro, A. & Callieri, D.A.S. (1986) *Biotechnol. Lett.* **8**: 501-504.
Margo, P., Varvaro, L., Chilosi, G., Avanzo, C. & Balestra, G.M. (1994) *FEMS Microbiol. Lett.* **117**: 1-6.
Markovic, O., Slezarik, A. & Labudova, I.(1985) *FEMS Microbiol. Lett.* **27**: 267-271.
McCarthy, R.E., Kotarsky, S.F. & Salyers, A.A. (1985) *J. Bacteriol.* **161**: 493-499.
McComb E.A, & McCready R.M. (1957) *Anal. Chem.* **29**: 819–821,
Membre, J.M. & Burlot, P.M. (1994) *Appl. Environ. Microbiol.* **60**: 2017-2022.
Meyer, L.H. (1987) *Food Chemistry,* LBS Publishers and Distributors, New Delhi, pp. 87-95.
Miller, L., & Mac Millan, J.D. (1971) *J. Bacterol.* **102**: 72-78.
Minjares-Carranco, A., Trejo-Aguilar, B.A., Aguilar, G. & Vniegra-Gonzalez, G. (1997) *Enz. Microb. Technol.* **21**: 25-31.
Minussi, R.C., Coelho, J.L.C., Baracat-Pereira, M.C. & Silva, D.O. (1996) *Biotechnol. Lett.* **18**: 1283-1286.
Miyazaki, Y. (1991) *Agric. Biol. Chem.* **55**: 25-30.
Nagel, C.W. & Hasegawa, S. (1967) *Anal. Biochem.* **21**: 411-415.
Nagel, W.C. & Vaughn, R.H. (1961) *Arch. Biochem. Biophys.* **93**: 344-352.
Nagel, W.C. & Wilson, T.M. (1970, *Appl. Microbiol.* 20: 374-383.
Nasser, W., Awade, A.C., Revercon, S. & Robert-Baudouy, J. (1993) *FEBS Lett.* **335**: 319-26.
Nasuno, S. & Starr, M.P. (1966) *Phytopathol.* **56**: 1414-1415.
Nasuno, S. & Starr, M.P. (1967) *Biochem. J.* **104**: 178-184.
Nurizzo, D., Turkenburg, J. P., Charnock, S. J., Roberts, S. M., Dodson, E. J., McKie, V. A., Taylor, E. J.,Gilbert, H. J. & Davies, G. J. (2002) *Nature Structural Biol.* **9**: 665-668.
Ofuya, C.O. (1984) *Curr. Microbiol.* **10**: 141-146.
Oriol, E., Raimbault, M., Roussos, S. & Viniegra, G. (1988) *Appl. Microbiol. Biotechnol.* **27**: 498-503.
Pandey, A. (1992) *Industrial Microbilogy* (ed. Malik, V. S. & Sridhar, P.) IBH and Oxford Publishing Co., New Delhi. pp. 525-537.
Pandey, A., Soccol, C.R. & Mitchell, D. (2000) *Process Biochem.* **35**: 1153-1169.
Paul, N.B. & Bhattacharya, S.K. (1979) *J. Textile Inst.* **12**: 512-517.
Perombelon, M.C.M. & Kelman, A. (1980) *Ann. Rev. Phytopathol.* **18**: 361-387.
Pitkänen, K., Heikinheimo, R. & Pakkanen, R. (1992) *Enz. Microb. Technol.* **14**: 832-836.
Reid, I. & Ricard, M. (2000) *Enz. Microb. Technol.* **26**: 115-123.
Rijssel, M.W., Gerwig, J.G.J. & Hausen, T.A. (1993) *Appl. Env. Microbiol.* **59**: 828-836.

Ros, J.M., Nunez, J.H., Saura, D., Samtron, M.C. & Laencina, J. (1991) *Biotechnol. Lett.* **13**: 287-290.
Rosemberg, J.A. & de Franca, F. P. (1967) *Appl. Microbiol.* **15**: 484-486.
Roussos, S., Olmos, A., Raimbault, M., Saucedo-Castaneda, G. & Lonsane, B.K. (1991) *Biotechnol. Lett.* **5**: 415-420.
Sakai T., Sakamoto T., Hallaert E. & Vandamme E.J. (1993) *Adv. Appl Microbiol.* 39:213-294
Sakai,T. & Sakamoto, T. (1990) *Agric. Biol. Chem.* **54**: 879-889.
Sakamoto, T., Yamada M., Kawasaki H. & Sakai T. (1997) *Eur. J. Biochem.* **245**: 708-14.
Sakellaris, G., Nikolaropoulos, S. & Evangelopoulos, A.E. (1989) *J. Appl. Bacteriol.* **67**: 77-85.
Salazar, L. and Jayasinghe, U. (1999) Techniques in Plant Virology. CIP Training Manual 5.0. Virus purification, International Potato Center, Peru, pp1-10.
Sharma, H.S.S. & Robinson, E. (1983) *Tech. Rep.* p.11
Sharma, H.S.S. (1987) *Appl. Microbiol. Biotechnol.* **26**: 358-362.
Sheiman, M.I., Macmillan, J.D., Miller, L. & Chase, T. Jr. (1976) *Eur. J. Biochem.* **64**: 565 –72
Shevchik, V.E., Boccara, M., Vedel, R. & Hugouvieux-Cotte-Pattat, N. (1998') *Mol. Microbiol.* **29**: 1459-1469.
Shevchik, V.E., Condemine, G., Hugouvieux-Cotte-Pattat, N. & Robert-Baudouy, J. (1996) *Mol. Microbiol.* **19**: 455-466.
Singh, S.A., Plattner, H. & Diekmann, H. (1999) *Enz. Microb. Technol.* 420–425.
Solis-Pereyra, S., Favela-Torres, E., Gutierrez-Rojas, M., Roussoss, S., Saucedo
Solis-Pereyra, S., Favela-Torres, E., Viniegra-Gonzalez & Gutierrez-Rojas, M. (1993) *Appl. Microbiol. Biotechnol.* 39: 36-41.
Steven, H.R., Brian, K., Carolyn, H.K., Fred, H. & Paul, S. (2001) *Plant Mol. Biol. Rep.* **19**: 353-359.
Sugiura, J., Yasuda, M., Kamimiya, S., Izaki, K. & Takahashi, H. (1984) *J. Gen. Appl. Microbiol.* **30**: 167-175.
Tanabe, H., Kobayashi, Y. & Akamatsu, I. (1986) *J. Ferment. Technol.* **64**: 265- 268.
Tanabe, H., Yoshihara, K., Tamura, K., Kobayashi, Y., Akamatsu, I., Niyomwan, N. & Footrakul, P. (1987) *J. Ferment. Technol.* **65**: 243 - 246.
Tardy, F., Nasser, W. & Robert-Baudouy, J. (1997) *J. Bacteriol.* **179**: 2503-2511.
Tardy, F., Nasser, W., Robert-Baudouy, J. & Hugouvieux-Cotte-Pattat, N. (1997) *J. Bacteriol.* **179**: 2503-2511.
Thornton, J., Ekman, R,, Holmbom, B. & Orsa, F. (1994) *J. Wood Chem. Technol.* **14**: 159-175.
van Houdenhoven, F.E., de Wit, P.J. & Visser, J. (1974) *Carbohydr. Res.* **34**: 233-239.
Versteeg, C. & Versl, L.O. (1979) *Agric. Res. Rep.* **892**: 1-109.
Vonzyakovskaya, Y.M., Avrova, N.P. & Andronikashvili, E. D. (1974) *Microbiol.* **43**: 357-360.
Ward, O.P. & Fogarty, W.M. (1974) *Appl. Microbiol.* **27**: 346-350.
Whistler R.L. & Smart, C.L. (1953) *Polysaccharide Chemistry*, Academic Press, New York, p. 161.
Wong, D.W.S. (1995) *Food Enzymes, Structure and Mechanism*, Chapman and Hall Publications. pp. 212-236.
Wood, P.J. & Siddiqui, I.R. (1971) *Anal. Biochem.* **39**: 418-428.

5

Properties, Uses and Directed Evolution of Glucansucrases Elaborated by Various *Leuconostoc mesenteroides*

Doman Kim[a,b*], Hee-Kyoung Kang [a,c], Jin-Ha Lee[a,c], Hwa-Ja Ryu[e], Eun-Seong Seo [a,d], Myung-Suk Ha [c]

[a] Laboratory of Functional Carbohydrate Enzymes and Microbial Genomics,
[b] Faculty of Applied Chemical Engineering,[c] Engineering Research Institute,
[d] Department of Materials and Biochemical Engineering,
Chonnam National University, Gwang-Ju, 500-757 (Korea)
[e] Department of Biochemistry, Biophysics and Molecular Biology,
Iowa State University, Ames, IA, 50011 (USA)

Dextrans make up a class of polysaccharides that are D-glucans of various structures with contiguous α-1, 6 glycosidic linkages in the main chains and α-1, 2, α-1, 3, or α-1, 4 branch glycosidic linkages, depending on the specificity of the particular dextransucrase. The enzymes responsible for the synthesis of these dextrans from sucrose are known as dextransucrases, glucansucrases, and glucosyltransferases. The enzymes that catalyze glucans synthesis from sucrose form covalent glucosyl- and glucanyl-enzyme complexes. They transfer the glucosyl unit to the reducing-end of the growing glucan chain. When other carbohydrates, in addition to sucrose, are present in the enzyme digest, the enzyme transfers glucose to the carbohydrate acceptors in the secondary reaction that diverts some of the glucose from incorporation into glucan. Many carbohydrate acceptors have been recognized and the products are dependent on the particular enzyme and the structure of the particular acceptor. Because of these unique catalytic characteristics, various dextransucrases have many important industrial and medical use. To improve the understanding of their action mode and extend their applications, this paper describes the historical background and structure of glucans, production and selection of mutants of *Leuconostoc mesenteroides* constitutive for glucansucrases, mechanism of glucan synthesis and potential industrial use of dextransucrases, and our recent findings on the structural, functional organization and directed evolution of the glucansucrases to offer for designing glucansucrases with improved properties.

HISTORICAL BACKGROUND AND STRUCTURE OF GLUCANS

In the nineteenth century, there were reports of mysterious thickening and something gelling of cane and beet sugar solutions. Pasteur reported in 1861, that these "viscous fermentations" of sucrose resulted from microbial action (Paul *et al.*, 1986). The material that produced such changes in the sucrose solution was isolated and found to be a polysaccharide that was called 'dextran' (Scheubler, 1874). Van Tieghem (1878) isolated and named the bacterium that produced the polysaccharide, *Leuconostoc mesenteroides*. Hehre (1941), reported the first cell-free synthesis of dextran from sucrose. Sucrose serves as a high energy donor of glucose units. Sucrose has a glycosidic bond energy equivalent to that of UDP-glucose (6-7 kcal/mole). Jeanes *et al.* (1954) reported the synthesis of several different kinds of glucans by 96 strains of *L. mesenteroides* (Kato and Kuramitsu, 1990). The polysaccharides were characterized by optical rotation, viscosity, periodate oxidation and physical appearance after alcohol precipitation (Table 1). These descriptions of the appearance of the alcohol precipitates provided an early suggestion that the different polysaccharides had different structures. Both water-soluble and water-insoluble polysaccharides were obtained, and some strains appeared to form more than one kind of polysaccharide, as judged by their water solubility and differences in the amount of alcohol needed to precipitate them.

Table. 1: Glucans synthesized from sucrose by glucansucrase from selected *Leuconostoc* and *Streptococci*.

Species and Strain No.[a]	Solubiliy Class[b]	Percent of Linkages					Description of the ethanol precipitate
		$1\rightarrow6$	$1\rightarrow3$	$1\rightarrow3Br^{C}$	$1\rightarrow2Br^{C}$	$1\rightarrow4Br^{C}$	
L.m. B-512F	L	95					translucent gel
L.m. B-742	L	87				13	heavy, opaque
L.m. B-742	S	50		50			fine[d]
L.m. B-1299	L	66		1	27		flocculent[d]
L.m. B-1299	S	65			35		fine[d]
L.m. B-1355	L	95		5			translucent gel
L.m. B-1355	S	54	35	11			heavy, opaque
L.m. B-1191	L	94		2		4	cohesive, stringy[d]
L.m. B-1308	L	95				2	pasty, crumbly
S.s. B-1526	I	83				17	fluid, stringy[d]
L.m. B-523	I		100				water-insoluble
L.m. B-1149	I		100				water-insoluble
S.m. 6715	S	64		36			heavy, opaque
S.m. 6715	I	4	96	2			water-insoluble
S.v. B-1351	S	89		11			short[d]

[a]*L.m.* = *Leuconostoc mesenteroides*; S.m. = *Streptococcus mutans*; S.s. = *Strep.* species; S.v. = *Strep. viridans*; B-numbers refer to the strain number in the Northern Regional Research Laboratory Collection (NRRL) of the USDA Laboratory, Peoria, IL. [b]L = less soluble referring to glucans precipitated by 34-37% ethanol; S = more soluble referring to glucans precipitated by 40-44% ethanol; I = water-insoluble. Br^{C} = branched linkage. Adapted from Table 1 in ref. [Robyt, 1996] and reprinted by permission of the Academic Press, Inc.

A dextran is now defined as a glucan that is glucopyranose units in the main chains. All known dextrans are branched. This is one of the important characteristics that distinguishes the different kinds of dextrans. Branch linkages have been found to be α-1→2, α-1→3, and α-1→4 and branches can be single D-glucose units and/or chains of α-1→6 linked D-glucose units (Robyt, 1996). Some strains of *L. mesenteroides* produce one type of glucansucrase and one type of dextran, whereas others produce multiple types of glucansucrases and dextrans (glucans). Differences in the proportions of types are evident from both heterogeneity in glucan solubility profiles and electrophoretic band patterns of glucansucrases (Kato and Kiramitsu, 1990). Dextransucrase from *L. mesenteroides* NRRL B-512F synthesizes a linear water-soluble dextran with 95% α-1→6 glucosidic linkages in the main chains and 5% α-1→3 branch linkages (Fig. 1). Some strains produce two different dextrans. *L. mesenteroides* NRRL B-742 produces two types of dextrans - fraction L dextran, which is precipitated by 38% alcohol, has about 14% α-1→4 branched linkages, and fraction S dextran, which is precipitated by 44% alcohol and has a very high degree (50%) of α-1→3 branched linkages that primarily consist of single glucose units (Fig.1) (Robyt, 1996). *L. mesenteroides* strain B-1142 dextran has a structure similar to that of B-742 fraction S dextran. The structure is essentially a B-512F dextran structure having some α-1→6 linked chains attached by an α-1→3 branch linkage, but having single glucose residues attached by α-1→3 branched linkages to every glucose residue in the α-1→6 glucosidic linked chains (Fig. 1). This structure results in the most highly branched dextran possible. *L. mesenteroides* NRRL B-1355 produces two sucrose–inducible extracellular glucansucrases. One produces a dextran (or fraction L dextran) and the other produces a glucan referred to as an alternan (or a fraction S dextran). Dextran has a structure very similar to that of B-512F dextran. Alternan has glucose residues linked alternately by aα-1→6 and aα-1→3 linkages. Fraction L dextran was considered the less water soluble of the two glucans because it precipitated at lower alcohol concentrations than did fraction S. A water-insoluble glucan which pelleted with the cells upon centrifugation was also produced. The water-insoluble fraction has not been considered distinct from fraction L. Other strains (B-1498, B-1501) also produce alternanusucrase (EC 2.4.1.140) that synthesizes alternan.

Various dextransucrases have many important industrial and medical uses (Grownwell and Ingelman, 1944; Grownwell and Ingelman, 1945; Ricketts *et al.*, 1950; Robyt and Taniguchi, 1976; Robyt *et al.*, 1992). Each enzyme is also important because of its theoretical and practical aspects in understanding the mechanism of glucan synthesis (Ebert and Schenk, 1962; Robyt and Taniguchi, 1976; Robyt *et al.*, 1974) and in its ability to synthesize a wide variety of oligosaccharides by glucosyl transfer reactions to acceptors (see below) (Robyt and Eklund, 1983; Robyt and Walseth, 1978). Also, different structure dextrans might be chemically or enzymatically modified to give new gums, gels, or films that could have applications in food preparations, pharmaceuticals, and drug or pesticide delivery systems. Before dextran, alternan, and/or other branched dextrans, however, can be marketed for food and pharmaceutical applications, their technologies for; mass production must be developed, such as isolation of hyper glucansucrase producing strains, development of relatively simple purification methods, and the development of continuous processes for polysaccharide production using enzyme reactors.

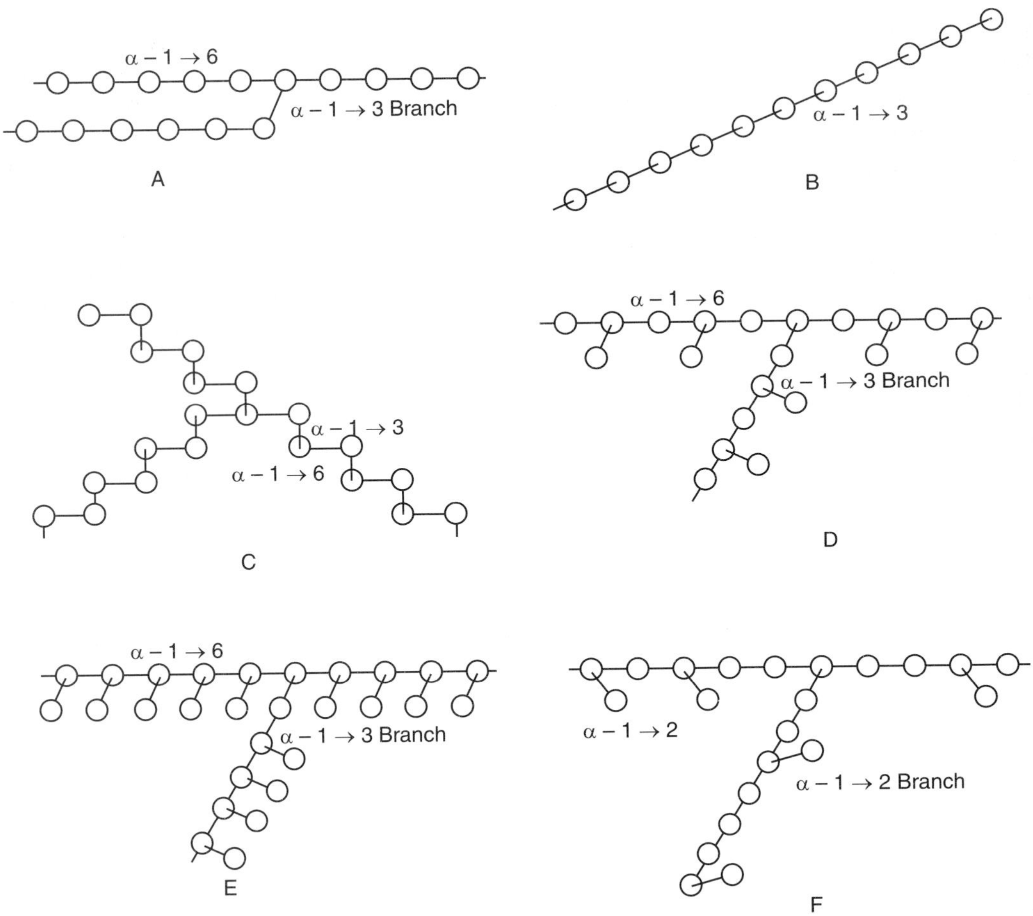

Fig. 1: Diagrammatic representation of the structures of the different glucans synthesized by different glucansucrase from various *L. mesenteroides* and *Streptococcus mutans*. A, *L. mesenteroides* B-512F dextran; B, S. mutans insoluble mutan; C, *L. mesenteroides* B-1355 alternan; D, S. mutans soluble dextran; E, *L. mesenteroides* B-742 S-dextran; and F, *L. mesenteroides* B-1229 dextran. Adopted from Fig. 1 (Robyt, 1996) and reprinted by permission of the Academic Press, Inc.

PRODUCTION AND SELECTION OF MUTANTS OF *LEUCONOSTOC MESENTEROIDES* CONSTITUTIVE FOR GLUCANSUCRASES

Selection of Mutants

The glucansucrases of *L. mesenteroides* strains are inducible and require sucrose in the culture medium for elaboration except several constitutive mutants developed (Kim and Day, 1994; Kim and Robyt, 1994; 1995a; 19995b). The enzymes are secreted into the culture broth during growth, the presence of sucrose results in production of glucan in the culture broth and creates a problem that the glucansucrases are contaminated with relatively large amounts of glucan that impart a high viscosity to the culture supernatant. This limits the concentration of sucrose that can be used in the medium for the production of glucansucrases to 2-3% (w/v). To purify the glucansucrases, a hydrolase treatment is needed to lower the viscosity and glucan content. Dextrans with a low degree of branching (5-10%), for example B-512F (MC) dextran, can be hydrolyzed by *Penicillium* endo-dextranase, but the more highly branched dextrans, such as B-1142, B-742, and B-1299 are resistant to endo-dextranase hydrolysis. The non-dextran glucans, such as B-1355 alternan and the α-1$\rightarrow$3 linked glucans are completely resistant to endo-dextranase hydrolysis. Hydrolases for these linked glucans have yet to be isolated and described. In addition, B-742 and B-1299 strains produce mixture of two dextransucrases that cannot be separated easily. To overcome difficulties, using ethyl methane sulfonate as a mutagen, several constitutive mutants have been isolated from *L. mesenteroides* (B-512FM, B-1142, B-1355, B-742 and B-1299) that produce glucans having different structures from each other (Kato and Kiramitsu, 1990) and that have potential of commercial use (Kim and Day, 1994; Kim and Robyt, 1994b; 1995a; 19995b). They have designated these glucansucrase constitutive mutants as B-5l2FMC, B-1142C, B-1355C, B-742C, B-742CA, B-742CB, and B-1299C, B-1299CA, and B-1299CB, respectively (Table 2).

Table 2: Comparison of glucansucrase activities of parents and mutants grown on glucose and sucrose.

Strains	Glucansucrase activity (U/ml)	
	Glucose (2%)	Sucrose (2%)
B-512F	0.00	0.05
B-512FMC	1.10	3.20
B-1355	0.00	0.10
B-1355C	0.34[a]	2.52[a]
B-1142	0.00	0.04
B-1142C	0.12	1.28
B-742	0.00	0.04
B-742C	0.14	0.29
B-742CA	0.43	0.65
B-742CB	0.10	0.15

[a]Contains both glucansucrase activities, alternansucrase and dextransucrase, as judged by the nature of the glucans produced.

The mutants produced glucansucrase(s) on glucose with higher activities (from 3 to 22 times) than those produced on sucrose by the wild types. The mutants, however, produced less enzymes when grown on a glucose medium than they did when grown on a sucrose medium. This may indicate that some induction for glucansucrase may still be present in the mutants. There is also the possibility that a complex of glucan with the glucansucrase stabilizes the glucansucrase may still be present in the mutants. Enzymes is a favorable conformation for the synthesis of glucan and thereby stimulates glucan synthesis, or that the glucan prevents subunit dissociation into less active or inactive subunits. Since the constitutive glucansucrase grown on glucose do not have glucan present, they may have less glucansucrase activity due to these possibilities. Also the mutants, grown on glucose or sucrose, produced a dextransucrase that synthesized both, a soluble and an insoluble dextran. Differences in water solubility, susceptibility to endo-dextranase hydrolysis, and the physical appearance of the ethanol precipitated dextrans elaborated by different mutants grown on glucose media and on sucrose media were also found.

Properties of *Leuconostoc mesenteroides* (Constitutive Dextransucrases)

After SDS-PAGE of B-512FMC glucose-grown culture supernatant, the gels were incubated with sucrose, followed by the periodic acid- Schiff stain. It showed one major activity band at 184 kDa and a minor band at 128 kDa. B-512FMC culture supernatant grown on sucrose showed one major band at 184 kDa and no minor bands. B-1142C culture supernatant grown on glucose showed one major band at 173 kDa and a minor band at 240 kDa. The sucrose grown B-1142C culture supernatant contained an additional band at the origin. B-1355C grown on glucose medium produced one band at 173 kDa. When grown on sucrose, there were additional activity bands at 184 kDa, 125 kDa, and the origin. This major 184 kDa activity band also exists for each of the dextransucrases from B-742 and B-1299 mutants as well as parental strains, even though each of the strains elaborated dextransucrase(s) that synthesized different dextrans. Thus, it is possible that many of the dextransucrases elaborated by the various strains of *L. mesenteroides* dextransucrases have originated from the same gene. Enzyme produced in glucose medium was adsorbed on Sephadex G-100 and G-200, but much less enzyme was adsorbed when it was produced in sucrose medium. Sephadex adsorption decreased when the glucose-produced enzyme was pre-incubated with dextrans of molecular size greater than 10 kDa. The release of dextransucrase activity from Sephadex by buffer (20 mM acetate, pH 5.2) was highest at 28°C - 30°C. The addition of dextran to the enzyme stimulated dextran synthesis but had very little effect on the temperature or pH stability.

MECHANISM OF GLUCAN SYNTHESIS AND POTENTIAL INDUSTRIAL USES OF DEXTRANSUCRASES

Two-site Insertion Mechanism and Activation for the Synthesis of Dextran by Dextransucrase

In earlier work on the initiation of dextran synthesis, Hehre (1951) suggested that the dextran associated with the enzyme could prime the polymerizing reaction. Germaine *et al.* showed that the addition of dextran to *Streptococcus mutans* dextransucrase digests increased the rate of dextran (Germaine *et al.*, 1974; Germaine and Schachtele, 1976). They found that the rate was dependent on the size of the dextran chain and reached a maximum when the average size of the added dextran was 30 glucose residues. The purified dextransucrases elaborated by both *L. mesenteroides* B-512F and *Streptococcus sp.* were stimulated by the addition of dextran, although both enzymes could synthesize dextran without the addition of dextran to the digests (Kobayashi and Matsuda, 1980; Kobayashi and Matsuda, 1986). In contrast, Robyt *et al.* (1974) using pulse and chase techniques with [^{14}C] sucrose and Bio-Gel P2-immobilized dextransucrase, showed that glucose and dextran were covalently attached to the enzyme during synthesis and that the glucose is added to the reducing end of the growing dextran chain by a two-site insertion mechanism. During dextran elongation, the glucosyl unit is transferred to the reducing end of the growing dextran chain by an insertion mechanism (Fig. 2).

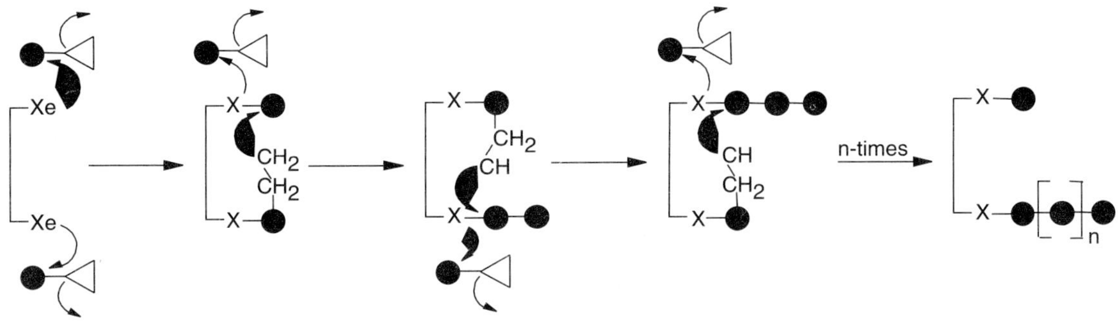

Fig. 2: Two-site insertion mechanism for the synthesis of *L. mesenteroides* B-512F dextran by dextransucrase. ●–◁ is sucrose, ● is glucose, ◁ is fructose, X represents an enzyme nucleophile, ●–● represents two glucose residues linked α-1?6. X orients the glucosyl units so that their C-6 hydroxyl groups can make an attack onto C-1 of the apposed glucosyl unit (Robyt, 1996).

The 6-hydroxyl group of one of these glucosyl intermediates nucleophilically attacks the C-1 of the other glucosyl intermediate to form an α-1→6 glucosidic linkage and an isomaltosyl-enzyme intermediate. The newly released nucleophile then attacks another sucrose molecule to give a new glucosyl-enzyme intermediate (this is shown in Fig. 2 as a concerted reaction that takes place as the nucleophile is displaced from the glucosyl unit). The 6-hydroxyl group of the new glucose intermediate then attacks the C-1 of the isomaltosyl intermediate to give a second α-1→6 glucosidic linkage and the formation of an isomaltotriosyl-enzyme intermediate. The process continues in a similar fashion between the two sites, affecting the synthesis of a dextran chain by the addition of glucose to the reducing end of the chain and the apparent insertion of glucose between the enzyme and the growing chain. In this mechanism, a dextran chain can be synthesized *de novo* in a continuous manner without the presence of any pre-formed dextran.

Fu and Robyt (1988) showed that two imidazolium groups of histidine were essential for dextran synthesis by chemical modification of the enzyme with diethyl pyrocarbonate and photo-oxidation with Rose Bengal dye. They postulated that two imidazolium groups donate their hydrogen ions to the leaving fructose units and that the resulting imidazole group, in a second step, becomes reprotonated by abstracting a proton from attacking 6-hydroxyl group of the glucosyl-enzyme intermediate, thereby facilitating the nucleophilic attack and the formation of the α-1→6 linkage. Thus the imidazole group also becomes reprotonated for the next reaction with sucrose. Mechanisms for the synthesis of other glucans, such as *L. mesenteroides* B-1355 alternan and *L. mesenteroides* B-742 regular comb-dextran can also be formulated by a two-insertion mechanism. For the synthesis of alternan the mechanism can be postulated to have the two glucosyl intermediates stereochemically positioned differently. On one site (the X site), the glucosyl intermediate is stereochemically positioned so that only its 6-hydroxyl group is in position to attack the C-1 of the opposite glucosyl-enzyme intermediate to give an α-1→6 linkage; on the other site (the Y site), the glucosyl-enzyme intermediate is stereochemically positioned so that 3-hydroxyl group attacks the C-1 of the opposite glucosyl intermediate to give an α-1→3 linkage. In this manner, the chain goes back and forth between the two sites, giving an alternating synthesis of α-1→6 and α-1→3 glycosidic linkages (Fig. 3).

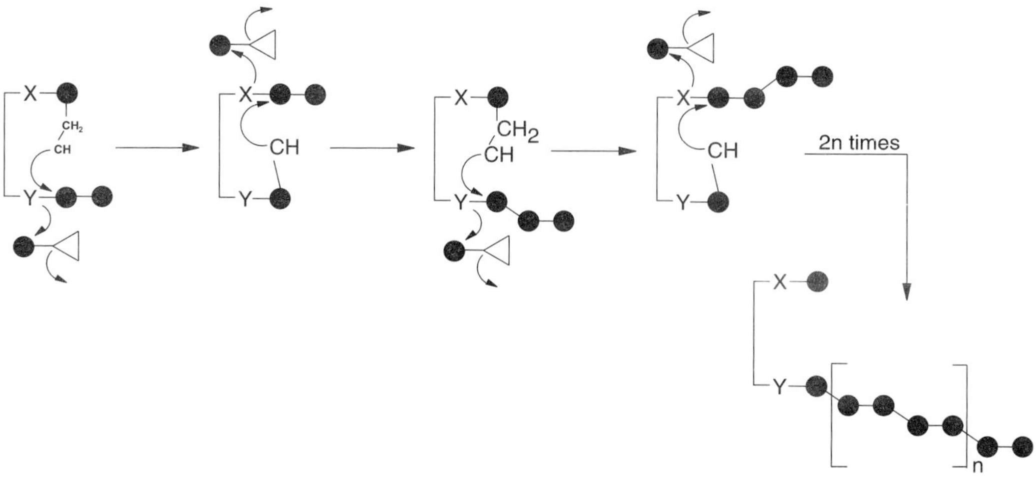

Fig. 3: Two-site insertion mechanism for the synthesis of *L. mesenteroides* B-1355 alternan by alternanuscrase. ●—◁ is sucrose, ● is glucose, ◁ is fructose, the two nucleophiles are X and Y, ●—●, represents two glucose residues linked α(1>6). X orients the glucosyl units so that its C-6 hydroxyl groups can attack at C-1 of the apposed glucosyl unit and Y orients its glucosyl unit so that its 3-hydroxyl group can attack at C-1 of the apposed glucosyl units (Robyt, 1996).

In two-site insertion mechanism, a primer is not necessary for the synthesis to occur. One of the reasons for the primer controversy was the uncertainty of complete dextran removal in purified *Leuconostoc* dextransucrase preparations. There was no carbohydrate found with the B-512FMC dextransucrase released from Sephadex gel using buffer, and this preparation could synthesize dextran (Kim and Day, 1994; Kim and Robyt, 1994b). This purified enzyme preparation does not have any factors that can compromise the experimental results for the study of the enzyme reaction mechanism. Thus, it is clear that the synthesis of dextran by *L. mesenteroides* B-512FMC dextransucrase does not require a dextran primer. Robyt *et al.* (1974) also investigated the mechanism of dextran activation of dextransucrase by studying the initial velocity of the synthesis of dextran as a function of the concentration of added dextran, using B-512FMC dextransucrases (Parnaik *et al.*, 1983). The results of the experiments are shown in Fig. 4 in which the initial velocities (mm mol of glucose incorporated into dextran/min) are plotted against the log of the concentration (g/ml) of added dextran T-40. The *L. mesenteroides* B-512FMC dextransucrase reaction gave sigmoidal curves as a function of the concentration of added dextran. It had an initial velocity lag between 1.25 and 3.12 g/ml of exogenous dextran. Thereafter, the initial velocity increased up to an exogenous dextran concentration of 40 g/ml, where a constant, maximum velocity was attained. In this digest, the initial velocities decreased slightly with an additional increase in the concentration of exogenous dextran. The sigmoidal curves obtained are not that of usual Michaelis-Menten kinetics. Sigmoidal curves are particularly characteristic of a cooperative binding of ligands (in this case dextran) to enzymes that have multiple binding sites. The sigmoidal increase in the reaction rate with increasing ligand concentration can be interpreted as due to the binding of the ligand at a non-catalytic site or allosteric site to give a more active enzyme. This positive allosteric effect often involves the joining together of subunits. This is very likely the mechanism of exogenous dextran activation of dextransucrases. It was found that *S. mutans* dextransucrase has a dextran-binding domain that is distinct and some distance from the active-site domain (Parnaik 1983; Moser and Iwaoka, 1989). Further, both the *S. mutans* dextransucrases and the *L. mesenteroides* B-512FMC dextransucrases bind to lightly crossed-linked dextran (Sephadex G-200) (Robyt *et al.*, 1974). The binding was inhibited when exogenous dextran were present. This non-catalytic, dextran-binding domain is very likely an allosteric site to which the dextran binds and induces a favourable conformation for the synthesis of dextran from sucrose. Further, there is evidence that the favourable conformation involves the formation of multiple subunit complexes. As indicated previously the molecular weights of the dextransucrases have ranged from 184 kDa to 64 kDa. Very high-molecular-weight aggregates also have been observed, especially when the enzymes are elaborated in a sucrose medium where dextran is synthesized. This indicates that the presence of dextran in relatively large amounts, with a high molecular weight, acts as a crosslinking matrix that can join many enzyme subunits together. As the concentration of enzyme is increased per unit volume, the concentration of inactive or less active subunits is increased, giving a higher probability of association. Thus, a lower concentration of dextran is required to induce the association into active units and give activation. The maximum degree of activation, however, remains the same as the relative proportion of subunits of associate.

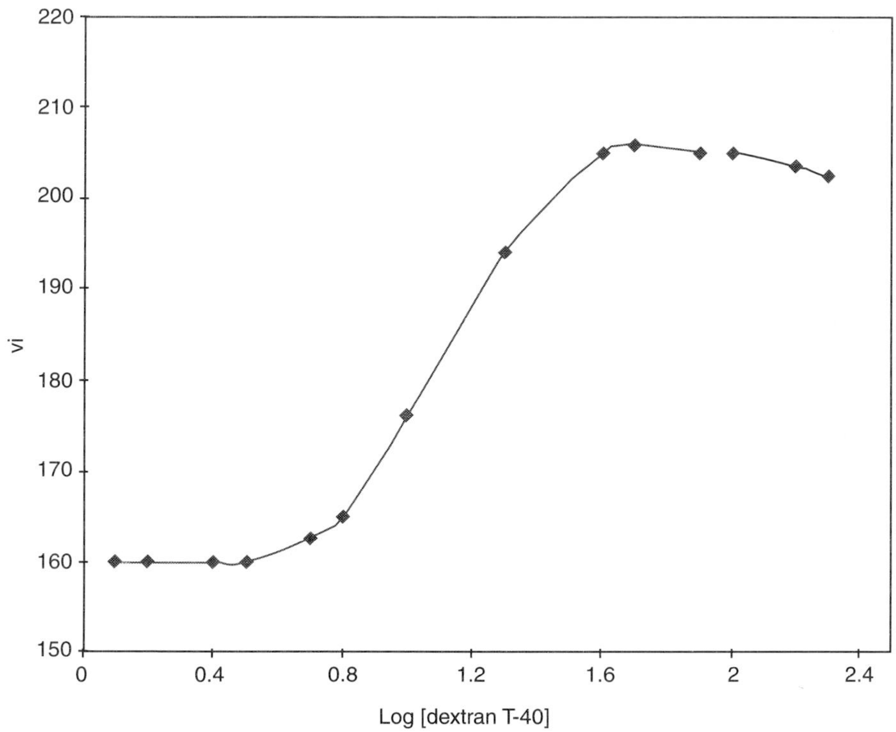

Fig. 4: Activation of dextransucrase by native dextran. The enzyme solution was incubated with sucrose and samples were taken at different times, giving various concentrations of dextran. Dextransucrase activity was determined on these enzyme solutions by the ^{14}C sucrose radioactivity assay and by the determination of the amount of fructose released by TLC analysis (Robyt, 1996). Vi; m moles Glucose/min; dextran T-40; mg/mL.

Acceptors and Acceptor Reaction

In addition to catalyzing the synthesis of dextran from sucrose, dextransucrase also catalyzes the transfer of glucose from sucrose to other carbohydrates that are present or are added to reaction mixture. The added carbohydrates are called 'acceptors' and the reaction is called 'acceptor reaction'. When the acceptor is a monosaccharide or disaccharide, a series of oligosaccharide acceptor-products is usually produced. This acceptor reaction was first described by Koepsell *et al.*, (1953). In the presence of sucrose, they observed that introduction into the reaction medium of molecules like maltose, isomaltose, *O*-α-methylglucoside shifted the pathway of glucan synthesis towards the production of oligosaccharides (Table 3). Moreover, glucan itself can act as an acceptor molecule to yield a branched polysaccharide. Robyt and Walseth (Robyt and Walseth, 1978) proposed that only one covalent glucosyl-enzyme complex was necessary and that acceptor molecules were incorporated at the reducing end of the glucan or the oligosaccharide produced. Oligosaccharide elongation might occur at the reducing end. According to the mechanism proposed by Robyt *et al.* (1974) for glucan synthesis, oligosaccharides may be synthesized by a nucleophilic attack of the hydroxyl group located at the non-reducing end of the acceptor to the C-1 of one of the two glucosyl-enzyme complexes. Various experiments were tended to show that the acceptor binding site was really unique and separated from the two active sites (Robyt and Taniguchi, 1976; Robyt and Eklund, 1983a), and till date there is no direct evidence that a separate acceptor binding site exists.

Table 3: Acceptor reaction products formed by dextransucrase [Kim & Robyt, 1994b].

Acceptors	First Acceptor Reaction Product
Maltose	Panose (6^2-α-glucosylmaltose)
Isomaltose	Isomaltotriose
Nigerose	6^2-α-D-glucopyranosyl nigerose
D-glucose	Isomaltose
Lactose	2^{Glc}-α-D-glucopyranosyl lactose
Cellobiose	2^1-α-D-glucopyranosyl cellobiose
D-fructose	5-α-D-glucopyranosyl-D-fructose
Raffinose	2^{Glc}-α-D-glucopyranosyl raffinose
D-mannose	α-D-glucopyranosyl-D-mannose
D-galactose	α-D-glucopyranosyl-D-galactose
α, α-trehalose[b]	—
α, α-trehalose	α-isomaltosyl-α-D-glucopyranoside
α-isomaltosyl-α-D-glucopyranoside	
α, α-trehalose	α-isomaltosyl-α-D-glucopyranoside
Maltotriose[c]	6^3-*O*-α-D-glucopyranosyl maltotriose
6^1-*O*-α-D-glucopyranosyl maltotriose	
Maltotetraose[c]	6^4-α-D-glucopyranosyl maltotetraose
6^1-α-D-glucopyranosyl maltotetraose	
Panose[d]	6^2- *O*-α-nigerosylmaltose

[a]The products are produced by *L. mesenteroides* B-512F dextransucrase.
[b]Not an acceptor
[c]Two products are formed by *L. mesenteroides* B-512F dextransucrase.
[d]Product from by an alternansucrase.

The addition of maltose in the reaction digest activates the reaction and accelerates the reaction velocity. It may be explained by a change in the limiting step occurring during the acceptor reaction (Paul *et al.*, 1986). In the presence of maltose, formation of glucosyl-enzyme complex becomes the limiting step instead of the polymer transfer in case of glucan synthesis reaction. Mayer *et al.* (1981) attributed this activator effect to a change in the enzyme conformation resulting of acceptor binding with dextransucrase. Broker *et al.* (1994) proposed that the presence of fructose may create a steric hindrance inhibiting the growth of glucan chain. Some acceptors, for instance disaccharides such as like maltose or oligosaccharides presenting an isomaltosyl residue at their non-reducing end, allow a series of oligosaccharides to be produced. When D-glucose, maltose, and isomaltose are the acceptors, the glucose from sucrose is transferred either to the 6-hydroxyl group of the monosaccharide or to the 6-hydroxyl group of the non-reducing end glucose residue of the disaccharides to give a series of isomalto-oligosaccharides of degree of polymerization (d.p.) of 2-7 attached to the acceptor (Fig. 5).

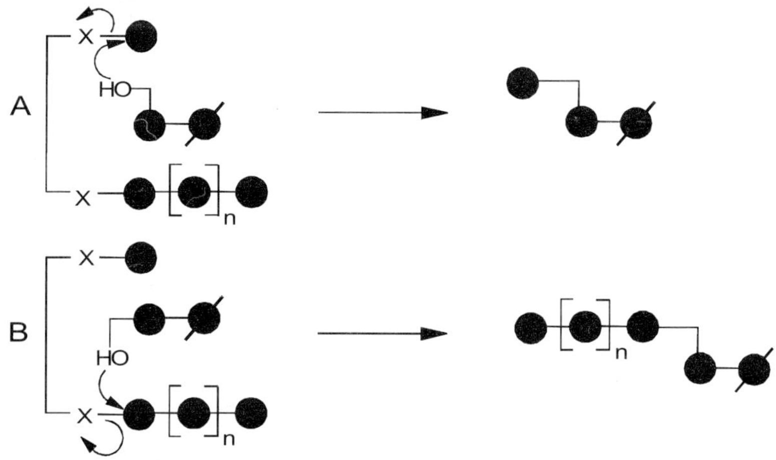

Fig. 5: Mechanism for the acceptor reaction of *L. mesenteroides* B-512F dextransucrase. A disaccharide acceptor binds in the acceptor binding site so that (A) its non-reducing 6-hydroxyl group can attack at C-1 of the glucosyl unit, releasing it from the active site to give a trisaccharide, or (B) its non-reducing 6-hydroxyl group can attack at C-1 of the glucosyl chain, releasing it from the active site (Kim & Robyt, 1994a).

Robyt and Walseth (1978) concluded that the acceptor reactions serve to terminate polymerization of dextran by displacing the growing chain from the active site, in contrast to previous ideas (Koepsell *et al.*, 1953) that acceptors were serving as primers for dextran synthesis. Later, Fu and Robyt (1990) studied the structures of the maltooligosaccharides, maltotriose to maltooctaose (G3-G8), acceptor products produced by *L. mesenteroides* B-512FM dextransucrase, by *S. mutans* dextransucrase (GTF-S) and mutansucrase (GTF-I) (Fu and Robyt, 1991). They found that B-512FM dextransucrase transfers D-glucose to the 6-hydroxyl group of both the non-reducing end and the reducing-end residues of G3-G8. Other acceptors, like fructose, allow only one acceptor product (leucrose) to be produced in which the glucosyl residue coming from the breakdown of sucrose is added to the acceptor. The reason why weak acceptor, such as fructose, inhibits the reaction remains mostly unclear. The leucrose synthesis reaction, in fact, becomes important at the end of the glucan synthesis reaction when the fructose concentration is high. Several acceptors may be shared into two main classes, according to their abilities to compete with glucan synthesis or their effects on the reaction velocity: (a) the strong acceptors, like maltose or isomaltose, have an activator effect on the reaction velocity and strongly inhibit the yield of glucan synthesis (Robyt and Eklund, 1983b); (b) the weak acceptors, like fructose or melibiose, have an inhibitory effect on the reaction (Koepsell *et al.*, 1953). Table 2 summarizes the major acceptors and their products.

Synthesis of Branch Linkages and Acceptor Reaction-Novel Process for Production of Oligosaccharides and Size-limited Dextrans

Oligosaccharides are the subject of an increasing number of research programmes due to both, their technological and physiological properties. The technological properties of oligosaccharides have been stressed in the food application field by the introduction of intense sweeteners and the need for stabilizers and/or bulking agents. Their viscosity and water-retention properties are keys to the development of new low-calorie diets. The physiological properties of oligosaccharides result from their physico-chemical and their specific biological characteristics (Mooser and Wong, 1988). The involvement of these oligosaccharides in specific recognition mechanisms (immunology, cell adhesion, etc.) has been underestimated for a while, but is of great interest for both, human and animal applications (Wong *et al.*, 1990). In biological systems, synthesis reactions generally involve cofactors (NAD(P)H, ATP etc.) which are consumed on a stoichiometric basis during the reaction. This raises the problem of cofactor regeneration which up to now limits the industrial application of this type of enzymatic synthesis reactions. Recent processes in enzymology have demonstrated that the alternative potential use of hydrolases in the catalysis of reverse reactions for the synthesis of glucooligosaccharides and galactooligosaccharides (Takemoto and Liebharber, 1962; Mitsuya *et al.*, 1988). However, the reaction yields and the limited specificity of most hydrolases in reverse reactions limit the application of this technology. Another enzymatic approach for oligosaccharide production consists of the use of transfer reactions by transferases. Industrial enzymatic processes are used to synthesize cyclodextrins (Ajisaka *et al.*, 1987) and fructooligosaccharides (Kitahata *et al.*, 1974). Glucooligosaccharides can be synthesized using glucosyltransferases (E.C. 2.4.1.5). These enzymes catalyze the transfer of glucose units from sucrose onto acceptor molecules, mainly sugars, resulting in the synthesis of glucooligosaccharides. In the absence of acceptors the main product of the reaction is dextran. The chemical structure of dextran was shown to be highly dependent on the glucosyltransferase producing strain. Different carbohydrate structures will bind at the acceptor binding-site with different affinities and therefore have different acceptor efficiencies (Castillo *et al.*, 1990). There are reports about the production of glucooligosaccharides via acceptor reactions using the glucosyltransferases from *L. mesenteroides* strains (Koepsell *et al.*, 1953; Robyt, 1996). The synthesis of oligosaccharides (or acceptor products) and controlled molecular weight dextran by the acceptor reaction using dextransucrase from *L. mesenteroides* NRRL B-512F(M) have been described (Kitahata and Okada, 1982) and the enzymatically synthesized dextrans were shown to have less branched linkages than the dextran fractions produced by acid hydrolysis (Lopez-Munguia *et al.*, 1990). These glucosyltransferases were shown to display specificity in the acceptor reaction close to that observed in high-molecular weight glucan synthesis. The efficiency of the acceptor reactions is dependent on the ratio of the concentrations of the acceptor to sucrose, the absolute concentration of acceptor and sucrose (Fig. 6). In general, when the concentration of the acceptor is higher than that of sucrose, the synthesis of the acceptor-product is favoured over the synthesis of glucan. Further, with acceptors that give multiple acceptor products, the number of acceptor products is decreased as the ratio of the concentration of acceptor to sucrose is increased. The formation of acceptor-products is also favoured when the concentrations of both, acceptor and sucrose, are relatively high. In the useful conditions of action onto sucrose, dextransucrase yields high molecular weight (10^7 - 10^8 daltons) dextran polymers. However, the industrial applications of this polysaccharide in the analytical field (chromatography supports) or in the medical field (blood plasma substitute) require lower molecular weight polymers (40,000 or 70,000 daltons and even 5,000 for iron- or sulphate-dextran derivatives) (Kim and Robyt, 1994a). Such molecules are obtained by acidic hydrolysis of high molecular weight dextran polymers, followed by alcohol-precipitation fractionation steps. On the other hand, the addition of efficient acceptors to the reaction medium results in the synthesis of low molecular weight oligosaccharides instead of high molecular weight

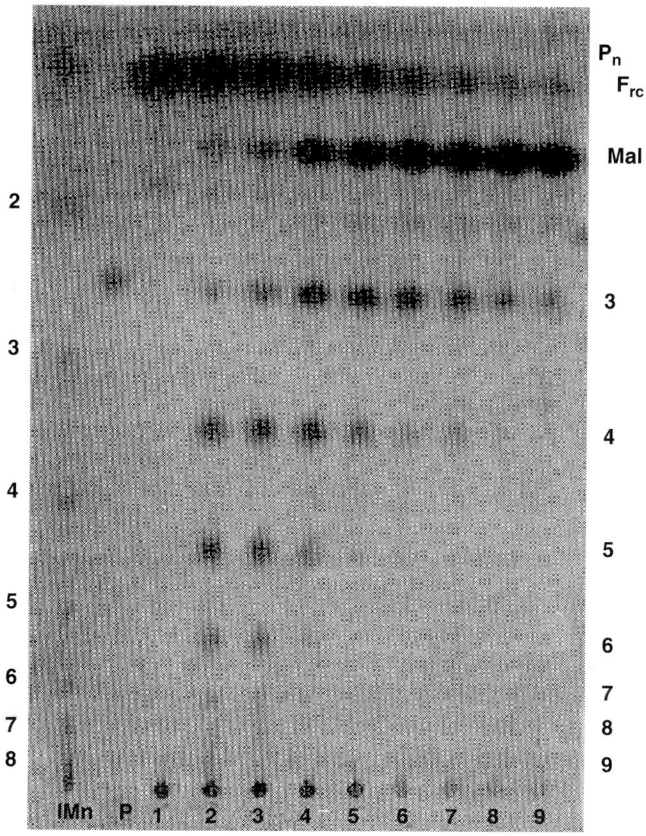

Fig. 6: Thin-layer chromatographic diagram of ^{14}C-sucrose acceptor reactions using different ratios of maltose to sucrose at 100 mM constant total carbohydrate and 120 mIU of B-742C prepared from glcose culture/ 100 μL. The chromatography was conducted on Whatman K5F plates using 3 ascents of 2:5:3 nitromethane - 1-propanol - water. Pn is a saccharide acceptor-product of dp=n, having an isomaltodextrin chain linked α-1→6 to the non-reducing end of maltose; Frc = D-fructose, Mal = maltose, P = panose, IMn = isomaltooligosaccharide standard (Robyt, 1996).

dextran polysaccharides. In addition, a mixed culture fermentation system was designed for the production of size-limited dextran (Day and Kim, 1993; Kim, 1992; Kim and Day, 1995). Low molecular weight dextrans have their greatest application in the pharmaceutical industry. Dextran 70 (average MW 70,000) and dextran 40 (average MW 40,000) have been used as blood plasma extenders and blood flow improvers respectively. This size range of dextran (MW 75,000 - 25,000) is named as clinical dextran. Anticoagulant activity with low toxicity has been obtained using the dextran derivative, dextran sulphate (MW 7,300) which also interacts specifically with the major envelope glycoprotein of HIV-1, inhibiting its ability to bind to $CD4^+$ cells (Mbemda *et al.*, 1992; Hidaka *et al.*, 1990). In addition, cross-linked dextran, mercaptodextran, and iron dextran have been used for various applications. An important factor in these applications is the size of the dextran derivative. Commercially, low-molecular weight dextran is produced by whole-culture fermentation of *L. mesenteroides* NRRL B-512(F), followed by controlled acid hydrolysis and organic solvent fractionation. Yields are relatively low due to losses during hydrolysis and fractionation. There have been several studies on ways to improve clinical dextran production using purified dextransucrase (Tanaka *et al.*, 1981). A concept for producing clinical dextran has been to use a maltose acceptor in conjunction with dextransucrase (Tanaka *et al.*, 1981). An enzymatic method replaces acid hydrolysis for clinical dextran production has also been patented, but neither of these methods is used commercially (Paul *et al.*, 1986; Su and Robyt, 1993).

A new, simpler and industrially practical method for producing size-limited dextrans has been developed (Kim and Robyt, 1994; Kim, 1992 Kim and Day, 1995; Day and Kim, 1993). It is the mixed culture fermentation system that require the coproduction of two microorganisms, an ascosporogenous yeast *Lipomyces starkeyi* (dextranase constitutive mutant assigned as ATCC 74054) *and L. mesenteroides* NRRL B-512. Acceptable conditions for cell growth, enzyme production and enzyme reactions for both species have been established. Both enzymes have similar pH optima for activity and stability between 5.2 and 5.5. A pH of 5.2 (±0.1) which was found to be optimum for product formation. The optimum temperature for maximum dextran yield was 28 (±0.5)°C. The highest yield of clinical dextran, 70% of theoretical yield, was obtained in mixed-culture fermentation with 0.9% yeast extract.

A single addition of sucrose (15%) produced more dextran than discontinuous addition, but clinical dextran yields were higher with discontinuous addition (Table 4). Clinical dextran formation in the mixed culture system appears to be the result of both dextranase action and acceptor (dextranase hydrolyzates) reactions. It is logical that the balance of sucrose to acceptor would be important. A ratio of less than 4 to 1 sucrose to acceptor in mixed culture resulted in high levels of clinical dextran. Practically, this means that the sucrose concentration at any given period should be less than 4 times the available acceptor concentration. Operationally there was less than 10% (w/v) sucrose in the fermentor at any given time. Several factors are known to affect dextran yields in *L. mesenteroides* fermentation (Kim, 1992). These are requirements for substrate dispersion, dextransucrase activity maintenance, viscosity control, and by-product (i.e. fructose) levels. Fructose is a poor acceptor for dextran formation. When fructose concentrations are high, glucose is transferred from sucrose to fructose molecules with the formation of leucrose. Leucrose is not a good acceptor for dextransucrase and decreases dextran yield. In mixed culture fermentation, *L. starkeyi* used fructose as a carbon source, producing dextranase and decreasing leucrose formation. In addition, concentrations of isomaltose and other oligosaccharides (dextran hydrolysis products by dextranase) were higher, displacing fructose as an acceptor. These oligosaccharides were incorporated into dextran and showed up as increased amounts of clinical dextran. Mixed culture fermentation maintained lower viscosity in the fermentation broth (less than 50 times as viscous as pure culture broths) (Koepsell *et al.*, 1953).

Table 4: Effect of discontinuous and single addition of dextran productivity.

	Sucrose concentration			
	15%		30%	
	Discontinuous	Single	Discontinuous	Single
Total Dextran[a]	0.35(±0.01)[b]	0.40(±0.005)	0.41(±0.01)	0.41(±0.005)
Clinical Dextran (%)[c]	94	79	94	73
Lm/Ls Ratio[d]	50.8(±1.6)	306(±15.6)	22.3(±1.3)	100.0(±2.2)

[a]Total dextran yield (Yp/s: kg dextran/kg sucrose); average value of total dextran production (100% of the theoretical yield of dextran is Y p/s = 0.48), standard deviation (±0.01) and (< ± 0.01) for pure and mixed culture, respectively; values are averages from results of gel permeation chromatography, phenol-sulfuric acid, and saccharimeter assays.
[b]Standard deviation (n = 4).
[c]Clinical dextran (MW 75,000 ± 25,000) % portion of the total dextran produced (±1.5).
[d]Cell number ratio of *L. mesenteroides* to *L. starkeyi*

Our process for the production of controlled-size dextrans differs in several ways from the traditional process. This process requires co-inoculation and establishment of a co-culture. Because of higher sucrose concentrations (over 30%), discontinuous sucrose addition is required. One of the most expensive steps in the traditional process, resulting in the highest dextran loss, is acid hydrolysis. The new process does not require acid hydrolysis, and because of this, deionization is also not required. The size-limited dextran produced in mixed culture broth has a small poly-dispersity index. The dextran size can be readily controlled simply by stopping the fermentation at the desired size. Since this process produces small-size dextrans that have lower viscosity, cell removal is easier. In addition, using different *L. mesenteroides* strains producing a dextransucrase, such as B-742CA or B-742CB, the new process can produce new structure low-molecular weight dextran that has different physico-chemical and biological characteristics.

Dextran Molecular Size and Degree of Branching as a Function of Sucrose Concentration, pH, and Temperature of Reaction of *Leuconostoc mesenteroides* B-512 FMCM Dextransucrase

The branching of dextran does not occur from the action of a separate dextran-branching enzyme as occurs in the biosynthesis of starch. Robyt and Taniguchi (1976) showed that dextran was branched by dextransucrase itself in an acceptor reaction in which a dextran chain is the acceptor that binds in the acceptor binding-site, where its C-3-OH group makes a nucleophilic attack onto C-1 of the glucopyranosyl- or dextranyl-groups, that are covalently linked at the active site, releasing them from the active site and forming an α-1→3 branch linkage with the

D-glucose or with the dextran chain. Alsop (1983) reported that as the sucrose concentration in a dextransucrase digest was increased from 2% to 20% (w/v), the amount of low molecular weight (LMW) dextran increased, and with 20% sucrose, there was 43.9% LMW dextran which was defined as polysaccharides and/or oligosaccharides with molecular weights below 5,000 Da. There is no report on the size distribution and the degree of branching of dextrans synthesized under different enzyme reaction conditions such as pH, temperature, and sucrose concentrations of 0.1 M – 4 M. We have studied the size distribution and the degree of branching of dextrans synthesized by *L. mesenteroides* B-512FMCM dextransucrase, using a relatively broad concentration of sucrose, 0.1 M – 4.0 M with different pH values of 4.5 to 6.0 and different temperatures of 4°C – 45°C.

Table 5: Molecular weight distribution of dextrans synthesized by *L. mesenteroides* B-512FMCM dextransucrase at varying sucrose concentrations, pH, and temperature.

Sucrose[a]	$10^3 – 10^4$	$10^4 – 10^5$	$10^5 – 10^6$	$> 10^6$ MWb
(M)	I	II	III	IV
0.1	0.12	0	0.8	99.8
0.3	1.27	0.78	0	98.0
1.0	5.8	18.6	17.9	57.7
1.5	8.7	30.9	3.8	56.6
2.0	61.5	0	0	38.5
3.0	68.4	0	0	31.6
4.0	69.9	0	0	30.1
pH[c]	I	II	III	IV
4.5	13.2	1.2	26.8	58.8
5.0	1.6	1.1	10.4	86.9
5.2	1.3	0	10.1	88.6
5.5	0.5	0	4.5	95.0
6.0	0.3	0	0	99.7
°C[d]	I	II	III	IV
4	0	0.4	16.4	83.2
15	4.2	0	20.6	75.2
23	6.1	0	9.1	84.8
28	2.3	7.8	15.4	74.5
37	1.3	0	33.4	65.3
45	0	0	15.9	84.1

[a]Reactions conducted at pH 5.5 and 28°C and different sucrose concentrations.
[b]Molecular weight divided into four divisions.
[c]Reactions conducted at 1.5 M sucrose and 28°C and different pH values.
[d]Reactions conducted at 1.5 M sucrose and pH 5.5 and different temperatures.

Table 5 shows the relative percentages of the different molecular weights of dextrans synthesized by *L. mesenteroides* B-512FMCM dextransucrase as a function of sucrose concentration, pH and temperature. The molecular weights of the synthesized dextrans were determined by Wyatt Technology multi-angle light scattering intensities. At 0.1 M sucrose, 28°C, pH 5.5, 99.8 % of the dextran had a molecular weight (MW) greater than 10^6 Da. Table 1 shows that as the concentration of sucrose was increased, the lower MW fractions were increased. At 1.0 M and 1.5 M sucrose, the intermediate MWs, 10^4 – 10^5 Da and 10^5 – 10^6 Da increased the most with a concomitant decrease of MWs greater than 10^6 Da. At 2.0 M, 3.0 M, and 4.0 M sucrose, the lower MWs (10^3 – 10^4 Da and 10^4 – 10^5 Da) increased and the intermediate MWs decreased. At these higher concentrations of sucrose, MWs was definitely bimodal with the lowest MW dextrans (10^3 - 10^4 Da) predominating at the higher concentrations of sucrose and the very high MW dextrans (> 10^6 Da) decreasing, but remaining relatively constant between 38.5 % and 30.1 %. The effect of pH was studied, using 1.5 M sucrose and 28 °C. The pH had much less effect on the MWs of the synthesized dextrans (Table 5) as stated by Lee *et al.* (1998) than the sucrose concentration or the temperature. At pH 4.5, the very high MW dextran predominates at 58.8 %. There were, however, significant amounts of most of the other MW dextran fractions of 13.2 % to 26.8 %, with the exception of the 10^4 – 10^5 Da fraction, which was only 1.2%. As the pH was increased above 4.5, the very high MW dextran (> 10^6 Da) increased and all of the lower MW fractions decreased, but especially the lowest MW fractions (10^3 - 10^4 Da and 10^4 - 10^5 Da) was synthesized. The effects of temperature were measured at 1.5 M sucrose and pH 5.5. At the very lowest temperatures studied (4°C and 15°C), the very high MW dextran (> 10^6 Da) was the major dextran synthesized (Table 5). As the temperature was increased, the very high MW dextran decreased, with modest increases in the intermediate MW dextrans being synthesized. At 37°C – 45°C, the predominant MWs were the very high (> 10^6 Da) and the high (10^5 - 10^6 Da) MW dextrans were the exclusive products. The optimum conditions for the synthesis of very high MW dextran in high yields would be obtained at relatively low sucrose concentrations (0.1 M – 0.3 M), high pH values of 5.5 - 6.0, and high temperature between 37°C and 45°C. The optimum conditions for the synthesis of low MW dextrans would be high sucrose concentrations of 3.0 M – 4.0 M, low pH (4.5), and intermediate temperatures of 23°C – 28°C. The effect of the concentration of sucrose on the degree of branching was studied at pH 5.5 and 28°C (Table 6). The data clearly shows that as the concentration of sucrose was increased from 0.1 M to 4.0 M, the percentage of branching significantly increased from 5.0 % to 16.6%. The high sucrose concentrations gave very high degrees of branching, with 4.0 M giving one α-1$\rightarrow$3 branch linkage out of every six D-glucose residues. Changes in the pH had very little effect on the degree of branching, using 1.5 M sucrose and 28°C. In going from pH 4.5 to 5.2, the degree of branching increased from 6.6% to 8.2%, and then decreased again to 6.5% at pH 6.0. The temperature, however, had a significant effect on the degree of branching, using 1.5 M sucrose and pH 5.5. At 4°C, the degree of branching was the lowest at 4.8%. As the temperature was increased, the degree of branching was increased to 5.3% at 15°C, 6.9% at 28°C, 8.8% at 37°C, and 14.7% at 45°C (Table 6). The effect of the increase in temperature, giving an increase in the degree of branching, might have been expected from the mechanism of branching dextran by dextransucrase. The linear dextran molecules bind at the acceptor-site and form the branch linkages by an acceptor transglycosylation reaction (Robyt and Taniguchi, 1976). The binding of dextran to the acceptor-site is controlled by the diffusion of the dextran molecules, and therefore, it is temperature dependent. The higher the rate of diffusion, the higher is the degree of branching due to greater interaction between dextransucrase, sucrose and acceptor dextran. Tanriseven and Robyt (1993) reported that when the sucrose concentration was increased over 200 mM, sucrose was bound to a third, low-affinity binding site of dextransucrase that allosterically changed the conformation of the active-site, preventing the synthesis of dextran. This could also be the cause here in both, decreasing the dextran molecular weight and increasing the degree of branching at high sucrose concentrations. Although the trends are apparent in this study for controlling both the molecular weight and the degrees of branching of the dextran synthesized by *L. mesenteroides* B-512FMCM dextransucrase, the study could be expanded, using a more complete combinational study of sucrose concentration and temperature to give a pattern that could be used to select a wide range of molecular weights and degrees of branching. This would lead to better controlling and selecting dextrans of different molecular weights with different degrees of branching.

Table 6: Dextranase/TLC determination of the degree of branching of dextrans synthesized by *L. mesenteroides* B-512FMCM dextransucrase at various concentrations of sucrose, pH, and temperatures.

Sucrose (M)[b]	Mono-Saccharide	Isomalto-dextrins	Branched Isomaltodextrins	Unhydrolyzed Dextran	Ratios (R) [BIM+UHD]	Percent Branching[a]
	[% M]	[% IMD]	[% BIM]	[% UHD]	[M+IMD]	
0.1	31.3	48.8	12.5	7.3	0.247	5.0
0.3	33.9	44.9	14.6	6.6	0.269	5.4
1.0	30.8	43.2	21.9	4.1	0.351	7.1
1.5	31.5	39.7	24.9	4.1	0.404	8.2
2.0	25.4	42.3	30.0	2.3	0.477	9.7
3.0	16.7	44.8	36.8	1.7	0.626	12.7
4.0	18.4	36.6	37.5	7.5	0.818	16.6
pH[c]	[% M]	[% IMD]	[% BIM]	[% UHD]	[BIM+UHD]	Percent Branching[a]
					[M+IMD]	
4.5	35.2	40.1	18.9	5.7	0.326	6.6
5.0	29.4	44.3	20.0	6.3	0.357	7.1
5.2	28.5	43.2	21.9	6.2	0.403	8.2
5.5	27.1	44.3	21.8	6.8	0.401	8.1
6.0	34.5	41.1	17.5	6.9	0.323	6.5
?[d]	[% M]	[% IMD]	[% BIM]	[% UHD]	[BIM+UHD]	Percent Branching[a]
					[M+IMD]	
4	23.4	57.6	14.5	4.5	0.235	4.8
15	19.8	59.3	16.0	4.9	0.265	5.3
23	17.9	57.6	18.5	6.0	0.325	6.6
28	16.4	58.3	18.9	6.4	0.339	6.9
37	16.5	53.1	20.7	9.7	0.437	8.8
45	8.7	49.2	26.6	15.5	0.727	14.7

[a]Branching for dextran synthesized from 0.1 M sucrose was assumed to be 5% from refs. 24 & 25. A branching conversion factor was obtained by dividing 5% by the ratio (R), 0.247, giving 20.24. The per cent branching for the other dextrans was then obtained by multiplying their ratios (R) by the conversion factor, 20.24.

[b]Reactions were conducted at pH 5.5 and 28?.

[c]Reactions were conducted at 1.5 M sucrose and 28?.

[d]Reactions were conducted at 1.5 M sucrose and pH 5.5.

Enzymatic Modification of 5-Cellulose Using *Leuconostoc mesenteroides* B-742CBM Dextransucrase

The mechanism for forming branch linkages is by an acceptor-reaction in which a glucan chain itself acts as an acceptor and the chain being synthesized is transferred to an acceptor chain, forming a branch linkage between the two chains. *L. mesenteroides* B-742CB dextransucrase could modify various starch granules or gelatinized starches and pullulan (Lee *et al.* 1998a; 1998b). Modified starches were more resistant against the hydrolysis of α-amylase, isoamylase, pullulanase and endo-dextranase than those to native starch. After modification of granule starch with B-742CB dextransucrase a soluble modified starch was produced. The products were modified by the addition of glucose to positions C3, C4 and C6 free hydroxyl group of glucose residues in the starch. The constitutive mutants for specific glucansucrases do not require sucrose in the growth media for glucansucrase elaboration and hence the enzyme preparations from these mutants are devoid of glucan (Kim and Day, 1994; Kim and Robyt, 1994b; 1995a). In a patent issued in 1964, it was proposed that cellulose and other polysaccharides could be modified using dextransucrase and sucrose. Attempts to obtain these products by other investigators, however, resulted in failure. These failures might have been due to not using the right dextransucrase (the dextransucrase was not specified in the patent) (Wilke-Douglas *et al.*, 1989). In 1964, it might not have been recognized that different *L. mesenteroides* strains elaborated different kinds of glucansucrases with different specificities. The successful modification of cellulose using sucrose and *L. mesenteroides* B-742CBM dextransucrase that has been developed from B-742CB for dextransucrase hyper-producing (Kim *et al.*, 1999). 5g of α-cellulose was suspended in 250 mL of buffer (20 mM Na-acetate, pH 5.2, 28°C) and dextransucrase (200 IU/reaction digest) was added. Modification reaction was initiated by the addition of 100 mM sucrose and a total of 100 ml was added at the rate of 0.2 ml per minute. The sucrose solution was continuously pumped into the digest. After finishing the addition of sucrose, the reaction was allowed to go an additional 30 minutes with stirring. After reaction any insoluble material in the digest was collected by centrifugation, washed with water (3-times) and titurated with acetone and ethanol. The supernatants from centrifugation were treated with two volumes of ethanol to precipitate any polysaccharide that had become soluble during the substitution reaction. These were also titurated with acetone and ethanol. The polysaccharides, thus, obtained was used in further reactions with enzymes and characterized as follows; the modified cellulose was hydrolyzed with cellulase and/or endo-dextranase. The hydrolysis products were analyzed by TLC using 2/5/1.5 (v/v/v) of Nitromethane/1-propanol/water. After reaction the yields of soluble and insoluble polysaccharides were 21(± 2.2) and 68% (± 5.1) of theoretical, respectively. 100% of the theoretical yield means 'the sum of gram cellulose and 0.48 gram sucrose added' in the reaction digest. The hydrolysis enzymes did not produce any products from B-742CBM dextran. The glucose was only produced from cellulose hydrolysis by cellulase. There were, however, many kinds of hydrolysis products from modified soluble glucans after hydrolysis enzyme treatments. Especially, oligosaccharides were produced by cellulase treatment. This means that the portion of soluble glucan is composed of cellulose like structure, β-1→4 linkage. This might be modified cellulose that has a β-1→4 linkage as well as an α-1→6 linkage. Kim *et al.* (1995) have reacted cellobiose and cellotriose with sucrose and B-742CMB dextransucrase. Interestingly, the efficiency of acceptor reaction was increased when the size of acceptor was increased. In the case of B-512FMCM dextransucrase, the proportion of cellobiose acceptor product was 25.5% of the total carbohydrates (excluding fructose and unreacted cellobiose) and that of cellotriose was 69.4%. Also, several kinds of oligosaccharides were produced as acceptor products. In the case of B-742CBM dextransucrase reaction, the proportion of cellobiose acceptor product was 41.9% of the total carbohydrates (excluding fructose and unreacted cellobiose) and that of cellotriose was 70.6%. Fu and Robyt (1990) showed that *L. mesenteroides* B-512FM dextransucrase could transfer glucose from sucrose to maltodextrins, maltotriose to malto-octaose. The percentage of acceptor reaction was decreased exponentially as the size of the maltodextrin increased. Similar results were obtained for *Streptococcus mutans* dextransucrase and mutansucrase with the exception that the reaction percentage decreased until with the maltopentaose acceptor reaction and started to increase again with maltohexaose and heptaose. It suggests the possibility that higher molecular glucan chains might react giving single glucose residue attached by α-1→6 linkage to the glucan chain. This might give possibility for cellulose modification by B-742CBM dextransucrase. To confirm the existence of cellulose structural materials in a soluble glucan, they did methylation and acid hydrolysis analysis (Table 7). *L. mesenteroides* B-742CBM glucan contained (1→3), (1→4), and (1→6) glycosidic linkages in the main chains, as shown by the formation of 2, 4, 6-tri-O-methyl-D-glucose, 2, 3, 6-tri-O-methyl-D-glucose, and 2,3,4-tri-O-methyl-D-glucose respectively. The latter was the predominant tri-O-methyl-D-glucose residue, indicating that the (1→6) linkage was dominant. Glucan was also highly branched through α-1→3 linkage, as shown by the formation of 2, 3, 4, 6-tetra-O-methyl- D-glucose from the non-reducing ends of the branched chains. It was quite interesteing that there was (1→4) linkage that did not exist in B-742CBM dextran. (1→4) linkage could be found in cellulose that was not soluble in water. Thus, the results indicated that the insoluble cellulose has been modified by dextransucrase and solubilized. It contained primarily (1→4) linkage as evidenced by the formation of large amounts of 2,3,6-tri-O-methyl-D-glucose, and also (1→6) linkage as evidenced by the formation of large amounts of 2, 3, 4-tri-O-methyl-D-glucose. The result indicated that there was a dextran-like structure in insoluble glucan. Thus, it might be possible that there was a modification of insoluble α-cellulose by dextransucrase and sucrose same as previous soluble glucan. It was, however, that the modification was less as evidenced by the formation of small amounts of 2, 3, 4, 6-tetra-O-methyl-D-glucose. Since there were still significant amounts of (1→6) linkage, it was possible that instead of high numbers of branching points there is longer branch linkages.

Table 7: Proportions of the methylated D-glucose in the acid hydrolysis products of the methylated cellobiose, B-742CBM dextran, modified soluble and insoluble glucan (Kim *et al.*, 1999).

	% of methylated glucose			
	Cellobiose	B-742CBM Dextran	Soluble Glucan	Insoluble Glucan
2,3,4,6-Me_4-Glc	50.6	8.2	13.6	4.3
2,3,6-Me_3-Glc	49.4	—	22.7	58.4
2,4,6-Me_3-Glc	—	12.5	6.0	-
2,3,4-Me_3-Glc	—	64.2	42.6	26.4
2,4/2,3-Me_2-Glc	—	15.1	15.1	10.9

Radioprotective Effect of Various Carbohydrates on DNA and Cells

DNA is the primary target for lethal radiation damage in organisms (Nataraj *et al.*, 1993; Rapine *et al.*, 1981). Radiation causes mutation or carcinogenesis by radiation-induced scission of DNA molecules. Mammalian cells and yeast cells were killed by using electron beams of various energies (Roots and Okada, 1972). X-ray irradiation of mammalian cells produces single-strand scissions in DNA. Therefore, there has been a search for substances that protect DNA from radiation-induced damage. One such substance is dimethyl sulfoxide (DMSO) which prevents single strand breaks of DNA (Stern and Lunder, 1972). Since DMSO is toxic and is required in large amounts, non-toxic protectors, effective in small concentrations were searched for. Previously, tea catechin was reported as a strong protectant (Stern, 2001). Another candidate would be trehalose, which is a naturally occurring disaccharide and known as a multipotent protector of living things from the stress of freezing, heating, drying, and irradiation (Roots and Okada, 1975). However, the value of other carbohydrates in protecting DNA and microorganisms from radiation-induced damage is not known. Therefore, an attempt was made to measure the protective effect of monosaccharides (glucose, fructose), disaccharides [sucrose, maltose, isomaltose, lactose, trehalose, gentiobiose (ß-D-glucopyranosyl-1→6-D-glucopyranose)] and the trisaccharides, oligosaccharides and polysaccharides [acarbose (4, 5, 6-trihydroxy-3-hydromethyl-2-cyclohexen-1-yl 4-amino-4, 6-dideoxy-D-glucopyranosyl maltose), maltotriose, maltopentaose, isomaltosyl maltooligosaccharides, isomaltosyl raffinooligosaccharides, dextran, levan, pullulan, xylan] (Ryu *et al.*, 2002). Branched oligosaccharides (BOS-9 and BOS-12) were prepared via a mixed culture fermentation of *Lipomyces starkeyi* and *Leuconostoc mesenteroides* (Heo *et al.*, 1999; Kim and Kim, 1999). *L. starkeyi* was grown with YS medium for 40 h at 30. The composition of YS medium was 0.5% yeast extract, 0.5% KH_2PO_4, 1% soluble starch and 2% sucrose. After fermentation of *L. starkeyi* for 40 h, *L. mesenteroides* B-512FMCM was inoculated. Then 9% sucrose (based on total volume) was added continuously to the fermentor and produced BOS-9. Through an analogous fermentation procedure, except that 12% sucrose was continuously added, BOS-12 was prepared. Following fermentation, the supernatant was separated from cells by centrifugation (6,000 x g). Fructose in the supernatant was removed by alginate immobilized *Saccharomyces cerevisiae*, and dextran was precipitated by the addition of ethanol up to 43%. The remaining supernatant was concentrated, and the branched oligosaccharides were prepared using a rotary vacuum evaporator (N-N series, Eyela, Japan). For the preparation of isomaltosyl maltooligosaccharides, maltose and sucrose were mixed 1:1 (200 mM: 200 mM), and B-512FMCM dextransucrase (20 U/ml) was added to this mixture with the volume ratio of 1:1 (v/v). For the preparation of isomaltosyl raffinooligosaccharides, the same conditions applied, except that B-1355 glucansucrase (10 U/ml) was used instead of B-512FMCM dextransucrase. Both enzyme reactions were carried out at 30°C until all the sucrose was consumed.

Ultrasoft X-ray was obtained from the LIGA beamline of the Pohang Accelerator Laboratory (South Korea) using a specially constructed chamber. An aluminum target was placed inside the exposure chamber, and the ultrasoft X-ray was generated by irradiating this target with a white beam. The chamber was filled with helium (He) gas that is effectively transparent to the ultrasoft X-rays. The binding energies of Al orbitals K (1), L(1), L(2), and L(3) were 1559.6 eV, 117.8 eV, 73.1 eV, and 72.7 eV, respectively. If an electron of the K-shell is taken out by the incident photon, an electron from a neighboring orbital moves to that empty orbital. The difference in the orbital energies is emitted as a photon of the energy range of the ultrasoft X-ray. In this experiment, the Al K ray was corresponded to energy of 1.487 KeV. BOS-12, BOS-9, isomaltosyl raffinooligosaccharides, isomosaccharides also showed DNA and cell protection from ultrasoft X-ray exposure (Table 8). BOS-12 showed the highest protective effect for both, DNA and cells. Protective effects for DNA and cells following ultrasoft X-ray exposure were 1.9 and 8.3 times higher than the control. Almost the same constant rate of 4×10^{-8} /M·s for the OH radical scavenging reaction by DNA and glucose was reported, suggesting that equimolar DNA and disaccharides or oligosaccharides can consume the radical at equivalent kinetic rates (Yoshinaya *et al.*, 1997). The DNA concentration in the experiment was 1×10^{-3} g/L, equivalent to 3.08×10^{-6} M of nucleotide (assuming an average molecular weight of 325). When the solution contained carbohydrates, the absorption of UV was increased. BOS-12 showed the highest absorption value among those solutions tested. Thus, this protective effect is potentially related to the absorption of UV or ultrasoft X-ray by carbohydrates during exposure. This protection was expected because the main factor for radiation-induced scission of DNA molecule is attributable to the OH radicals formed as a result of decomposition of the surrounding water molecules, and the protection of carbohydrates may be derived from scavenging by the OH radicals (Roots and Okada, 1972). With the addition of carbohydrates to CC form DNA, ultrasoft X-ray converted 50% of CC form DNA to OC form after 7.7 minutes

Table 8: Time required for the conversion of 50% CC form DNA to OC form DNA after ultrasoft X-ray exposure (Ryu *et al.*, 2002).

Carbohydrates	Time for 50% conversion (min)
No sugar (control)[a]	7.7
Glucose†	6.6
Fructose†	13.0
Sucrose†	8.6
Trehalose†	17.0
Gentiobiose†	13.0
Lactose†	13.0
Maltose†	10.5
Maltotriose†	12.6
Acarbose†	23.7
Maltopentaose†	13.6
á-Cyclodextrin†	10.9
â -Cyclodextrin†	7.4
Isomaltosyl malto-oligosaccharides‡	14.7
Isomaltosyl raffinooligosaccharides‡ rraffino-oligosaccharidesraffino-oligosaccharides	24.8
Amylodextrin‡	9.9
BOS-9[b]‡	57.3
BOS-12[c]‡	63.8
Pullulan‡	15.2
Levan‡	7.3
Xylan‡	7.6
Dextran T-10‡	13.2
Dextran T-40‡	6.3

[a] No addition of any saccharide; [b, c] See Ref. [Djouzy *et al.*, 1995]

†, The final concentration was 100mM.

‡, The final concentration was 2.5 % (w/v).

of exposure under the experimental conditions (Table 8). However, by the addition of BOS-12, BOS-9, isomaltosyl raffino-oligosaccharides, and acarbose, the time required for the 50% conversion was extended to 63.8, 57.3, 24.8, and 23.7 minutes, respectively.

Psoralen is used as an ingredient of cosmetics and suntan products. 4'-hydroxymethyl-4, 5', 8-trimethylpsoralen (HMT), 5-methoxypsoralen, and 8-methoxypsoralen are the psoralen derivatives. A combination of psoralen and ultraviolet A radiation, commonly referred to as 'PUVA', is widely used in the treatment of psoriasis and for tumor photochemotherapy (Dalla and Marciani, 2001; Honigsmann *et al.*, 1979). This therapy consists of oral or tropical administration of 8-methoxypsoralen (8-MOP) followed by exposure to long wave UVA radiation (320 - 400 nm) (Honigsmann *et al.*, 1979). However, the widespread use of PUVA therapy has made its potential long-term side-effects an issue of concern and debate (Nataraj *et al.*, 1993; Shiroza *et al.*, 1987). PUVA follow-up studies revealed increased risks of the development of squamous cell carcinoma in patients receiving PUVA therapy, although European studies appeared to be at odds with these reports (Hanada and Kiramatsu, 1988). The addition of MOP was shown to protect against DNA conversion to the open form following UV exposure (Table 9). The best protective effects were associated with the use of branched oligosaccharides when compared with commercially available fructooligosaccharides (FOS) and maltooligosaccharides (MOS). In the case of FOS, the% conversion of CC form DNA to OC form DNA was 81% and that of MOS was 73%. BOS-12 and BOS-9 showed the % conversion rates of 46 and 29, respectively. The addition of MOP to oligosaccharides further increased the protective effect. FOS or MOS, in combination with MOP decreased the conversion rate to 23% and 22%, respectively. This protection was even higher when MOP was added to branched oligosaccharides, decreasing the % conversion (CC form to OC form DNA) to 0 - 4%. Thus, the potential side-effects of MOP as an ingredient of cosmetics and suntan products can be reduced by the addition of carbohydrates with increased protection effects, especially by branched oligosaccharides.

Table 9: Protection of DNA from UV damage by various oligosaccharides and MOP (Ryu *et al.*, 2002).

Treatment	Conversion (%)
DNA (No UV exposure)	0
DNA + FOS (2.5%)	81
DNA + MOS (2.5%)	73
DNA + BOS-12 (2.5%)	46
DNA + BOS-9 (2.5%)	29
DNA + MOP	32
DNA + FOS (1%) + MOP	23
DNA + MOS (1%) + MOP	22
DNA + BOS-12 (1%) + MOP	4
DNA + BOS-9 (1%) + MOP	3
DNA + FOS (2.5%) + MOP	19
DNA + MOS (2.5%) + MOP	24
DNA + BOS-12 (2.5%) + MOP	0
DNA + BOS-9 (2.5%) + MOP	0

FOS, fructo-oligosaccharide; MOS, malto-oligosaccharide; BOS-9 and BOS-12, See Ref. [18]; MOP, 8-methoxypsoralen (5 mM)

Values in parenthesis represent the final concentration of each sample.

The protective effect by carbohydrates of DNA and cells from radiation damages will be useful in developing methods of protecting living cells from harmful environmental radiation such as in outer space, or leakage of radioactive materials from nuclear power plants. Carbohydrates can readily be added as ingredients in cosmetics and suntan products for protection from UV damage while reducing potential MOP side-effects.

STRUCTURAL, FUNCTIONAL ORGANIZATION AND DIRECTED EVOLUTION OF THE GLUCANSUCRASES

The isolation and sequencing of nineteen genes coding for glucansucrase were initiated more than 10 years ago allowed from oral streptococci involved in the cariogenesis process, *S. mutans* GS5 or LM7, *S. sobrinus* serotype h (*S. downei* Mfe28), serotype g (*S. sobrinus* 6715), serotype d (*S. sobrinus* OMZ176), by *S. salivarius* ATCC 25975 and by *L. mesenteroides* NRRL B-512F, its constitutive mutant B-512FMCM, B-1299, B-742CB and B-1355. (Table 10) (Gilpin *et al.*, 1985; Aoki *et al.*, 1986; Ferretti *et al.*, 1987; Pucci *et al.*, 1987; Russel *et al.*, 1987; Shiroza *et al.*, 1987; Hanada and Kuramitsu, 1988; Ueda *et al.*, 1988; Hanada and Kuramitsu, 1989; Pitty *et al.*, 1989; Wilke-Douglas *et al.*, 1989; Gilmore *et al.*, 1990; Honda *et al.*, 1990; Abo *et al.*, 1991; Giffard *et al.*, 1991; 1993; Hanada *et al.*, 1991, 1993; Sato *et al.*, 1993; Banas *et al.*, 1994; Simpson *et al.*, 1995; Vickerman *et al.*, 1997; Monochois *et al.*, 1996, 1997; 1998; Arguello-Morales, 2000; Kim *et al.*, 2000).

Isolated glucansucrase genes are for all large enzymes with an average Mr of 160,000. *Streptocci* glucansucrases synthesize primarily α-1$\rightarrow$3 rich mutant polysaccharides and α-1$\rightarrow$6 rich dextran polysaccharides. *L. mesenteroides* glucansucrases produce a-1?6 rich dextran (Mooser *et al.*, 1991; Kato *et al.*, 1992; Devulapalle *et al.*, 1997; Cope and Mooser, 1993; Arguello-Morales *et al.*, 2000). However, genes coding for glucansucrases forming branched glucan through α-1$\rightarrow$2 or α-1$\rightarrow$4 glucosidic bonds have not yet been isolated. Analysis and comparison of the different protein sequences show that these enzymes are closely related and have a common structure (Monchois *et al.*, 1999). They are composed of four distinct structural domains (Giffard *et al.*, 19991; Banas *et al.*, 1990; Simpson *et al.*, 1995; Russell, 1990). Except for DSR-A from *L. mesenteroides* NRRL B-1299 (Kato *et al.*, 1992), their N-terminal end begins with a signal peptide of 32-34 aa followed by a stretch of 123-129 amino-acids which is highly variable. The non-conserved region located just downstream of the signal peptide seems to have no important role in the enzyme mechanism. Its deletion does not affect the enzyme activity (Abo *et al.*, 1991). Moreover, DSR-A, isolated from *L. mesenteroides* NRRL B-1299 is an active enzyme that does not possess this variable region (Kato *et al.*, 1992). The significance of this variable region also remains unknown. It was also presented a highly conserved core region of about 1000 amino-acids. This domain, also named as a catalytic domain, is the sucrose binding domain, capable of binding and cleaving sucrose. The C-terminal end of glucansucrases covering about 500 a is composed of a series of tandem repeats and constitutes the glucan binding domain.

Structure-function Relationships of *N*-terminal Catalytic Domain

Identification of the catalytic site: The research group of Mooser has demonstrated that a site making a covalent link with the glucosyl residue coming from sucrose breakdown is present inside the N-terminal domain. Mooser and Iwaoka (1989) have isolated from *S. sobrinus* GTF-S a covalent glucoyl-enzyme complex, while Mooser *et al.* (1991) could isolate a peptide from GTF-I and GFT-S of *S. sobrinus* presenting the active site. The corresponding sequences were very similar and contained three aspartic acids. Mass spectrometry analysis revealed that the Asp residue inside the peptide coming from GTF-I (Asp-Ser-Ile-Arg-Val-**Asp**-Ala-Val-Asp) made an ester bond with the glucosyl residue. This peptide was located at 449 α from the N-terminal end of GTF-I. Site directed mutagenesis experiments have confirmed the importance of this amino acid. Replacement of homologous Asp residues in the GTF-B sequence (Asp-451) by Thr, Asn, and Gln (Kato *et al.*, 1992) and in the GTF-I sequence from *S. downei* Mfe28 (Asp-453) by Asn (Devulapalle *et al.*, 1997) completely suppressed the enzymes, activities. Change of Asp-551 in Asn for DSR-S also abolished the activities of glucan and oligosaccharide synthesis. The mutations do not affect glucan binding ability of these two enzymes GTF-B and DSR-S (Devulapalle *et al.*, 1997; Kato *et al.*, 1992). Other mutations affecting Asp residues near Asp451 from GTF-B had no significant effect on enzyme activity. The functional importance of this region has also been confirmed by the fact that an antibody directed against a peptide of sequence homologous to this region inhibited glucansucrase activity (Devulapalle *et al.*, 1997; Cope and Mooser, 1993; Loloi *et al.*, 1996; Smith *et al.*, 1994; Taubman *et al.*, 1995).

With respect to an action mechanism similar to the glucosidase action mechanism, it was suggested that at least one more amino-acid has to be involved in the catalytic process to facilitate the fructose release by playing the role of proton donor (Mooser *et al.*, 1991). The involvement of histidines in the enzymatic mechanism has been illustrated by photo-oxidation experiments where only histidine residues were modified (Koga and Inouc, 1981) or by using diethylpyrocarbonate (DEP) (Fu and Robyt, 1988). They also concluded from inhibition curves with DEP that two histidines were involved in the action mechanism of glucansucrases (Fu and Robyt, 1981). Site-directed mutagenesis experiments have enabled functionally important conserved His to be located. The replacement of His-661 of DSR-S to Arg 101 and its equivalent in GTF-B (His-561) to Gly (Tsumori *et al.*, 1997) led to a very weak active enzyme.

Identification of other important regions for catalytic activity: Funane *et al.* (1993) reported the existence of a second active site for the dextransucrase produced by *L. mesenteroides* NRRL B-512F in an region with a sequence homologous to the 435-453 GTF-C region (corresponding to the 509-527 DSR-S region). They showed that the modification of carboxylic groups with 1-ethyl-3- (3-dimethylamino)-propylcarbdodiimide (EDC) and glycine ethyl ester (GEE), the latter compound playing the role of nucleophile, abolished glucansucrase activity. Since addition of sucrose delayed the enzyme inactivation, they have concluded that these modified carboxyl groups may be involved in the binding with the substrate. The presence of functional carboxyl groups in this region has been confirmed by site-directed mutagenesis experiments. Substitution of Asp-511 and Asp-513 of DSR-S in Asn resulted, respectively, in a complete suppression and a strong decrease in the glucan and oligosaccharide synthesis activities (Cope and Mooser, 1993). The substitution of the similar amino-acids (Asp-411 and Asp-413) of GTF-B in Asn showed that Asp-413 was essential, but not Asp-411, a still active enzyme being obtained after mutation (Tsumori *et al.*, 1997). The two GTF-B variants exhibited K_m values similar to that of the wild-type GTF-B, suggesting that the two Asp residues were not directly involved in the binding of sucrose. Inhibition studies of dextransucrase produced by *L. mesenteroides* NRRL B-512F with *o*-phthalaldehyde (OPA) also suggested that lysines are essential for activity (Funane *et al.*, 1995; Goyal and Katiyar, 1995). One of these essential lysines may be close to the active site (Mitsuya *et al.*, 1988). Replacement of four conserved Tyr residues at position 169-172 in GTF-B suggested that this region, presenting a significant homology with repeating units thought to be involved in glucan-binding, may play a role in the enzymatic mechanism (Tsumori *et al.*, 1988).

Site-directed mutagenesis experiments have also allowed the identification of amino-acid residues influencing the structure of the glucan produced. By comparing GTF-I, GTF-S, GTF-B, GTF-C and GTF-D sequences, Shimamura *et al.* (1994) have identified positions where amino acid residues are conserved for the enzymes producing an insoluble glucan. Ile-448, Asp-457, Asp-567, Lys-779 and Lys-1014 present in GTF-B, an enzyme-producing insoluble glucan, have been chosen to be mutated in the corresponding amino-acid present in GTF-D, an enzyme producing a soluble glucan: Val-462, Asn-471, Thr-589, Gln-810 and Thr-1046, respectively. Only the mutation D567T enabled a change in the structure of the glucan produced by GTF-B to be achieved; the percentage of α-1$\rightarrow$3 linkages is 38% instead of 76% with the wild-type enzyme. The mutation, T589D in GTF-D, has also an effect on the glucan structure, the mutated enzyme producing an insoluble one. However, even though this study clearly shows the direct involvement of these residues in the glucan structure determination, the mechanism defining their involvements remains to be elucidated.

Structure-function Relationships of *C*-terminal Glucan Binding Domain

***Structure of C-terminal domain*:** Sequence analysis of different glucansucrases showed that all these enzymes presented a C-terminal domain composed of a series of repeated units. It contains different homologous direct repeats named A, B, C, D and N repeats on the basis of their sequence similarities (Giffard *et al.*, 1991; Monchois *et al.*, 1998; Vickerman *et al.*, 1996; Simpson *et al.*, 1995; Koya and Inoue, 1981). The number and distribution of these repeats is specific to each enzyme (Table. 11). However, it appears that D repeated units are specific to enzymes produced by *S. salivarius* ATTC 25975. "A" units are always present, often in an A-C pattern. The C-terminal ends of DSR-S and DSR-B are characterized by the presence of small repeated units, not very homologous to each other (Mooser *et al.*, 1991; Devulapalle *et al.*, 1997; Vickerman *et al.*, 1997). The exact involvement of these different patterns in glucan structure determination has not been elucidated. However, except for GTF-L, enzymes producing an insoluble glucan possess the same pattern, A-A-C-A-C-A-C (Table 11). Similar repeated units composed of A-C motif are present in a glucan binding protein produced by *S. mutans* (Banas *et al.*, 1990). They can also be found in the C-terminal part of a dextranase inhibitor protein produced by *S. sobrinus* UAB 108 acting like a glucan-binding protein.

Wren showed that 'A' repeats found in the C-glucan binding domain of glucansucrases and in *C. difficile* toxins always present three conserved amino-acid residues a tyrosine-phenylalanine dipeptide and a glycine located ten residues downstream (Wren, 1999). Sequence comparison enabled a consensus sequence for these repeated units to be established. According to von Eichel-Streiber *et al.* (1992), the repeating units are composed of the same amino acid residue pattern consisting of an aromatic amino acid residue stretch, sometimes including a tyrosine, surrounded by conserved residues. The following consensus sequence has been proposed: IDG**YY**D + N + G. More recently, another consensus motif named YG repeat has

Table 11: Comparison of the pattern of repeated units composing the C-terminal domain of glucosyltransferases (Kitahata *et al.*, 1974).

Strain	Gene	Repeating Units	References
S. mutans GS5	*gtf-B*	A-A-C-A-C-A-C-A-C-A-C-A-C	Shiroza *et al.* (1987)
	gtf-C	A-A-C-A-C-A-C-A	Ueda *et al.* (1988)
	gtf-D	A-A-A-A-A	Honda *et al.* (1988)
S. downei Mfe28	*gtf-I*	A-A-C-A-C-A-C-B-A-C-B-A-C	Ferretti *et al.* (1987)
	gtf-S	A-A-C-A-C-A-C	Gilmore *et al.* (1990)
S. sobrinus 6715 (serotype g)	*gtf-Ia*	A-A-C-A-C-A-C-A-C-A-C	Abo *et al.* (1991)
S. sobrinus OMZ176	*gtf-T*	A-A-C-A-C-C-A-A-C	Hanada *et al.* (1993)
(serotype d)	*gtf-Is*	A-A-C-A-C-A-C-A-C-A-C	Sato *et al.* (1993)
S. salivarius ATCC	*gtf-J*	A-D-A-D-A-D	Giffard *et al.* (1991)
25975	*gtf-K*	A-D-A-A-A-D-A-D-A-D-A-D	Giffard *et al.* (1993)
	gtf-L	A-A-C-A-C-A-C	Simpson *et al.* (1995b)
	gtf-M	A-C-A-C-A-A-C-A	Simpson *et al.* (1995b)
S. gordonii (*S. sanguis*)	*gtf-G*	A-A-C-A-C-A-C-A-C-A-C	Vickerman *et al.* (1997)
L. mesenteroides NRRL B-512F	*dsr-S*	A-C-C-A-A-N-C-N-N-N-N	Wilke-Dougles *et al.* (1989), Scheubler (1874)
L. mesenteroides NRRL B-1299	*dsr-A*	A-A-A-C-A-C-A-C	Monchois *et al.* (1996)
	dsr-B	A-C-C-A-A-N-C-N-N-N-N	Monchois *et al.* (1998)
L. mesenteroides B-742CB	*dsrB742*	A-C-C-A-A-N-C-N-N-N-N	Monchois *et al.* (1997)

Consensus sequences are the followings; A repeated units, WYYFNXDQAATGLQTIDGQTVFDDNGXQVG; B repeated units, VNGKTYYFGSDGTAQTQANPKGOTFKDGSVLRFYNLEGQYVSGSGWY; C repeated units, GKIFFDPDSGEVVKNRFV; D repeated units, GGVKNADGTYSKY.

been presented (Giffard and Jacqueus, 1994): NDGYYFxxxGxxH°x(G/N)x—H°H°H° (x, non-conserved amino-acid rsidue; H°, hydrophobic amino-acid residue). It is found both within the A-D repeats and also outside them. It includes an aromatic amino-acid stretch surrounded by neutral or polar amino-acid residues, such as a conserved glycine residue. Its end is composed of hydrophobic residues (Mooser and Iwaoka, 1989).

Role of the C-terminal domain in glucan binding: The C-terminal domain is responsible for the binding of glucan. Conserved amino-acid residues may be involved in binding with glucosidic units. The clustered aromatic residues (tyrosine, tryptophane and phenylalanine) may stabilize the binding between sugar and protein by interacting with sugar unit (Quiocha, 1986). The polar (tyrosine, serine, asparagines and glutamine) or acid (aspartic acid) residues may allow the creation of hydrogen bonds with hydroxyl residues of the sugar. The presence of amino-acid residues, like lysine, glycine, asparagines or serine, able to introduce flexibility into the protein structure, may allow the glucosyl residue to be correctly orientated to the binding sites.

Singh *et al.* (1993) showed by using different chemical inhibitors, that tyrosine, tryptophan, histidine and aspartic (or glutamic) acid residues were important for the binding abilities of glucanscrases. After the action of *o*-phthaldehyde (OPA) on the glucansucrase of *L. mesenteroides* NRRL B-512F in the presence of T-10 glucan and mild tryptic digestion, resultant peptides contained lysines that were protected from OPA because of the binding with glucan (Funane *et al.*, 1995). The minimum number of these repeated units necessary to ensure glucan binding properties was investigated. For GTF-D, the loss of only one C-terminal A repeated unit or of two N-terminal 'A' repeated units is sufficient to suppress its binding capacity (Lis *et al.*, 1995). Thus all the units are not required for glucan binding. Moreover, it appears that the number of required units is different for enzymes producing a soluble glucan than for those producing an insoluble one; the latter enzymes appearing less sensitive to deletions. The presence of only the first two N-terminal repeated units are sufficient for GTF-I and GTF-Ia to bind glucan (Ferretti *et al.*, 1987; Abo *et al.*, 1991; Kato and Kuramitsu, 1990). Deletions of three C-terminal units (A-C) are necessary to affect binding capacity of GTF-G (Vickerman *et al.*, 1996).

***Role of the C-terminal domain in glucansucrase activity*:** The presence of the C-terminal glucan binding domain seems to be necessary to keep an active enzyme. As for glucan binding capacities, the activity of enzymes producing a soluble glucan appears to be more affected by deletions of the C-terminal domain than is the activity of enzymes producing an insoluble one. For GTF-D the loss of the C-terminal 'A' repeated unit is sufficient to produce an activity decrease of 90% (Lis *et al.*, 1995). The truncation of the last 85 amino-acid residues of the C-sterminal domain of DSR-S resulted in only about 25% of the initial activity being retained (Monchois *et al.*, 1999). In contrast, deletion of the three C-terminal units (A-C) of GTF-G is necessary to produce an activity decrease of 85% (Vickerman *et al.*, 1996). Only the two N-terminal A repeated units are necessary to keep GTF-I or GTF-Ia significantly active. The importance of the C-terminal domain in the glucansucrase catalytic mechanism remains mostly unknown. However, the fact that with some deleted enzymes, hydrolytic activity remains but glucan binding and synthesis properties disappear, suggests that the C-terminal domain may be important for the polymer chain growth. C-terminal truncations of DSR-S did neither modify the K_m for sucrose nor the optimum pH and energy of activation of this enzyme both, for dextran synthesis reaction and for the oligosaccharide synthesis reaction; it is thus supposed that the C-terminal domain is not directly involved in the catalytic process for dextran formation, but may make release of products from the catalytic site easier (Monchois *et al.*, 1998). The glucansucrase glucan-binding domain seems also to be required for the synthesis of low molecular weight oligosaccharides. C-terminal truncations of DSR-S resulted in a strong decrease in oligosaccharide synthesis velocity, whatever the acceptor molecule used (maltose or fructose). Moreover, the fact that the C-terminal domain of DSR-S modulated the size of oligosaccharides produced in the presence of maltose, without affecting the reaction yield and the activator effect of maltose, seems to show again that its role may be to facilitate the release of products from the catalytic site without being directly involved in the catalysis.

The C-terminal domain also seems be involved in glucan structure determination. Deletion of the three C-terminal units (A-C) of GTF-G led to an enzyme being obtained which produced an α-1$\rightarrow$6 and α-1$\rightarrow$3 glucosidic linkages (Nakano and Kuramitsu, 1992). The in-frame fusion of the C-terminal domain of GTF-B, an enzyme producing an insoluble glucan, with the N-terminal catalytic domain of GTF-D, an enzyme producing a soluble glucan, resulted in the formation of an enzyme producing an insoluble glucan (Nakano and Kuramitsu, 1992). However, the inverse in-frame fusion of the C-terminal domain of GTF-B did not yield an enzyme producing soluble glucan. This shows that the C-terminal domain may influence the structure of the produced dextran, but is not the unique determinant.

Directed Evolution of a Dextransucrase for Increased Constitutive Activity and Synthesis of a Highly Branched Dextran

Kim *et al.* (2000), cloned and sequenced a dextransucrase gene from B-742CB (*dsrB742;* GenBank Accession No. AF294469). Unlike other genes, the *dsrB742* gene encodes an extracellular dextransucrase in *E. coli.* However, the dextransucrase primarily synthesizes an α-1→6 linked dextran with 3–5% α-1→3 branch linkages. The nucleotide sequence of *dsrB742* shows one open reading frame (ORF) composed of 4,524 bp encoding the dextransucrase. The amino acid sequence of B-742CB dextransucrase (DSRB742) shows a 50% similarity with DSRA from *L. mesenteroides* B-1299, which synthesizes a dextran containing 87% α-1→6 linkages and 13% α-1→3 branch linkages, a 70% similarity with *L. mesenteroides* B-512FMCM dextransucrase from (Ryu *et al.*, 2000) and a 45-56% similarity with *Streptococcal* glucansucrases (GTFs) (Banas *et al.*, 1990; Mooser *et al.*, 1991, Ferrette *et al.*, 1987). The expressed and purified enzyme (DSRB742) from the clone showed similar enzymatic properties, such as acceptor reactions, molecular size, optimum pH and temperature to that of B-742CB dextransucrase, but with the inability to form a significant amount of α-1→3 branch linkages compared to that of *L. mesenteroides* B-742CB dextransucrase (Park *et al.*, 2001). Until now, there has not been a report of the isolation or development of a mutant or clone that elaborates a constitutive, extracellular dextransucrase, capable of synthesizing a highly branched dextran. In a previous study, Kim *et al.* (1997) developed a novel mutation method for the hyper-production of dextransucrase by using vacuum ultraviolet radiation or ultrasoft X-rays. The electromagnetic radiation that lies between the UV and X-ray is called ultrasoft X-ray and its energy ranges from 10 eV to 2 keV (Koch, 1983). They also elaborate a novel extracellular dextransucrase gene (*dsrB742ck*) after ultrasoft X-ray irradiation, producing a dextransucrase of increased activity and synthesis of a highly branched dextran. To develop novel mutants, *pdsrB742* was irradiated with ultrasoft X-ray. Ultrasoft X-ray was obtained from the LIGA (Lithography, Galvanoformung-electroplating, Abformung-replication) beam-line in the Pohang Accelerator Laboratory (Pohang, Korea) using a specially constructed chamber. An aluminum target was placed inside the exposure chamber and the ultrasoft X-ray was generated by irradiating this target with a white beam. The chamber was filled with helium (He) gas which is effectively transparent to the ultrasoft X-rays. The binding energies of Al orbital K(1), L(1), L(2), and L(3) were 1559.6 eV, 117.8 eV, 73.1 eV, and 72.7 eV respectively. If an electron of the K-shell is taken out by the incident photon, an electron from a neighbouring orbital moves to that empty orbital. The difference in the orbital energies is emitted as a photon of the energy range of the ultrasoft X-ray. In this experiment, the Al K ray was corresponded to energy of 1.487 keV. The DNA (20 μl of a 50 μg/ml concentration) was allocated in the well of a 96-well micro plate (Nalge NUNC International, USA), and ultrasoft X-ray was evenly exposed. In this irradiation, 6.4×10^6 photons/s were irradiated to each sample in a well. After irradiation, the *dsrB742* in each well was transformed to *E. coli* strain DH5a and transformants were plated onto LB medium supplemented with ampicillin (50 μg/ml) and 2% sucrose (w/v).

Among the glucansucrases, *L. mesenteroides* NRRL B-742 produces two extracellular dextrans from sucrose, namely, L-dextran, which is precipitated at an ethanol concentration of 39%, and S-dextran, which is precipitated at an ethanol concentration of 45%. Fraction S-dextran contained 50% α-1→3 branches and fraction L-dextran contained 14% α-1→4 branch linkages and about 1% α-1→3 branch linkages. One dextransucrase gene from *L. mesenteroides*

NRRL B-742CB has already been cloned (Park *et al.*, 2001). Unlike other genes, the *dsrB742* gene encodes an extracellular dextransucrase in *E. coli*. However, it synthesizes a dextran with a relatively low degree of branching containing 95-97% of α-1→6 linkages and 3-5% α-1→3 branch linkages.

The ultrasoft X-ray irradiation affected the viability of dextransucrase gene (*dsrB742*); the longer the exposure to irradiation, the higher was the percentage of conversion of supercoiled DNA to relaxed DNA. Ninety per cent conversion was detected after 25 minutes exposure with 1.487 keV energy with an aluminum target. The exposed DNA was transformed and

```
dsrB742CK  GGTAAAGAATTGGTAATCGTTTGTGAAATGATATAAGTTCGTAATTTTATAGCTTATAAT    660
dsrB742     GTTAAAGAATTGTTAATCGTTTGTGAAATGATATAAGTTCGTAATTTTATAGCTTATAAT   660
- 3 5                                                                   - 1 0
*  * * * * * * * * * *   * * * * * * * * * * * * * * * * * * * * * * * * * * * * * * * * * * * *

dsrB742CK  TCATTTGTTTCAAAGAATAATCACCAATGGAGGAGAGAATGTTT--GATTAA-GAGAGAA   717
                                                    M   F     D  *   E  R
dsrB742    TCATTTGTTTCAAAGAATAATCACAAATGGAGGAGAGAATGTTTATGATTAAAGAGAGAA   720
R B S             M      F      M     I     K       E      R              7
* * * * * * * * * * * * * * * * * * * * * * *   * * * * * * * * * * * * * * * * * * *   * * * * * *   * * * * * * *

dsrB742ck  ATGTACGAAAAAAGCTCTACAAGTCTGGTAAGAGTTGGGTTATTGGGGGACTCATTTTAT   777
dsrB742    ATGTACGAAAAAAGCTCTACAAGTCTGGTAAGAGTTGGGTTATTGGGGGACTCATTTTAT   780
N  V  R  K  K  L  Y  K  S  G  K  S  W  V  I  G  G  L  I  L                2 7
          * * * * * * * * * * * * * * * * * * * * * * * * * * * * * * * * * * * * * * * * * * * * *

dsrB742ck  CGACAATTATGCTGTCTATGACCGCTACTTCACAAAATGTTAATGCAGATAGCACAAACA   837
                       M L S M T A T S Q N V N A D S T N                  17
dsrB742    CGACAATTATGCTGTCTATGACCGCTACTTCACAAAATGTTAATGCAGATAGCACAAACA   840
S  T  I  M  L  S  M  T  A  T  S  Q  N  V  N  A  D  S  T  N                4 7
* * * * * * * * * * * * * * * * * * * * * * * * * * * * * * * * * * * * * * * * * * * * * * * * * * * * * * *

dsrB742ck  ATCAATACATTACTGACCAAACCGGGACCGCCTATTACTTCCAGAATGATGGCACAATGG  4437
N  Q  Y  I  T  D  Q  T  G  T  A  Y  Y  F  Q  N  D  G  T  M              1 2 1 7
dsrB742    ATCAATACATTACTGACCAAACCGGTACCGCCTATTACTTCCAGAATGATGGCACAATGG  4440
N  Q  Y  I  T  D  Q  T  G  T  A  Y  Y  F  Q  N  D  G  T  M              1 2 4 7
* * * * * * * * * * * * * * * * * * * * * * * * *   * * * * * * * * * * * * * * * * * * * * * * * * * * * * * * * * *

dsrB742ck  ATATTCCCTGATTAGCTAATAACACCTTCACTTCTGGCATTTTTAAAACAGGTGGTTCTA  5517
dsrB742    ATATTCCCTGATTAGCTAATAACACCTTCACTTCTGGCATTTTTAAAACAGCTGCTTCTA  5520
* * * * * * * * * * * * * * * * * * * * * * * * * * * * * * * * * * * * * * * * * * * * * * * * * * * * *   * * * * * * * *

dsrB742ck  ATTCAAACGACTTAGCTTGCTGTGCAACTGAGCCCAACCATTTCTATCATCATGATAATG  5577
dsrB742    ATTCAAACGACTTAGCTTGCTGTGCAACTGAGCC-AAACATTTCTATCATCATGATAATG  5580
* * * * * * * * * * * * * * * * * * * * * * * * * * * * * * * * * * *   * *   * * * * * * * * * * * * * * * * * * * * * *

dsrB742ck  ATTGCTTCAATAGCAACTAACCCGGGTAATAATTTGTGCGATCATTTTATACGGTCTCTT  5637
dsrB742    ATTGATTCAATAGCAACTAACCCGG-TAATAATTTGTGCGATCATTTTATACTGTCTCTT  5640
* * * *   * * * * * * * * * * * * * * * * * * * * * * *   * * * * * * * * * * * * * * * * * * * * * * * * * *   * * * * * * *

dsrB742ck  TTTGTAATATAATCAGCGAGTGTGGTTAAATCTTTACCATTAGTATCCAACCAACCGGGT  5817
dsrB742    TTTGTAATATAATCAGCGAGTGTGGTTAAATCTTTACCATTAGTATCCAACCAACCGG-T  5820
       ************************************************************ *
```

Fig. 7: Base substitutions of the mutant (pDSRB742CK) developed using the irradiated DNA. Boxed letters indicate changed nucleotides. Underlined nucleotides represent the -35 and -10 regions for presumptive promoter. Double underlined nucleotides correspond to the putative ribosome-binding sites. The nucleotide sequence of the dsrB742ck dextransucrase gene has been deposited in the GenBank database under accession No. AY280636.

about 500 colonies were found, forming mucous dextran on an agar plate containing sucrose. These were selected and analyzed for the production of dextransucrase on a sucrose or glucose medium. A constitutive and hyper-producing dextransucrase *E. coli* transformant (pDSRB742CK) was obtained. The purified DSRB742CK dextransucrase showed 2.3-times higher activity per mg protein, compared to that of the parent clone, DSRB742 dextransucrase. The nucleotide sequence of *dsrB742* shows one ORF composed of 4,524 bp encoding dextransucrase (GenBank database under accession No. AF294469). The deduced amino-acid sequence gave a calculated molecular mass of 168.4 kDa. The nucleotides sequence of the *dsrB742ck* dextransucrase gene showed seven DNA base differences comparing to that of *dsrB742* dextransucrase gene (Fig. 7); three nucleotide deletions (at nucleotides 705, 706, and 713) and one nucleotide substitution (at nucleotide 4406; from G to T). Therefore, the start codon appears at 30 amino-acids downstream of DSRB742 dextransucrase. There are two nucleotide sequence differences at the promoter region (at nucleotides 602 and 613) and one nucleotide sequence difference at the RBS site at nucleotide 685. DSRB742 dextransucrase is produced without isopropyl-β-D-galactose (IPTG) induction on sucrose medium, but not on glucose medium. pDSRB742CK did produce dextransucrase on glucose medium (0.019 U/ml) as well as on a sucrose medium (0.035 U/ml). These values are higher than that of IPTG-induced cell extract (0.004 U/ml), yet the approximate specific activity of each preparation was the same. Two general mechanisms are suggested for the introduction of protein into the liquid medium of *Escherichia coli* (Shokri *et al.*, 2003). One is a two-stage translocation using active transporters in the cytoplasmic membrane followed by passive transport through the outer membrane. Passive transport is achieved through either external or internal destabilization of the *E. coli* structural components (Cornelis, 2000; Yohata *et al.*, 1999). The other is transplantation (type I secretion/ABC transport, fusion protein/α-auto transport, specific signal peptide and over expression) of proteins capable of active transport over one or both of the membranes. This involves the transplantation of secretion mechanisms into the *E. coli* cell from pathogenic *E. coli* as well as from other species. A detailed study of the potential secretion mechanism and involving component of DSRB742 and DSRB742CK dextransucrase is in progress.

It has been reported that a constitutive dextransucrase mutant, *L. mesenteroides* B-512FMCM, produced 13-times higher activity than that of the parent strain, *L. mesenteroides* B-512FM. Furthermore, 1000-times higher amount of dextransucrase was produced than that of the industrial-dextran producing strain, *L. mesenteroides* B-512F. *L. mesenteroides* B-512FMCM is a completely constitutive mutant for the production of dextransucrase. There is a single nucleotide change in the promoter region and two amino-acid changes in the structural gene from that of the parent *L. mesenteroides* NRRL B-512F [Ryu *et al.*, 2000]. Whether the single substitution in the promoter affects the high production of dextransucrase in *L. mesenteroides* B-512FMC, remains to be determined. The change in the nucleotides in the promoter region in *dsrB742ck* could affect the pattern of expression and the amount of protein produced. A detailed study of the role of *dsrB742ck* promoter is in progress.

The dextran synthesizing activity of DSRB742CK dextransucrase was 2.3 times higher than that of DSRB742 dextransucrase based on Western blot analysis using anti-dextran antibody. The slight increase in protein production in culture supernatant of pDSRB742CK (1.07 times) compared to that of pDSRB742 clone and the activity increase of 2.3-times suggests that the 30 amino-acid in the N-terminal region significantly affects the specific activity of the dextransucrase.

Electrophoresis analyses and *in situ* activity assays were performed with sucrose culture supernatants. The molecular weights of the active bands of DSRB742CK and DSRB742 dextransucrases were 164.1 and 168.6 kDa, respectively. Even when we applied the same amount of each protein, the intensity of each active band was significantly different. For DSRB742CK, there were higher molecular weight activity bands. These higher forms are likely the aggregated forms of a dextransucrase, as previously reported from *L. mesenteroides* B-512F, B-1299, B-1355 and B-742 (Devulapalle *et al.*, 1997; Djaizy *et al.*, 1995; Ebert and Schenk, 1962; Ferretti, 1987). The DSRB742CK dextransucrase, prepared using glucose, gave a single activity band of 164.1 kDa (data not shown). Thus, dextransucrase, prepared on sucrose culture had bound dextran that could play a role in the formation of multiple aggregated dextransucrase forms.

The structure, size, and the position and degree of branch linkages are dependent on the specific glucansucrases that is elaborated by a specific gene. The genome also determines whether the organism requires an inducible medium with sucrose as the carbon source to elaborate the enzyme or elaborates the enzyme in a constitutive medium that can use glucose or fructose as the carbon source. The structure, size, and the position and degree of branching in the glucans are dependent on the specific glucansucrases that is elaborated by a specific gene. Glucansucrase encoding genes from *L. mesenteroides* code for enzymes that have an average molecular mass of 170 kDa. Key elements in the structure of the enzyme that determine the structure of the glucans have not been determined and remain unclear. The cloning and development of new and unique glucansucrases genes and the determination of the structures of the glucans products as well as the structures of the acceptor products and the catalytic parameters of the enzymes are necessary for improving and understanding the reaction mechanisms of the *Leuconostoc* and *Streptococcus* glucansucrases.

The dextran prepared from DSRB742CK was hydrolyzed with *Penicillium* dextranase and the hydrolyzate composition was compared with that of DSRB742 dextran (Table 12). Both hydrolyzates were primarily composed of glucose, isomaltose, and branched isomaltodextrins.

Table 12: Dextranase/TLC determination of the degree of branching of dextrans synthesized by DSRB742 and DSRB742CK dextransucrases*

	DSRB742 dextran	DSRB742CK dextran
Mono-saccharide (M, %)	35.4	29.5
Isomaltodextrins (IMD, %)	42.2	27.1
Branched Isomaltodextrins (BIM, %)	19.1	23.8
Unhydrolyzed Dextran (UHD, %)	3.3	19.6
[BIM+UHD] Ratio		
[M+IMD] Per cent		
Branching **	0.29	
5.9%	0.77	
15.6%		

* Hydrolysis reactions were conducted at 37° C for 3 hrs.

** Branching for dextran synthesized from 0.1 M sucrose was assumed to be 5% from refs. 23 & 24. A branching conversion factor was obtained by dividing 5% by the ratio, 0.247, giving 20.24. The per cent branching for DSRB742 and DSRB742CK dextrans was obtained by multiplying their ratio by the conversion factor.

DSRB742CK dextran contained 15.6% branching and showed 2.7-times higher resistance to *Penicillium* endodextranase hydrolysis compared to that of DSRB742 dextran. Thus, DSRB742CK dextran contains much higher degrees of branching than previously reported before for dextrans synthesized by dextransucrase-producing clones (Botchway *et al.*, 1995; Valette *et al.*, 1993).

When various carbohydrates, in addition to sucrose, are present in glucansucrase reaction digests, some of the glucosyl groups are transferred to the carbohydrates to make branched oligosaccharides and are diverted from the formation of glucan. Branched oligosaccharides resist hydrolysis by digestive enzymes in animals and humans because of the configuration of their glycosidic bonds. They selectively stimulate intestinal microflora such as *Bifidobacterium* sp., *Lactobacillus* sp. or *Bacteroides* sp. as prebiotic agents (Djouzy *et al.*, 1995; Vallette *et al.*, 1993) and can have uses in dermocosmetic applications (Monsan and Paul, 1995).

The branch linkage was analyzed by using ^{13}C-NMR. The ^{13}C-NMR spectrum signals of DSRB742 and DSRB742CK dextrans were assigned as shown in Fig. 4. No signal was observed for C3-OS in DSRB742 dextran, while C3-OS signals for α-(1$\rightarrow$3) linkages were observed in DSRB742CK dextran. Until now different clones, such as DSRB742 [108], DSRS (from *Leuconostoc mesenteroides* NRRL B-512F dextransucrase), FMCMDS (from *Leuconostoc mesenteroides* B-512 FMCM) and DSRB (from *Leuconostoc mesenteroides* B-1299) (Monchois *et al.*, 1998), produced primarily linear dextrans that were composed of α-(1$\rightarrow$6) linkages when grown in a sucrose containing media. A clone, DSRA (from *Leuconostoc mesenteroides* B-1299), synthesizes a dextran containing 13% α-(1$\rightarrow$3) linkages (Monchois *et al.*, 1996) using cell extract and sucrose reaction after IPTG induction. pDSRB742CK is the first clone that makes highly branched dextran in sucrose culture supernatant without IPTG induction or in glucose medium and is evolved from a gene producing a dextransucrase of linear dextran. The DSRB742 dextransucrase amino-acid sequence is very different from those of DSRS, FMCMDS, DSRA and DSRB dextransucrases; the amino-acid sequence similarities of DSRB742 dextransucrase with DSRS, FMCMDS, DSRA and DSRB dextransucrases were ranged from 50 to 70%. As of yet, there is no detailed molecular explanation and information for the synthesis of branch linkages by different dextransucrases. Studies on DSRB742CK dextransucrase the structure of crystalline enzyme could give a more complete picture of branch formation and their regioselectivity.

To determine which specific nucleotide(s) of the promoter region is related to the constitutivity of dextransucrase and which amino-acids are involved in branch formation, molecular evolution and/or site-directed mutagenesis of dextransucrase gene and the determination of a crystal structure must be performed.

ACKNOWLEDGEMENTS

This work was partially supported by Korea Science and Engineering Foundation through Korea-Japan Joint Research Project (20016-211-01-2), Republic of Korea.

REFERENCES

Abo, H., Matsumura, T, Kodama, T., Ohta, H., Fukui, K., Kato, K. & Kagawa, H. (1991) *J. Bacteriol.* **173**: 989-996.

Ajisaka, K., Nishida, H. and Fujimoto, H. (1987) *Biotechnol. Lett.* **9**: 243 - 248.

Alsop, R. M. (1983) In *Progress in Industrial Microbiology* (Ed. Bushell, M. E.), Elsevier, Amsterdam, pp.1-44.
Aoki, H., Shiroza, T., Hayakawa, M., Sato, S. & Kuramitsu, H. K. (1986) *Infec. Immun.* **53**: 587-594.
Argüello-Morales, M. A., Remaud-Simeon, M., Pizzut, S., Sarcabal, P., Willemot, R. M. & Monsan, P. (2000) *GenBank Accession No.* AJ250172.
Aubin, F., Puzenat, E., Arveux, P., Quencez, E. & Humbert, P. (2001) *Br. J. Dermatol.* **144**: 1204-1206.
Banas, J. A., Russell, R. R. B. & Ferretti, J. J. (1990) *Infect. Immun.* **58**: 667-673.
Banas, J. A., Simon, D., Williams, L. K., Ferretti, J. J. & Russell, R. R. B. (1994). FEMS Microbiol. Lett. **123**: 349-354.
Botchway, S.W., Stevens, D. L., Hill, M. A., Jenner, T. J. & O'neill, P. (1997) *Radiat. Res.* **148** : 317-321.
Broker, M., Jordening, H. & Buchholz, K. (1994) *Biotechnol. Bioeng.* **43**: 392-394.
Castillo, E., Iturbe, F., Lopez-Munguia, A., Pelenc, V., Paul, F. & Monsan, P. (1990) *Ann. NY. Acad. Sci.* **672**: 425-430.
Cope, P. A. & Mooser, G. (1993) *Infect. Immun.* **61**: 4814-4817.
Biochemistry 32: 13696-13702.
Cornelis, P. (2000). *Curr. Opin. Biotechnol.* **11**: 450.
Dalla, V. L. & Marciani, M. S. (2001) *Curr. Med. Chem.* **8**: 1405-1408.
Day, D.F. & Kim, D. (1992) *Ann. NY. Acad.* **672**: 573 - 576.
Day, D. F. & Kim, D. (1993) *US Patent.* 5,229,227.
Devulapalle, K. S., Goodman, S. D., Gao, Q., Hemsley, A. & Mooser, G. (1997) *Protein Sci.* **6**: 2489-2493.
Djouzy, Z., Andrieux, C., Pelenc, V., Somarriba, S., Popot, F., Paul, F., Monsan, P. & Szylit, O. (1995) *J. Appl. Bacteriol.* **79**: 117.
Ebert, K.H. & Schenk, G. (1962) *Z. Naturforsch.* **17**: 732 - 741.
Ferretti, J. J., Gilpin, M. L. & Russell, R. R. B. (1987) *J. Bacteriol.* **169**: 4271-4278.
Fu, D. & Robyt, J. F. (1988) *Carbohydr. Res.* **183**: 97-109.
Fu, D. & Robyt, J. F. (1990) *Arch. Biochem. Biophys.* **283**:379-387.
Fu, D. & Robyt, J. F. (1991) *Carbohydr. Res.* **217**: 201-211.
Funane, K., Shiraiwa, M., Hashimoto, K., Ichishima, E. & Kobayashi, M. (1993) *Oyo. Tos. Kaga.* 40: 17-25.
Funane, K., Arai, T., Chiba, Y., Hashimoto, K., Ichishima, E. & Kobayashi, M. (1995) *Oyo. Tos. Kaga.* 42: 27-35.
Germaine, G.R., Chludzinski, A.M. & Schachtele, C.F. (1974) *J. Bacteriol.* **120**: 287 - 294.
Germaine, G.R. & Schachtele, C.F. (1976) *Infect. Immun.* **13**: 365 - 372.
Gilmore, K. S., Russel, R.R.B. & Ferretti, J.J. (1990) *Infect. Immun.* **58**: 452-2458.
Giffard, P. M., Simpson, C. L., Milward, C. P. & Jacques, N. A. (1991) *J. Gen. Microbiol.* **137**: 2577-2593.
Giffard, P.M., Allen, A.M., Milward, C.P., Simpson, C.L. & Jacques, N.A. (1993) *J. Gen. Microbiol.* **139**: 1511-1522.
Giffard, P. M. & Jacques, N. A. (1994) *J. Dent. Res.***73**: 1133-1141.
Gilpin, M. L., Russell, R. R. B. & Morrissey, P. (1985) *Infect. Immun.* **49**: 414-416.
Goyal, A. & Katiyar, S. S. (1995) *Biochem. Mol. Int.* **36**: 579-585.
Gronwell, A. J. T. & Ingleman, B. G. A. (1944) *Acta Physiol. Scand.* **7**: 97-107.
Gronwell, A. J. T. & Ingelcican, B. G. A. (1945) *Acta Physiol. Scand.* **9**: 1-27.
Hanada, N. & Kuramitsu, H. K. (1988) *Infect. Immun.* **56**: 1999-2005.
Hanada, N. & Kuramitsu, H. K. (1989) *Infect. Immun.* **57**: 2079-2085.
Hanada, N., Yamashita, Y., Shibata, Y., Sato, S., Katayama, T., Takehara, T. & Inoue, M. (1991) *Infect. Immun.* 59: 3434-3438.
Hanada, N., Isobe, Y., Aizawa, Y., Katayama, T., Sato, S. & Inoue, M. (1993) *Infect. Immun.* **61**: 2096-2103.
Hehre, E.J. (1941) *Science* 93: 237-238.
Hehre, E.J. (1951) *Adv. Enzymol.* **11**: 297 - 337.
Heo, S. J., Kim, D., Lee, I. S. & Chang, P. S. (1999) *Kor. J. Appl. Microbiol. Biotechnol.* **27**: 304-310.
Hidaka, H.,Hirayama, M., Tokunaga, T. & Eida, T. (1990) *New Developments in Dietary Fibre* (Ed. Furda, I. and Brine, C. J.), Plenum Press, New York, pp.105 - 117
Honda, O., Kato, C. & Kuramitsu, H. K. (1990) *J. Gen. Microbiol.* **136**: 2099-105.
Honigsmann, M., Jaschke, E., Gscnait, F., Brenner, S., Fritsch, W. P. O. & Wolff, K. (1979) *Br. J. Dermatol.* **101**: 369-378.

Jeanes A.W., Haynes, A.W., Williams, C.A., Rankin, J.C., Melvin, E.H., Austin, M.J., Cluskey, J.E., Fisher, B.E., Tsuchiya, H.M. & Rist, C.E. (1954) *J. Am. Chem. Soc.* **76**: 5041-5047.
Kato, C. & Kuramitsu, H. K. (1990) *FEMS Microbiol. Lett.* **72**: 299-302.
Kato, C., Nakano, Y., Lis, M. & Kuramitsu, H. K. (1992) *Biochem. Biophysic. Res. Communi.* **189**: 1184-1188.
Kim, D. (1992) *Ph.D. Thesis*, Louisiana State Univ. USA
Kim, D. & Day, D. F. (1994) *Enzyme Microbial. Technol.* **16**: 844 - 848.
Kim, D. & Robyt, J. F. (1994a) *Enzyme Microb. Technol.* **16**: 659-664.
Kim, D. & Robyt, J. F. (1994b) *Enzyme Microb. Technol.* **16**: 1010-1016.
Kim, D. & Robyt, J. F. (1995a) *Enzyme Microb. Technol.* **17**: 689-695.
Kim, D. & Robyt, J. F. (1995b) *Enzyme Microb. Technol.* **17**: 1050-1056.
Kim, D. & Day, D. F. (1995) *Lett. Appl. Microbiol.* **20**: 268 - 270.
Kim, D., Kim, D.W., Lee, J. H., Park, K. H., Day, L. M. & Day, D. F. (1997) *Biotechnol. Techn.* **11**:319-321
Kim, D. & Kim, D. W. (1999) *J. Microbiol. Biotechnol.* **9**: 219-222.
Kim, D., Kim, Y. M., Park, M. R. & Park, D. H. (1999) *J. Micorbiol. Biotechnol.* **9**: 704-708.
Kim, H. S., Kim, D., Ryu, H. J. & Robyt, J. F. (2000) *J. Microbiol. Biotechnol.* **10**: 559-563.
Kim, D., Robyt, J. F., Lee, S.Y., Lee, J. H. & Kim, Y. M. (2003) *Carbohydr. Res.* **338**: 1183–1189.
Kitahata, S., Tsuyama, N. & Okada, S. (1974) *Agri. Biol. Chem.* **38**: 387 - 393.
Kitahata S. & Okada S. (1982) *J. Jap. Soc. Starch Sci.* **29**: 7 -12.
Kobayashi, M. & Matsuda, K. (1980) *Biochem. Biophysics Acta* **614**: 46 - 62.
Kobayashi, M. & Matsuda, K. (1986) *J. Biochem.* **1986**: 100: 615 - 621.
Koch, E. E. (1983) *Handbook on synchrotron radiation*, 3rd ed., New York, North Holland
Koepsell, H. J., Tsuchiya, H. M., Hellman, N. N., Kazendo, A., Hoffman, C. A., Sharpe, E. S. & Jackson, R. W. J. (1953) *J. Biol. Chem.* **200**: 793-801.
Koga, T. & Inoue, M. (1981) *Carbohydr. Rcs.* **183**: 97-109.
Laloi, P., Munro, C. L., Jones, K. R. & Macrina, F. L. (1996) *Infect. Immun.* **64**: 28-36.
Lee, J. H., Kim, D., Ryu, H. J., Heo, S. J., Jhon, D. Y., Han, N. S. & Robyt, J. F. (1998). *Kor. J. Appl. Microbio. Biotechnol.* **26**: 264-268.
Lee, J. H., Kim, D., Baek, J. S., Park, K. H., Han, N. S. & Robyt, J. F. (1998) *Kor. J. Appl. Microbiol. Biotechnol.* **26**:143-150.
Lis, M., Shiroza, T. & Kuramitsu, H. K. (1995) *Appl. Env. Microbiol.* **61**: 2040-2042.
Lopez-Munguia, A., Paul, F., Monsan, P. & Biton, J., *et al.* (1990) *Ann. NY. Acad. Sci.* **613**: 717 - 722.
Mayer, R. M., Metthews, M. M., Futerman, C. L., Parnaik, V. K. & Jung, S. M. (1981) *Arch. Biochem. Biophys.* **208**: 278-287.
Mbemda, E., Cham, V., Gluckman, J. C. & Kiatzmann, D. *et al.* (1992) *Biochem. Biophy. Acta* **1138**: 62 - 67.
Mitsuya, H., Looney, D., Kuno, S. & Ueno, R. *et al.* (1988) *Science* **240**: 646 - 649.
Monchois, V., Willemot, R. M., Remaud-Simeon, M., Croux, C. & Monsan, P. (1996) *Gene.* **182**: 23-32.
Monchois, V., Remaud-Simeon, M., Russell, R. R. B., Monsan, P. & Willemot, R. M. (1997) *Appl. Microbiol. Biotechnol.* **48**: 465-472.
Monchois, V., Remaud-Simeon, M., Monsan, P. & Willemot, R. M. (1998) *FEMS Microbiol.* Lett. **159**: 307-315.
Monchois, V., Reverte, A., Remaud-Simeon, H., Monsan, P. & Willemot, R. M. (1998) *Appl. Environ. Microbiol.* **64**: 1644-1649.
Monchois. V., Willemot, R. M. & Monsan, P. (1999) *FEMS Microbiol. Rev.* **23**: 131-151.
Monsan, P. & Paul, F. (1995) *FEMS Microbiol. Rev.* **16**: 187.
Mooser, G. & Wong, C. (1988) *Inf. Immun.* **56**: 880 - 884.
Mooser, G. & Iwaoka, K. R. (1989) *Biochem.* **28**: 443-449.
Mooser, G., Hefta, A., Paxton, R. J., Shively, J. E. & Lee, T. D. (1991) I. **266**: 8916-8922.
Nakano, Y. J. and Kuramitsu, H. K. (1992). J. Bacteriol. **174**: 5639-5646.
Nataraj, A. J., Black, H. S. & Ananthaswamy, H. N. (1993) *Proc. Natl. Acad. Sci. U. S. A.* **93**: 7961-7965.
Nataraj, A. J., Black, H. S. & Ananthaswamy, H. N. (1993) *Proc. Natl. Acad. Sci. U. S. A.* 93: 7961-7965.
Park, M.R., Ryu, H.J., Kim, D., Choe, J.Y. & Robyt, J.F. (2001) *J. Microbiol. Biotechnol.* **11**: 628.
Parnaik, V. K., Luzio, G. A., Grahame, D. A., Ditson, S. L. & Mayer, R. M. (1983) *Carbohydr. Res.* **121**: 257-268.

Pasteur, L. (1861) *Bull. Soc. Chim. Paris*, 30-36.

Paul, F., Oriol, E., Auriol, D. & Monsan, P. (1986) *Carbohydr. Res.* **149**: 433-441.

Pitty, L. J., Giffard, P. M., Gilpin, M. L., Russell, R. R. B. & Jacques, N. A. (1989) *J. Dent. Res.* **68**: 1681-1682.

Pucci, M. J., Jones, K. R., Kuramitsu, H. K. & Macrina, F. L. (1987) *Infect. Immun.* **55**: 2176-2182.

Quiocho, F. A. (1986) *Annu. Rev. Biochem.* **55**: 287-315.

Repine, J. E., Pfenninger, O. W., Talmage, D. W., Berger, E. M. & Pettijohn, D. E. (1981) *Proc. Natl. Acad. Sci. U. S. A.* **78**: 1001-1003.

Ricketts C. R., Lorenz, L. & Maycock, D.A. (1950) *Nature* **165**: 770 - 771.

Robyt, J.F., Kimble, B.K. & Walseth, T.F. (1974) *Arch. Biochem. Biophy.* **165**: 634 - 640.

Robyt, J.F. & Taniguchi, H. (1976) *Arch. Biochem. Biophy.* **174**: 129 - 135.

Robyt, J.F & Walseth, T.E. (1978) *Carbohydr. Res.* **61**: 433 - 445.

Robyt, J.F. & Eklund, S.H. (1983) *Carbohydr. Res.* **121**: 279 - 286.

Robyt, J. F. & Eklund, S. H. (1983) *Carbohydr. Res.* **68**: 95-111.

Robyt, J. F. (1995) *Adv. Carbohydr. Chem. Biochem.* **51**: 133-168.

Robyt, J. F. (1996) In *Progress in Biotechnology*,(Eds. Park, K. H,, Robyt, J. F. and Choi, Y. D.), Elsevier, 12:1-22.

Roots, R. & Okada, S. (1972) *Int. J. Radiat. Biol.* **21**:329-342.

Roots, R. & Okada, S. (1975) *Radiat. Res.* **64**: 306-320.

Russell, R. R. B., Gilpin, M. L., Mukasa, H. & Dougan, G. (1987) *J. Gen. Microbiol.* **133**: 935-944.

Russell, R. R. B. (1990) *Arch. Oral. Biol.* 35: 53S-58S.

Ryu, H. J., Kim, D., Kim, D. W., Moon, Y. Y. & Robyt, J. F. (2000) *Biotechnol. Lett.* 22: 421-425.

Ryu, H. J., Yi, K. E., Kim, D. W., Jung, Y. D., Chang, S. S., Seo, E. S., Lee, K. Y., Day, M. L. & Kim, D. (2002) *J. Microbial. Biotechnol.* **12**: 598-602.

Sato, S., Masaku, I., Hanada, N., Aizawa, Y. & Isobe, Y. (1993) *DNA Seq.* **4**: 19-27.

Scheubler, C. Z. (1874). Ver. Dtsch. Zuker-Ind. **24**: 309-335.

Shimamura, A., Nakano, Y. J., Mukasa, H. &Kuramitsu, H. K. (1994) *J. Bacteriol.* **176**: 4845-4850.

Shiroza, T., Ueda., S. & Kuramitsu, H. K. (1987) *J. Bacteriol.* **169**: 4263-4270.

Simpson, C. L., Cheetham, N. W. H., Giffard, P. M. & Jacques, N. A. (1995a) *Microbiol.*. **141**: 1451-1460.

Simpson, C. L., Giffard, P. M. & Jacques, N. A. (1995b) *Infect. Immun.* **63**: 609-621.

Singh, J. S., Taylor, K. J. & Doyle, R. J. (1993) *Carbohydr. Res.* **244**: 137-147.

Smith, D. J., Taubman, M. A., King, W. F., Eida, S., Powell, J. R. & Eastcott, J. (1994) *Infect. Immun.* **62**: 5470-5476.

Stern, R. S. & Lunder, E. J. (1998) *Arch. Dermatol.* **134**: 1582-1585.

Stern, R. S. (2001) *J. Am. Acad. Dermatol.* **44**: 755-761.

Shokri, A., Sanden, A.M. & Larsson, G. (2003) *Appl. Microbiol Biotechnol*. **60**: 654.

Su, D. and Robyt, J. F. (1993). Carbohydr. Res. **248**: 339 - 348.

Takemoto, K.K. & Liebharber, H. (1962) *Virol.*17: 499 - 501.

Tanaka, T., Yacicamoto, S. & Oi, S. (1981) *J. Biochem* (*Tokyo*) **90**: 521 - 526.

Tanriseven, A. & Robyt, J. F. (1993) *Carbohydr. Res.* **245**: 97-104.

Taubman, M. A., Holmberg, C. J. & Smith, D. J. (1995) *Infect. Immun.* **63**: 3088-3093.

Tieghem, P. V. (1878) *Ann Sci. Nat. Bot. Biol. Veg.* **7**:180-203.

Tsumori, H., Minami, T. & Kuramitsu, H. K. (1997) *J. Bacteriol.* **179**: 3391-3396.

Ueda, S., Shiroza., T. & Kuramitsu, H. K. (1988) **69**: 101-109.

Valette, P., Pelenc, V., Djouzy, Z., Andrieux, C., Paul, F., Monsan, P. & Szylit, O. (1993) *J. Sci. Food Agric.* **62**: 121.

Vickerman, M. M., Sulavik, M. C., Minick, P. E. & Clewell, D. B. (1996) *Infect. Immun.* **64**: 5117-5128.

Vickerman, M. M., Sulavik, M. C., Nowak, J. D., Gardner, N. M., Jones, C. W. & Clewell, D. B. (1997) *DNA Seq.* **7**: 83-95. von Eichel-Streiber, C., Saueborn, M. & Kuramitsu, H. K. (1992) *J. Bacteriol.* **174**: 6707-6710.

Wilke-Douglas. M., Perchorowicz, J. T., Houck, C. M. & Thomas, B. R. (1989) *PCT Patent WO* 89/12386

Wong, C., Hefta, S.A., Paxton, R.J. & Shively, J.E. *et al.* (1990) *Infect. Immun.* **58**: 2165 - 2170.

Wren, B. W. (1991) *Mol. Microbiol.* **5**: 797-803.

Yokota, N., Kuroda, T., Matsuyama, S. & Tokuda, H. (1999) *J. Biol. Chem.* **274**: 309-355.

Yoshinaga, K., Yoshioka, H., Kurosaki, H., Hirasawa, M., Uritani, M. & Hasegawa, K. (1997) *Biosci. Biotech. Biochem.* **61**: 160-161.

6

Biotechnological Relevance of Ergot Alkaloids with Future Perspectives

P. Mehta and A. Mehta

Microbial Technology Laboratory, Department of Botany

Dr HS Gour University, Sagar-470 003 (India)

In nature the ergot alkaloids are produced mainly by *Claviceps* species. Ergot alkaloids are indole compounds having a tetracyclic ring called ergolin which are formed from tryptophan, mevalonic acid and methionine. Clavines and simple lysergic acid derivatives are easier to manufacture than ergopeptines. Mathematical modeling extended by physiological studies can increase the production of alkaloids in bioreactors. The ergot alkaloids and its semi-synthetic derivatives are widely used as potential therapeutic agents in many diseases. The series of ergot alkaloids can be widened by their semi-synthetic analogues for use in medicine by exploiting *Claviceps* biotechnologically as the production of ergot alkaloids thrives on new knowledge generated by molecular biology, genetics, microbiology and biochemical engineering.

INTRODUCTION

Ergot alkaloids belong to the group of compounds produced by fungi, which are referred to as secondary metabolites. Ergot alkaloids (ergoline-derived toxic metabolites) are a group of pharmacologically active compounds produced by a number of fungi mainly of *Claviceps* sp. but they have been also found in other fungi and higher plants. Ergot (sclerotium of the pyrenomycete *Claviceps purpurea*) develops on plants of rye, cereals and in florets of grasses. In early days, the medieval midwives used to collect the fungus from naturally infected plants and used it in the induction of childbirth and in the control of postpartum bleeding. Many additional uses of the ergot alkaloids have been discovered since then.

The role of ergot (*Secale cornutum*) has undergone important changes from a dreaded toxic parasite to an important source of biologically effective substance. *Claviceps purpurea*, apart from yeast, is the first fungus which was biotechnologically exploited without its existence being known. The beginning of modern ergot research dates back to the extraction of the first crystalline alkaloid (ergotinine) from sclerotia in 1875, to the major pharmacological discoveries of ergotoxine in 1906 and the discovery of the first clinically used compound, ergotamine in 1918. In the beginning of the fifties the biosynthesis of ergot alkaloids received much attention from a subject of fruitful speculation to the region of experimental research. Toxic effects of ergot compounds have been recognized since the discovery that consumption of cereals infected by parasitic fungus, *Claviceps* could cause severe illness (ergotism). The disease is due to the effects of alkaloids produced by the fungus and contained in the ergot body. It is of particular interest that even before the ergot was identified as a cause of ergotism, it was evidently used in obstetrics, to enhance muscle contraction and was also used for abortions. However, because during these early days of medicinal use, neither its physiological action nor the appropriate dosage was understood, many deaths were brought about by the ergot extracts. At the end of 19th century ergot was banned as a drug in Germany. Later, with increasing understanding of pharmacology, it was found that chemical alterations of various alkaloids improved their medical efficiency. Further, after a correlation between clinical indications and therapeutical dosage was developed, pure ergot alkaloids and their semi-synthetic derivatives are being used as drugs at low concentrations extensively in medicine at present.

Since the supply of naturally grown ergots is not sufficient, the biotechnological production of ergot alkaloids has gained importance. The microbial technology of ergot alkaloids has two approaches: (i) the acquisition of knowledge for producing microorganism, (ii) and the development of large-scale processes to exploit such knowledge. Initially, chemical properties of the alkaloids were studied while biologists studied the physiological aspects of the synthesis of alkaloids from a practical point of view *i.e.* for both agronomists who were trying to obtain grain with lowest possible alkaloid content and for pharmacologists, who, on the other hand were interested in high alkaloid yields. The dependence of pharmacological effectiveness on the alkaloid chemical structure, enabling preparation of substances for specific purposes, was an important stimulus for the development of ergot alkaloid research. The chemistry, biosynthesis, physiology, biochemistry, genetics, biotechnology, and therapeutic applications of ergot alkaloids have been extensively studied (Berde and Sturmer, 1978; Mukherjee and Menge, 2000; Rehacek and Mehta, 1991, 1993Tudzynski *et al.*, 2001). The present review gives an overview of biotechnological relevance of ergot alkaloids with a perspective for the future.

Sources

In nature, primarily the fungus *Claviceps* forms the ergot alkaloids. More than fifty different alkaloids have been isolated from this ascomycetous genus, which has about fifty species. All *Claviceps* species form ergot sclerotia on various plants. Atleast another hundred plants can be attacked by ergot to form the sphacelial but not the sclerotial stage of the fungus (Rehacek and Sajdl, 1990). Ergot alkaloids have also been found from organisms other than *Claviceps* (Flieger *et al.*, 1997; Kozlovsky, 1999).

Fungi imperfectii: *Aspergillus clavatus, A.flavus, A.japonicus, A. tamarii, A. versicolor, A. oryzae, A. fumigatus, A.nidulans, Penicillium chermesinum, P. concavorugulosum, P. fumigatus, P.viridicatum, P. sizovae, P. rugulosum, P. rubrum, P. kapuscinskii, P.palitans, P.aurantiovirens, P.camembertii, P.clavigerum,P. crustosum, P. griseofulvin, P. roquefortii*

Hypocreales: *Balansia claviceps, B. epichloe, B. obtecta, B. strangulans, Epichloe typhinna, Neotyphodium coenophialum, N. lolii, Hyphomyces aurantius, Sepedonium sp.*

Zygomycetes: *Rhizopus arrhizus, R. nigricans, Cunninghmella blakesleana, Mucor hiemalis.*

Basidiomycetes: *Corticum caeruleum, Lenzites trabea, Pellicularia filamentosa*

Connvolvulaceae: *Argyreia nervosa, Ipomoea argyrophylla, I. Hildebrandtii, I.ruberosa, I.violacea,*

I. rubrocaerulea, I. piurensis, Rivea corymbosa, Stictocardia tiliifolia

Chemistry

The ergot alkaloids constitute the largest known group of nitrogenous fungal metabolites. The common part of chemical structure of ergot alkaloids is the tetracyclic ergoling ring system (Fig. 1). Ergoline, a common part of ergot alkaloids is formed from tryptophan (Taber and Vining, 1959), mevalonic acid (Groger *et al.*, 1961) and methionine (Baxter *et al.*, 1961).

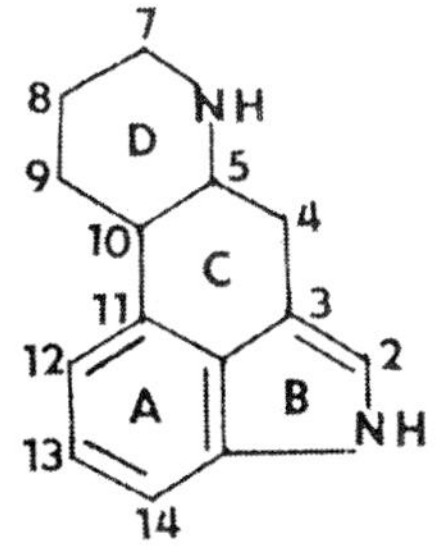

Fig. 1: Chemical structure of ergoline.

Classification

The ergot alkaloids can be separated structurally into four main groups: (i) clavine alkaloids, (ii) lysergic acids, (iii) simple lysergic acid amides and (iv) peptide alkaloids (Fig. 2).

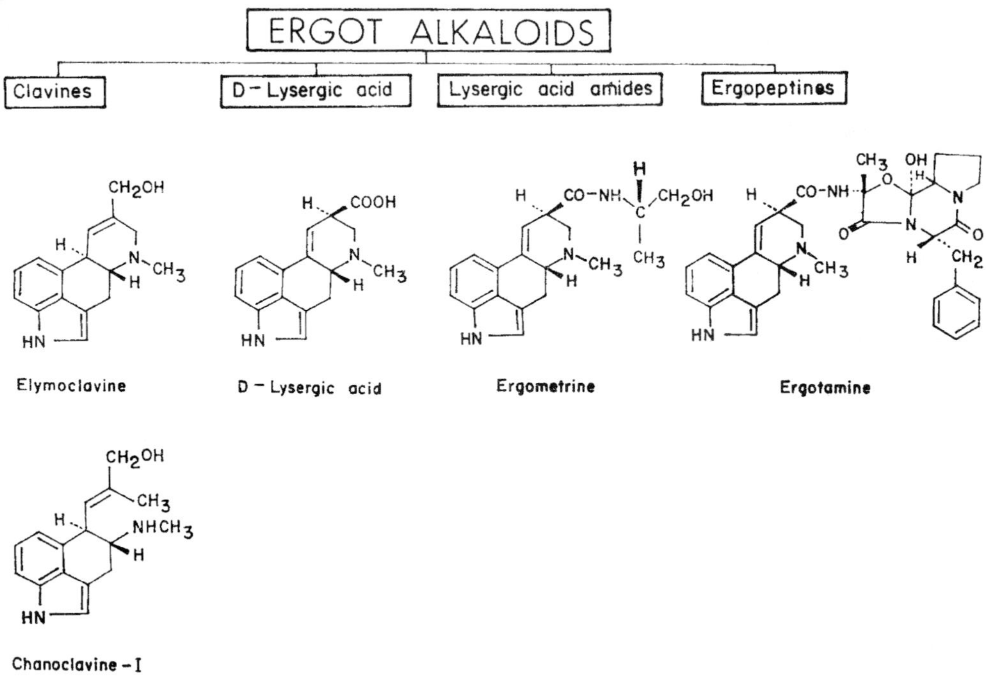

Fig. 2: Characteristic representative of the main groups of ergot alkaloids.

Clavines: The clavines are (Fig. 3) substituted 6, 8-dimethylergolines and include a few members, namely the chanoclavines, with a 6, 7-seco D ring. The degree of oxidation is a criterion for further differentiation in the group of clavine alkaloids. Some examples of clavine alkaloids are as follows:

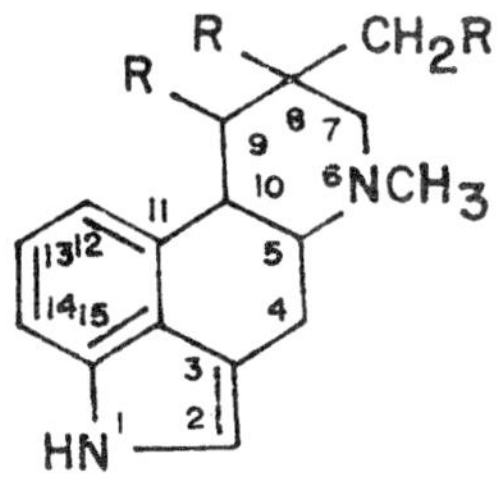

Fig. 3: Clavine.

Chanoclavine- I (secaclavine), Chanoclavine-II, Isochanoclavine-I, Norchanoclavine-I, Norchanoclavine-II, Chanoclavine-I acid, Paliclavine, Paspaclavine, Dihydrochanoclavine-I, Isodihydrochanoclavine-I, Rugolovasine A, Rugolovasine B, Clavicipitic acid, Agroclavine, Elymoclavine, Molliclavine, Elymoclavine-O-β-D-fructofuranoside, Lysergine, Lysergol, Isolysergol, Lysergene, Setoclavine, Iisosetoclavine, Norsetoclavine, Penniclavine, Isopennniclavine, Festuclavine, Costaclavine, Pyroclavine,Dihydrolysergol-I, Dihydrosetoclavine, Fumigaclavine A, Fumigaclavine B, Roquefortine A (Isofumigaclavine A), Roquefortine B (Isofumigaclavine B), Cytoclavine, Aurantioclavine

Lysergic acid and its simple derivatives: The derivatives of lysergic acid are mostly amides, in which the amide part is a small peptide or a simple alkylamide. Non-peptide amides of lysergic acid found in ergot are ergometrine, lysergic acid 2- hydroxyethylamide, lysergic acid amide (ergine) and paspalic acid (Fig. 4). Ergometrine (ergonovine and ergobasine) was also isolated from many laboratories. *Claviceps paspali* could also produce paspalic acid. Some examples of lysergic acid derivatives are: Ergometrine (ergonovine, ergobasine), Lysergic acid 2- hydroxyethylamide, Lysergic acid amide (ergine), Lysergic acid etc.

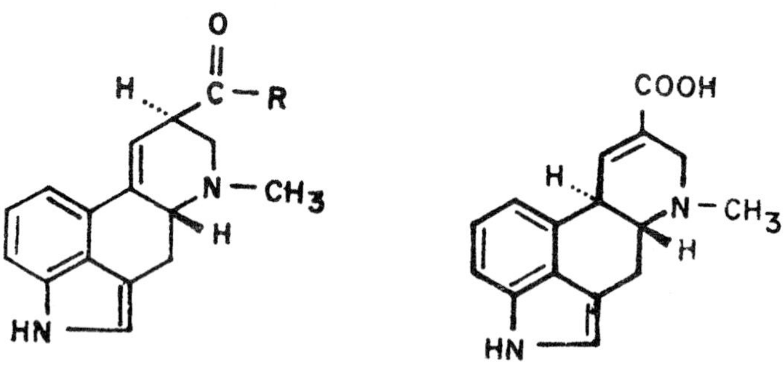

Fig. 4: Simple derivatives of lysergic acid (A) and paspalic acid (B).

Name	R
Ergometrine (ergonovine, ergobasine)	CH_3 | –HN–CH | CH_2OH
Lysergic acid 2-hydroxyethylamide	CH_3 | –HN–CH | CH_2OH
Lysergic acid amide (ergine)	$-NH_2$
Lysergic acid	– OH

Ergopeptines: Ergotamine was the first alkaloid to be isolated from ergot crop and became commonly used compound for therapeutics. Ergotamine and other classical peptide ergot alkaloids (ergopeptines) are composed of lysergic acid and an L- proline containing complex tripeptide moiety. Ergopeptines are the only natural substances shown to have cyclol structure. Some examples are shown below:

Table1. Some new natural ergot alkaloids (Rehacek and Sajdl, 1990).

Alkaloid	Source
Chanoclavine- II	*Claviceps* sp.
Isochanoclavine- I	*Claviceps* sp.
Aurantioclavine- I	*Penicillium aurantiovirens*
Cycloclavine	*Ipomoea hildebrandtii* *Aspergillus japanicum*
Cividiclavine	*Penicillium citreoviride*
Dyhidrosetoclavine	*Claviceps paspali*
6- Norsetoclavine	*Claviceps* sp.
Elymoclavine- O- β-D fructofuranoside	*Claviceps* sp. SD 58
Epoxyagroclavine	*Penicillium corylophium*
Paliclavine	*C. paspali*
Paspaclavine	*C. paspali*
Clavicipitic acid	*C. fusiformis*
Ergoheptine	*Claviceps* sp.
Ergohexine	*Claviceps* sp.
Paspalic acid	*C. paspali*
Paspalicine	*C. paspali*
Paspaline	*C. paspali*
Ergocornom	*C. purpurea*
Ergobutine	*C. purpurea*
Ergobutyrine	*C. purpurea*
8- Hydroxyergotamine	*C.purpurea*
Ergosecaline	*C.purpurea*
Ergosine	*Ipomoea argyrophylla*
Dihydroergosine	*Sphacelia sorghi*
Ergostine	*C.purpurea*
Roquefortine A, B	*Penicillium roqueforti*
Rugulovasine A, B	*P. Concavorugulosum*

Ergotamine, Ergosine, Ergovaline, Ergostine, Ergoptine, Ergonine, Ergobutine, Ergocrystine, Ergocryptine, Ergokryptine, Ergocornine, Ergobutyrine etc.

New Alkaloids: Relatively few new natural alkaloids have been described recently (Table 1), many of these have only biogenetic importance.

Biogenesis

Ergot alkaloid is derived from tryptophan, mevalonic acid and methionine (Birch *et al.*,1960; Groger *et al.*, 1960; Taylor and Ramsted, 1960). The origin of ergoline part from tryptophan and dimethylallyapyrophosphate was established at an enzymic level (Fig. 5). The cell-free bioynthesis of chanoclavines-I and II, agroclavine and elymoclavine from precursors has been reported by several workers (Cavender and Anderson, 1970; Sajdl and Rihacek, 1975). The cell-free synthesis of ergot peptides from their constituent amino acids and D-lysergic acid has not yet been accomplished. To obtain such a suitable cell-free system for this purpose protoplasts might be used for actively synthesizing ergopeptines. Tryptophan has also been found to be a factor in the induction and depression of enzymes catalyzing alkaloid formation (Krupinski *et al.*, 1976). The first enzyme of alkaloid biosynthesis is dimethylallytryptophan synthetase (DMATS) (Heinstein *et al.*, 1971) and the coding gene was cloned from *C. fusiformis* and *C. purpurea* (Tai *et al.*, 1995; Tudzynski *et al.*, 1999). Recent data (Tudzynski *et al.*, 2001) show that in *C. purpurea* all genes involved in alkaloid biosynthesis are organized in a cluster and are regulated by phosphate and pH level. This finding verified the previously published data on the negative role of phosphate in the induction of enzymes catalyzing the alkaloid synthesis (Krupinski *et al.*, 1976) and on inhibition of alkaloid synthesis by high concentration of phosphate in the culture medium (Menge *et al.*, 2001; Pazoutova *et al.*, 1981). Gene for peptide synthetase homologous to that of *C. purpurea* was also detected in *Epichloe* and *Neotyphodium* species (Annis *et al.*, 1998).

Ergot alkaloid synthesis requires a change in differentiation, which is considered to be a result of different expression of the genome under various conditions and at different times. Morphological differentiation is preceded by synthesis of membrane, i.e. protein and lipids. Normally, the metabolism of the early sequences of fungal differentiation is characterized by an increase in the relative importance of the pentose phosphate route and that of the glyoxylate cycle, NADP dependent functions predominate and there is a decrease in oxygen consumption. The vegetative fungal cells initially form a polarisation gradient in the cytoplasm where respiration plays an important role in maintaining polarised growth. In *Claviceps purpurea* the formation of conidia is inversely dependent on the synthesis of alkaloids for those saprophytic strains which partially retain parasitic development, i.e. differentiation of the sphacelial phase to the conidial or the sclerotial phase or both (Mehta, 1984). In contrast, in those strains for which the conidial phase, the phase of vegetative growth and the alkaloid phase are not clearly separated, the conidia formation occurs simultaneously with alkaloid synthesis. Physiological conditions necessary for conidiation and germination of conidia are different from conditions controlling alkaloid synthesis (Rehacek *et al.*, 1974). The morphological development of the mycelium in submerged *Claviceps* cultures show significant differences between high-yielding and degenerated cultures (Fleiger *et al.*, 1982). The character of fat hyphae and sclerotia like cells of submerged mycelium is reminiscent of the plectenchymatic structure of production sclerotia of parasitic cultures. Such sclerotia like cells influence the alkaloid production in *C. fusiformis* (Dickerson *et al.*, 1970).

Cultural conditions suitable for metabolism of primary substances are not suitable for alkaloid metabolism (Taber and Vining, 1963). To obtain over production of alkaloids, high citrate or Krebs cycle intermediate level in the medium is required (Pazoutova *et al.*, 1981). The inability of submerged *Claviceps* cultures to grow on hexoses in absence of Krebs cycle intermediates comes from the parasitic way of life where the level of citrate and malate in the host plant phloem sap is high. The rate of oxidative metabolism of saccharides and the activity of alkaloid synthesis were proved to be proportional (Pazoutova *et al.*, 1981). Another important fact is that clavine alkaloids are extracellular products with high solubility in the culture medium and feedbacks their own biosynthesis (Fleiger *et al.*, 1988). Ergopeptines are mostly intracellular and are accumulated in lipid droplets (oleosomes) with no influence on the metaboloism (Neumann *et al.*, 1979).

Clavine alkaloids lie at the beginning of the biogenetic order and classic ergot peptide alkaloids at the end. The oxidation order agroclavine - elymoclavine- lysergic acid is the main biogenetic pathway (Fig. 5). A variety of clavine alkaloids is formed due to action of various enzymes which direct elymoclavine and agroclavine from the main biosynthetic route. Such type of shunt activity is particularly evident in strains that lack the complete pathway and are unable to synthesize subsituted lysergic acid (Vining, 1980).

The biosynthesis of ergot alkaloids is correlated with distinct genotypes (Esser and Tudzynski, 1979). Ergot alkaloids are produced under conditions of substrate limitation other than by carbon. Most of the enzymes and co-enzymes of basic metabolism and alkaloid biosynthesis are common. In submerged culture, alkaloid synthesis takes place under conditions of decreased proliferation and clearly defined degradation and resynthesis of nucleic acids and proteins. Many researchers (Garve *et al.*, 1998; Rehacek *et al.*, 1972; Tenberge *et al.*, 1999) have isolated an extracellular catalase, and β-1, 3 glucanase in *C. purpurea* during infection on rye. The rate of alkaloid synthesis and the activity of Krebs cycle were proved to be inversely proportional. The level of the Krebs cycle enzymes decreased in favour of glyoxylate enzymes. Citrate is an intermediate, which has an important role in metabolic regulation. Therefore, in parasitic strains the impairment of mentioned regulatory relationships has not been excluded during the *Claviceps* evolution as a selection disadvantage.

The region of formation of ergot alkaloids is not generally the same as where alkaloids are found in increased concentrations. In *Ipomoea* the alkaloids are formed in leaves while they are found accumulated in seeds which do not produce alkaloids. The clavine alkaloids synthesis takes place in endoplasmic reticulum which later forms vacuoles containing lipoproteins (Vorisek and Rehacek, 1978). Clavines and simple derivatives of lysergic acid are found primarily in the medium while peptide ergot alkaloids are mostly intracellular. Ergoline alkaloids are mainly stored in lipid droplets (oleosomes) where they are accumulated without any influence on the metabolism (Neumann *et al.*, 1979).

Ergot alkaloids are not waste physiologically inert products of cell metabolism (Neumann *et al.*, 1979). Their biosynthesis plays a role in the physiology of the production organism by maintaining certain mechanisms of the metabolism and cell divisions in an operative state, especially under cultivation of the limited growth of the organism. Ergometrine added to a submerged culture of *Claviceps paspali* increases the alkaloid production by 120%. Ergotamine

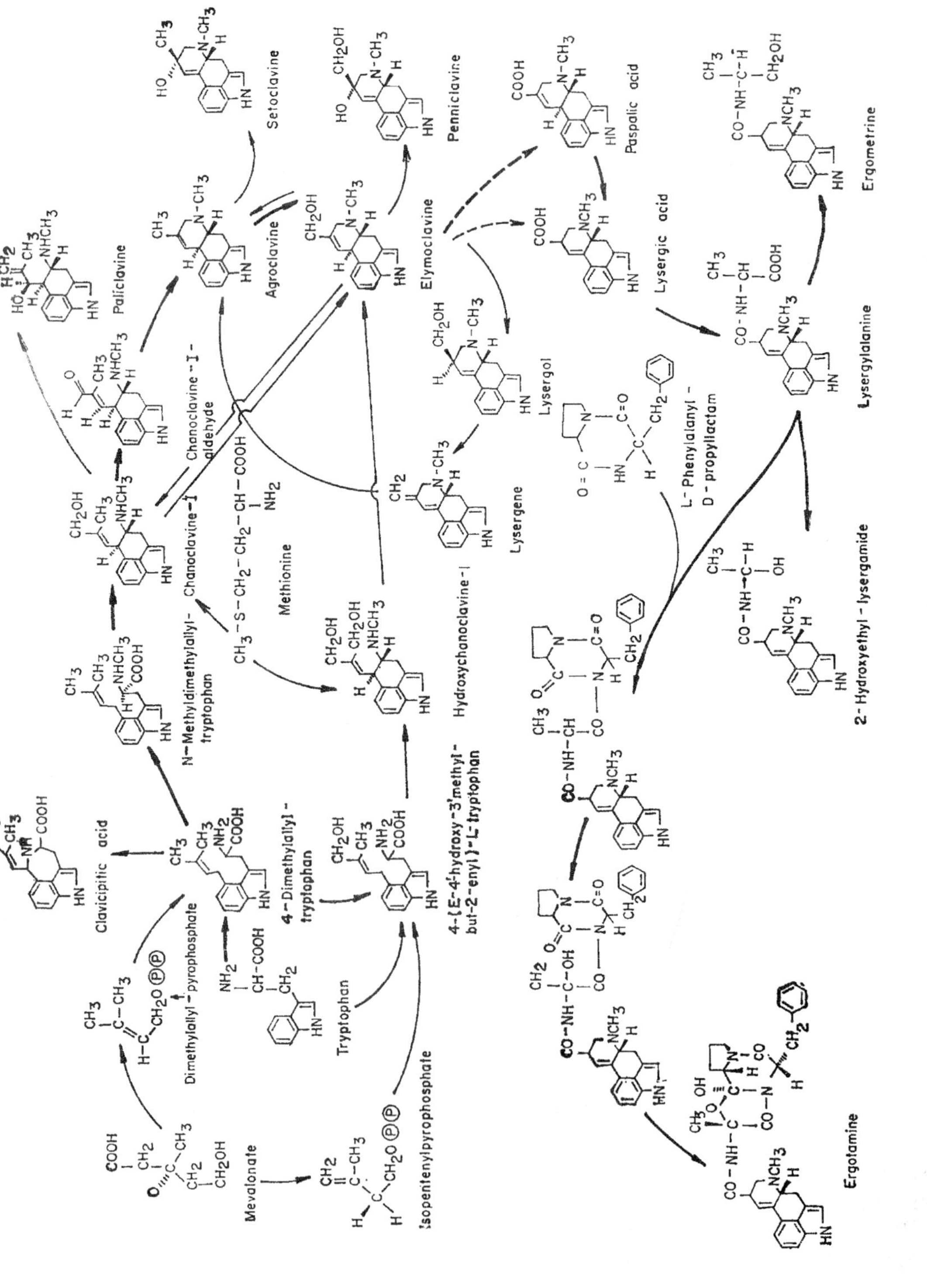

Fig. 5: Biogenesis of ergot alkaloids.

and ergometrine reduce the intensity of conidia formation in a productive culture of *Claviceps purpurea* (Rehacek *et al.*, 1974). The effect of tryptophan on the enzyme system in the ergot alkaloid synthesis (Robbers *et al.*, 1972; Taber and Vining, 1963) suggests participation of the alkaloids in the feedback inhibition and repression of the enzyme-forming system. The formation of clavine alkaloids is a part of endogenous metabolism releasing the energy required for maintenance for the integrity and viability of cells under conditions of limited growth (Rehacek, 1979). Clavine alkaloids probably may form a reservoir, which may lead to formation of peptide alkaloids.

SOME INDUSTRIAL PERSPECTIVES

Strain

Like other fermentation processes the key to successful production of ergot alkaloids is in obtaining the proper strain of the fungus. There are three main processes used for the preparation of *Claviceps* strains for saprophytic culture: (i) Plating of plectenchymatic tissue from the surface sterilized sclerotia on an agar growth medium (Mantle, 1969), (ii) Plating of honeydew drops containing conidia formed at early stage of *Claviceps* infection (Janardhana and Husain, 1984; Pazoutova *et al.*, 2002), (iii) Trapping of sexual ascospores ejected from fruiting bodies on germinated sclerotia. By this method monosporic culture can be obtained (Vasarhelyi *et al.*, 1980). In *Claviceps* sp. high alakloid production is accompanied with thick walled round having full cytoplasm with long, thin vacuolated cells (Mehta, 1984).

Saprophytic Cultivation of *Claviceps* sp.

The first saprophytic cultivation of *Claviceps* on artificial nutritional media dates back to 1922 (Bonns, 1922). The first attempt to industrial production of ergot alkaloids was isolation of clavine alkaloids from submerged cultures of different *Claviceps* spp. (Abe *et al.*, 1952; Abe and Yamatodani, 1954, 1955, 1956). Later, conditions for saprophytic production of simple derivatives of lysergic acid were developed using different strains of *C. paspali* (Arcamone *et al.*, 1960, 1961). Only five years later a new isolate of *C. purpurea* was found to produce ergotamine under submerged conditions (Amici *et al.*, 1966). Since that time all types of ergot alkaloids for direct use as therapeutic agents or precursors for the preparation of semi-synthetic drugs could be obtained by fermentation. Malinka (Malinka, 1999) described the successful submerged fermentations for the production of D-lysergic acid, the simple D-lysergic acid amides and D-lysergic acid peptides. Chemical synthesis of the natural D-lysergic related compounds such as ergometrine and ergopeptines as well as of new semisynthetic derivatives used in therapy (methylergometrine, methysergide or cabergolin) has led to D-lysergic acid fermentations being preferred to that of its derivatives. Since D-lysergic acid rarely occurs in free form it is obtained indirectly either via paspalic acid or via the simple D-lysergic acid amides such as D-lysergic acid α-hydroxyethylamide, ergine/erginine and ergometrine. In some cases, ergopeptines themselves have been reported as a source of D-lysergic acid. All of the D-lysergic acid amides can be hydrolyzed by alkaline hydrolysis to D-lysergic acid in good yields (Rucman, 1976).

Production in Parasitic Culture

Bekesy developed the first large technology for the parasitic production of ergot on rye in 1934 (Bonns, 1922). For a long time due to severe problems in early fermentation processes, the field cultivation was the major technology in Europe. Recently, the methods of cultivation, inoculation and harvesting have been improved (Nemeth, 1999). Presently, hybrid rye based on cytoplasmic-male-sterile (CMS) lines (Geiger and Bausbach, 1979) is used as host because the infection rate and the alkaloid yield is significantly higher than with normal rye cultivars (probably because the florets stay 'open' longer), also, a major advantage is that spraying conidial suspension instead of pricking can perform infection. The field production technology can reach 1,000 – 2,000 kg sclerotia/ha (Nemeth, 1999). Since the alkaloid content in sclerotia of commercial strains averages 1% and more, an alkaloid yield of 8 – 10 kg per hactare can be achieved, even if losses during the purification process are taken into account (Udwardy, 1980). The main technical problem in this technology is the generation and storage of high quality inocula, which is mostly achieved by using deep-frozen conidia suspensions. However, several disadvantages of this method have led to a significant reduction of parasitic alkaloid production: (*i*) production is too dependent on environmental parameters, (*ii*) the areas in which the climate is adequate for such production (and where it is allowed: protection of other cereals) are limited, (*iii*) protection of the valuable production strains is difficult; and (*iv*) the purification procedures are expensive. Therefore the present proportion of alkaloids produced in parasitic culture worldwide is low, although exact figures are not available (Tudzynski *et al.*, 2001).

Strain Improvement

The classical methods (selection pressure, mutagenesis and recombination) used for the strain improvement are, to some extent, more complicated with *Claviceps* due to incomplete information on cell nucleus. Strains used for saprophytic cultivation might be heterokaryotic and homokaryotic (Didek-Brumec *et al.*, 1991; Mantle and Nisbet, 1976). Recently, it has been reported that the number of chromosomes in *C. purpurea* is variable so that haploid as well as aneuploid nuclei may be encountered. Mutagenesis followed by subsequent selection of strain is an important technique in increasing the yield of alkaloids (Didek-Brumec *et al.*, 1987). An ergocristine producing *C. purpurea* strain showed 180-fold increase in alkaloid production after eight-step mutation-selection with different mutagens (Kobel and Sanglier, 1987). Mutagenesis of sporulating strain results in monosporic isolates. To increase the mutation frequency the protoplasts prepared from spores of selected strains were used (Olasz *et al.*, 1982; Zalai *et al.*, 1990). More complicated situation is with mutagenesis of asporogenic strains. Hyphal fragments are rather unsuitable for mutagenesis due to the higher number of nuclei. Even protoplast formation from young mycelium and subsequent regeneration without any mutagenic treatment yielded strains with different properties (Schumann *et al.*, 1987). Protoplast fusion technique is beginning to find useful applications either in producing improved mutant strains by intraspecific crosses or in formation of novel spectrum of products by interspecific hybrids (Socic and Gaberc-Porekar, 1992). Relevant structural and regulatory genes of the alkaloid biosynthesis in *C. purpurea* form cluster of about 50 kbp in length (Tudzynski *et al.*,

1999), therefore, isolation and cloning of the entire pathway for more rapidly growing fungus would be difficult.

Maintenance and Long-term Preservation

Cultures are the basic ingredient of microbiological work and at present there is much concern regarding the need to conserve genetic resources. Selection of a preservation method and the stability of desired properties of a microorganism are correlated with that microorganism's identity, properties, use, restoration and growth conditions. High yielding strains of *Claviceps* (Kobel, 1969) normally show degeneration and loss of production capabilities. Conservation and systematically performed selection of the isolates is the only way to eliminate the biological effects given by transfer of cultures, ageing, and other influences. Methods applied for long-term preservation have been recently reviewed (Hunter-Cervera and Belt, 1996). Two main methods are applied for preserving sporulating cultures, they are: (*i*) deep-freezing and (*ii*) maintaining of cultures on rye grains or agar slants placed in refrigerator. Non-sporulating strains due to higher sensitivity to conservation procedure are frequently preserved as cultures on the agar plates. The most universal technique applied for preservation is keeping of lyophilized cultures and cultures frozen in liquid nitrogen (Baumert *et al.*, 1979; Pertot *et al.*, 1990).

Genetic Engineering

Many field isolates of *C. purpurea* produce significant amounts of ergot alkaloids only in the parasitic phase and not in axenic culture. Thus the screening of different ergot strains from field surveys or various culture collections over the years has yielded interesting strains producing different specific alkaloids (or desired mixtures) at moderate rates in plants, but they were only suited (if at all) for parasitic production and not for submerse production of alkaloids. Therefore, ergot biotechnology strain improvement programmes were launched, aimed at an increased capacity of strains to produce high amounts of a specific alkaloid. This was mainly achieved, as in other strain improvement programmes of secondary-metabolite-producing fungi, by a stepwise mutagenesis selection process. However, due to the lack of antibiotic or their activity that can be tested in quick bioassay, the screening is more time consuming than normal and requires a great deal of analytical work in order to select among the numerous clones the desired ones. Preselection methods to narrow down the number of interesting isolates involve growing natural isolates or mutated clones under culture conditions that enforce sclerotia-like cell growth, such as by the addition of salt to the media (Upwardy, 1980) or by strictly using media imitating the natural physiology of parasitic culture (Strandova *et al.*, 1986). Molecular genetic techniques have recently been successfully applied to *Claviceps* sp. (Pazoutova and Parbery, 1999; Tudzynski *et al.*, 1999) and have considerably widened the options for genetic research. They have been used for: (*i*) detailed characterization of strains for evolutionay studies (*ii*) for a molecular analysis of the biosynthetic pathway and (*iii*) for detailed regulatory studies have been initiated. These molecular approaches will allow the design of recombinant strains with improved production properties (e.g. by enhancing expression of genes of interest or by creating defined detection mutants) or that produce new compounds, which are not accessible by chemical synthesis.

Ergot alkaloids are synthesized by the enzymatic transformation of small precursor molecules and hence they are not direct gene products. In biogenesis of *Claviceps* the relevant structural and regulatory genes have not been mapped, therefore, isolation and cloning of the entire pathway of structural genes for alkaloids would be difficult. Gene transfer via protoplast fusion has opened up new possibilities in both the cellular and molecular aspects of *Claviceps* genetic engineering. There are three potentially profitable avenues which may help to bridge the gap between academic and industrial genetics of *Clavicep,* they are: (*i*) the construction of a plasmid that replicates autonomously in *Claviceps* and carries easily selectable markers, (*ii*) the development of transformation systems with drug resistance markers for use in species lacking available auxotrophic mutants and (*iii*) the study of genes controlling the regulation of entire pathway. The recent indentification of a cluster of genes (Tudzynski *et al.*, 1999) involved in ergot alkaloid biosynthesis in *C. purpurea* and the availability of molecular genetic techniques allow the development of strategies for rational drug design of ergoline-related drugs by enzymic engineering and by biocombinatorial approaches (Tudzynski *et al.*, 2001). Cultural conditions affect the fermentation broth directly as well indirectly (Fig. 6). However,

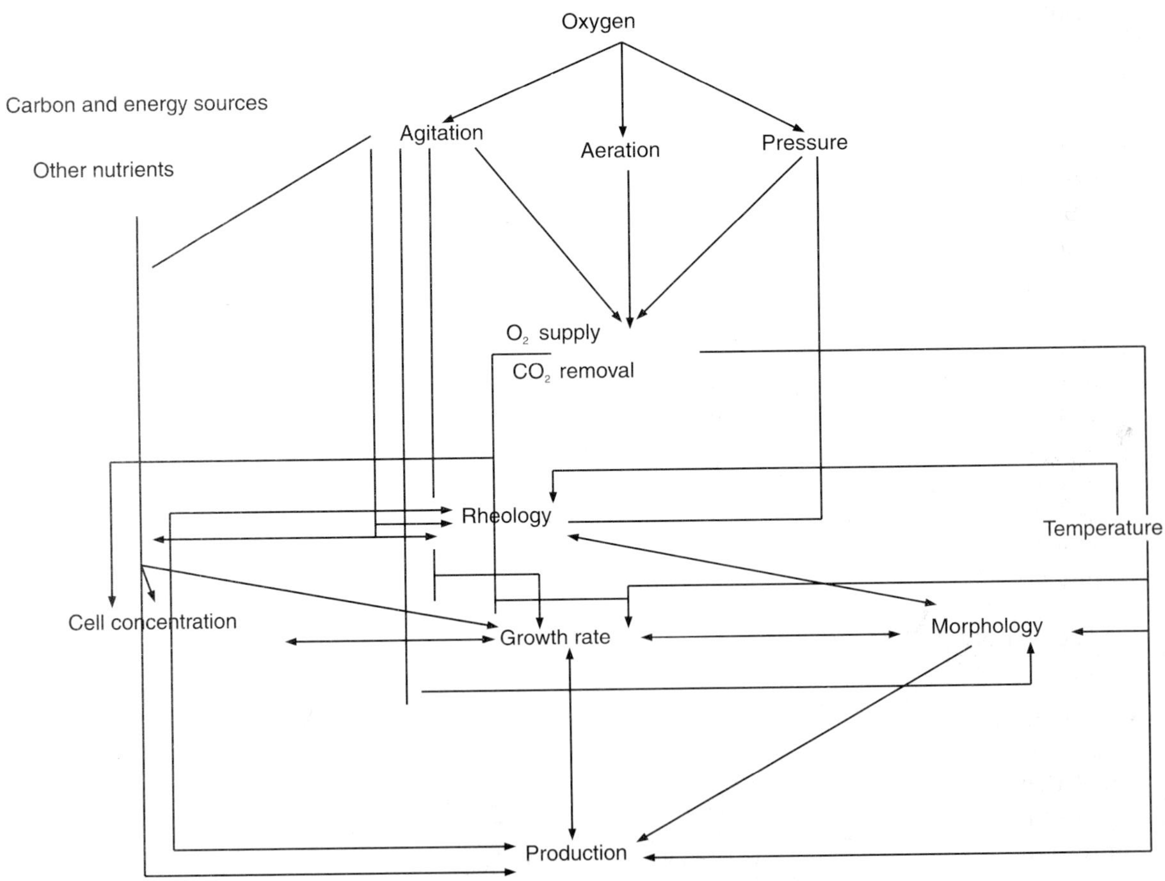

Fig. 6: Effect of cultivation on the culture growth rate and metabolite production (82).

the effects of rheological characteristics have not really been considered appropriately in any of the published transport correlation (Rehacek, 1983).

Fermentation

All technologies developed for industrial scale have the same aims i.e. maximum production of desired ergot alkaloids, minimum amount of accompanying contaminants (other alkaloids, toxins, etc.), the shortest possible time of fermentation, minimized cost of medium, energy, equipment, and labour (Fig. 6). For the production of ergot alkaloids, different fermentation technologies have been employed (*i*) stationary cultivation on liquid or solid medium (kybal and Vlcek, 1976; Trejo-Hernandez and Lonsane, 1993), (*ii*) submerged cultivations (Kobel and Sanglier, 1986) also adapted to semicontinuous or continuous processes (Kopp and Rehm, 1984). In some cases the immobilized microorganisms were used for the production of ergot alkaloids under condition of submerged fermentation (Komel *et al.*, 1985; Kopp and Rehm, 1983; Kren, 1991).

Stationary surface cultivation: In the beginning of sixties, the development of processes for the production of ergot alkaloids under conditions of stationary cultivation on liquid media started (Adams, 1962). The stationary surface cultivation on agar slants was commonly used for preparation of starting cultures. When transformed to industrial scale this technology showed some limitations mostly due to difficulties with control of aseptic conditions of large surfaces. Vlcek and Kybal (Vlcek and Kybal, 1974) developed technology for stationary cultivation of *C. purpurea* in plastic bags partially filled with inoculated liquid medium. This procedure was used for production of ergotoxins and later adapted to production of asexual spores of *C. purpurea* for field production.

Submerged cultivation: Many laboratories have got success in isolating strains of *Claviceps purpurea* and their cultivation under submerged conditions to produce ergot peptide alkaloids (ergopeptines), simple derivatives of lysergic acid and clavine alkaloids (Vorisek and Rehacek, 1976). Submerged fermentation in laboratory scale, i.e. shaker cultivation, is a primary step in getting knowledge of production microorganism physiology, biosynthesis, sporulation, stability, and influence of medium composition on production of alkaloids. In industrial scale, submerged fermentation in shaker culture is mostly used for preparation of sufficient amount of inoculum for further cultivation step. During the inoculum preparation an optimal state of the culture for biosynthesis of ergot alkaloids in the production step can be established (Socic *et al.*, 1985, 1986). Usually, the whole industrial process consists of four steps : (*i*) shaker culture, (*ii*) pre-inoculating fermentation, (*iii*) seed fermentation, and (*iv*) production fermentation followed by downstream processing (Malinka, 1999). Industrial production of clavine alkaloids, agroclavine and elymoclavine by submerged fermentation of different strains of *Claviceps* sp. was reported in a number of patents (Kren *et al.*, 1988; Rehaeek *et al.*, 1986; Trinn *et al.*, 1990), reaching the maximum production of alkaloidal mixture about 6 g/L (Pazoutova, 1987; Pazoutova and Tudzynski, 1999). In contrast to high production and well elaborated procedure of submerged cultivation the industrial production of clavine alkaloids did not receive much attention due to limited amount of procedures leading to preparation of desirable final products. Production of lysergic acid and its simple derivatives is of high technological and industrial interest due to relatively simple procedure of chemical modifications to semisynthetic ergot preparations (Table 2). The main producers of these alkaloids are strains of *C. paspali*. The basic studies on biosynthesis, physiology and production of simple derivatives of lysergic acid were done on relatively limited number of strains including *C. paspali* 31 (Rosazza *et al.*, 1967), *C. paspali* ATTC 13892 (95), and *C. paspali* MG-6 (Bumbova-Linhartova *et al.*, 1991). The concentration of produced alkaloids reached nearly 3 g/L of fermentation broth (Pertot *et al.*, 1990) with the strain *C. paspali* L-52. One of the very interesting features of *C. paspali* strains is their ability to convert clavine alkaloids added to the cultivation medium to the simple derivatives of lysergic acid (Mothes *et al.*, 1962). Flieger *et al.* (1989 a, b) and Harazim *et al.* (1989) used this capability of *C. paspali* CCM 8061 and developed technology of aggressive bioconversion of clavine alkaloids to simple derivatives of lysergic acid with total production of nearly 6 g/L in batch cultivation and about 3 g/L in industrial fermentor. Different strains of *C. purpurea,* as the only producers of ergopeptines, have been described for their submerged production. Ergotamine producing strains and their cultivation are the best-studied processes due to direct therapeutical use of ergotamine.

Table 2. Industrial submerged production of ergot alkaloids (Rehacek, 1983).

Alkaloid	Company			
	A	B	C	D
Ergocornine	s	s	s	s
Ergocrystine	s	s	-	s
Ergokryptine	s	s	s	s
Ergotamine	-	s	s	-
Ergometrine	-	-	s	-
Lysergic acid	-	s	-	-
Paspalic acid	s	-	-	s

A Biochemie GmbH, Kundl/Tirol, Austria, C Gedeon Richter, Budapest, Hungary, D Sandoz-Wander AG, Basal, Switzerland

Bioreactor design: Many ergot laboratories have used different bioreactors from time to time. In scaling up a 'conventional' stirred tank fermentor must provide for at least and the lowest possible cost and should satisfy all requirements (good heat, mass transfer, perfect sterilization, aseptic procedures, reliable foam control, good bulk mixing, simple, rapid and thorough cleaning methods, reliable and appropriate monitoring control systems, proven construction methods, appropriate materials of construction and multiple product application).

Solid substrate fermentation: Solid substrate fermentation describes the microbiological transformation of biological materials in their natural state, in contrast with liquid of submerged fermentation, which is carried out in dilute solutions or slurries. Recently Robinson *et al.* (2001) have proposed a solid-state (substrate) fermentation (SSF) for the production of enzymes and secondary metabolites to meet the global demand. Production of ergot alkaloids by SSF was found 3.9 times greater than that produced by *C. fusiformis* in SmF (Hernandez *et al.*, 1993). Products of secondary metabolites by submerged liquid fermentations (SmF) have been found to be less efficient than SSF. The use of antifoam chemicals and the sheer stresses caused by stirring prevented high yields in SmF. Better air circulation can also be achieved in SSF increasing ergot alkaloid yields further (Balakrishnan and Pandey, 1996). The solid substrate fermentation has been shown to produce a more stable product requiring less energy in smaller fermenters with easier downstream processing measures. Also, by removing the cost and trouble associated with anti-foaming chemicals and by maximizing yield production, SSF may be seen as a viable option for industrial scale production of ergot alkaloids.

Optimization of ergot alkaloid fermentation: The following facts should be taken into account for optimization and scaling up of fermentation processes for the production of ergot alkaloids (Kobel and Sanglier, 1978): (*i*) the production strains should be continually tested to maintain the optimal quality of selected production strain and thus maximally eliminate the biological effects given by transfer of cultures, ageing and other external influences, (*ii*) long cultivation period in absence of antibiotics require very high standard of sterility in operation and equipment, (*iii*) balanced aeration and stirring are required due to the sensitivity of the *Claviceps* cultures to the stress and high oxygen tension, (*iv*) the use of antifoam agents could cause considerable loss in alkaloid yield. Recently the application of oxygen vectors to *Claviceps purpurea* cultivation has been published (Menge *et al.*, 2001). The classical problem of the large-scale fermentations is an optimal supply of oxygen to growing *Claviceps* sp. High oxygen demand in the exponential growth phase can be met by addition of different hydrocarbons (Gilmanov *et al.*, 1996) or perfluorocarbons (Menge *et al.*, 2001) to shake flask and/or stirred reactors. Perfluorocarbons were also applied successfully in cultivations of other microorganisms (Lowe *et al.*, 1998). Besides these technological aspects other techniques to optimize the fermentation process have also been described.

Scale-up

There are three basic ways to get ergot alkaloids, and they are: (*i*) by isolation from the sclerotia of *Claviceps purpurea* collected from the field (95% of the world production), (*ii*) by extraction of saprophytic cultures of other ascomycetes especially of the *Claviceps* genus and (*iii*) by partial/ total synthesis. Sclerotia of *Claviceps purpurea* collected from rye may contain 0.8% peptide alkaloids on dry weight basis which is an equivalent of about 500 mg/L alkaloid level in a fermentation broth of 6% mycelial dry weight (Udwardy, 1980). Since the naturally grown ergot is not sufficient, also as its production is labour intensive and wheather dependent, the biotechnological production of ergot alkaloids has gained importance. In some cases, the saprophytic production of ergot alkaloids is more economical than parasitic cultivation on rye plants. In bioreactors the process is more precisely controlled, not dependent on specific season and the production and isolation of alkaloids, can take place on the same site and is possible to apply some manipulations, which are not feasible in field conditions and above all clavines and simple lysergic acid derivatives are easier to manufacture than ergopeptines. In bioreactors the alkaloid yield may exceed 20% of the mycelial dry mass while natural sclerpotia yield is normally less than 1% or even less than 0.1% in some cases. For optimization and scale-up process certain facts should be taken into account (Kobel and Sanglier, 1986), they are: (*i*) as bulk of the peptide alkaloids are mainly confined in mycelial cells, therefore large mycelial mass with high alkaloid content should be obtained, (*ii*) long cultivation period and absence of any antibiotic-form in *Claviceps* require very high standards of sterility in operation and equipment, (*iii*) looking to the sensitivity of the organism against stree and necessity for high oxygen tension require balanced aeration and stirring, (*iv*) alkaloids production varies with the composition of media, especially with antifoam agents which cause considerable loss in alkaloid yield, (*v*) keep the production strains under constant observation as *Claviceps* cultures soon lose their production activity.The saprophytic *Claviceps* strains are prerequisite for modern production of ergot alkaloids. Some saprophytic strains of *Claviceps* for industrial production of ergot alkaloids are worth mentioning:

Clavines : *C. fusiformis* B 35, *C. fusiformis* w1, *C. fusiformis* F 27,
C. purpurea Pepty 695/ch, *C. purpurea* 129, *C. purpurea* CP 7 and
C. paspali SD 58

Simple lysegic acid derivatives: *C. paspali* NRRL 3027, *C. paspali* Li 342/SE 60

Ergopeptines:

(*i*) ergotamine: *C. purpurea* ATCC 15383, *C. purpurea* FI 275, *C. purpurea* OKI 56/1970
(*ii*) ergosine: *C. purpurea* MUT 168
(*iii*) ergocristine: *C. purpurea* ATCC 20 103 (IMI 131669, FIS 40), *C. purpurea* Ech 420
(*iv*) ergocornine and ergosine: *C. purpurea* FI 43/14
(*v*) ergotamine and ergotryptine: *C. purpurea* FI 32/17, *C. purpurea* IC/39/20
(*vi*) ergocornine and ergokryptine: *C. purpurea* Exy 20, C. purpurea OKI 00022, *C.purpurea* DM 838 (IMET PA 13), *C. purpurea* Pepty 695
(*vii*) dihydroergosine: *Sphacelia sorghi*

Processing and Cost Engineering

The commercialization of ergot alkaloids and their production made possible through biotechnology requires a coordinated coupling of unit operations in order to develop efficient processes. There are many factors involved in scaling down the solids and liquid handling processes. After fermentation to septate cell and broth membrane filters are widely in use in recent time in addition to conventional processes like centrifugation and filtration. New developments in membranes as well as improved process configuration continue to make this an attractive tool in biochemical process synthesis. Electrodialysis is an important operation for purification and concentration of low molecular weight salts. Improvements in process configuration and membrane design make this technique attractive in a variety of applications.

Modern Technology

Modern research into ergot alkaloids can draw on a very considerable number and variety of truly new means and methodologies, which are in continuous rapid evolution. A full range of its features involves (*i*) computerization of documentation (*ii*) automation of various experimental activities (*iii*) mathematical modeling (*iv*) calculation of atomic, molecular and crystal structures and their relevant physical, chemical, biological and therapeutical properties (*v*) statistical design of experimentation and the interpretation thereof with translation into mathematical models.

Mathematical Modeling

The growth of the fungus and alkaloid formation in submerged batch fermentation can be described by mathematical model, which can be used in automatic process control and optimization of an industrial process. Grm *et al.* (1980) while working with *C. purpurea* proposed a growth model based on morphological features of a cell population during the fermentation process. Pazoutova *et al.*, (1981), however, proposed another model accentuating the antagonistic effects of phosphate on growth and alkaloid production. Another possibility for process optimization lies in feeding the culture, and thus, prolonging the idiophase of the fermentation process (Milicic *et al.*, 1987). This process which seems to be one of the perspectives in modern alkaloid production (Socic and Gabere- Porekar, 1992). The activation-inhibition kinetics of clavine alkaloids production was evaluated for two *Claviceps fusiformis* strains (Flieger *et al.*, 1988) and it was found that feedback inhibition could be eliminated by combination of fermentation and separation units in a closed loop. Increased efficiency (more than 100 %) of the fermentation process was also found when inducers of cytochrome P-450 were used (Rylko *et al.*, 1988).

THERAPEUTIC APPLICATION

The ergot alkaloids have many potential therapeutic uses, some of which have been realized and others have not survived the critical test of time. Ergot alkaloids are of prime interest to pharmacologists because of their multiple actions on utrine and vascular muscle, inhibitors of prolactin secretion, as a tool for studying the mechanism of sympathetic nervous system to produce stimulation of dopaminergic receptors and in the treatmnt of parkinsonism. The ergot and its semi-synthetic derivatives have been widely used in many diseases: acromegaly, amenorrhea-galactorrhea, suppression of postpartum lactation, treatment of breast cancer and possibly cancer of prostrate, angina pectoris, enterics (Demain, 1999) and in infectious diseases like glaucoma and herpes zoster. Within the last few years new aspects of pharmacology of ergot derivatives have been recognized (Berde and Sturmer, 1978). A new group of alkaloids has been reported which interferes with DNA synthesis in mammalian cells. Ergot alkaloids can be chemically manipulated to form derivatives that possess altered pharmacologic effects. Preparations containing ergot alkaloids or their semi-synthetic derivatives are also applied for the treatment of migraine (ergotamine), high blood pressure (dihydrocornine, dihydrokryptine) and obstetrics (ergometrine). Ingestion of isoergine and lysergic acid diethylamide (LSD) are known to cause marked hallucinations (De Groot *et al.*, 1998). Ergot alkaloids are not simple pharmacological or therapeutical entity as they also have wide range of biological effects (Berde, 1984): (*i*) on different receptor sites (*ii*) exert both agonistic and antagonistic actions on the same receptor and in many cases (*iii*) one and the same compound acts on all the three types of receptors but in different ways, i.e. as agonist and / or antagonist. The clavine alkaloids cause apathy, drowsiness and sedation in many mammalian systems and have antimicrobial activity against a series of human pathogenic bacteria and *Candida albicans* (Eich, *et al.*, 1984), hypotensive and oxytoxic properties and are able to inhibit prolactin secretion Ref. Several other ergot derivatives influence the secretion of several hormones and act at the central nervous system. Ergot has three main types of actions: (*i*) central (*ii*) peripheral and (*iii*) neurohormonal. LSD is pharmacologically inactive but its various derivatives display pharmacological actions which include: (*i*) sympatholytic effect-inhibition of the hormones adrenalin, noradrenalin and serotonin and thereby a dilation of veins and decrease in blood pressure and (*ii*) effect on uterus via contraction of smooth muscle. The ergot and its derivatives have been further used for various purposes. Psychogeriatry is directed at treating and improving or reversing particular symptoms which are the consequence of the ageing process. Nicergolin has been found to be of therapeutic value in senile dementia. Ergot alkaloids are mainly employed in peripheral vascular disorder, hypertension, and cervical syndromes and in the treatment of migraine. They have the ability to improve tissue metabolism by normalizing and stabilizing the phosphatases. Ergometrine and its semi-synthetic derivatives like methylergometrine have been widely used in postpartum to prevent hemorrhage. Serotin antagonist lisuride (lysenyl), a derivative of isolysergic acid, has been used against hypertension and migraine. Since early time ergot alkaloids have shown their effects on the reproduction processes of animals, therefore ergot derivatives (ergotoxine and ergocornine) have been tried as antifertility agents in rat and can also terminate early pregnancy in both, rat and mice (Sommer and Buchaman, 1955). Lisuride could show effects on neurohormonal processes which are controlled by the hypothalamic pituitary system particularly in the release of prolactin from the pituitary gland (Cassydy and Floss, 1977). Natural ergot alkaloids and its semi-synthetic derivatives bring inhibition of pathological changes in the microcirculation. Terguride and lisuride (lysenyl compounds act on dopamine (DA) receptors in the central nervous system and also affect serotonin and noradrenaline-mediated neural transmission. Lisuride was also introduced into therapy as an antimigraine drug. Ergot alkaloids play an important role in the pharmacotherapeutic spectrum of drugs in neurology. Bromokryptine with dopaminergic effects on DA receptors are used in the therapy of Parkinson's disease and for curing acromegaly. Pergolide, a synthetic ergoline, has been found to have a positive effect on parkinsonism (Liberman *et al.*, 1979). Lisuride, another promising antiparkinsonian drug, has a high dopaminergic activity. With the increasing understanding of pharmacology it was found that chemical alterations of various alkaloids improved their medical efficiency.

The ergot alkaloids can be chemically manipulated to form derivatives that possess altered pharmacologic effects. Moreover, after obtaining a correlation between clinical indications and therapeutic dosage, pure ergot alkaloids and their semi-synthetic derivatives can be used in medicine extensively in future.

CONCLUSION

In nature, the ergot alkaloids are formed primarily by *Claviceps* sp. Ergot alkaloids are indole compounds having a tetracyclic ring called ergoline which are formed from tryptophan, mevalonic acid and methionine. Saprophytic production of ergot alkaloids is more economical than parasitic cultivation on rye plant, as in bioreactors the process is more precisely controlled, the cultivation is not limited to the specific season and the production and isolation of alkaloids can take place on the same site, also possible to apply some manipulations which are not practicable under field conditions. Clavines and simple lysergic acid derivatives are easier to manufacture than ergopeptines. Production of ergot alkaloids is due to action of multiple enzyme complexes. Biosynthesis of ergot alkaloids lies at the level of genetic information apparatus and continues at the level of physiological realization. The ergot alkaloids and its semi-synthetic derivatives in addition to antimicrobial and biological activity have been widely used as potential therapeutic agents in obstetrics and also to control various diseases (parkinsonism, acromegaly, cancer, blood pressure and migraine etc.). Ergot alkaloids can be chemically manipulated to form derivatives that possess altered pharmacologic effects. Ergot alkaloids are not waste, physiologically inert, products of cell metabolism. Genes involved in alkaloid biosynthesis are under rigorous metabolic control. It is possible to understand the genetic control of alkaloid production and to find the genes responsible for specific alkaloids. Mathematical modeling extended by physiological studies can increase the production of alkaloids in bioreactors. The series of ergot alkaloids can be widened by their semi-synthetic anologues for use in medicine by exploiting *Claviceps* biotechnologically as the production of ergot alkaloids thrives on new knowledge generated by molecular biology, genetics, microbiology and biochemical engineering.

Future Perspectives

Future research on biotechnological aspects of ergot fungi will most probably focus on the analysis of the biosynthetic pathway at both, the enzymatic and genetic level. The recent cloning of the ergot peptide alkaloid gene cluster will allow a functional analysis of all genes involved in ergoline ring synthesis and identification of the corresponding steps by enzymatic analysis of the gene products. The characterization of the various enzymes in terms of substrate specificity and structure will facilitate the development of strategies for rational drug design of ergoline-related drugs by enzyme engineering and by biocombinatorial approaches e.g. by introducing unnatural precursors into the ergoline ring system which as yet have failed to become incorporated into the ergoline ring backbone. Improvement of the efficiency of the pathway could also be achieved by the targeted alteration of some of the enzymatic steps in the processing of DMAT to D-lysergic acid. Alternatively, construction of recombinant peptide synthetases containing the D-lysergic acid-activating module should be a means to produce recombinant ergot drugs. Preferable, in these compounds 2D-lysergic acid should be combined with substituents that were difficult to obtain by chemical synthesis, thus leading to new or improved ergoline-derived compounds.

For biotechnological production of ergot alkaloids and its semi-synthetic derivatives the following few points need more attention: (*i*) shortening of the initial non-productive phase of alkaloid fermentation (*ii*) immobilization of cells (*iii*) construction of a plasmid which can carry selectable markers (*iv*) the development of transformation systems with drug resistance markers (*v*) study of membrane processes and vacuoles in the productive organism (*vi*) use of molecular genetic techniques (*vii*) genetic control of alkaloid production and to find the genes responsible for specific alkaloids (*viii*) concerted breeding of the strains concerned which requires a prerequisite knowledge of the life cycle and ability to manipulate it in order to get controlled genetic recombination (*ix*) use of mathematical modeling for description of the phenomenon observed during culture growth and alkaloid production, and (*x*) possibility to use solid-state fermentation, perfluorocarbons as oxygen vectors, and modern analytical methods (capillary electrophoresis and 2D fluoresence spectroscopy) in *Claviceps* fermentation.

The biotechnological relevance of ergot alkaloids due to their therapeutic use is unquestionable. Ergot alkaloids have gained considerable agricultural importance not only due to contamination of cereals with sclerotia of the genus *Claviceps*, but also because of ergopeptine-producing endophytes that are involved in the intoxification with ergot drugs of animals in many parts of the world. These endophytes, mainly of the genus *Acremonium* and

Neotyphodium cause serious problems in livestock breeding. It is therefore of great importance to develop new strategies for rational designing of ergot alkaloid based drugs and to understand and control the production of ergot alkaloids by endophytic fungi.

In short, detailed biochemical and molecular genetic research covering the various aspects of ergot alkaloid-based drug development and control of toxin production in the environment will continue to be of great importance.

REFERENCES

Abe. M. & Yamatodani, S. (1954) *Bull. Agric. Chem. Soc.* (Japan). **18**:501-502.

Abe. M. & Yamatodani, S. (1955). *Bull. Agric. Chem. Soc.* (Japan). **19**:161-162.

Abe. M., Yamano, T., Kozu, Y., & Kusumoto, M. (1952) *J. Agric. Chem. Soc.* **25**:458- 459

Abe. M., Yamatodani, S., Yamano, T. & Kusumoto, M. (1956) *Bull. Agric. Chem. Soc.* (Japan). **20**:59-60.

Adams. R.A. (1962) *US Patent No.* 3117917.

Amici. A.M., Minghetti, A., Scotti, T., Spalla, C. & Tognoli, L. (1966) *Experimentia.* **22**:415-418.

Annis. S.L. & Panaccione, D.G. (1998) *Can. J Microbiol.* **44**:80-86.

Arcamone. F., Bonino, C., Chain, E.B., Ferreti, A., Pennela, P., Tonolo, A. & Vero, I. (1960) *Nature* (London). **187**:238-239.

Arcamone. F., Chain, E.B., Ferreti, A., Minghetti, A., Pennela, P., Tonolo, A. & Vero I. (1961) *Proc. Royal Soc.* B. **155**:26-54.

Balakrishnan. K. & Pandey, A. (1996) *J. Sci. Ind. Res.* **55**:365-372.

Baumert. A., Erge, D., Groeger, D., Maier, W., Schmauder, H.P., Schumann, B., Breuel, K. & Hoehne, I. (1979) DD Pat 234171.

Baxter. R.M., Kandel, S. & Okany, A. (1961) *Tetrahedron Lett.* 596-600.

Berde. B. & Sturmer, E. (1978). *Ergot alkaloids and related compounds.* (Eds. Berde, B & Schild, O.), Berlin: Springer Verlag, pp 1-28.

Berde. B. (1984) *J. Royal Soc. Med.* **77**: 5.

Birch. A.J., McLoughlin, B.J. & Smith, H. (1960) *Tetrahedron Lett* .**7**:1-3.

Bonns. W.W. (1922) *Am. J. Bot.* **9**:339-353.

Bumbova-Linhartova, R., Flieger, M., Sedmera, P. & Zima, J. (1991) *Appl. Microbiol. Biotechnol.* **34**:703-706.

Cassady, J.M. & Floss, H.G. (1977). *Lloydia.* **40**: 90-106.

Cavender, F.L. & Anderson, J.A. (1970) *Biochem. Biophys Acta.* **208:** 345-348.

De Groot, A., Dongen, P.W.J.V., Vree, T.B., Hekster, Y.A. & Roomalen, J.V.(1998) *Drugs.* **56**: 523-535.

Demain, A.L. (1999) *Appl. Microbiol. Biotech.* **52**: 455-463.

Dickerson, A.G., Mantle, P.G. & Szezyrbak, C.K. (1970) *J. Gen. Microbiol.* **60**: 403- 415.

Didek-Brumec, M., Gaberc-Porekar, V., Alacevic, M., Druskovic, B. & Socic, H. (1991) *J. Basic Microbiol.* **31**:27-35.

Didek-Brumec, M., Puc, A., Socic, H. & Alacevic, M. (1987) *Food Technol. Biotechnol. Rev.* **25**:103-109.

Edwardson, J.A. (1968) *Brit. J. Pharma.* **33**: 215.

Eich, E., Eichberg, D. & Muller, W.E.G. (1984) *Biochem. Physiol..* **33**: 523-526.

Esser, K. & Tudzynski, P. (1979) *Theor. Appl. Genet.* **53**: 145-149.

Fleiger, M., Sedmera, P., Vokoun, J., Ricicova, A. & Rehacek, Z. (1982) *J. Chromatogr.* **236**: 453-459.

Flieger, M., Linhartova, R., Rehacek, Z., Saidl, P., Stuchlik, J., Malinka, Z., Harazim, P., Cvak, L. & Bremek, J. (1989a). CCM 8061 CZ Patent 276430 (In Czech)

Flieger, M., Linhartova, R., Rehacek, Z., Saidl, P., Stuchlik, J., Malinka, Z., Harazim P., Cvak, L. & Bremek, J. (1989b). *CZ Patent* 281196 (In Czech)

Flieger, M., Votruba, J., Kren, V., Pazoutova, S., Rylko, V., Sajdl, P. & Rehacek, Z. (1988) *Appl. Microbiol. Biotechnol.* .**29**:181-185.

Flieger, M., Wurst, M. & Shelby, R. (1997) *Folia Microbiol.* **42**:3-30.

Garve, V., Tenberge, K.B. & Eising, R. (1998) *Phytopath.* **88**: 744-753.

Geiger, H.H. & Bausbach, G.A. (1979) *Z. Pflanzenzuchtg.* **83**: 163-175.

Gilmanov, V., Flieger, M., Dymshits, V. (1996) *J. Appl. Bacteriol.* **81**:678-680.

Grm, B., Mele, M. & Kremser, M. (1980) *Biotechnol. Bioengg.* **22**:255-270.

Groger, D., Mothes, K., Simon, H., Floss, H.G. & Weygand, F. (1960) *Z. Naturforsch.* **15**:141-143.

Groger, D., Tyler, V.E.Jr & Dusenberry, J.E. (1961) *Lloydia* **24:97**-102.

Harazim, P., Malinka, Z., Stuchlik, J., Cvak, L., Bremek, J., Flieger, M., Linhartova, R., Rehacek, Z. & Sajdl, P . (1989) CCM 8063.*CZ Patent* 276032 (In Czech)

Heinstein, P.F., Lee, S.L. & Floss, H.G (1971). *Biochem. Biophys. Res. Commu.* **44**:1244-1251.

Hernandez, M.R.T., Lonsane, B.K., Raimbault, M. & Roussos, S. (1993). *Process Biochem.* **28**:23-27.

Hunter-Cervera, J.C. & Belt, A. (1996) *Maintaining Cultures for Biotechnology and Industry*. New York, New York.

Hüsgen, U., Buttner, P., Müller, U. & Tudzynski, P. (1999) *J. Phytopathol.* **147**:591- 597.

Janardhanan, K.K. & Husain, A. (1984) *Proc. Indian Nat. Sci. Acad* . **B 50**:438-440.

Jungehulsing, U. & Tudzynski, P. (1997) *Mycol. Res.* **101**: 1-6.

Kobel, H. & Sanglier, J.J. (1978) *Antibiotics and other secondary metabolites,* (Eds. Hutter, R., Leisinger, T., Nuesch, J. & Wehrli, W.), Academic Press, New York, pp. 233- 242.

Kobel, H. & Sanglier, J.J. (1986) *Biotechnology* Vol. 4, (Eds. Rehm, H.J. & Reed, G). Weinheim, Germany, pp. 569-609.

Kobel, H. (1969) *Pathologia Microbiol.* **34**:249-251.

Komel, R., Rozman, D., Puc, A.& Socic, H. (1985) *Appl. Microbiol. Biotechnol.* **23**:106-109.

Kopp, B. & Rehm, H.J. (1983) *Eur. J. Appl. Microbiol. Biotechnol.* **18**:257-263.

Kopp, B. & Rehm, H.J. (1984) *Appl. Microbial Biotechnol.* **19**:141-145.

Kozlovsky, A.G. (1999) *Medicinal and aromatic plant-industrial profile. Ergot, the genus Claviceps,* (Eds. Kren, V. & Cvak, L.), Harwood Academic Publ, The Netherlands, pp. 479-499.

Kren, V. (1991) *Advances in Biochemical Engineering/Biotechnology*, (Ed. Fiechter, A.), Springer Verlag, Germany, pp. 123-144.

Kren, V., Sajdl, P., Rehacek, Z., Malinka, Z. & Bremek, J. (1988) *CZ Patent* No. 272562 (In Czech).

Krupinski, V.M., Robbers, J.E. & Floss, H.G. (1976) *J. Bacteriol* .**125**:158-165.

Kybal, J. & Vlcek, V. (1976) *Biotechnol. Bioeng.* **18**:1713-1718.

Liberman, A.N., Neibowitz, M., Neophytides, A., Kuperschmit, M., Mehl, S., Kleinberg, D., Serby, M. & Goldstein, M. (1979) *Lancet.* **2**: 1129-1130.

Lowe, K.C., Davey, M.R. & Power, J.B. (1998) *Trends Biotechnol.* **16**:272-277.

Malinka, Z. (1999) *Medicinal and aromatic plant-industrial profile. Ergot, the genus Claviceps,* (Eds. Kren, V. & Cvak, L.), Harwood Academic Publ. The Netherlands, pp. 321-371.

Mantle, P.G. & Nisbet, I.J. (1976) *J. Gen. Microbiol.* **93**:321-324.

Mantle, P.G. (1969) *Trans. Br. Mycol. Soc.* **52**:381-392.

Mehta, P. (1984) *Submerged fermentation of clavine alkaloids.* Ph.D. Thesis, Czech Academy of Sciences, Prague, Czech Republic, pp.202.

Menge, M., Mukherjee, J. & Scheper, T. (2001) *Appl. Microbiol. Biotechnol.* **55**:411-416.

Milicic, S., Kremser, M., Gaberc-Porekar, V., Didek-Brumec, M. & Socic, H. (1987) *Appl. Microbiol. Biotechnol.* **27**:117-120.

Minghetti, A. & Crespi-Perellino, N. (1999) *Medicinal and aromatic plant-industrial profile. Ergot, the genus Claviceps,* (Eds. Kren, V. & Cvak, L.) Harwood Academic Pub., The Netherlands, pp. 1-24

Mothes, K., Winkler, K.,Groger, D., Floss, H.G., Mothes, U. & Weygand, F. (1962) *CH Patent* 455820.

Mukherjee, J. & Menge, M. (2000) *Adv. Biochem. Engg. Biotech.* **68**:1-20.

Nemeth, E. (1999) *Medicinal and aromatic plant-industrial profile. Ergot, the genus Claviceps,* (Eds. Kren V, & Cvak, L.) Harwood Academic Chur, pp. 303-319

Neumann, D., Losecke, W., Maier, W. & Groger, D. (1979) *Biochem. Physiol. Pflzen.* **174**:504-508.

Olasz, K., Gaal, T. & Zalai, K. (1982) *Acta. Biochim. Biophy. Acad. Sci.* Hungary. **17**:126.

Panaccione, D.G., Johnson, R.D., Wang, J., Young, C.A., Damrongkool, P., Scott, B. & Schardl, C.L. (2001) *Proc. Natl. Acad. Sci.* **98**:12820-12825.

Pazoutova, S. & Parbery, D. (1999) *Medicinal and aromatic plant-industrial profile. Ergot, the genus Claviceps, (Eds.* Kren V. & Cvak, L.). Harwood Academic Press, The Netherlands, pp. 57-77

Pazoutova, S. & Tudzynski, P. (1999) *Mycol. Res.* **103**:1044-1048.

Pazoutova, S., Flieger, M., Rylko, V., Kren, V. & Sajdl, P. (1987) *Curr. Microbiol.* **15**: 97-101.

Pazoutova, S., Johnson, N. & Rajasab, A.H. (2002) *J. Phytopathol* .**150**:1-4.

Pazoutova, S., Slokoska, L.S., Nikolova, N. & Angelov, T.I. (1983) *Eur. J. Appl. Microbiol. Biotechnol.* **16**:208-211.

Pazoutova, S., Votruba, J. & Rehacek, Z. (1981) *Biotech. Bioengg.* **23**:2837-2849.

Pertot, E., Gaberc-Porekar, V. & Socic, H. (1990) *J. Basic Microbiol.* **30**:51-56.

Rehacek, Z. & Mehta, P. (1991) *The Genet Engineering and Biotechnol,* London. **11**:6: 19- 21.

Rehacek, Z. & Mehta, P. (1993) *Fungal Ecology and Biotechnology,* (Eds. Rai, B., Arora, D.K., Dubey, N.K. & Sharma, P.D). Rastogi Publications, Meerut, India, pp. 275-287.

Rehacek, Z. & Sajdl, P. (1990) *Ergot alkaloids: Chemistry, Biological Effects, Biotechnology.* Czech Academy of Sciences, Academia, Praha, Czech Republic, pp. 376.

Rehacek, Z. (1974) *Zbl. Bakt II.* **129**: 20-49.

Rehacek, Z. (1983) *Process Biochem.* 22-33.

Rehacek, Z., Kozova, J., Sajdl, P. & Vorisek, J. (1974) *Can. J. Microbiol.* 20: 1323-1329.

Rehacek, Z., Pazoutova, S., Kren,V., Rylko,V., Kozova, J. & Sajdl, P. (1986) Tul 59 CC 5/86. *CS Patent* 252603 (In Czech).

Rehacek, Z., Sajdl, P., Kozova, J. & Ricicova, A. (1972) *Folia Microbiol.* **17**: 308-315.

Robbers, J.E., Robertson, L.W., Horneman, K.M., Jindra, A. & Floss, H.G. (1972) *J. Bacteriol.* **112**: 791-796.

Robinson, T., Singh, D. & Nigam, P. (2001) *Appl. Microbiol. Biotech.* **55**:284-289.

Rosazza, J.P., Kelleher, W.J. & Schwarting, A.E. (1967) *Appl. Microbiol.* **15**:1270- 1283.

Rucman, R. (1976) *J. Chromat.* **121**:353-360.

Rylko. V., Flieger, M., Sajdl, P., Rehacek, Z., Malinka, Z., Harazim, P. & Stuchlik, J. (1988) *CS Patent* 4725-88.

Sajdl, P. & Rehacek, Z. (1975) *Folia Microbiol.* **20**:365-367.

Schumann, B., Maier, W. & Groeger, D. (1987) *Z. Naturforsch.* **42**:381-386.

Socic, H. & Gaberc-Porekar, V. (1992) *Handbook of Applied Mycology, Fungal Biotechnol.* Vol. 4 (Eds. Arora, D.K., Elander, E.P. & Mukerjee, K.G.) Marcel Dekkar, Inc, New York, pp. 475-515.

Socic, H., Gaberc-Porekar, V. & Didek-Brumec, M. (1985) *Appl. Microbiol. Biotechnol.* **21**:91-95.

Socic, H., Gaberc-Porekar, V., Pertot, E. Puc, A. & Milicic, S. (1986) *J. Basic Microbiol.* **26**:533-539.

Sommer, A.F. & Buchanan, A.R. (1955) *Am. J. Physiol.* **18**: 296-300.

Strandova, K., Kybal, J., Svoboda, E., Spacil, J., Krajicek, A., Malinka, Z. & Harazim, P. (1986). CCM8043. *CS Patent No.*267573.

Taber, W.A. & Vining, L.C. (1959) *Chem. Ind.,* London. 1218-1219.

Taber, W.A. & Vining, L.C (1963) *Can. J. Microbiol.* **9**:1-14.

Taylor, E.M. & Ramsted, E. (1960) *Nature.* **188**: 494-495.

Tenberge, K.B., Brockmann, B. & Tudzynski, P. (1999) *Mycol. Res.* **103**: 1103- 1118.

Trejo-Hernandez, M.R. & Lonsane, B.K. (1993) *Process Biochem.* **28**:23-27.

Trinn, M., Manczinger, L., Polestyukne, N.A., Kordik, G., Pecsne, R.A., Zalai, K., Beszedics, G., Ferenczy, L., Nagy, L., Robicsek, K. & Szegedi, M. (1990) *HU Patent No. 209325* (In Hungarian).

Tsai, H-F., Wang, H., Gebler, J.C., Poulter, C.D.& Schardl, C.L. (1995) *Biochem. Biophy. Res. Comm.* **216**: 119-125.

Tudzynski, P., Correia, T. & Keller, U. (2001) *Appl. Microbiol. Biotech.* **57**:593-605.

Tudzynski, P., Holter, K., Correia, T., Arntz, C., Grammel, N. & Keller, U. (1999) *Mol. Genet.* **261**:133-141

Udwardy, N.E. (1980) *Process Biochem.* **15**: 5-8.

Vasarhelyi, G., Tetenyi, P., Zambo, I., Kinicky, M., Balassa-Barkanyi, I., Kiss, J., Pecs-Razso, A. & Vining, L.C. (1980) *Economic Microbiology,* (Eds. Rose, A.H.), Academic Press, New York, pp. 523-573.

Vining, L.C. (1980) *Economic Microbiology* (Ed. Rose, A.H.), Academic Press, New York, pp. 523-573.

Vlcek, V. & Kybal, J. (1974) *CS patent* 172552 (In Czech).

Vorisek, J. & Rehacek, Z. (1976) *Arch. Microbiol.* 107: 321-327.

Vorisek, J. & Rehacek, Z. (1978) *Arch. Microbiol.* 117: 297-302.

Zalai, K., Kordik, G., Manczinger, I., Pecsne, R.A., Beszedics, G., Polestyukne, N.A., Ferenczy, I., Olasz, K., Szegedi, M.& Trinn, M. (1990) *HU Patent* 209324 (In Hungarian).

7

Production of Value-added Hydroxy Fatty Acids by a Bacterial Strain *Pseudomonas aeruginosa* PR3

Hakryul Kim[1] and Sun Chul Kang[2]
[1]*Department of Animal Science and Biotechnology, Kyungpook National University, Daegu 702-701 (Korea)*
[2]*Department of Chemical Engineering and Biotechnology, Daegu University, Kyungsang, Daegu (Korea)*

Hydroxy fatty acids are known to be produced in nature mainly from plant systems. The hydroxyl group of fatty acid is well known to give a fatty acid special properties, such as higher viscosity and reactivity compared with other normal fatty acids and are used in a wide range of industrial products. Microbial conversion of unsaturated fatty acids by different microbial strains and different substrates have been widely exploited to produce new, value-added hydroxy fatty acids. Among those microbial strains, *Pseudomonas aeruginosa* NRRL strain B-18602 (PR3) originally isolated from a water sample of a pig farm located in Morton, Illinois, USA was well studied as a bio-catalyst for the conversion of various unsaturated fatty acids into various hydroxy fatty acids. Among those unsaturated fatty acids used for the production of hydroxy fatty acids by PR3, oleic acid, linoleic acid, and linolenic acid were well studied as substrates to produce mono-, di-, and trihydroxy fatty acids. Important environmental conditions for their production were also well studied. Based on these successful data, further studies should be focused on the investigation of their biological properties for their industrial applications.

INTRODUCTION

Hydroxy fatty acids are known to be produced in nature mainly from plant systems. The hydroxyl group of fatty acid is well known to give fatty acid special properties, such as higher viscosity and reactivity compared with other normal fatty acids and are used in a wide range of industrial products. Those products include resins, waxes, nylons, plastics, lubricants, cosmetics and additives in coatings and paintings. At present, imported castor oil and its derivatives are the only commercial source of these hydroxy fatty acids. Those castor oil derivatives are used in the synthesis of high performance polymers for military equipment and classified by the United State Department of Defense as strategic and critical materials. However, the fluctuating supplies and price of castor oil render scientists to seek alternative raw materials.

Since Wallen *et al.* (1962) reported the first bioconversion of oleic acid to 10-hydroxystearic acid by a Pseudomonad, microbial conversion of unsaturated fatty acids have been widely exploited to produce new, value-added hydroxy products. Microbial oxidation of unsaturated fatty acids was recently reviewed (Hou, 1995). Unsaturated fatty acids are the major components of corn and soyabean oils, representing more than 85 % of fatty acids present in triglycerides. Many works have been done to investigate microbial conversion of vegetable oils and their component fatty acids to value-added products.

Microbial conversion of unsaturated fatty acids by different microbial strains and different substrates have been widely exploited to produce new, value-added products. Among those unsaturated fatty acids used for microbial production of hydroxy fatty acids, oleic acid, linoleic acid, and linolenic acid were well studied as substrate to produce mono-, di-, and trihydroxy fatty acids. In this review, we introduce production of hydroxy fatty acids from their correspondent fatty acid substrates by a bacterial strain *Pseudomonas aeruginosa* PR3 (Hou *et al.*, 1993) and optimized environmental conditions for their production. *Pseudomonas aeruginosa* NRRL strain B-18602 was originally isolated from a water sample of a pig farm located in Morton, Illinois, USA.

PRODUCTION OF MONOHYDROXY FATTY ACID

Production of 10-hydroxystearic Acid

Production of 10-hydroxystearic acid (10-HSA) was first reported by Wallen *et al.* (1962). They found that a Pseudomonad isolated from fatty material hydrated oleic acid at the cis-9-double bond resulting in the production of 10-hydroxystearic acid with a 14% yield (Fig. 1). The 10-HSA is optically active (Schroepfer and Block, 1963; 1965) and has the D-configuration (Schroepfer and Block, 1965). Incubation of this organism with oleic acid in a medium enriched with deuterium oxide yields 10-HSA containing one deuterium atom is ideal. (Schroepfer and Block, 1965). Moreover, this deuterium was shown to be on carbon atom 9 and in the L-configuration. By using soluble cell-free extract of a Pseudomonad, Niehaus and Schroepfer (1965) were able to demonstrate the conversion of oleate to 10-HSA under anaerobic conditions and the reversibility of the reaction. These findings, coupled with the observed stereospecific uptake of one atom of solvent hydrogen into 10-HSA and the lack of conversion of either the *cis*- or *trans*-9, 10-epoxystearic acid to 10-HSA, are compatible with a mechanism involving hydration of the double bond of oleic acid and rule out an epoxide intermediate (Schroepfer, 1966). The same enzyme preparation was later found to catalyze both the hydration of *cis*- and *trans*-9, 10-epoxystearic acids to yield *threo*- and *erythro*-9, 10-dihydroxystearic acid, respectively (Niehaus and Schroepfer, 1967). Niehauls *et al.* (1970) demonstrated the interconversion of oleic acid and 10-HSA by a soluble enzyme preparation from a Pseudomonad. This further ruled out the possible intermediate role of an epoxide in the overall conversion for two reasons. First, the enzymatic conversion of oleate to 10-HSA was observed to proceed under anaerobic conditions, a feature not characteristic of enzymatic epoxidations of olefins. Second, neither the DL-*cis*-9,10-epoxystearate nor the DL-*trans*-9,10-epoxystearate served as precursors of either oleate or 10-hydroxystearate under the conditions studied.

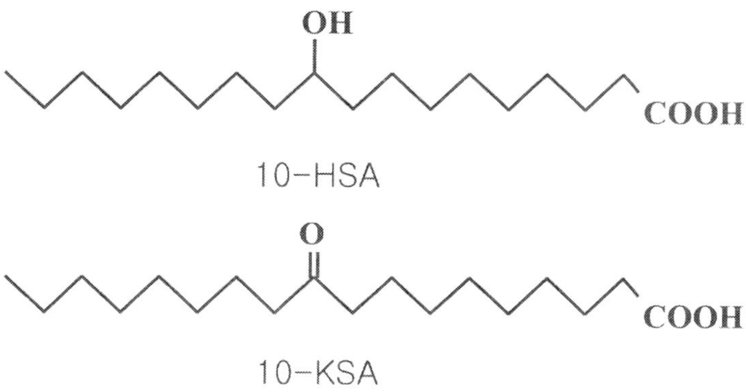

Fig. 1: Structure of 10-hydroxystearic acid (10-HSA) and 10-ketostearic acid (10-KSA).

Production of 10(S)-hydroxy-8(E)-octadecenoic Acid

10(S)-hydroxy-8(E)-octadecenoic acid(HOD) was produced as an intermediate in the conversion of oleic acid to 7,10-dihydroxy-8(E)-octadecenoic acid (DOD) (Kim *et al.*, 2000a). The production of DOD from oleic acid is unique in that it involves an introduction of two hydroxyl groups at carbon number 7 and 10 and a rearrangement of the double bond from carbon 9 to 8. Based on structure complexity, DOD formation seems to involve more than one step. In an effort to elucidate the metabolic pathway leading to DOD from oleic acid, Hou *et al.* (1992) reported isolation of two more products from the reaction mixture of which one was identified as phenazine-1-carboxylic acid and the other was identified to be a plausible intermediate in the bioconversion of oleic acid to DOD, 10-hydroxy-8-octadecenoic acid (HOD) in which the 8-double bond was possibly in *cis* form. However, the rearranged double bond of HOD turned out to be *trans* rather than *cis* by NMR and IR data and the carbon number 10, on which a hydroxyl group was introduced, represented the *S*-configuration with a minor isomer of *R*-form (Kim *et al.*, 2000b).

Structure determination: The electron impact (EI) GC/MS data of TMS derivative of the methylated HOD were, m/z (% relative intensity): 73(48.5), 75(22), 95(7.1), 107(7.1), 121(8.2), 129(17.3), 143(5.5), 149(8.5), 169(0.7), 215(0.8), 241(7.5), 271(100), and 369(M-CH_3, 0.7). This was consistent with the TMS derivative of a methylated C18 hydroxy monoenoic fatty acid with a molecular mass of 384. The location of hydroxyl group was apparent from the fragments observed in the EI spectrum of the TMS derivative of the methylated fatty acid. The intense fragment arising from alpha cleavage to the derivatized hydroxyl group to give a fragment containing both the TMS and the double bond was observed at m/z 271. This fragment, together with the m/z 241 fragment, located the hydroxyl group at either C10 or C8 and established that the hydroxyl group is alpha to a double bond (Hou, 1995). In experiments described below, the MCO derivative of HOD was oxidatively cleaved by $KMnO_4$ followed by methyl esterification. The cleavage products, methyl 2-MCO-decanoate and dimethyl octadecenoate unequivocally established the position of the 10-hydroxyl and 8-unsaturation.

Determination of stereochemistry: Purified HOD was subjected to chiral analysis by the MCO method. Two major peaks were detected by GC at retention time 10.65 min. and 20.2 min. under the temperature programming from 65°C to 260°C at a rate of 10°C/min. with a post-holding for 5 minutes. The first eluting peak was determined to be dimethyl octanedioate from the standard MS library and the second peak to be methyl 2- MCO-decanoate by comparison with standard racemic methyl 2(*R*, *S*)-MCO-decanoate. After changing temperature programming to 100°C - 260°C at the rate of 2°C/min to enhance the separation and elucidate the stereo-configuration, the methyl 2-MCO-decanoate appeared at 60.15 min. with a minor peak at 60.37 min. (about 3 %). It is known from previous work that methyl 2(*S*)-alkanoates elute earlier than methyl 2(*R*)-alkanoates (Fahlstadius and Hamberg, 1989; Hamberg, 1971; Hamberg *et al.*, 1986), specifically including methyl 2-MCO-decanoate (Fahlstadius and Hamberg, 1989). The injection of standard racemic methyl 2(*R*, *S*)-MCO-decanoate represented two peaks, one [methyl 2(*S*)-MCO-decanoate] at 60.17 min. and the other [methyl 2(*R*)-MCO-decanoate] at 60.38 min. and coinjection of the racemate with the MCO derivative from HOD demonstrated that the derivatives from HOD were the same compounds as the racemic standard. The isomers of methyl 2-MCO-decanoate afforded virtually identical mass spectra, regardless of its origin from HOD, 2(*S*) including the minor amount of 2(*R*) or the racemic methyl 2-hydroxydecanoate, 2(*R*,*S*) as follows: m/z (% relative intensity) 247(4), 215(1.5), 185(3), 153(6), 139(55), 138(100), 123(29), 109(8), 95(52), 83(68), 81(51), 69(32), 55(41). From these results it was clear that the first major peak of the MCO derivative from HOD was the methyl 2(*S*)-MCO-decanoate and the second minor peak the methyl 2(*R*)-MCO-decanoate suggesting that the hydroxyl group at C10 of HOD contained mainly *S*-configuration with a minor isomer of *R* configuration (about 3%).

Based on the results illustrated above and the work previously done (Kim *et al.*, 2000b), the overall bioconversion pathway of oleic acid to DOD by strain PR3 was postulated as shown in figure 2. Substrate (oleic acid) is first converted to HOD by PR3 during which one hydroxyl group is introduced at C10(*S*) and a double bond is shifted from C9 *cis* to C8 *trans*. Although Hou (1995a,b) reported that a C10 position-specific and *cis*-specific hydratase was involved in the hydration of unsaturated fatty acid by *Flavobacterium* sp. DS5 in which C10 hydroxyl group was introduced with the removal of C9 *cis* double bond, typical of hydration reaction of fatty acids, it is unlikely that a hydratase is involved in this reaction by PR3 in that the double bond at C9 of substrate retained as a shifted *trans*-configured form during the hydroxylation. The hydration of oleic, linoleic, and linolenic acid to their corresponding 10-hydroxy products have been reported from various microbes as reviewed by Hou (1995b). The resulting plausible intermediate HOD is then subject to another hydroxylation at C7(*S*) by possibly the same enzyme(s) involved in the first hydroxylation reaction resulting in DOD formation.

PRODUCTION OF DIHYDROXY FATTY ACID

Production of 7,10-dihydroxy-8(E)-octadecenoic Acid

7,10-dihydroxy-8(E)-octadecenoic acid (DOD) was produced by bioconversion of oleic acid by PR3. Hou *et al.* (1991) reported the discovery of a new compound, 7,10-dihydroxy-8(*E*)-octadecenoic acid (DOD) which was produced from oleic acid at a greater than 60 % yield by

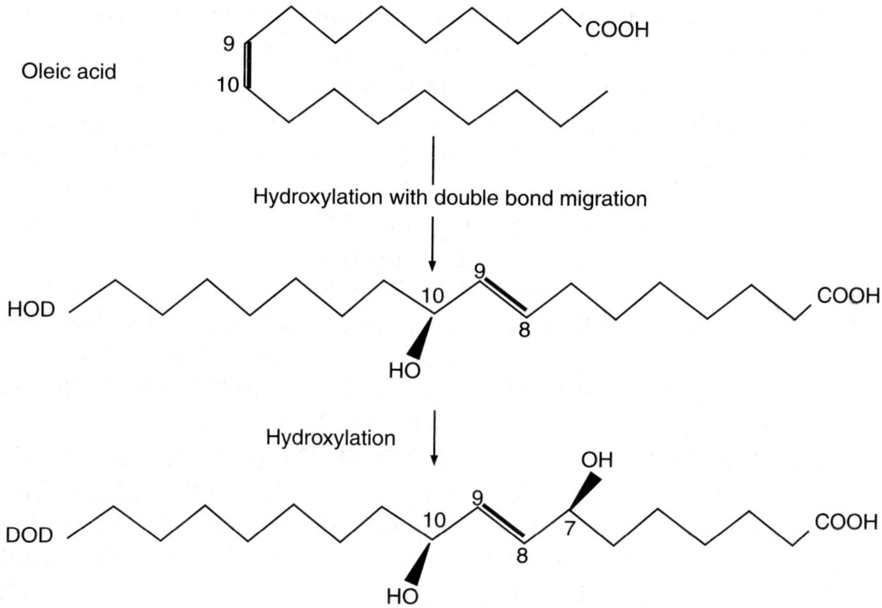

Fig. 2: Schematic diagram of the production of DOD[7,10-dihydroxy-8(E)-octadecenoic acid] from oleic acid by microbial strain PR3. HOD represents 10(S)-hydroxy-8(E)-octadecenoic acid.

a new bacterial isolate *Pseudomonas* sp. PR3 (Hou and Bagby, 1991)(Fig. 2). Recently the yield of DOD production by strain PR3 was improved to more than 80 % yield through modifying the culture medium and reaction parameters (Kou *et al.*, 1998). The production of similar compound by *Pseudomonas* 42A2 was reported and studied (De Andres *et al.*, 1994; Mercade *et al.*, 1988). The production of DOD from oleic acid is unique in that it involves an introduction of two hydroxyl groups at carbon number 7 and 10 and a rearrangement of the double bond from carbon 9 to 8.

Production of 10,12-Dihydroxy-8(*E*)-Octadecenoic Acid

10,12-Dihydroxy-8(*E*)-Octadecenoic Acid (DHOD) was produced as an intermediate in the conversion of ricinoleic acid to 7,10,12-Trihydroxy-8(*E*)-Octadecenoic Acid by *Pseudomonas aeruginosa* PR3 (Kim *et al.*, 2000c). Recently, Kuo *et al.* (1998) found that strain PR3 converted ricinoleic acid to 7,10,12-trihydroxy-8(E)-octadecenoic acid (TOD). The reaction mechanism involved in TOD production from ricinoleic acid looked similar to that of DOD production from oleic acid in that both the substrates (oleic acid and ricinoleic acid) contained C 18 fatty acid backbone with a *cis*-double bond at carbon 9 except for a hydroxyl group pre-introduced at C 12 of ricinoleic acid. In an effort to elucidate the metabolic pathway leading to TOD from ricinoleic acid, plausible compound from the reaction mixture was purified and characterized. The structure of the new compound was determined by GC/MS, FTIR, and NMR to be 10,12-dihydroxy-8(E)-octadecenoic acid (DHOD).

Structure determination: The electron impact (EI) GC/MS data of TMS derivative of the methylated DHOD were, m/z (% relative intensity): 73(88.7), 75(23), 103(9.9), 107(5.8), 121(7.1), 129(14.9), 147(28.1), 149(12.4), 187(58.1), 239(100), 261(7.3), 271(85.5), 297(19.3), 387(1.5), and 457(M-CH_3, 0.3). This was consistent with the TMS derivative of a methylated C18 dihydroxy monoenoic fatty acid with a molecular mass of 472. The locations of hydroxyl group were apparent from the fragments observed in the EI spectrum of the TMS derivative of the methylated fatty acid. The intense fragment arising from alpha cleavage to the derivatized hydroxyl group to give a fragment containing the TMS and both the TMS and the double bond were observed at m/z 187 and m/z 271, respectively.

The purified free fatty acid was subjected to IR analysis. The presence of hydroxyl group was indicated by the broad strong IR absorption at 3600 cm^{-1} region. An absorption observed at 1702 cm^{-1} represented carbonyl group and the absorption at 966 cm^{-1} indicated that the unsaturation was *trans* (Method, 1981). The data obtained from GC/MS, FTIR confirmed that the purified compound was 10,12-dihydroxy-8(*E*)-octadecenoic acid.

Optimum conditions for the production of DHOD: The reaction was carried out at 28°C for the time specified after ricinoleic acid (200mg) was added to the culture. The amount of DHOD and TOD produced in the culture along with the amount of the remained substrate were determined (Fig. 3). The amount of the produced DHOD and TOD started to sharply increase after 13 hrs incubation and reached a maximum at 30 and 37 hrs, respectively, while the amount of the remained ricinoleate decreased in the same manner during that period. After peaked at 30 hrs, the production of DHOD decreased sharply up to 60 hrs. However, the production of TOD further increased beyond 30 hr and reached a maximum at 37 hr after which it remained a plateau up to 84 hrs strongly suggesting that DHOD might be a possible intermediate of TOD production from ricinoleate. Further incubation reduced the TOD content in the medium indicating that PR3 metabolized TOD.

The optimum pH for the production of DHOD appeared to be 6.5, which was same as that reported for the production of HOD, an intermediate of DOD production from oleate by the same strain (Kim *et al.*, 2000). However, at pH 7.0 DHOD was also produced at a level of 88 % of the maximum value at pH 6.5. Optimal temperature for the production of DHOD was 25°C which was different from that reported for the production of HOD (Kim *et al.*, 2000b) and DOD (Hou and Bagby, 1991) by the same strain in that 30°C was the optimum temperature for their production.

Based on the structural similarity between DHOD and TOD and the results from this study, the overall bioconversion pathway of ricinoleic acid leading to TOD by strain PR3 was postulated as shown in Fig. 4. Substrate (ricinoleic acid) is first converted to DHOD by PR3 during which one hydroxyl group is introduced at C10 and a double bond is shifted from C9 *cis* to C8 *trans*. The resulting plausible intermediate DHOD is then subject to another hydroxylation at C7 by possibly the same enzyme(s) involved in the first hydroxylation reaction resulting in TOD formation. Although Hou (Hou, 1995a,b) reported that a C10 position-specific and *cis*-specific hydratase was involved in the hydration of unsaturated fatty acid by *Flavobacterium* sp. DS5 in which C10 hydroxyl group was introduced with the removal of C9 *cis* double bond, typical of hydration reaction of fatty acids, it is unlikely that a hydratase is involved in this reaction by PR3 in that the double bond at C9 of substrate retained as a shifted *trans*-configured form during the hydroxylation. On the other hand, the pathways involved and the structural similarities of substrates (oleate and ricinoleate for DOD and TOD, respectively), which represent C18 fatty acid with a *cis*-double bond at carbon 9 except a hydroxyl group introduced at C12 of ricinoleate, suggest that the enzyme system involved in this reaction might be the same as that involved in the reaction producing DOD from oleate by the same strain PR3 with HOD being a plausible intermediate (Kim *et al.*, 2000b).

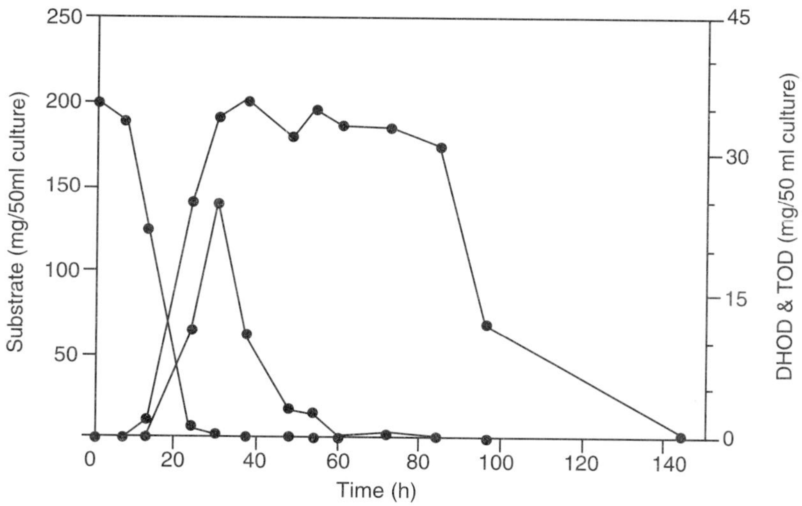

Fig. 3: Time course production of DHOD and TOD from ricinoleic acid by PR3. ● : TOD ▲ : DHOD, ■ : ricinoleic acid.

PRODUCTION OF TRIHYDROXY FATTY ACID

Production of 7,10,12-trihydroxy-8(E)-octadecenoic Acid from Ricinoleic Acid

Strain PR3 converts oleic acid to form a surface active compound, 7,10-dihydroxy-8(E)-octadecenoic acid (DOD) (Hou and Bagby, 1991). Based on the structural similarity between oleic acid and ricinoleic acid except a hydroxyl group pre-introduced at carbon 12, ricinoleic acid was used as a substrate for the bioconversion by PR3. Kuo *et al.* (2001) described the ability of strain PR3 in converting ricinoleic acid to a novel trihydroxy unsaturated fatty acid 7,10,12-trihydroxy-8(E)-octadecenoic acid (TOD)(Kou *et al.*, 1998).

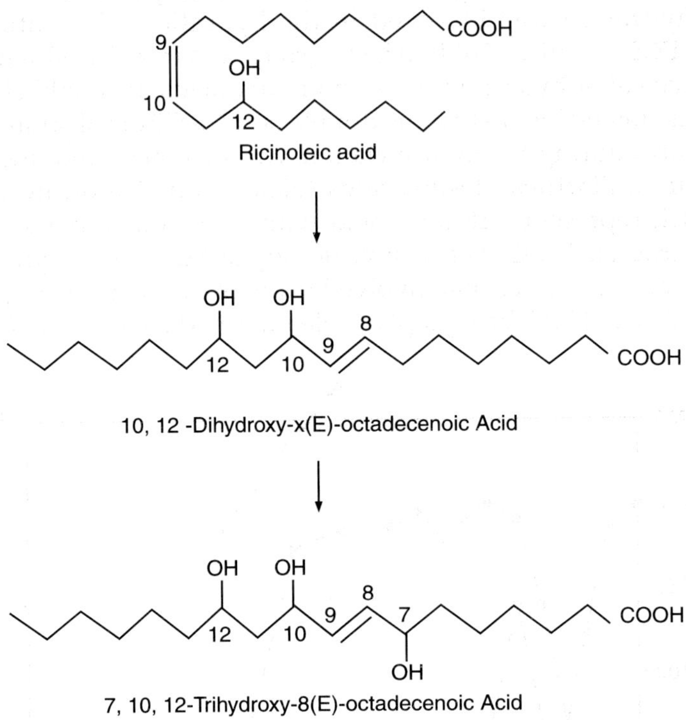

Fig. 4: Schematic diagram of TOD[7,10,12-trihydroxy-8(E)-octadecenoic acid] production from ricinoleic acid by PR3.

Time coursed production of TOD: Time coursed study of TOD production by strain PR3 was carried out at 28°C for the time specified after ricinoleic acid was added to the culture. During the conversion, the amount of TOD increased almost linearly in the first 36 hrs of reaction and reached its maximum levels between 36 and 48 hrs after adding the substrate (Figure 5). Thereafter, the amount of TOD decreased rapidly to approximately 60% of the maximum level after 60 hrs and to 35% after 72 hrs. This indicates that strain PR3 actively metabolizes TOD, a phenomenon similar to the formation of DOD from oleic acid (Hou and Bagby, 1991). The results indicate that control of reaction time is important to achieving a maximum yield.

Optimum conditions for TOD production: The effect of carbon source, substrate concentration, temperature, and reaction pH on the production of TOD by PR3 was studied. Strain PR3 utilized glucose, glycerol, raffinose, stachyose, and lactose as carbon sources for growth and could convert ricinoleic acid to TOD. However, glucose consistenly provided the best yield. Substrate concentration for maximal TOD production was 1% of working culture volume. Optimal temperature and pH value for TOD production by PR3 were 25°C and pH 7.0, respectively.

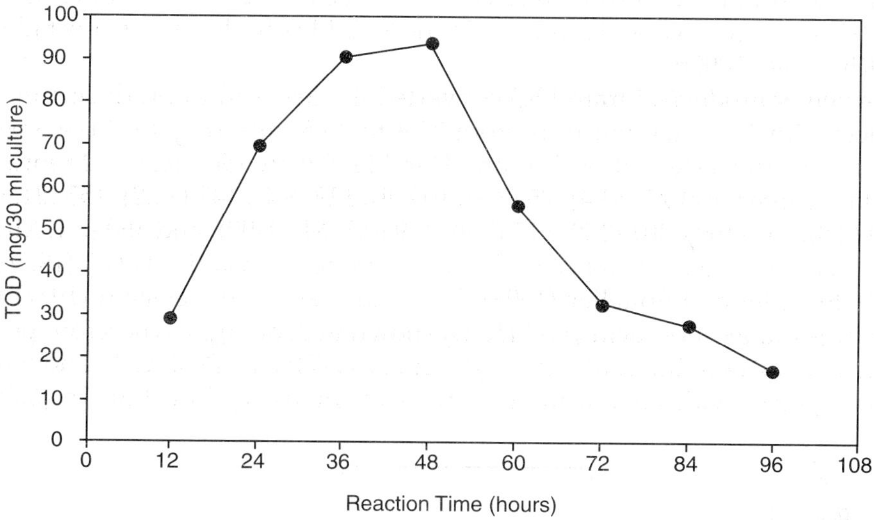

Fig. 5: Time coursed production of TOD.

Production of isomeric 9,10,13 (9,12,13)-trihydroxy-11*E* (10*E*)-octadecenoic Acid

There are some reports about the production of certain 18 carbon-trihydroxy fatty acids from plant (Baur *et al.*, 1977; Dix and Marnett, 1985; Esterbauer and Schauenstein, 1977), animal (Funk and Powell, 1983) and chemical methods (Gardner *et al.*, 1984). Of these trihydroxy fatty acids, 9,10,11-trihydroxy-12-octadecenoic acid (12-THOD), 9,10,13-trihydroxy-11(*E*)-octadecenoic acid (11-THOD), and 9,12,13-trihydroxy-10(*E*)-octadecenoic acid (10-THOD) have gained special attention. Kato *et al.* (1984) reported that the mixed hydroxy fatty acids isolated from the *Sasanishiki* variety of rice plant suffering from the rice blast disease exhibited a strong inhibition activity toward germination and elongation of germ tube of the conidia of rice blast fungus. Their structures were identified as 9*S*,12*S*,13*S*-trihydroxy-10-octadecenoic acid and 9*S*,12*S*,13*S*-trihydroxy-10,15-octadecadienoic acid (1985;1986). 10-THOD was also isolated from tubers of taro (*Colocasia antiquorum*) inoculated with *Ceratocystis fimbriata* and showed activity against black rot fungus (Masui *et al.*, 1989). Funk and Powell (1983) reported the formation of 10-, 11-, 12-THODs from endogenous substrate during the incubation of slices of faetal calf aorta in physiological medium and suggested the possibility that some of these products or their hydroperoxy precursors might have some biological significance. Isomeric mixtures of 11-THOD and 10-THOD were also isolated from onion (*Allium cepa*) and showed prostaglandin E-like activity in *in vitro* bioassay tests.

However, although the biological significance of these THODs were reported, their productions in nature are rare and restricted mostly in plant system, and the compounds reported so far were all produced in trace amounts. No microbial system was reported on the production of trihydroxy unsaturated fatty acids. Previously, Kim *et al.* (2000c) reported that a bacterial strain *Pseudomonas aeruginosa* PR3 was able to convert unsaturated fatty acids to dihydroxy fatty acids (Hou *et al.*, 1991; Hou and Bagby 1991; Kim *et al.*, 2000b; 2000c). They also reported that strain PR3 also converted linoleic acid to two new compounds: 9,10,13-trihydroxy-11(*E*)-octadecenoic acid (11-THOD) and 9,12,13-trihydroxy-10(*E*)-octadecenoic acid (10-THOD)(Kim *et al.*, 2000b).

Identification of products: Strain PR3 converted linoleic acid to produce a mixture of six major products with GC retention time from 37.8 to 40.5 min. (Fig. 6). Mass spectra of the TMS-derivatized methyl esters of peak A and D in Fig. 6 were identical and represented ions at m/z (% relative intensity) 73 (53.4), 75 (7.8), 109 (8), 129 (8.4), 147 (12.2), 155 (21.6), 173 (7.1), 211 (5.9), 230 (3.8), 259 (100), 301 (2.7), 317 (2.9), 439 (0.3, M-31-90), and 460 (2.1, M-100). These mass spectra were in good accordance with those obtained by Galliard *et al.* (1975), Funk and Powell (1983), Spiteller and Spiteller (1998), Esterbauer and Schauenstein (1977) for 9,10,13-trihydroxy-11(*E*)-octadecenoic acid (11-THOD) shown in A of Fig. 7. Therefore peak A and D were considered to be two different stereo isomers of 11-THOD. Peak C, E, and F represented identical mass spectra with ions at m/z (% relative intensity) 73 (51.3), 75 (6.7), 103 (8.6),

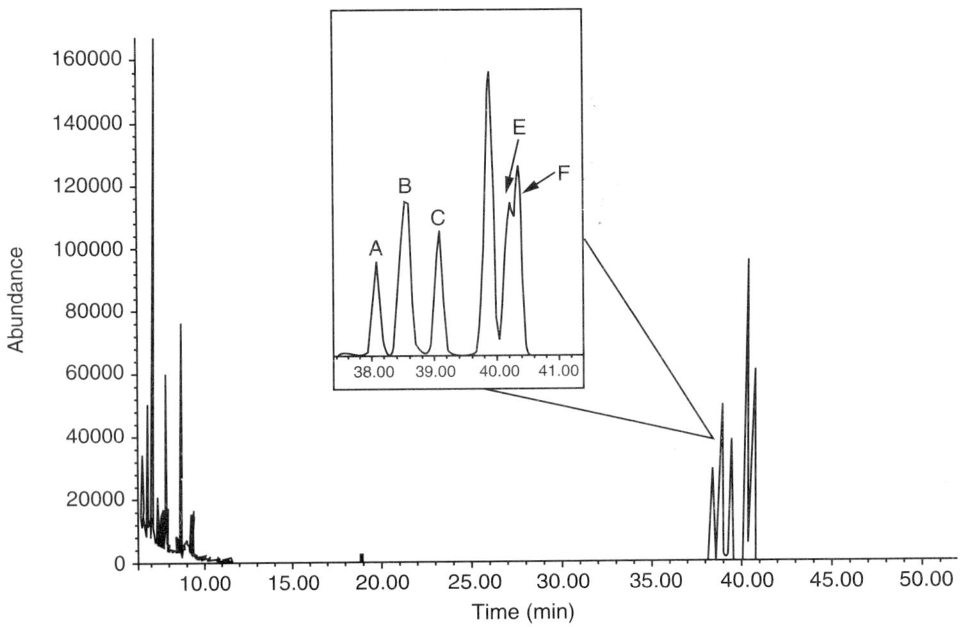

Fig. 6: Gas chromatogram of the partially purified trihydroxy fatty acid from the bioconversion of linoleic acid by PR3.

129 (3.8), 147 (12), 155 (3.6), 173 (100), 230 (5), 259 (12.4), 297 (1.7, 387-90), 317 (1.1), 387 (2.4), 439 (1.2, M-31-90), and 460 (7, M-100). These mass spectra were in good accordance with those reported in the literature (Esterbauer and Schauenstein, 1977; Funk and Powell, 1983) for 9,12,13-trihydroxy-10(*E*)-octadecenoic acid (10-THOD) shown in B of Fig. 7. Accordingly peak C, E, and F were considered to be different stereo-isomers of 10-THOD. Predominating base peak ions at m/z 259 (A of Fig. 7) and at m/z 173 (B of Fig. 7) well explained the different locations of the vicinal hydroxy groups on carbons 9, 10 and 12, 13 for 11-THOD and

THOD, respectively as shown in Fig. 7. However, the mass spectra of peak B of Fig. 6 were quite different from other two cases mentioned above with ions at m/z (% relative intensity): 73 (81.5), 75 (10.9), 103 (10.2), 109 (5.7), 129 (7.9), 147 (15.5), 155 (15.9), 173 (100), 212 (3.8), 230 (6.3), 259 (70.9), 297 (1.2, 387-90), 301 (1.4), 317 (2.5), 387 (1.7), 439 (0.8, M-31-90), and 460 (6.4, M-100). These were similar to those reported in the literature for the mixture of 11-THOD and 10-THOD (Graveland, 1970; Naughton, 1974; Spiteller and Spiteller, 1998) with 10-THOD relatively predominating. Based on the results obtained above, it was concluded that the six major peaks of GC chromatogram in Fig. 6 represented the different stereo-isomers or their mixture of trihydroxy octadecenoic acid with a *trans* double bond on either carbon 10-11 or 11-12. Even though the ratio of each peak remained constant, trial to further separate the distinct isomer peaks from peak B was failed resulting in unavailability of calculation of the amount of two separate THODs from peak B.

(A)

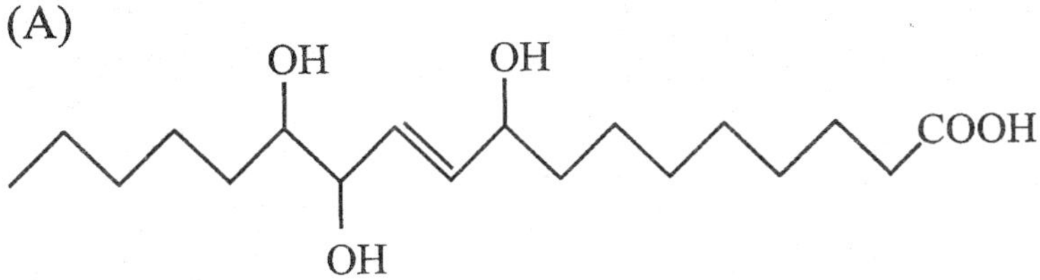

(B)

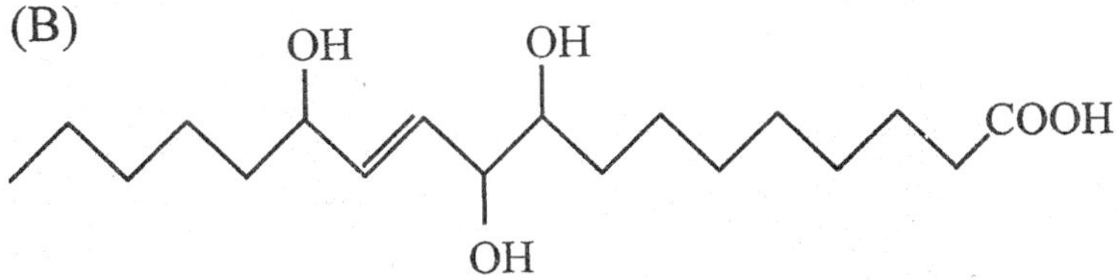

Fig. 7: Structure of 9,12,13-THOD(A) and 9,10,13-THOD(B).

Time course THOD production: The reaction was carried out with standard medium at 28°C for the time specified after the addition of linoleic acid to the 24 hr-old culture. The total amount of THODs produced in the culture increased with time up to 72 hrs. After peaked at 72 hrs with a maximum THOD production (220 mg/50 ml culture, 44% weight yield), the amount of THOD in the medium remained virtually unchanged up to 240 hrs indicating that THODs were not further metabolized by strain PR3. This result was different from that reported for the production of 7,10-dihydroxy-8(*E*)-octadecenoic acid (DOD) from oleic acid by the same strain PR3 (Hou and Bagby, 1991) in which DOD was metabolized by PR3 after peaked at 48 hrs. However, the production of 12,13,17-9(Z)-octadecenoic acid (THOA) from linoleic acid by *Clavibacter* sp. ALA2 (Hou *et al.*, 1996; Hou *et al.*, 1997) showed similar results that THOD was not further metabolized. Therefore, THOD production was standardized at 72 hr whenever other variables were evaluated.

Conditional optimization for THOD production: Various conditional factors were investigated for the maximal production of THOD by PR3. THOD production increased with the substrate concentration and peaked at 500 mg after which decreased slightly and remained a plateau up to 1 g. The effect of reaction temperature on THOD production was studied between 15°C and 45°C representing that THOD production increased significantly in parallel with temperature up to 30°C followed by saturation up to 40°C and sharp decrease at 45°C. THOD production was relatively sensitive to the reaction pH. The optimum pH for the production of THODs was determined to be 7.0 out of the range between 5.0 and 8.5 studied. Various organic and inorganic nitrogen compounds were tested independently for the production of THODs. Of those tested nitrogen sources (yeast extract, peptone, tryptone, NH_4NO_3, NH_4Cl, $NaNO_3$, urea, sodium glutamate), sodium glutamate was most effective to produce THOD but not sufficient compared to the control. Maltose, starch, sucrose, and glucose were effective for THOD production. Effects of glucose concentration on cell growth and THOD production were studied. THOD production and cell growth increased in parallel with glucose concentration up to 0.3 %. However, as glucose concentration continued to increase, cell growth reached saturation but THOD production declined significantly while the remaining substrate increased. These results suggested that PR3 utilized linoleic acid as a substituted carbon source for their growth when carbon source was not sufficient in the medium and THODs were produced as non-metabolizable by-products in this microbial process.

Effect of metal ions on the production of THODs by PR3: It has been reported that lipid peroxide molecules (LOOH) were decomposed to radicals in the presence of bivalent metal ions, in particular, iron or copper since these ions were known to be required to generate intermediate radicals (LO) leading to the formation of hydroxyl group on fatty acid (Gardner, 1989). To address the effect of metal ions on the production of THODs by PR3, we prepared medium in which all the defined metal ions in the standard medium were replaced with 70 mM of single metal ion-sulphate salt and tested each medium to compare THOD production under the standard condition. As shown in Table 1, Fe^{+2} and Cu^{+2} ions among the eight metal ions tested were sufficiently effective to produce THOD compared to the control medium. In the case of copper ion at a given concentration, THOD production was about 20% higher than that with control medium. However single use of iron ion at a given concentration was less effective by 30 % for THOD production compared to the control except for Mn^{+2} ion which showed 6% relative productivity of THODs. Other ions tested produced no THODs.

Table 1. Effect of metal ions on THOD production by PR3.

Metal ions (70 μM)[a]	Total THOD[c] (mg/50ml culture)	Cell density (O.D.@540 nm)
Control[b]	18.6	1.29
Fe^{+2}	13.0	1.54
Cu^{+2}	21.9	1.58
Zn^{+2}	nd	1.40
Mn^{+2}	1.1	1.30
Mg^{+2}	nd	1.48
Ca^{+2}	nd	1.67
Na^{+1}	nd	1.67
K^{+1}	nd	1.69

[a] All metal ions tested were supplied as sulphate salt
[b] Standard medium was used as control
[c] nd indicates not detectable

Effect of Fe^{+2} ion concentration on the production of THODs by PR3: Since Fe^{+2} ion was shown to be required for THOD production from linoleic acid by PR3, we investigated the effect of varied Fe^{+2} ion concentrations on the production of total THODs from linoleic acid by PR3. THOD production showed non-linear increment in relation to the increase of Fe^{+2} ion concentration (Fig. 8).

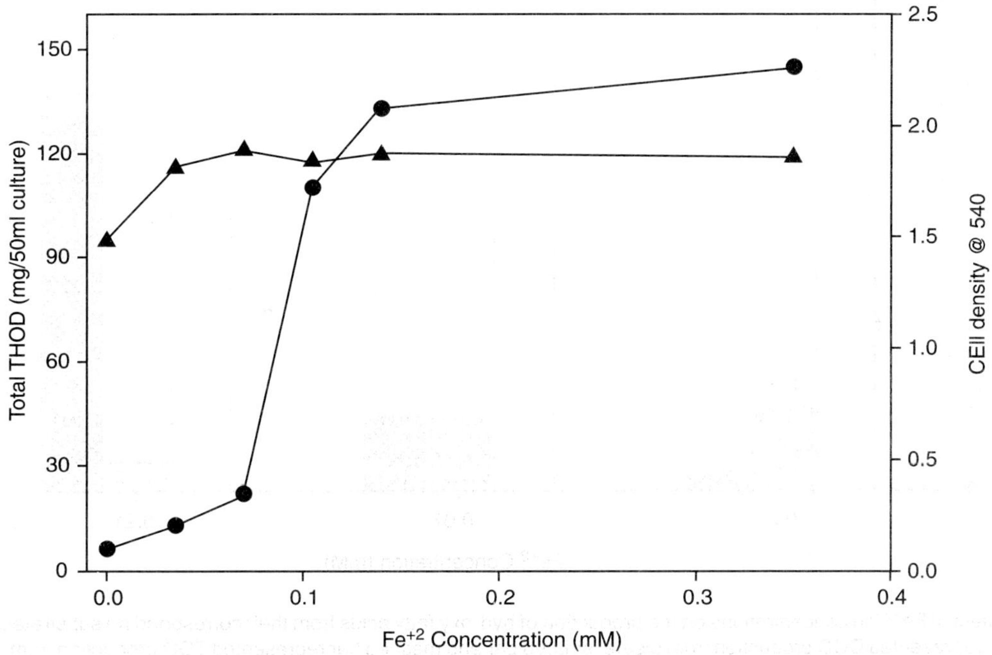

Fig. 8: Effect of Fe^{+2} ion concentrations on the production of total THODs from linoleic acid by PR3. Circle and triangle represented total THODs and cell density, respectively

THOD production increased steadily resembling lag phase of cell growth curve up to 0.07 mM of Fe^{+2} ion concentration after which the production entered exponential increase up to 0.14 mM of Fe^{+2} ions followed by saturation of THOD production. However, microbial growth was not significantly influenced by the concentration of Fe^{+2} ions tested. Even in the absence of defined metal ions at all, cell density marked up to 80 % of the maximal density. These results represented that the threshold concentration level of Fe^{+2} ions for the production of THODs from linoleic acid by PR3 existed between 0.07 mM and 0.14 mM. This concentration range accounted for the substrate/Fe^{+2} ion molar ratio of from 5100 to 2550. It is highly plausible to assume that existence of the threshold concentration level of Fe^{+2} ion for THOD production could possibly be raised by the preference of two iron ions Fe^{+2}/Fe^{+3} in the equilibrium for discrete reactions by the enzyme system involved in this event.

Effect of Fe^{+2} ion concentration on the production of hydroxy fatty acids from various substrates by PR3: PR3 used in this study has been reported to convert oleic acid and ricinoleic acid into 7,10-dihydroxy-8(E)-octadecenoic acid (DOD) and 7,10,12-trihydroxy-8(E)-

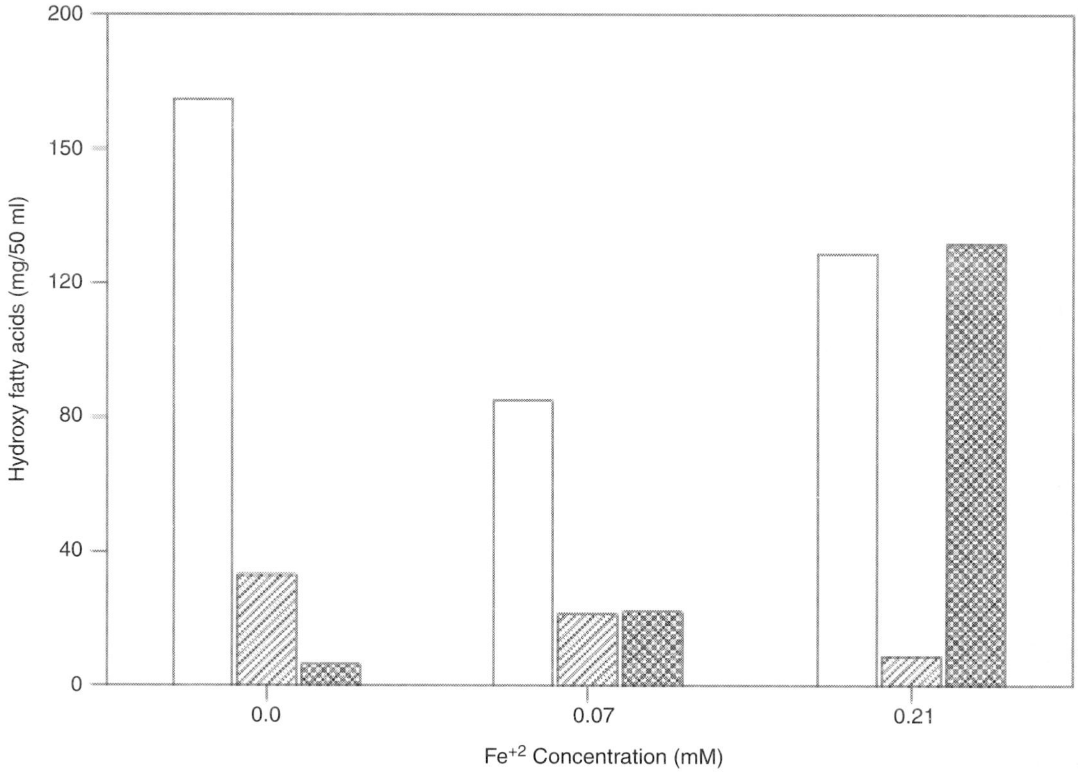

Fig. 9: Effect of Fe^{+2} ion concentrations on the production of hydroxy fatty acids from their corresponding substrates by PR3. Open bar represented DOD production from oleate, striated bar and meshed bar represented TOD production from ricinoleic acid and total THOD production from linoleic acid, respectively.

octadecenoic acid (TOD), respectively (Hou and Bagby, 1991; Kou *et al.*, 1998). In previous reports, it was postulated that oleic acid and/or ricinoleic acid-specific lipoxygenase rather than hydratase would be involved in the production of DOD and TOD from their corresponding substrate by pR3 (Kim *et al.*, 2000a,c). Hence, we investigated the effect of Fe^{+2} ions on the production of DOD and TOD from their corresponding substrates in comparison with THOD production from linoleic acid by PR3 (Fig. 9). When Fe^{+2} ions were added to at two different concentrations of 0.07 mM and 0.21 mM, THOD production increased greatly as expected. However, TOD production from ricinoleic acid linearly decreased with relatively low production level and DOD production from oleic acid fluctuated with its production in the absence of Fe^{+2} ion being even higher than that in the presence of iron. These results demonstrated that the enzyme(s) involved in the production of THOD from linoleic acid must be different from the enzyme system involved in the formation of DOD and TOD from their corresponding substrates by PR3 and that the enzyme could be lipoxygenase. This was supported by the report that the key element of the substrate for the non-heme iron containing lipoxygenase was a 1,4-*cis,cis*-diene unit present in the structure of linoleic acid (Veldink and Vliegenthart, 1984). This structure is different from the single *cis* double bond found on C9, 10 of oleate and ricinoleic acid. However, further research is needed to identify the lipoxygenase in PR3.

CONCLUSION

Various unsaturated fatty acids were used for the production of various hydroxy fatty acids by a bacterial strain *Pseudomonas aeruginosa* PR3. Among those unsaturated fatty acids used for microbial production of hydroxy fatty acids, oleic acid, linoleic acid, and linolenic acid were well studied as substrate to produce mono-, di-, and trihydroxy fatty acids. Important environmental conditions for their production were also well studied. Based on these successful data, further studies should be focused on the investigation of their biological properties for their industrial applications.

REFERENCES

Baur, C., Grosch, W., Wieser, H. & Jugel, H. (1977) *Z. Lebesm. Unters-Forsxh* **164**: 171-176.
De Andres, C., Mercade, E., Guinea, J. & Manresa, A. (1994) *W. J. Microbiol. Biotechnol.* **10**: 106-109.
Dix, T. A. & Marnett, L. J. (1985) *J. Biol. Chem.* **260**: 5351-5357.
Esterbauer, H. & Schauenstein, E. (1977) *Monatshefte. Für. Chemie.* **108**: 963-972.
Fahlstadius, P. & Hamberg, M. (1989) *Chem. Phys. Lipids* **51**: 15-22.
Funk, C. D. & Powell, W. S. (1983) *Biochim. Biophys. Acta.* **754**: 57-71.
Galliard, T., Phillips, D. R. & Matthew, J. A. (1975) *Biochem. Biophys. Acta.* **409**: 157-171.
Gardner, H. W. (1989) *Free. Radic. Biol. Med.* **7:** 65-86.
Gardner, H. W., Nelson, E. C., Tjarks, L. W. & England, R. E. (1984) *Chem. Phys. Lipids.* **35**: 87-101.
Graveland, A. (1970) *J. Am .Oil. Chem. Soc.* **47**: 352-361.
Hamberg, M. (1971) *Anal. Biochem.* **43**: 515-526.
Hamberg, M., Herman, R. P. & Jacobsson, U. (1986) *Biochem. Biophys. Acta.* **879**: 410-418.
Hou, C. T. & Bagby, M. O. (1991) *J. Ind. Microbiol.* **7**: 123-130.
Hou, C. T. & Bagby, M. O. (1991) *J. Ind. Microbiol.* **7**: 123-130.
Hou, C. T. & Bagby, M. O. (1992) *J. Ind. Microbiol.* **9**: 103-107.
Hou, C. T. (1995) In: *Advances in Applied Microbiology: Microbial Oxidation of Unsaturated Fatty Acids*, (Eds. Laskin, A. I.) Academic Press, Orlando, pp 1-23.
Hou, C. T. (1995) *J. Ind. Microbiol.* **14**: 31-34.
Hou, C. T. (1996) *J. Am. Oil. Chem. Soc* **73**: 1359-1362.
Hou, C. T., Bagby, M. O., Plattner, R. D. & Koritala, S. (1991) *J. Am. Oil. Chem. Soc.* **68**: 99-101.
Hou, C. T., Brown, W., Labeda, D. P., Abbott, T. P. & Weisleder, D. (1997) *J. Ind. Microbiol. Biotechnol.* **19**: 34-38.
Hou, C. T., Nkamura, L. K., Weisleder, D., Peterson, R. E. & Bagby, M. O. (1993) *W. J .Microbiol. Biotechnol.* **9**: 570-573.
Hou, C.T. (1995) *J. Am. Oil. Chem. Soc.* **72**: 1265-1270.
Kato, T., Yamaguchi, Y., Abe, N., Uyehara, T., Nakai, T., Yamanaka, S. & Harada, N. (1984) *Chem. Lett.* **25**: 409-412.
Kato, T., Yamaguchi, Y., Abe, N., Uyehara, T., Namai, T., Kodama, M. & Shiobara, Y. (1985) *Tetrahedron. Lett.* **26**: 2357-2360.
Kato, T., Yamaguchi, Y., Ohnuma, S. I., Uyehar, T., Namai, T., Kodama, M. & Shibara, Y. (1986) *Chem. Lett.* 577-580.
Kim, H., Gardner, H. W. & Hou, C. T. (2000a) *J. Am. Oil. Chem Soc.* **77**: 95-99
Kim, H., Gardner, H. W. & Hou, C. T. (2000b) *J. Ind. Microbiol. Biotechnol.* **25**: 109-115
Kim, H., Kuo, T. M. & Hou, C. T. (2000c) *J. Ind. Microbiol. Biotechnol.* **24**: 167-172

Kleiman, R. & Spencer, G. F. (1973) *J. Am. Oil. Chem. Soc.* **50**: 31-38.
Kuo, T. M., Kim, H. & Hou, C. T. (2001) *Curr. Microbiol.* **43**: 198-203
Kuo, T. M., Manthey, L. K. & Hou, C. T. (1998) *J. Am. Oil. Chem. Soc* **75**: 875-879.
Masui, H., Kondo, T. & Kojima, M. (1989) *Phytochem.* **28**: 2613-2615.
Mercade, E., Robert, M., Espuny, M. J., Bosch, M. P., Manreesa, M. A., Parra, J. L. & Guinea, J. (1988) *J. Am. Oil. Chem. Soc.* **65**: 1915-1916.
Method Cd 14-61. (1981) Official and tentative methods of AOCS, 3rd edn (Ed. Walker Ro), AOCS, Champaign, IL.
Naughton, F. C. (1974) *J. Am. Oil. Chem. Soc.* **51**: 65-71.
Niehaus, W. G.. & Schroepfer, G.. J. Jr. (1965) *Biochem. Biophys. Res. Commun.* **21**: 271-275.
Niehaus, W. G.. & Schroepfer, G.. J. Jr. (1967) *J. Am. Chem. Soc.* **89**: 4227-4228.
Niehaus, W. G.., Kisic, A., Torkelson, A., Bednarczyk, D. J. & Schroepfer, G.. J. Jr. (1970) *J. Biol. Chem.* **245**: 3790-3797.
Schroepfer, G.. J. Jr. & Block, K. J. (1963) *J. Am .Chem. Soc.* **85**: 3310-3315.
Schroepfer, G.. J. Jr. & Block, K. J. (1965) *J. Biol. Chem.* **240**: 54-65.
Schroepfer, G.. J. Jr. (1966) *J. Biol. Chem.* **241**: 5441-5447.
Spiteller, P. & Spiteller, G. (1998) *Biochim. Biophys. Acta* **1392**: 23-40.
Veldink, G. A., Vliegenthart, J. F. (1984) *Adv. Inorg. Biochem.* **6**: 139-61
Wallen, L. L., Benedict, R. G. & Jackson, R. W. (1962) *Arch. Biochem. Biophys.* **99**: 249-253.

8

Bioactive Potential of Actinomycetes from India and Saudi Arabia

N. N. Nawani, M. S. Shejul, A. Y. Al-Mahdi and B. P. Kapadnis

Department of Microbiology, University of Pune, Pune 411007 (India)

Actinomycetes are known for their bioactive potential which includes production of several enzymes and antibiotics. Nevertheless, the search for new and potent bioactive compounds is mandatory due to emergence of new infections and recurrent problems of environmental pollution. In view of this, actinomycetes with an ability to grow optimally at acidic, neutral and alkaline pH were isolated from soil and water samples from two geographically distant countries, India and Saudi Arabia. A comparative account of the microbiological and bioactive characteristics of isolates from two regions indicated considerable diversity as well as certain similarities suggesting differential divergence during evolution and probable impact of environmental conditions. Overall, the isolates possessed several enzymatic and antimicrobial activities competitive with the existing reported strains. Enzymes like chitinases, xylanases, pectinases, cellulases, amylases, lipases and proteases active at different pH and temperature values were produced by many isolates. An acidophilic and thermophilic chitinase active at pH 3.0 and 60°C was reported for the first time from *Microbispora* sp. V2 isolated from India. Several isolates also produced antibacterial and antifungal compounds particularly non-polyene in nature. Morphologically distinct genera like *Nocardioides, Kitasatosporia, Micromonospora, Microbispora, Nocardia, Actinomadura, Catellatospora* and *Streptomyces* were obtained from both regions. The knowledge of this generic and bioactive diversity can also be a useful inception in correlation of intra- and interspecific diversity.

ACTINOMYCETES: OCCURRENCE AND CHARACTERISTICS

Actinomycetes are a diverse group of heterotrophic prokaryotes forming hyphae at some stage of their growth (Gottlieb, 1973), hence referred to as filamentous prokaryotes. The existence of the actinomycetes has been recognized for over hundred years where earlier they were recognized as an exotic group with affinities to both bacteria and fungi, but determination of their fine structure and chemical composition confirmed their prokaryotic nature. They have different morphological, cultural, biochemical and physiological characters (Gattlieb, 1973).

This group is a potential producer of many enzymes, enzyme inhibitors, growth promoting substances and antibiotics (Lechevalier and Lechevalier, 1967).

Actinomycetes are Gram-positive bacteria belonging to the order Actinomycetales (Lechevalier, 1981), characterized by the formation of substrate and aerial mycelium on solid media, presence of spores and a high GC content of the DNA (60-70 mol%). The composition of the growth media can profoundly affect the growth and stability of substrate and aerial mycelium (Kalakoutskii and Arge, 1976). The majority of actinomycetes are soil bacteria that play an important role in mineralization processes in nature. These bacteria can be separated into different genera on the basis of morphological, physical and chemical criteria. Actinomycetes possess cell-wall type I to IV, depending on the presence of L-diaminopimelic acid (DAP) and glycine (type I), meso-DAP and glycine (type II), meso-DAP (type III), or meso-DAP, arabinose and galactose (Type IV) (Goodfellow, 1989).

Actinomycetes are a successful group of bacteria that occur in a multiplicity of natural and man-made environments (Goodfellow and Williams, 1989). The ubiquity of actinomycetes in the natural environment is due to two main factors: metabolic diversity and the evolution of specific mechanisms for dispersal. Most actinomycetes are strict saprophytes, but some of them form parasitic or mutualistic associations with plants and animals. Since the actinomycete group constitutes varied genera physiologically and nutritionally distinct, they have been detected in all types of ecosystems. Their ability to degrade almost all natural polymers makes them ubiquitous group. Their slow growth rate is a limiting factor for their presence in all types of niches. However, this has been overcome due to the production of several secondary metabolites inhibitory to the bacteria and fungi, which make them a successful group amongst all microorganisms. They have been found in all ecological niches such as soil, water (marine and fresh), compost, extreme environments, marshy places etc. Extreme environments such as alkaline, acid, high temperature, high pressure, complex ecological system, polluted environments, rhizospheres can be looked for screening new actinomycetes. The recovery of acidophilic forms from soil of low pH (Hagedoen, 1976, Khan and Williams, 1975) and identification of a novel actinomycete population colonizing a salt marsh are good examples of the ability of actinomycetes to adapt to range of environmental factors. The actinomycetes have received comparatively little attention from aquatic habitats. The genera most frequently present in water are *Actinoplanes, Rhodococcus, Thermactinomyces* and *Streptomyces*. A variety of genera have been isolated from, sea-water and marine sediments including *Micromonospora, Microbispora, Nocardia* and *Streptomyces*.

Actinomycetes are able to produce extracellular enzymes (Morosoli *et al.*, 1997) which can decompose various materials. Actinomycetes have been implicated in the biodeterioration of silicone and natural rubber, wool, PVC plastic, cotton, jute fibre, wood, hay and grain. Out of 50 rubber degrading actinomycetes, 33 were identified as streptomycetes which have been studied. Mesophilic actinomycetes, many of which belong to the genus *Streptomyces*, have been frequently isolated from all kinds of building materials, especially from ceramic materials, paints and glues (Hyvarinon *et al.*, 2002). They are able to grow and proliferate on plasterboards (Murtoniemi *et al.*, 2001).

Actinomycetes constitute a significant component of the microbial population in most soils. It has been estimated that counts of actinomycetes over 1 million per gram are commonly obtained. Over twenty genera have been isolated from soil. Lechevalier and Lechevalier found that 95% isolates belonged to streptomycetes (Lechavalier and Lechavalier, 1967). Environmental factors influence the type and population of actinomycetes in soil. Most actinomycete isolates behave as neutrophiles capable of growth range at pH ranging from 5.0 to 9.0 with an optimum around 7.0. The pH is a major environmental factor determining the distribution and activities of soil actinomycetes. Neutrophiles occur in less number in acidic soils below pH 5.0, whereas acidophilic and acidoduric streptomycetes are numerous in acidic soils (Goodfellow and Williams, 1983). However, there are few reports of basophilic streptomycetes. Most actinomycetes behave as mesophiles in the laboratory with optimum growth temperature between 25 and 30°C. Many mesophilic actinomycetes are active in compost. However, the capacity of self-heating during decomposition often provides ideal conditions for obligate or facultative thermophilic actinomycetes which are able to grow at temperatures above 40°C.

It has been observed that the actinomycetes in soil grow attached to surfaces, such as plant residues or fungal hyphae (Mayfield *et al.*, 1972), and they may have an important ecological role in the degradation of litter in soil (Goodfellow and Williams, 1983, Khan and Williams, 1975). Actinomycetes play an important role in decomposition of plant and other substances especially in the degradation of complex and relatively recalcitrant polymers (Lacey, 1973). They degrade lignin, cellulose, lignocelluloses, hemicellulose, pectin, keratin, chitin and fungal cell-wall materials (Waksman, 1961). Actinomycetes from rhizosphere suppress the growth of pathogens and can also synthesize gibberellin-like substances. Some actinomycetes produce indole acetic acid in culture and when inoculated into the soil (Lechavaliar, 1988) indicating their growth promoting nature. They also have an antagonistic role as reduced incidence of root infection has been correlated with an increase in number of antibiotic producing *Streptomyces* in the rhizosphere (Lechavaliar, 1988). The strains of genus *Frankia* are nitrogen-fixing endophytes in root nodules of various non-leguminous shrubs and trees. There have been surprisingly few quantitative studies, but R-S ratios based on viable counts are often low in fertile soils, while those in sand dunes ranged from 16 to 50.

Many mesophilic actinomycetes are active in compost in the initial stages of decomposition. However, the capacity for self-heating during decomposition provides ideal conditions for obligate or facultative thermophilic actinomycetes. The predominance of actinomycetes is well established, and can even be directly observed as a surface bloom, termed 'fire fang', in composts. Some genera like *Thermoactinomycetes* and *Saccharomonospora* are strictly thermophilic. Thermophilic actinomycetes grow well on animal manure and are active in fermentation and deodorization of pig faeces (Goodfellow and Williams, 1988). *Thermomonospora* species particularly grow during the second indoor phase of preparation of manure for mushroom cultivation, whereas *Streptomyces diastaticus* and *Thermoactinomyces vulgaris* predominate in the spent steamed compost and its dust (Lacey, 1973). *Thermomonospora curvata* was shown to be active in decomposition of municipal waste compost and to produce thermostable cellulase (Goodfellow and Williams, 1983).

Actinomycetes are widely distributed in aquatic habitats and can be recovered readily from fresh water, sea-water and sediment samples. Taxonomic diversity comparable to that in soil is usually encountered and this leads to the suggestion that isolates are merely the result of ' wash-in '. Thus they are part of the indigenous microflora or possibility of wash-in forms from surrounding terrestrial habitats is a matter of repeated retrospection of the water samples.

Actinoplanes, *Micromonospora*, *Rhodococcus*, *Streptomyces* and the endospore forming *Thermoactinomyces* can readily be isolated from freshwater sites. Evidence of the existence and growth of actinomycetes in fresh water habitats has been studied by Cross (1981). *Actinoplanes* and related organisms are common in soils, rivers and lakes and can be grown on plant litter in rivers. The *Micromonospora* were considered to be indigenous to freshwater ecosystem and have a role in the turnover of cellulose, chitin and lignin (Nawani *et al.*, 2002). Several workers confirmed the presence of *Micromonospora* in streams, rivers and lake sediments (Burman, 1973, Nawani and Kapadnis, 2003).

Streptomycete spores are also being continually washed into freshwater and marine habitats. Some investigations have claimed the existence of aquatic streptomycetes and growth on the chitinous exoskeletons of *Procambarus versutus* immersed in a woodland stream (Cross, 1981). Actinomycetes change taste and odour of potable water rendering it unpalatable. Earthy taste and odour is caused because of compounds like geosmin and methyl iso-borneol are produced by *Streptomyces*. Plant and animal debris at water margins can provide substrates for actinomycetes growth and geosmin production (Jiang and Xu, 1996, Niemi *et al.*, 1982).

Evaluation of the actinomycete load in air uses a cascade impactor and liquid impinger (Gregory & Lacey, 1963) and particular attention to sampling regimes is required to obtain quantitative data on them. The ability of actinomycete spores to act as respiratory allergens has provided the main impetus for studies on the recovery of actinomycetes from air. Their further implication in some cases of humidifier fever or ventilation pneumonitis has extended beyond the traditional agricultural or industrial environments. This relationship between actinomycete spores and respiratory disease has generated a substantial amount of information on numbers and species of actinomycetes found in air (Lacey and Crook, 1988).

CHARACTERISTICS OF ACTINOMYCETES

Actinomycetes are heterotrophic in nature. Most of them are strict saprophytes, while some form parasitic or mutualistic associations with plants and animals (Goodfellow and Williams, 1983). They are aerobic and most of them readily grow on the common bacteriological media like nutrient agar, trypticase agar, blood agar, brain heart infusion agar and starch casein agar. *Sporoactinomycetes* require special media to allow differentiation and the development of characteristic spores and pigments. Some of these media are not available commercially and must be prepared in the laboratories using colloidal chitin, soil extract and decoctions of plant materials. Nocardiform actinomycetes exhibit fragmentation, the hyphae break up into rods and cocci and form soft or friable colonies (Holt, 1994). The most diversely studied genus *Streptomyces* gives pale, shiny, hard colonies on nutrient agar which can be transformed into bright colonies with a powdery white mycelium on media like oatmeal or inorganic salts starch agar (Williams, 1989). Outgrowths from spore or fragments of mycelium develop into hyphae that penetrate the agar (substrate mycelium) and hyphae that branch repeatedly and become cemented together on the surface of the agar to form a tough, leathery colony. The density and consistency of the colony depend on the composition of the medium. The aerial mycelium can be initially white but later forms different colours when spore formation begins. Thus the colonies appear powdery or velvety and can be readily distinguished from the more typical bacterial colonies (Kalakoutskii and Agre, 1976; Krasilnkov, 1960). *Streptomyces* species have chains of spores on the aerial mycelium, which are normally absent on the substrate mycelium. These spores are arthrospores, regular segments of hyphae with a thickened spore-wall surrounded by a hydrophobic sheath that may bear spines or hairs. *Thermomonospora* species can have clusters of single spores, aleuriospores, on both aerial and substrate mycelium. *Rhodococcus* species have a rapidly fragmenting mycelium, the segments become rounded and the colony usually consists of a mass of coccoid elements. *Actinoplanes* species form globose sporangia on the surface of the colony, these sporangia liberate motile zoospores (Williams, 1989). *Dactylosporangium* species have tubular sporangia containing few potentially motile spores together with aleuriospores on the substrate mycelium.

Other than the media used for the growth of actinomycetes, other physiological characteristics are also important for their growth. Most of them are neutrophilic however, some acidophilic and alkalophilic actinomycetes have been detected (Nawani and Kapadnis, 2003). A branching mycelium growing at the surface of transparent agar may appear in 3 to 4 days. Formation of mature aerial mycelium with spores may require 7 to 14 days and few slow growing strains require one-month incubation. The growth temperatures are close to 28 to 30°C. The thermophilic strains require 45 to 55°C temperature (Iwai and Omura, 1982). Their primary carbon sources in the soil are insoluble, polymeric substrates. They have the ability to penetrate and solubilize these polymers whether of plant (lignocellulose) or animal (chitin) origin (McCarthy and Williams, 1990). The conditions used for their growth in the laboratory are a replica of the natural environments. Thus media made from decoctions of plant material support the growth of isolates that may not be able to grow on the conventional media.

An evaluation of acidophilic actinomycetes has shown that those which grow between pH 3.5 and 6.5 form a diverse group and they are numerous and widespread in natural and man-made acidic habitats (Williams *et al.*, 1971). They are probably active in decomposition processes in such soils and produce diastases (Williams and Flowers, 1978) and chitinases (Williams & Robinson, 1981) adapted to function at lower pH than those of neutrophilic actinomycetes. Members of the neutrotolerant acidophilic actinomycete group are poor in their ability to use a range of carbon sources than their acidophilic counterparts (Seong *et al.*, 1993, Miyadoh, 1993).

BIOACTIVITY AND INDUSTRIAL POTENTIAL

Antibiotics

More than 50% of the known natural antibiotics produced are from actinomycetes (Berdy, 1984, Miyadoh, 1993). Studies of primary metabolism in actinomycetes have been very limited but would provide important information about the switch between primary and secondary metabolism and for further improvement of industrial processes for the production of secondary metabolites. As mentioned, actinomycetes have been described as the greatest source of antibiotics. Two-thirds of the present-day antibiotics are of actinomycete origin like aminoglycosides, anthracyclines, chloramphenicol, macrolides, tetracyclines, nucleosides, peptides, polyenes and polyethers. Till 1974 antibiotics of actinomycete origin were almost exclusively confined to *Streptomyces*. Now-a-days efforts are being made to explore rare actinomycetes like *Actinomadura, Actinoplanes, Ampullariella, Actinosynnema, Dactylosporangium* and *Kibdilosporangium* etc. for the search of the antibiotics. A search for novel antibiotic producing actinomycetes in new ecological niches with a target directed approach is currently being pursued. Molecular biology techniques have helped on large scale in finding new antibiotics from actinomycetes (Austin, 1989).

Enzymes

Actinomycetes have been found to be a promising source of wide range of important enzymes, some of which are produced on an industrial scale, but many other remained to be harnessed. Presently, enzymes of microbial origin are widely used in food processing, detergent manufacturing, textile and pharmaceutical industries, medical therapy, bioorganic chemistry and molecular biology.

Cortisone reductase from *Streptomyces hydrogenans* has been used in model studies on the effect of substituents introduced to the steroid nucleus and the nature of the steroid-binding site on the substrate structure relationship. Cholesterol oxidase from *Nocardia* and *Streptomyces* (Smith, 1984) has been used for determination of cholesterol in blood serum. Besides this, attempts have been made to apply this enzyme to lower the blood cholesterol in mammals (Slobodskaya *et al.*, 1980). Choline oxidase from *Streptomyces nigrifaciens* (Manyu Seijaku, 1982) has been applied in tests used for the determination of levels of phospholipids in blood serum of patients with liver complaints such as obstructive jaundice.L-Phenyl alanine dehydrogenase catalyses the reversible formation of phenyl pyruvate from L-phenyl alanine. L-phenyl alanine from *Rhodococcus* sp. M4 is employed in the production of a dipeptide sweetener, aspartame that is used in food industry (Kieslich, 1976). Urate oxidase from *Streptomyces cyanogenus* has been attempted to treat patients suffering from hyperuraemia and primary gout (Brogard *et al.*, 1972).

Hydrolases like lipases, phospholipases, amylases, cellulases, endonucleases, xylanases, chitinases and peptide hydrolases are produced by actinomycetes. Many actinomycete enzymes are or could be successfully applied in biotechnology, clinical chemistry and medical therapy. They are more attractive than enzymes from other sources because of their high stability and unusual substrate specificity (Czoch and Mordarski, 1988). Glycosidases of all kinds are exploited for the degradation of plant biomass. Proteases from actinomycetes serve as additives to detergents or are used in tanning industry. Glucose isomerases have been successfully used to obtain fructose-heavy syrups. Bacteriolytic or mycolytic enzymes from actinomycetes may be used for beer and wine clarification or serve as non-toxic food preservatives (Czoch and Mordarski, 1988).

Transformation of Xenobiotics

Transformation of xenobiotics is defined as the structural modification of components foreign to an organism's metabolism, which occur in its chemical environment. The most characteristic reactions in transformation of xenobiotics are oxidative, reductive, hydrolytic, dehydrative and condensation. The ability of actinomycetes to perform a variety of microbial conversions of organic compounds is an important factor in the complicated processes of biodegradation of pollutants in soil and water and in the biodegradation of biopesticides, oil spills, chemical waste decomposition and the like. Members of the genera *Nocardia* and *Streptomyces* have the ability to perform highly selective chemical modification of complicated compounds of natural and synthetic origin. *Nocardia* strains have been found to degrade aromatic hydrocarbons, first by hydroxylation, followed by other reactions (Kieslich, 1976). Actinomycetes indigenous to soil and water are probably the first line of attack on hydrocarbon molecules, as these compounds are not broken down by the majority of microorganisms. In general actinomycetes have the ability to hydroxylate aliphatic chains of hydrocarbons in the terminal and subterminal positions and subsequently followed by shortening of the transformed chains (Golovlev *et al.*, 1978). Actinomycetes are able to degrade certain pesticides. The herbicide, dalapon, 2,2, - dichloropropionic acid, was degraded by selected *Nocardia* strains isolated from soil (Hirsch & Alexander, 1960). DDT (1,1,1-trichloro-2, 2-Di-4-chlorophenyl-ethane) is dechlorinated by *Streptomyces aureofaciens, Streptomyces cinnamoneous* and *Nocardia erythropolis* (Chacko *et al.*, 1966). The application of actinomycetes in bio-organic chemistry to perform microbial transformation of organic compounds has become a field of great interest and of commercial value. The ability to effect subtle modifications of complicated organic compounds has been employed in the production of several interesting products, some of which have been subjected to further chemical synthesis to provide valuable pharmaceutical agents.

Enzyme inhibitors

In 1972, Umezawa reported a low molecular weight enzyme inhibitor from a streptomycete strain. Since then more than 60 inhibitors have been reported, eg., inhibitors for papain, plasmin, trypsin, chymotrypsin and cathepsin B (Umezawa *et al.*, 1972). Enzyme inhibitors are finding possible use in cancer treatment, eg. revistin, an enzyme inhibitor, from *Streptomyces* sp. inhibit reverse transcriptase (Numata *et al.*, 1975). Streptonigrin and retrostatin synthesized by *Streptomyces* inhibit reverse transcriptase (Nisho *et al.*, 1983). The inhibitors are now-a-days used as therapeutic agents (Ellaiah *et al.*, 2002). The actinomycetes synthesize inhibitors of low molecular weight having various structures and pharmacological activities (Wu *et al.*, 2002).

Immunomodifiers

Low molecular weight compounds have been isolated from culture filtrates of actinomycetes, which enhance immune responses, such agents are called immunomodifiers. Inhibitors of enzymes located on the surface of cells involved in immunity may bind to such cells and augment immune responses. Moreover, immunity augmenting peptides have also been found. Bestatin (Umezawa *et al.*, 1972) from *Streptomyces olivoreticuli*, amastatin (Aoyagi *et al.*, 1978, Tobe *et al.*, 1982) from *Streptomyces* species ME 98-M3, phenicine (Aoyagi *et al.*, 1972) from *Streptomyces lavendulae*, enhanced immune responses in mice. Immunosuppressive agents such as FC-900506 reported by Fujisawa pharmaceutical Company, produced by *Streptomyces tsukubaensis*, shows stronger inhibition against interleukin-2 production, mixed lymphocyte reaction, interferon, cytotoxic-T cells and platelet activating factor- C induction.

Agriculture

It is well known that actinomycetes carry out numerous activities in soil including degradation of organic matter, inhibition or stimulation of other microorganisms and plants, transformation of chemical compounds such as herbicides and other agriculturally useful compounds (Ktause *et al.*, 1985, Lechevalier, 1981). Lignocellulose is converted to humus by *Streptomyces* imparting desirable properties to soil (Kaluskar, 1995). Actinomycetes with activity against plant pathogen and antagonistic actinomycetes have been detected in the rhizosphere of diseased plants (Kundu and Nandi, 1985). Plant growth has been enhanced by free-living actinomycetes. Actinomycetes isolated from rhizosphere of pine, produced B vitamin and are known to enhance the growth of plants. It was proposed that actinomycetes may thus indirectly contribute to plant growth stimulation (Strzelczyk and Leniarsko, 1985). There are also recent reports suggesting the use of whey for propagation of actinomycetes with potential to promote plant growth (Kanavade, 2003, Kanavade and Kapadnis, 2003).

Biotechnology

The attention given to the actinomycetes in biotechnological applications is a natural result of the great metabolic diversity of these organisms and their long association with the environment and needs of humans. Other beneficial activities of actinomycetes related to biotechnology is their gene expression activity which is becoming an active research area. Several different species of RNA polymerase holoenzyme have been isolated from *Streptomyces coelicolor*. A limited number of streptomycete promoters have been characterized, but a consensus sequence has not been formulated. Actinomycetes contain one major (circular) linkage group and a variety of extrachromosomal elements. The genome size in streptomycetes is thought to be 1.5 to 2 times that of *Escherichia coli*. *Streptomyces* DNA has a high GC content (70-74 %) which may have selective advantages in nature. On the other hand, mycobacterial DNA has a lower GC content and a broader range. Repetitive DNA sequences appear widespread in streptomycetes. Strain improvements to increase application or productivity can be utilized in industrial fermentations.

Waste Management

Wastes from food processing operations contain significant amounts of starch (Lemmet *et al.*, 1979). Amylolytic actinomycetes might be used for bioconversion of such wastes. Several mesophilic *Streptomyces* strains have been shown to produce amylase that hydrolyses starch and raw starch granules to maltose (Fairbairn *et al.*, 1986). The amylases of thermophilic actinomycetes have more commercial potential than those of mesophilic *Streptomyces* eg. *Thermoactinomyces vulgaris* is a good producer of a heat stable, highly active amylase. Cyanide and other toxic metals are produced on industrial scale for use in the metal extraction, electroplating and steel industry etc. In nature cyanide can be assimilated or detoxified by plants. Certain microorganisms also can degrade cyanide to formamide, ammonia and carbon dioxide or cyaloalanine. *Streptomyces lavendulae* found to be a degrader of cyanide. It has a rhodanase enzyme that transfers sulphur from sodium thiosulphate to cyanide, forming thiocyanate. Other streptomycetes such as *Streptomyces phaeoviridae* and *Micromonospora braunna* isolated from soil have been reported to degrade cyanide (Shete, 1994).

Members of many genera of actinomycetes have potential for use in the bioconversion of under-utilized agricultural and urban wastes into high value chemical products (Crawford *et al.*, 1984, Mackenzie *et al.*, 1984). Lignin is an aromatic polymer linked to cellulose in lignocellulose. The economic production of useful chemicals from lignin components is possible and lignin products could be used to produce biodegradable plastics. A solid-state fermentation of lignocellulose by lignolytic *Streptomyces* strains to produce APPL has been developed on a laboratory scale (Ball *et al.*, 1989, Crawford, 1981, Crawford *et al.*, 1984). APPL was produced by *Streptomyces sp.* LDSt1, LDSt2 and LDSt3 (Kaluskar, 1995).

Fermentation Technology

As actinomycetes form a major source of enzymes and antibiotics, so, the scaling up of these production processes is mandatory. Understanding the process and sequence of morphological development in actinomycetes colonies is accompanied by biochemical differentiation, which has important consequences in biotechnology as it often involves the formation of secondary metabolites. The appearance of glycogen storage material and antibiotic takes place when the rate of biomass accretion starts to decay (Bushell, 1988). Although surface growth is the means by which actinomycetes grow in nature, submerged culture is the method of choice for industrial fermentations. In such liquid cultures, growth can occur in different modes eg. pellets, mycelial fragments and well-dispersed mycelia. Pellet size is a function of inoculum size, dissolved oxygen, shear and pH. The well-dispersed mycelial type of growth is usually best for product formation. Branching is true in *Streptomyces* and *Thermoactinomyces*, whereas in actinomycetes such as *Nocardia* and *Rhodococcus* fragmentation follows branching.

Few genera of actinomycetes are fermentative, growing under low concentration of oxygen eg. *Actinomyces, Agramyces, Arachnia, Corynebacterium, Oerkovia* and *Rothia* (Okazaki and Okami, 1976). All other genera are oxidative requiring high levels of oxygen transfer. Good mixing is critical especially when broth becomes viscous. In such cases, growth is limited by transfer of oxygen to the cell surface rather than by oxygen solubility (Iwai and Omura, 1982). A number of fermentor configurations are available for free and immobilized cells to improve oxygen transfer. Most industrial fermentations are conducted as fed-batch cultures.

APPROACHES TO ISOLATION OF ACTINOMYCETES

Isolation and screening are rate-limiting steps in the search and discovery of previously unknown, natural products from microorganisms (Gottlieb, 1973). Even modern isolation and screening programmes for organisms with biochemical potential are often limited to the examination of aerobic, heterotrophic actinomycetes that grow well at neutral pH and ambient temperature, these organisms represent only a tiny fraction of the gene pool found in both, natural and man-made extreme habitats. Thus, innumerable members of a relatively small number of well-known genera and species are isolated and screened.

Selection of Sampling Locations

Many thousands of actinomycetes have been isolated from the environment, but relatively little is known about the ecological or geographical distribution of even better-known neutrophilic streptomycetes (Goodfellow and Williams, 1983). Choice of natural materials, such as soils, may be based on the assumption that samples from widely different locations are more likely to yield novel isolates and therefore, hopefully, novel metabolites. Alternatively, the search may be directed towards habitats in which the microbial populations are adapted to relatively extreme environmental pressures (Williams and Willington, 1982).

Pretreatment of Samples

Selectivity may be enhanced by chemical or physical pretreatment of samples before plating onto selective media (Cross, 1982). Enrichment procedures to increase actinomycete number are generally necessary and since bacilli dominate the reminder of the bacterial population, heat treatments to reduce their number are effective. For air dispersal techniques, samples must be completely dried. A variety of pretreatments (Table 1) and media (Table 2) can be selectively used for the isolation of specific genera of actinomycetes (Table 3).

Table 1: Modes of pre-treatment of samples for isolation of actinomycetes.

Treatment Physical	Sample	Actinomycetes
Heating:		
55°C for 6 min.	Soil, water and dung	*Rhodococcus, Micromonospora,*
40°C for 2 to 16 h	Soil and roots	*Streptomyces*
100°C for 1 h.	Soil	*Actinomadura, Microbispora*
Membrane filtration	Water	*Micromonospora*
	Water	*Thermoactinomyces*
Centrifugation	Sea-water and mud	*Streptomyces*
Agitation in sedimentation chamber	Moldy hay	*Thermophilic actinomycetes*
Chemical		
Enrichment with chitin	Soil	*Streptomyces*
$CaCO_3$ to increase pH	Soil	*Streptomyces*
Baiting:		
Paraffin-coated rod in carbon-free broth.	Soil	*Nocardia*
Pollen	Soil and water	*Actinoplanaceae*
Snake skin	Soil	*Ampullariella*
Human hair	Soil	*Pilimelia*

Source: *Principles and Problems of Selective Isolation of Microbes,* Williams and Wellington, 1982, Academic press, London, UK.

Table 2: Media used for selective isolation of actinomycetes.

Media	Actinomycetes recovered	Reference
Colloidal chitin, mineral salts	*Streptomyces, Micromonospora*	Lingappa and Lockwood (1962) Hsu and Lockwood (1975)
Half-strength nutrient medium	*Thermophilic actinomycetes*	Gregory and Lacy (1963)
Starch Casein, mineral salts	*Streptomyces, Micromonospora*	Kuster and Williams,(1964) Williams and Davies (1965)
Glucose-aspargine, mineral salts	*Actinomadura, Microbiospora, Streptosporangium*	Nonomura and Ohara (1969, 1971)
Diagnostic sensitivity test medium	*Nocardia*	Orchard and Goodfellow (1974) Orchard *et al.* (1977)
Mineral salts, sodium propionate, thiamin (M3 medium)	*Rhodococcus, Micromonospora*	Rowbotham and Cross (1977)

Source: *Principles and Problems of Selective Isolation of Microbes,* Williams and Wellington, 1982, Academic press, London, UK.

Table 3: Antimicrobial compounds used in media for the isolation of actinomycetes.

Actinomycete	Antibiotic(s) (mg ml^{-1})	Reference
Micromonospora	Bruneomycin (0.5-2)	Preobrazhanskaya *et al.* (1975)
		Ivanitskaya *et al.* (1978)
	Gentamycin (1-10)	Wakisaka *et al.* (1982)
	Tunicamycin (25-50)	Savshnikova *et al.* (1976)
	Novobiocin (25-50)	Tomita *et al.* (1980)
Streptomyces	Nitrofurazone (25)	Yoshioka (1952)
	Mycostatin (50)	Williams and Davies (1965)
	Cycloheximide (50)	
Actinomadura	Kanamycin (25)	Chormonova (1978)
	Streptomycin (0.5-2)	Probrazhanskaya *et al.* (1975)
	Bruneomycin (0.5-2)	
	Rubomycin (5-20)	Lavrova *et al.* (1972)
Actinoplanes	Tellurite	Willoughby (1971).
Acidophilic actinomycetes	Cycloheximide (50)	Goodfelow *et al.* (1988)
	Nystatin (50)	
Thermoactinomyces vulgaris	Cycloheximide (50)	Cross and Johnson (1971)
	Novobiocin (25)	
Nocardia	Cycloheximide (50)	Orchard *et al.* (1977)
	Mycostatin (50)	
	Chlortetracycline (45)	
	Methacycline (10)	

Source: *Search and Discovery of Industrially-Significant Actinomycetes,* Goodfellow and Donnell (1988) Academic press, London, UK.

Conditions of Incubation for Growth of Actinomycetes

An acidic pH is useful in isolation of acidophilic and acido-duric actinomycetes. Previous studies of the pH requirements of soil actinomycetes have distinguished between neutrophilic which grow from pH 5.0 to 9.0 with optima close to pH 7.0, and acidophiles which grow from 3.5 to 6.5 with optima from 4.5 to 5.5 (Flowers and Williams, 1977, William *et al.*, 1971). The temperature and length of incubation are also major variables which help in selective isolation of actinomycetes. Most of these are mesophilic and grow well between 25°C and 30°C (McCarthy and Williams, 1990). Thermophiles are isolated at 45°C to 55°C and psychrophiles at 4°C to 10°C, though little attention has been paid to the later group. The commonly isolated mesophilic genera, such as *Streptomyces* and *Micromonospora*, can be isolated in 7 to 14 days, common thermophiles, such as *Thermoactinomyces* require only 1-2 days (Williams *et al.*, 1971). However, novel or unusual isolates may be overlooked unless incubation periods are extended.

Approaches to identification of Actinomycetes

Morphological, physiological, biochemical, chemotaxonomic and molecular approaches are used for the identification of actinomycetes. Numerical taxonomy is 'an objective method of taxonomic analysis, whose first aim was to sort individual strain of bacteria into homogeneous group, and which would also assist in the arrangement of species into genera and higher grouping'. Probability identification systems are widely used to identify unknown actinomycete strains (Williams *et al.*, 1984). Probability matrix has been assembled for the identification of unknown actinomycetes (Saddler *et al.*, 1991) based upon several characters drawn from 150 unit characters used in the original classification. The actinomycete strains representing the major clusters defined by numerical taxonomic study on the basis of morphological and chemotaxonomic characteristics is confirmed by 16S rRNA gene sequencing.

Bioactive actinomycetes from India and Saudi Arabia

There is a pressing need for safe and effective antifungal antibiotics and industrially important enzymes with special activity and stability features. Actinomycetes as potential producers of many enzymes, enzyme inhibitors, growth promoting substances and antibiotics have been described as the greatest source of industrially important metabolites.

In order to explore the diversity of actinomycetes and their bioactive potential from two geographically distant regions, extensive sampling and isolation programme was sketched which lead to the collection of diverse actinomycetes with varied bioactivities. Table 4 gives the details of sampling sites and their characteristics.

Table 4: Sampling sites and their characteristics.

Sites and Region	Sample Type	pH	Temperature (°C)
1	2	3	4
INDIA			
Mula River, Pune	Soil	7.3	30.5
Mula River, Pune	Water	7.0	30.2
Mutha River, Pune	Soil	6.3	29.8
Mutha River, Pune	Water	6.7	29.7
Pavana River, Thergaon, Pune	Water	6.9	31.2
Pavana River, Thergaon, Pune	Soil	7.2	31.2
Pashan Lake, Pune	Water	8.5	29.9
Pashan Lake, Pune	Soil	8.3	30.1
Laundry Spot, Golibar Maidan, Pune	Soil	8.5	32.6
Laundry Spot, Southern Command, Pune	Soil	8.1	32.4
Agricultural land, Pune	Soil	6.6	26.6
Agricultural land, Pune	Soil	6.8	26.6
Compost, Pune	Solid Compost	6.5	49.2
Pune University campus,Pune	Soil	7.1	31.2
Sinhgad,Pune	Soil	6.9	32
Alibag	Soil	7.2-7.7	28.0
Aurangabad	Soil	6.9-7.3	39.8
Ahmednagar	Soil	7.5	31.0
Ulsoor lake,Bangalore	Water	7.4	29.6
Cubbon Park,Bangalore	Soil	7.0	29.8
Lalbagh botanical garden, Bangalore	Soil	7.2	30
Bharat river, Malapuram	Water	7.2	30.2
Nilambur teak plantation,Malapuram	Soil	6.8	29.6
Mahablashwar	Soil	5.4-5.7	18.7
Lonar Lake, Maharashtra	Water	10.0	29.7
20 feet from Coastline of Lonar Lake	Soil	9.8	29.3
Hipparga Lake, Solapur	Water	8.0	22.0
Ganeshpuri Hot Springs, Akloli, Thane	Water	6.0-6.5	57.9-61.6

1	2	3	4
Vrajreshwari Hot Springs, Akloli, Thane	Water	5.8-6.8	42.8-51.1
Upland area of Ranchi,	Soil	5.5	31.1
Kambar Lake, Solapur	Water	8.6	25.0
Sassoon Docks, Colaba, Mumbai	Water	7.1	31.3
Sasson Docks, Colaba, Mumbai	Soil	7.2	31.2
SAUDI ARABIA			
Sana, Yemen	Soil	5.7	20.0
AL-Shab Garden, Coastal regions of Red Sea, Hudaydah, SA	Soil	7.0	27.4
AL-Hudaydah Garden, Coastal regions of Red Sea, Hudaydah, SA	Soil	6.8	27.8
AL-Hodieda Beach, Coastal regions of Red Sea, Hudaydah, SA	Sandy Soil	6.5	27.4
Gold Beach, Aden, Coastal regions of Arabian Sea, Yemen	Sandy Soil	6.6	29.4
Kormaksar Beach, Aden, Coastal regions of Arabian Sea, Yemen	Sandy Soil	6.8	29.2
Kormaksar Garden, Aden, Coastal regions of Arabian Sea, Yemen	Soil	6.8	29.7
AL-Abdia, Mecca, Saudi Arabia	Soil	7.2	30.5
AL-Azizia, Mecca, Saudi Arabia	Soil	6.9	30.8
AL-Shate Beach, Jeddah, Saudi Arabia	Sandy Soil	6.7	30.4
Jeddah University Garden, Jeddah	Soil	6.9	30.5
South Jeddah, Jeddah	Soil	6.9	30.2

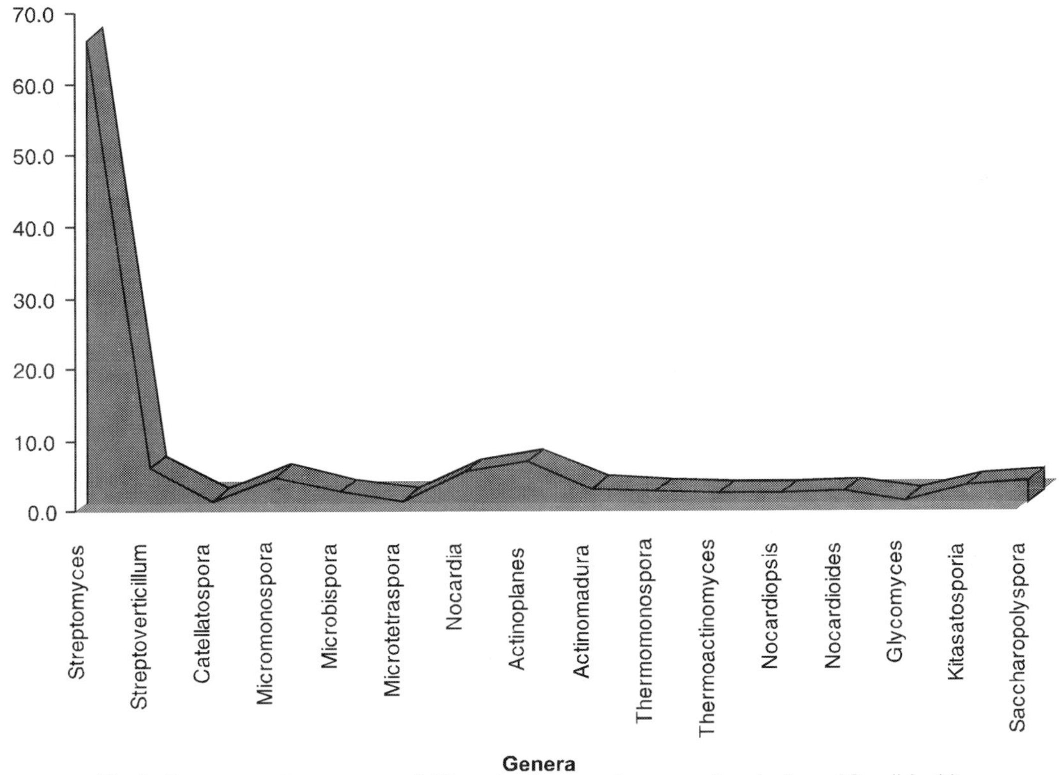

Fig 1: Frequency of occurrence of different actinomycete genera from India and Saudi Arabia.

As seen from Table 4, the sampling location characteristics are remarkably diverse. The isolation of actinomycetes on selective media yielded a large number of morphologically distinct genera with diverse biochemical activities. Overall, the frequency of getting extremophiles and mesophils depends on the ecological niche although some extremophilic isolates could be obtained from moderate locations as well.

Starch casein medium and chitin medium showed good growth of isolates from all samples. Identification of actinomycetes and screening for various enzyme and antimicrobial activities enabled the classification of the isolates based on frequency of their occurrence in particular extreme or moderate habitats. In general, actinomycetes from India and Saudi Arabia were equally competent in their bioactivities.

Fig. 1 shows the frequency distribution of actinomycetes obtained from both countries. As seen, the genus *Streptomyces* is highly predominant in all the samples occupying nearly 68% of the collection. However, this does not solely indicate their relatively higher occurrence but could also be due to their relatively easy isolation. The isolates belonging to other genera could also be isolated but with a lower frequency as seen in the graphical representation. Fig. 2 shows the enzyme activity profile of the isolates from both countries. Almost all the isolates displayed cellulase, chitinase, gelatinase and amylase activities and nearly 50-60% isolates also displayed pectinase and keratinase activities. Lipase, lecithinase, protease and xylanase

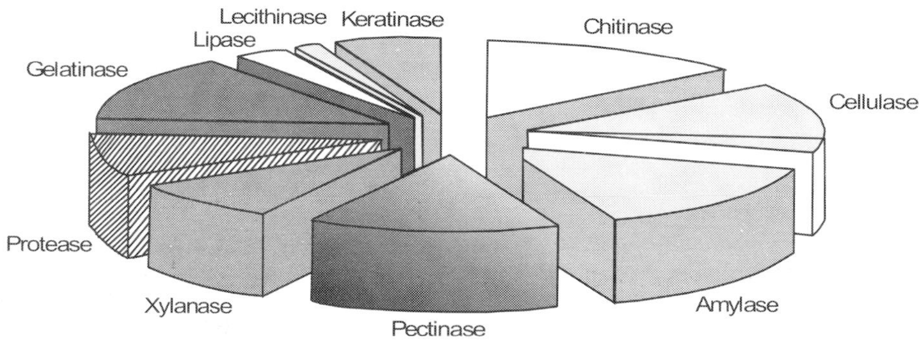

Fig. 2: Enzyme activity profile of actinomycete isolates from India and Saudi Arabia.

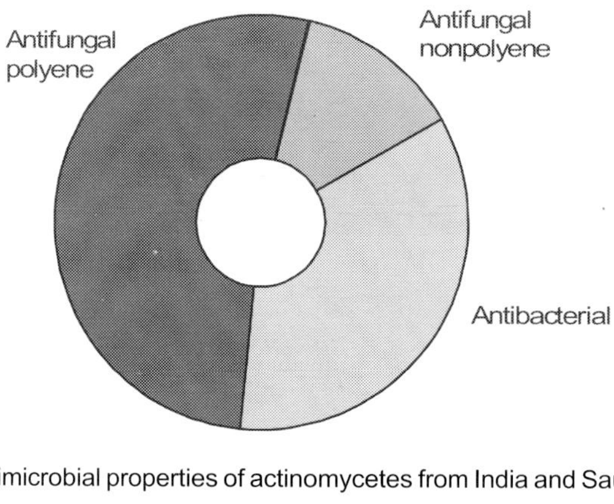

Fig. 3: Antimicrobial properties of actinomycetes from India and Saudi Arabia.

activities were relatively lesser than the above activities. A crucial primary and secondary screening programme for the chitinase producers was done, considering the importance of this enzyme in varied fields like biotechnology, waste management, biocontrol and medicine (Bhushan Moondal, 1999; Sampson and Gooday, 1998; Wiwat *et al.*, 1996).

Fig. 3 shows the antimicrobial properties of the isolates and the incidence of antifungal isolates is higher than antibacterial ones. The actinomycetes were shown to produce antifungal agents of both polyene and non-polyene nature.

Antibiotics from two of the potent isolates of the collection were purified and characterized. *Catellatospora citrea* AK49 produces an antibiotic which is non-polyene in nature. This acidophilic actinomycete produces substrate mycelium with straight and spirales short spore chains arising on the colony surface. The spore mass is orange coloured and it produces a diffusible pigment, the photomicrograph of the isolate with branched aerial and substrate mycelium is shown in Fig. 4. The isolate could hydrolyse the polymers chitin, xylan, pectin and cellulose. The antimicrobial agent it produces is active against wide range of targets, *Aspergillus niger, Cryptococcus humicolus, Candida albicans, Micrococcus luteus, Bacillus subtilis, Streptococcus mutans and Serratia marcescens.* The antibiotic could inhibit germination inhibition and elongation of hypha of *Aspergillus niger*. In the case of *Candida albicans* and *Cryptococcus humicolus* inhibition of budding was seen (Augustine *et al.*, 2005, Augustine *et al.*, 2004).

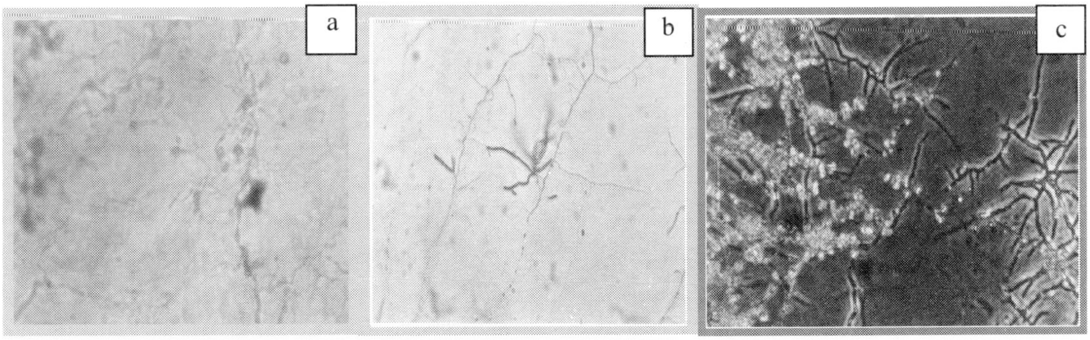

Fig. 4: Photomicrographs of antibiotic producers *Catellatospora citrea* AK49(a) and *Streptomyces* sp. PU23(b) and thermophilic, acidophilic chitinase producing *Microbispora* sp. V2(c).

The antibiotic from *Streptomyces* sp. PU23 was fairly stable for a period of twelve months at 4°C. The antifungal effect of the antibiotic from *Streptomyces* sp. PU 23 and hamycin is enhanced when the two drugs are used in combination, thus indicating the synergistic effect. The mode of action of the antibiotic is probably by binding to the ergosterol present in the fungal cell-membrane resulting in the leakage of intracellular material and eventually death of the cell. The antibiotic probably is a non-proteinic, straight chain, non-polyene, polyhydroxy polyether compound with a single double bond. These spectra differ from the known antifungal antibiotics i.e., hamycin, nystatin, Amphotericin B, cycloheximide, nikkomycin, griseofulvin etc. The properties of both antifungal antibiotics are given in Table 5. The antibiotics from *Catellatospora citrea* AK49 and *Streptomyces* sp. PU 23 seem to be novel, as they do not fit into any of the known classes of antifungal antibiotics as per our knowledge. The present research work can be an opening for a new field of research in the microbiological and pharmaceutical industry and academics.

Table 5: Physico-chemical properties of antibiotics from *Catellatospora citrea* AK49 and *Streptomyces* sp. PU 23.

Attribute	Character of antibiotic from *Streptomyces* sp. PU 23	Character of antibiotic from *Catellatospora citrea* AK49
Colour	Brown	Light orange
Nature	Amorphous	Amorphous
Melting point	218-220°C	213°C
R_f value	0.78	0.5, 0.3
Elemental analysis (%)	Carbon 37.09, Hydrogen 6.42	C(51.06%) H(6.43%) N(19.85%) O(22.67%)
UV λ_{max} (nm)	211, 216, 218	213, 261
IR (KBr) cm^{-1}	1031.8, 1639, 3296	3786, 3408,2923, 2852, 2279, 1625, 1550, 1382, 1321, 1022, 933, 673, 624
Molecular formula	$C_{16}H_{32}O_{19}$	$C_{12}H_{18}N_4O_4$ 282 Da

Table 6: Properties of chitinase from *Microbispora* sp. V2 .

Attribute	Chitinase from *Microbispora* sp. V2
Type	Chitobiosidase
Molecular weight	35 kDa
Optimum pH for activity	3.0
Optimum temperature for activity	60°C

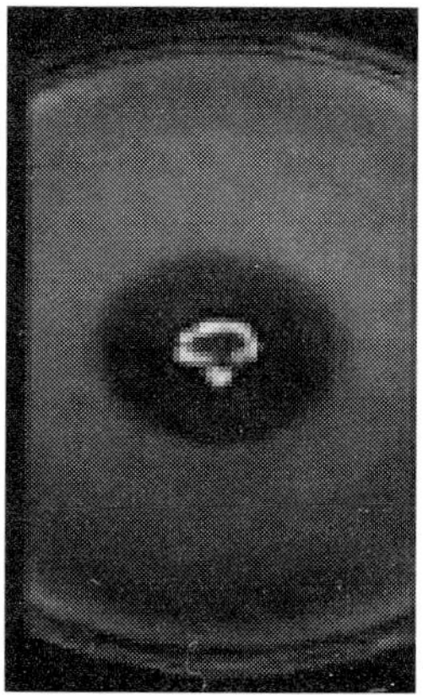

Fig 5: Chitin degradation ability of *Microbispora* sp. V2.

As regards to the chitinase activity, the collection yielded a *Microbispora* strain designated hereafter as *Microbispora* sp. V2 which produces an acidophilic and thermophilic chitinase (Nawani *et al.,* 2002). This genus has hitherto not been reported for a detailed chitinase production and characterization. The isolate was obtained from hot springs at Vrajreshwari, India which indicates adaptability of the strain and thus a thermophilic nature of its chitinase. Its chitin degradation ability has been shown in Fig 5 and Fig 4(c) shows the photomicrograph of the strain with typical spore pattern that is a characteristic feature of the genus *Microbispora*. The properties of chitinase from *Microbispora* are given in Table 6. A higher predominance of chitin degraders in non-extreme regions than in extreme locations was seen which could be either, since the extreme habitats are deviating from the normal ones in terms of pH and temperature, so they allow the survival of limited organisms capable of thriving under these conditions. The second reason could be a relatively lesser availability of chitin in these environments. Most of the actinomycetes can tolerate pH up to 5.0 and being cultivable region, the nutrient availability of the soils under cultivation must be higher. Besides, an acidic pH can lead to proliferation of fungi that would indirectly contribute to the chitin content of the soil thus allowing the predominance of actinomycetes.

In general, a higher predominance of actinomycetes was seen in most of the samples. An exception to this were the soil and water samples obtained from extreme type of locations like saline lakes, acid soils and hot springs. It was evident that acid soils have predominance of acidophiles, alkalophiles and neutrophiles predominate in alkaline and neutral habitats respectively. It could also be evaluated that neutrophiles do exist in acidic and alkaline habitats, but how far do they contribute in polymer recycling is a matter of introspection *in situ*. The relative abundance, in general, provides a clue to understanding the effect of nutrients and environmental factors on a particular phenotype which may not be restricted only to it. Large increase or decrease in the pH of the soil is generally not expected but it may happen in some localized sites of decomposition which explains the occurrence of neutrophiles in acid and alkaline habitats. Alkalophilic bacteria are found with highest frequency in alkaline soils, although, they are also present in neutral and acidic soils but in limited numbers. Most enzymes from alkalophiles are characterized by an optimum pH on the alkaline side. All these diverse isolates can be ecologically important as well as industrially important, eg., chitinases with alkaline pH optima and stability can be applicative in biological control. The peritrophic membrane of insect gut lining is chitinous and the alkali active chitinases can be used in synergism with other biocontrol agents. Enzymes, active and stable at extreme pH values, are expected to have a better half-life at extreme pH values in comparison to enzymes from neutrophiles or acidophiles. Soils and water bodies with a high organic content and thereby a higher competition seem to favour the predominance of isolates expressing good bioactivity.

FUTURE PERSPECTIVES

A balanced ecosystem is the need of the era with all the industrialization at the peak. The disinterest in conservation of nature is a major consequence of rapid economization. It has thus become very important to explore the nature for its diversity and to conserve the same. It is necessary to maintain a natural balance of the eco-cycles if a healthy life has to be sustained. The extension of diversity of actinomycetes to a molecular level will be of great use in understanding the complexity of the pathways in production of enzymes and antibiotics and if their replicas can be made in fast growing genera like *E. coli*. Despite the production of varied bioactive elements by this diverse group, the commercial and clinical applications of enzymes and antibiotics from actinomycetes have been limited. The need of rapid, thoughtful, directed approaches would yield many significant but unexplored strains, as the culturable isolates comprise of a very low percentage of the population existing in nature. So methods to detect the non-culturable strains are essential in order to explore the hidden diversity in them. Many more biochemists and physiologists are needed to take part in the screening for physiologically active substances from microorganisms. Human beings earnestly desire medicines effective against many diseases such as rheumatism, heart and hypertension, as well as antimicrobial agents effective against fungi, many kinds of antibiotic-resistant bacteria and parasites. Thus the search and discovery of new enzymes and antibiotics from actinomycetes requires interdisciplinary approach involving microbiologists, biochemists, geneticists and chemists.

REFERENCES

Aoyagi, T., Tobe, H., Kojima, F., Hamada, M., Takeuchi, T. & Umezawa, H. (1978) *J. Antibiot.* **31**: 636-638.

Aoyagi, T., Yamamoto, T., Kojiri, K., Kojima, F., Hamada, M., Takeuchi, T. & Umezawa, H. (1972) *J. Antibiot.* **31**: 244-246.

Augustine, S., Bhavsar, S. & Kapadnis, B. P. (2005) *J. Biosciences.* **30**: 201-211.

Augustine, S., Bhavsar, S., Baserisalehi, M. & Kapadnis, B. P. (2004) *Indian J. Exp. Biol.* **42**: 928-932.

Austin, B. (1989) *J. Appl. Bacteriol.* **67**: 461-470.

Ball, A. S., Betts, W. B. & Mc Carthy, A. J. (1989) *Appl. Environ. Microbiol.* **55**: 159-162.

Berdy, J. (1984) *Chinese Journal of Antibiotics.* **7**: 272-290.

Bhushan, B. & Hoondal, G. S. (1999) *World J. Microbiol. Biotechnol.* **15**: 403-404.

Brogard, J. M., Coumaros, D., Frankhauser, A. J., Stahl, A. & Stahl, J. (1972) *Res. Eur. Stud. Clin. Biol.* **17**: 890-895.

Burman, N. P. (1973) *Actinomycetales: Characteristics and practical importance* (Eds. G. Sykes & F.A Skinner) Academic Press, London, pp. 219-230.

Bushell, M. E. (1988) *Actinomycetes in Biotechnology* (Eds. M. Goodfellow, S.T. Williams & M. Mordarski) Academic press, London, pp 186-214.

Chacko, C. I., Lockwood, J. I. & Zabik, M. (1966) *Science.* **154**: 893-895.

Crawford, D. L. (1981) *Biotechnol. Bioeng. Symp. Series.* **11**: 275-291.

Crawford, D. L., Pometto, A. L. & Crawford, R. L. (1984) *Biotechnol. Advances.* **2**: 217-232.

Cross, T. (1981) *J. Appl. Bacteriol.* **50**: 397-423.

Cross, T. (1982) *Developments in Industrial Microbiology.* **23**: 1-18.

Czoch, W. P. & Mordarski, M. (1988) *Actinomycetes in Biotechnology* (Eds. Goodfellow, M., Williams, S.T. & Mordarski, M) Academic Press, London, pp. 219-283.

Ellaiah, P., Raju, K.V., Adinarayana, K., Adinarayana, G., Saisha,V., Madhavi, S. & Premkumar, J. (2002) *Hind. Antibiot. Bull.* **44**: 8-16.

Fairbairn, D. A., Priest, F. G. & Stark, J. R. S. (1986) *Enz. Microbiol. Technol.* **8**: 89-92.

Flowers, T. H. & Williams, S. T. (1977) *Microbios.* **18**: 223-228.

Golovlev, L., Golovleva, L. A., Eroshina, N. V. & Skryabin, G. K. (1978) *Zentralblatt Bacteriologie, Parasitenkunde, infections krankheiten and hygiene. 1. Abteilung, supplement.* **6**: 269-283.

Goodfellow, M. & Simpson, K. E. (1987) *Frontiers in Applied Microbiology.* **2**: 97-125.

Goodfellow, M. & Williams, S. T. (1983) *Ann. Rev. Microbiol.* **73**:189-216.

Goodfellow, M. (1989) *Bergey's Manual of Systematic Bacteriology*, Volume 4 (Eds. S. T. Williams) Baltimore: Williams and Wilkins Company, pp. 2333-2347.

Goodfellow, M., Williams, S.T. & Mordarski, M. (1988) *Actinomycetes in Biotechnology.* Academic Press, London.

Gottlieb, D. (1973) *Actinomycetales, characteristics and practical importance.* (Eds. M. Goodfellow, M. Mordarski & S. T. Williams) Academic Press, London, pp. 1-6.

Gregory, P. H. & Lacey, M. E. (1963) *J. Gen. Microbiol.* **30**: 75-88.

Hagedorn, C. (1976) *Appl. Environ. Microbial.* **432**: 368-375.

Hirsch, P. & Alexander, M. (1960) *Can. J. Microbiol.* **6**: 241-249.

Holt, J.H. (1994) *Bergey's Manual of Determinative Bacteriology*, Ninth edition. Williams & Wilkins, London.

Hyvärinen, A., Meklin, T., Vepsäläinen, A. & Nevalainen, A. (2002) *Int Biodeterior Biodegradation.* **49**: 27-37.

Iwai, Y. & Omura, S. (1982) *J. Antibiot.* **35**: 123-141.

Jiang, C.L. & Xu, L. H. (1996) *Appl. Environ. Microbiol.* **62**: 249-253.

Kalakoutskii, L.V. & Agre, N. S. (1976) *Bacteriol. Rev.* **40**: 469-524.

Kaluskar, V. M. (1995) Biodegradation of native lignin by actinomycetes and wood rotting fungi. Ph. D. Thesis, University of Pune, Pune.

Kanavade, V.L. & Kapadnis, B.P. (2003) *Maharashtra J. Agricul.* **27**: 168-172.

Kanavade, V.L. (2003) Use of bio-industrial waste for production of microbial biomass with potential in environmental management, Ph.D. Thesis, University of Pune, Pune.

Khan, M. R. & Williams, S. T. (1975) *Soil Biol. Biochem.* **7**: 345-348.

Kieslich, A. (1976) *Microbial transformation of non-steroid cyclic compounds.* John Wiley Interscience Publishers, New York.

Klausner, A. (1985) *Biotechnology.* **3**: 301-307.

Krasilnikov, N. A. (1960) Rules for the classification of antibiotic producing actinomycetes. *J. Bacteriol.* **79**: 75-80.

Krause, A., Hancock, W. G., Minard, R. D., Freyer, A. J., Honeycutt, R. C., LeBaron, H. M., Paulsen, D. L., Liu, S. Y. & Bollag, J. M. (1985) *J. Agri. Food. Chem.* **33**: 584-589.

Kundu, P. K. & Nandi, B. (1985) *Plant Soil.* **33**: 357-362.

Lacey, J. & Crook, B. (1988) *Ann. Occup. Hyg.* **32**: 515-533.

Lacey, J. (1973) *Actinomycetes, characterisation and practical importance* (Eds. G. Sykes & F. A. Skinner) Academic Press, London, pp. 231-251.

Lechevalier, H. A. & Lachevalier, M. P. (1967) *Ann. Rev. Microbiol.* **21**: 71-100.

Lechevalier, M. P. (1988) *Actinomycetes in Biotechnology.* (Eds.M. Goodfellow, S. T. Williams & M. Mordarski) Academic Press, London. pp 327-348.

Lechevalier, M.P. (1981) *Actinomycetes* (Eds. Schaal, K.P. & Pulverer, G.). Gustav Fischer Verlag, Stuttgart, pp 159-166

Lemmel, S. A., Heimsch, R. C. & Edwards, L. L. (1979) *Appl. Env. Microbiol.* **37**: 227-233.

MacKenzie, C. R., Bilous, D. & Johnson, K. G. (1984) *Biotechnol. Bioeng.* **26**: 590-594.

Manyu Seijaku, K. K. (1982) *Patent Japanese Kukai Tokyo Koho IP.* **57**: 132 880.

Mayfield, C. I., Williams, S. T., Ruddick, S. M. & Hatfield, H. L. (1972) *Soil Biol. Biochem.* **4**: 79-91.

McCarthy, A. J. & Williams, S. T. (1990) *Met. Microbiol.* **22**: 534-562.

Miyadoh, S. (1993) *Actinomycetologica.* **7**: 100-106.

Morosoli, R., Shareck, F. & Kluepfel, D. (1997) *FEMS Microbiol. Lett.* **146**: 167-174.

Murtoniemi T., Nevalainen, A., Suutari, M., Toivola, M., Komulainen, H., Hirvonen, M.R. (2001) *Inhal. Toxicol.* **13**: 233-247.

Nawani, N. N. & Kapadnis, B.P. (2003) Chitin degrading potential of bacteria from extreme and moderate environment. *Indian J. Exp. Biol.* **41**: 248-254.

Nawani, N. N., Kapadnis, B.P., Das, A.D., Rao, A.S. & Mahajan, S.K. (2002) *J. Appl. Micrbiol.* 93: 965-975.

Niemi, R. M., Knuth, S. & Lunstrom, K. (1982) *Appl. Environ. Mircobiol.* **43**: 378-388.

Nisho, M., Kuroda, A., Suzuki, M., Ishimaru, K., Nakamura, S. & Nomi, R. (1983) *J. Antibiot.* **36**: 761-769.

Numata, M., Nitta, K., Utahara, R., Maeda, K. & Umezawa, H. (1975) *J. Antibiot.* **38**: 993-996.

Okazaki, T. & Okami, Y. (1976) *Actinomycetes: The boundary Microorganisms* (Ed. T. Arai). Toppan Company Limited, Tokyo, Singapore, pp. 123-161.

Saddler, G. S., Tavecchia, P., Lociuro, S., Zanol, M., Colombo, L. & Selva, E. (1991) *J. Microbiol. Methods.* **14**: 185-191.

Sampson, M. N. & Gooday, G. W. (1998) *Microbiol.* **144**: 2189-2194.

Seong, C. N., Goodfellow, M., Ward, A. C. & Hah, Y. C. (1993) *Kor. J. Microbiol.* **31**: 355-363.

Seong, C. N., Park, S. K., Goodfellow, M., Kim, S. B. & Hah, Y. C. (1995) *J. Microbiol.* **33**: 95-102.

Shete, H. G. (1994) Degradation of cyanide by actinomycetes, M. Phil Dissertation, University of Pune, India.

Slobodskaya, V. V., Titiov, A. V. & Tereshin, J. M. (1980) *Voprosy Meditechieskoj Khimii.* **26**: 735-738.

Smith, L. K. (1984) *Biotechnology* (Eds. H. J. Rehm & G. Reed) Verlag Chemie, Weinheim, pp. 31-78.

Strzelczyk, E. & Leniarska, V. (1985) *Plant soil.* **86**: 387-394.

Tobe, H., Morishima, H., Aoyagi, T., Umezawa, H., Ishiki, K., Nakamura, K., Yoshiyoka, T., Shimauchi, Y. & Unui, T. (1982) *Agricul. Biol. Chem.* **46**: 1865-1872.

Umezawa, S., Tatsuta, K., Fujimoto, K., Tsuchiya, T., Umezawa, H. & Naganawa, H. (1972) Structure of antipain, a new Sakaguchi- Positive product of *Streptomyces. J. Antibiot.* **25**: 267-270.

Waksman, S. A. (1961) *The Actinomycetales, Volume II. Classification, Identification and Descriptions of Genera and Species*. Williams & Wilkins, Baltimore.

Williams, S. T. & Flowers, T. H. (1978) *Microbios*. **20**: 99-106.

Williams, S. T. & Robinson, C. S. (1981) *J. Gen. Microbiol.* **127**: 55-63.

Williams, S. T. & Wellington, E. M. H. (1982) *Bioactive Microbial Products: Search and Discovery* (Eds. J. D. Bulock, L. J. Nisbet & D. J. Winstanley) Academic Press, London, pp. 9-26

Williams, S. T. (1989) *Bergey's Manual of Systemic Bacteriology,* Vol. 4. Williams & Wilkins, Baltimore, U.S.A.

Williams, S. T., Davies, F. L. & Khan, M. R. (1971) *Soil Biol. Biochem.* **3**: 187-195.

Williams, S. T., Lanning, S. & Wellington, E. M. H. (1984) *The Biology of the actinomycetes.* (Eds. M. Goodfellow, M. Mordarski & S. T. Williams) London: Academic press, pp. 481-528.

Wiwat, C., Lertcanawanidrakul, M., Siwayapram, P., Pantuwatana, S. & Bhumiratana, A. (1996) *Gene.* **179**, 119-126.

Wu, H., Shimoi, H. & Ito, K. (2002) *Biosci. Biotechnol. Biochem.* **66**: 2515-2519.

9

Potential of Entomopathogenic Fungi for Management of Insect Pests

Sardul Singh Sandhu, Pallapotu Vikrant and $Ram Charan Rajak*

Fungal Biotechnology and Invertebrate Pathology Laboratory,
Department of Biological Sciences,
R. D. University, Jabalpur- 482001(India)
**Director (Academics), Seedling Academy of Design,*
Technology and Management, Seedling Campus, Jagatpur
Jaipur- 302 025 (India)

The continued use of insecticides, insect resistance and their environmental impact have necessitated the development of alternative means of biological control. The use of microbial control agents ranging from bacteria, viruses, fungi, and protozoas to nematodes render safety for humans and other non-target organisms, reduction of pesticides residues in food, preservation of other natural enemies and increased biodiversity in managed ecosystems. Entomogenous fungi differ from other microbial agents in both their mode of parasitism and survival in nature that are strongly influenced by abiotic factors. This unique feature of microclimate dependency is of major importance to the formation, germination and survival of various types of infective spores of conidia that attach to the host cuticle, germination and subsequently penetrate the insect cuticle.

Therefore, the present article deals with *in vitro* culturing of entomopathogenic fungi along with the infection process in their hosts. Persistence of entomopathogenic fungi in nature, light in relation to conidiation, spore discharge and survival, atmospheric and soil temperature and moisture in relation to growth and development, pathogenicity and survival, have been described. Natural and artificial antagonists of these fungi were also mentioned. Apart from these, large scale production of entomogenous fungi by solid state fermentation and liquid fermentation, their formulation and application strategies have been detailed. Finally, registration for use of these entomopathogenic fungi and the practical consideration.

INTRODUCTION

Recent concern over the continued use of insecticides, insect resistance and particularly, their environmental impact have necessitated the development of alternative means of biological control (Sandhu and Vikrant, 2004). Invertebrate pathology is a recently organized discipline, and its roots can be traced to ancient history with reference to solutions for preventing disease in honeybees and silkworms (Steinhaus, 1956; Vikrant and Sandhu, 2005). Notable early pioneers in invertebrate pathology such as Agostino Bassi, Louis Pasteur and Elie Metchnikoff proposed the application of microorganisms for control of insect pests that are important regulatory factors in insect populations (Lacey *et al.*, 2001). Entomopathogens range from bacteria, viruses, fungi, and protozoas to nematodes. Many species are employed as biological control agents of insect pests (Burges, 1981; Tanada and Kaya, 1993; Lacey and Kaya, 2000). The use of microbial control agents render safety for humans and other non-target organisms, reduction of pesticide residues in food, preservation of other natural enemies and increased biodiversity in managed ecosystems.

Fungi comprise a large heterogenous and ubiquitous group of heterotrophic eukaryotic organisms (Alexopoulos, 1979). Approximately 750 species are facultative or obligate pathogens on one or more developmental stages of insects commonly found in aquatic, terrestrial and subterranean habitats (Tanada and Kaya, 1991; McCoy *et al.*, 1988). Fungi differ from other entomopathogens, in that, both their mode of parasitism and survival in nature are strongly influenced by abiotic factors. This unique feature of microclimate dependency is of major importance to the formation, germination and survival of various types of infective spores or conidia that attach to the host cuticle, germinate and subsequently penetrate the insect cuticle (Carruthers and Hural, 1990). Both hyphal and spore survival associated with a dead host, and/or the surrounding substrate are also influenced by different abiotic and biotic factors (Gottwald and Tedders, 1984).

The term 'entomopathogenic fungi' refers to fungi which induce disease symptoms in insect hosts. Fungi, which could be considered to be normally present in the mycoflora of healthy hosts such as insect-associating members of the Labouleniomycetes, are not included (Rajak *et al.*, 1991). This domain does include the range of fungi from quick killers to absolute parasites that cause disease symptoms in the host. 'Absolute parasites' are considered fungi, which live in association with a host and benefit at the host's expense. It is important to note that facultative saprobes can become pathogenic only if other protagonistic forces, biotic or abiotic, compromise insect's immune system. The high virulence and epizootic efficiency of some of the entomopathogenic fungal strains towards certain insects has been used to develop commercial mycoinsecticides for the biological control of insect-pests (Agarwal *et al.*, 1990). Fungi such as *Beauveria, Nomuraea, Metarhizium, Verticillium, Acremonium, Fusarium* and *Paecilomyces* are capable of infecting and killing a variety of insects (Sandhu *et al.*, 1993a; Sandhu *et al.*, 2001; Thakur *et al.*, 2005). Fungal entomopathogens are found in every major taxonomic group except the higher basidiomycetes and the demataceous Hyphomycetes. As many as 700 species in 90 genera have been documented. Many of these entomopathogenic fungi are studied for use as pest control agents and some of them have been successfully marketed as mycoinsecticides. Nearly 700 species of entomopathogenic fungi have been reported, but only 10 of these have been or are currently being developed for insect control (Hajek and St. Leger, 1994; Sandhu and Vikrant, 2004). A broad range of biotic, from obligate parasites to opportunistic pathogens that can survive saprophytically in the absence of living hosts, have been documented for the entomopathogenic fungi (Lacey *et al.*, 2001).

IN VITRO CULTURING

Entomogenous fungi have now been cultured and are maintained *in vitro* and at industrial levels (Urbanczyk *et al.*, 1992). Samson (1982) and Bartlet and Jaronski (1988) have reported inoculum preparation and growth. Different scientists reported that these organisms can grow on different diets (Sandhu *et al.*, 1993b). The optimum temperature for *in vitro* growth and sporulation was found 25-28°C. The Entomophthorales are more nutritionally demanding, and often, are best growing on egg-based media such as Sabouraud's egg milk. The optimum temperature for growth and sporulation may be lower than that of Deuteromycetes that generally grow at 25-28°C. Cultures can be held for many months between subcultures on slopes at 4°C. For preservation without genetic alterations, it is recommended that cultures be stored under 10% glycerol in liquid nitrogen (-175°C). Most entomopathogenic fungi can be frozen without any special precautions (Khachatourians, 1991).

INFECTION

Insect pests are susceptible to infection by naturally occurring, insect pathogenic fungi. These fungi are very specific to insects, often to particular species, and do not infect animals or plants. Fungal growth is favored by moist conditions but fungi also have resistant stages that maintain infection potential under dry conditions. Fungi have considerable epizootic potential and can spread quickly through an insect population and cause their collapse. Since, fungi penetrate the insect body, they can infect sucking insects such as aphids and whiteflies that are not susceptible to bacteria and viruses (Sandhu, 1994).

The process of fungal infection involves different steps; each step is influenced by a range of integrated intrinsic (fungal) and extrinsic (host, environmental) factors, which ultimately determine whether the pathogen breaches the host cuticle. Virtually all entomopathogenic fungi enter the host through the cuticle. Unless spores, germ-lines or appressoria are attached to the host surfaces it cannot induce pathogenesis in the insect. Contact must first be made by the spore on the host surface. The airborne spore of entomopathogenic fungi makes contact where they land on the insect surface. In comparison motile spores, the zygotes of *Coelomomyces* and the zygospores of *Aphanomyces*, actively locate their host by chemotaxis. It might be presumed that oomycete zoospores and the zygospores of the Mucorales member Sporodiniella also exhibit chemotaxis towards their insect hosts (Burges, 1981). Adhesion involves the chemical and physical interactions of the insect epicuticle and the spore. Lipids, lipoproteins, polyphenols and proteins have been identified in insect epicuticles. The epicuticle is not impermeable as once thought. Many small channels do cross this layer. Waxes, sugars and proteins are exuded through them (Sandhu *et al.*, 1993). Adhesion is normally achieved through the secretion of mucilage which interact with and modifies epicuticular waxes as indicated by the production of imprints of invasive propagules on cuticle. This is caused by action of cuticle-degrading enzymes secreted with the mucilage and probably serves in host recognition as well as cementing the pathogen to a suitable substratum. Secretion of mucilage is not a passive response because some fungi will not adhere to certain surfaces. The surface of the spore varies with the type of spore. The spores of many members of the Entomophthorales are covered with glycoproteinaceous mucous. Motile spores of *Culicinomyces* may actively synthesize binding compounds during encystment after recognizing a host (Sweeney, 1985). Adhesion has been shown to be an important prerequisite to infection.

A wide range of factors influence spore germination and behaviour, water, ions, fatty acids, nutrients, the biota on the cuticle surface and the physiological state of the host. Water maintains cell turgor and provides a medium for the uptake of dissolved nutrients including breakdown products of the cuticle. Nutrients accelerate germination, growth and development (Hassan *et al.*, 1989). It is entirely possible that microclimates could exist on the insect's epicuticle due to respiration through the spiracles which could also enable germ tube formation when a low humidity is recorded in the ambient macroclimate. A number of compounds have been found on the cuticle which stimulate or inhibit germination. Not all compounds of the cuticle favour germination (Fargues, 1981; Latge *et al.*, 1987). Conidia may adhere to non-host surfaces but may not then receive the appropriate stimulus for germination. Actually solvent extracts of host and non-host insects stimulate spores of these fungi to germinate. Fatty acids have a profound effect on the spore germination and differentiation either being toxic, fungistatic or stimulatory (Sandhu, 1995).

The exact mechanism for entry is usually peculiar to that species. A range of cuticle degrading enzymes is produced during penetration. The three most important classes being lipases, proteases and chitinases. These are produced sequentially, reflecting the order of substrates they encounter, i.e. epicuticular waxes, followed by the protein-chitin matrix (Smith *et al.*, 1981). Great variation in the type and quantity of enzymes exists between species and within species. The importance of chitinases is obscure. The virulent isolates consistently contain significantly higher levels of activity at the time of penetration. Proteases have been shown to play a key role in the penetration process, and a wide range have been identified including trypsin, chymotrypsin, elastases, collagenases and chymoelastases (St. Leger *et al.*, 1988; Bidochka and Khachatourians, 1988). Also changes in curvature of the cuticle have been documented suggesting that penetration pegs are applying pressure to the cuticle. Resistance to penetration could be single most important defense to infection. Wounded hosts are much more susceptible to weak pathogens and to facultative saprophytes. More penetration is concentrated at the naturally thinner areas of the integument such as the setae and the intersegmental membranes.

There are several defense mechanisms in insects, which should prevent the penetration and the growth of the fungus. The most common reaction is melanization of the cuticle at the infection site (Gunnarson, 1988). However, in most cases this defense mechanism is not sufficient, since melanization is initiated too late or this process cannot stop the quickly penetrating hyphae (St. Leger *et al.*, 1991). When the fungal hyphae reach the haemolymph, further defense mechanisms have to be overcome. For example, in locusts the cellular defense reactions start during the penetration process. In this case haemocytes are accumulated at the infection site. The insect's defense system in the hemocoel employs phagocytosis and the secretion of antagonistic compounds, namely, quinines and melanines. First the insect must recognize the fungus as foreign. Glucans compounds on the fungus' cell-wall seem to trigger this immune response (Latge and Boucias, 1984). The fungus retaliates by rapid reproduction to overwhelm the immune system. This is usually accomplished by hyphal fission producing separate hyphal bodies. These hyphal bodies are not as antigenic as hyphae (Pendland and Boucias, 1986). Coelomomyces and members of the Entomophthorales produce wall-less protoplasts in the hemocoel, which are unrecognizable by the host. Toxins are another offensive waged by the fungus. Most of the toxin producers are Deuteromycetes. Much of the research has been conducted on the silkworm pathogens, *Beauveria* and *Metarhizium*. These *in vitro* assays showed the toxins to be cyclodepsipeptides (Roberts, 1966).

The toxin producers tend to be quick killers and consequently secrete antibiotics so that they can continue their feeding frenzy saprobicly free of competition. Eventually all organs of the insect are consumed and replaced with hyphae. Some of the members of the Entomophthorales are astute at savouring their vital meals (*i.e.* almost parasitic). The later stages of mycosis induce physiological symptoms such as convulsions and lack of coordination. Some members of the Cordicepitales and Entomophthorales induce altered states of insect behaviour. Victims of Entomophthoramycosis assume lofty positions. Infected winged insects fly towards the sun. Earthbound victims elevate themselves by climbing trees or foliage and then firmly affix themselves to an apex with their mandibles. This behaviour is followed shortly thereafter by death which usually occurs between 3:00 and 7:00 P.M. Grasshoppers have been witnessed to die with wings outstretched (McCleod, 1963). These behavioural alterations facilitate spore dispersal and are incredibly adaptive for the fungus. Curiously, Cordyceps might have some psychopharmacological effects on vertebrates too. Chinese folk healers prescribe insect cadavers infected with *Cordyceps sinensis* and boiled with pork to reduce the side effects of opium over-indulgence.

Upon favourable conditions hyphae repenetrate the cuticle and produce conidiophores on the outside. Resting spores can also be produced both inside and outside of the cadaver. Entomophthoralean members forcibly eject their conidia. The cadaver can serve as protection from the elements (Sandhu, 1995). After the death of insects infected with entomopathogenic fungi, fungal outgrowth from the insect body and coincidentally, the production and dispersal of spores to new host and the environment occurs. The conidiospores are released upon maturation; however, with many entomophthoralean species, host death and spores discharge occurs under unique conditions of time, daylight, relative humidity, temperature, etc. the conidial discharge level were correlated with relative humidity for previous 2 hours and with leaf wetness for the previous 3 hours (Hajek and Soper, 1992). Greater than 90% of cadavers discharges maximum numbers of conidia between 10 – 15 days. In this case, 1-day-old cadavers sporulated within 1.5 - 2.5 days of being placed in the field, and almost 72% produced conidia.

PERSISTENCE OF ENTOPATHOGENIC FUNGI

Many environmental factors, representative of different habitats either directly or indirectly, affect both persistence and survival of various fungi in nature (Roberts and Campbell, 1977; Fuxa, 1987; McCoy *et al.*, 1988). Temperature and moisture in some form are influential in all habitats. Light in the form of solar radiation is of major importance in terrestrial habitats. In aquatic habitats, temperature, organic pollutants, oxygen concentration and salinity have been identified as key factors influencing the persistence of pathogens of mosquitoes (Jaronski and Axtell, 1982; Lord and Roberts, 1985). It is safe to say that natural and artificial antagonism can be a factor in all habitats (Steinhaus, 1960). Generally, the natural mortality caused by abiotic and biotic factors can be very high among fungal species, resulting in sporadic spatial distribution on or in most natural substrates. To compensate this natural mortality, many species produce copious quantities of infective spores and/or form modified adaptive structures such as resistant hyphae, chlamydospores, resting spores, sclerotia, and macrocyclic conidia to maximize survival (Pendland, 1982). Fungi do persist in nature and frequently cause catastrophic decline in insect populations (Fuxa, 1987; Tanada and Kaya, 1993). These epizootics require a favourable environment interacting with host and pathogen in many intricate ways. Since mycelia and various infective structures of many species of entomopathogenic fungi also can be mass-produced, larger quantities of inoculum have been and are being applied experimentally in managed ecosystems as microbial biopesticides for pest control (McCoy, 1990). An understanding of the environmental factors common to different habitats that will affect the persistence and survival of fungi, is imperative to their use in biological control (Roberts and Campbell, 1977).

Effect of Light

Light plays an important role in biology of most fungi, particularly as it relates to conidiation, spore discharge and survival (Callaghan, 1969; Sakamoto *et al.*, 1988). For example, a number of hyphomycetes such as *Cordyceps subilifera* produce stromata or synemata that are attracted to light as a sporulation stimulus. On the other hand, Coelomycetes such as species within the genus *Aschersonia,* produce conidiogenous cells within highly pigmented pycnidia for some protection from light. Natural sunlight in the range of 290 to 400 nm will affect persistence of entomopathogenic fungi occurring on foliage and to a lesser extent on other substrates (Fuxa, 1987). In natural and cultivated niches that offer shade or specific photoperiodicity, they can survive and flourish under warm, humid conditions (Daoust and Pereira, 1986). Spores produced by the more primitive species of Phycomycetes and most species found in the Entomophthorales are more sensitive to simulated or natural solar effect than the typical conidia produced by the Ascomycotina and the Dueteromycotina. Within a given species, however, different strains show variable persistence to solar exposure. The half-life of spores of most entomogenous fungi exposed to simulated sunlight range from about one to four hours, while survival of spores measured in the field have longer half-lives (Ignoffo *et al.*, 1977). Carruthers *et al.* (1988) working with *Entomophaga grylli* found that both duration and intensity of simulated light were interactive, resulting in an exponential decay curve for spore viability. In the field, they found a 700- fold difference in loss of spore viability between open and shaded grassland. Brokyn *et al.* (1985) found that *Erynianeo aphidis* on bean leaves in the field showed a linear decline in sporulation over time, with some survival at 14 days post-inoculation. Plant shading was also important in preventing inactivation by sunlight. Even leaf surface influenced survival, with spores surviving longer on the abaxial compared to adaxaial surface. A comprehensive field study on the effects of sunlight was completed by Farques *et al.* (1988). Working with *Nomuraea rileyi* in field plantings of pigeon bean and cabbage, they determined the effect of different levels of sunlight UV on conidial persistence in the upper plant canopy using shade cloth and glass screening treated with UV inhibitors. In full sun, conidial half-life was 3.6 hours compared to more than 40 hours under shade cloth. By blocking wave length above 400 nm, they found that the half-life of conidia was increased about four times compared to exposure to UV in the range of 320 to 2500 nm. Using a predictive equation, they concluded that conidial viability was reduced 10,000-fold after seven days in full sun compared to 100 times during a cloudy seven-day period. Ignoffo and Garcia (1992) reported that the conidia of fungal species that are pigmented are more persistent than less-pigmented conidia when exposed to simulated sunlight. Conidial half-life of *Aspergillus niger* with dark green to black pigmentation was significantly greater than that for five other species with less conidial pigmentation. This result suggests that persistence of some fungi may be improved by selecting or genetically engineering for highly pigmented strains. Sunlight may directly affect chromosomal DNA within the cell or indirectly inactivate the entomopathogenic fungi via the production of reactive radicals. In laboratory studies, different pathotypes of *Hirsutella thompsonii* were sensitive to far-ultraviolet light from 200 to 300 nm (Tuveson and McCoy, 1982). However, the fungus was found to photoreactivate in visible light suggesting that it contains enzymes that bind to the pyrimidine dimers in DNA induced by the near-UV in the range of 290 to 400 nm wavelengths found in sunlight.

Atmospheric Temperature and Moisture

Entomopathogenic fungi are generally mesophiles, i.e. optimum temperatures for growth and development, pathogenicity and survival in nature generally are between 10 and 30°C (Roberts and Campbell, 1977; McCoy *et al.*, 1988). Temperature-dependent processes such as rate of disease development and level of disease are higher at the optimum temperature than at lower temperatures, although in some cases reduced temperatures increase the time from infection to mortality without reducing total mortality. Optimal temperatures for fungal survival differ among taxa, for example, spores of entomophthorales appear to be more sensitive than conidia of most deuteromycetes. Temperatures above 35°C inhibit growth and development of entomopathogenic fungi. Conidia of *Beauveria bassiana* and *Metarhizium anisopliae* have a thermal death point of less than 15 minutes at 40°C. Resting spores of the entomophthorales are most heat resistant, tolerating 80°C for one hour or longer. Below 4°C, fungal cells generally survive but rarely grow. Conidia of most deuteromycetes remain viable for years when stored at temperatures of –20°C or lower. Studies by Pell and Wilding (1992) suggest that survivorship of *Erynia radicans* as hyphae within the dead host is highest in the field and that capilliconidia appear to be more temperature tolerant than primary conidia. Ignoffo, (1985) determined that dry and wet conidia of *Nomuraea rileyi* were non-infectious within one day of exposure at 50°C. At intermediate temperatures, dry conidia were about twice as stable as wet conidia suggesting that once the germination process begins conidia are less persistent. In a recent series of papers, Cammon and Rath (1994) found that conidial germination rates and ranges for *Metarhizium anisopliae* differed widely among isolates collected from different temperate regions ranging from 2.5 to 37°C. Using canonical variate analysis, these authors separated 70 isolates into heat-active and meso-thermoactive groups. High relative humidity and/or free water is generally required by entomopathogenic fungi for germination of infective propagules and the formation of reproductive structures outside the host. Moisture also affects the persistence and survival of the fungus in other ways. The effects of temperature and humidity are intimately related; some fungi may tolerate temperature extremes if there is more moisture in the air or when condensation readily occurs and water loss is minimized. Wind can significantly alter the microclimatic moisture and greatly influence fungal behaviour. Millstein *et al.* (1983) working with *Erynia* spp. determined that maximum spore discharge by the fungus was closely associated with saturated still air. Wind at 4 m/min with a relative humidity (RH) of 90 per cent had an almost identical effect on spore discharge repression as still air at 75.6 per cent RH. Moisture conditions favouring epizootics of entomophthorales generally require 8 to 10 hours of relative humidity higher than 90% with prolonged periods of dew or fog occurring over a period of two to three days. Steinkraus and Slaymaker (1994), studying the temporal pattern of sporulation, spore discharge and formation of capilliconidia of *Neozygifesfresenii,* found that moisture affected host rehydration and subsequent conidial discharge. Matewele *et al.* (1994) developed a method where water activity of a culture medium for *Metarhizium anisopliae* was reduced through evaporation, thereby allowing for the selection of mutants capable of germinating and growing at lower relative humidity than parental strains. Some selected mutants were more virulent against green leafhopper than the parental strains.

Effect of Soil, Temperature and Moisture

Soil is a natural reservoir for fungal-infected insects that died in the soil and for insects killed on foliage which subsequently fall to the ground (Keller and Zimmermann, 1989). Fungi in soil generally are protected from desiccation, ultraviolet radiation and extreme temperature; however, their prevalence is seasonal and reflects an aggregated spatial distribution that approaches a negative binomial. This distribution suggests that they do not widely proliferate in the soil but remain in close proximity to their host at relatively low densities under natural conditions. The predominant soil-inhabiting entomopathogenic fungal species produce hydrophobic conidia as infective propagules that bind readily to the insect epicuticle upon contact. The texture of soil can influence sporulation and the efficacy of the fungal conidium in contacting its host and its persistence and density within the soil profile (Li and Holdum, 1993). If the source of infection remains only on the surface of the soil, only insects moving on or coming in contact with the surface will be exposed. Conversely, if the fungal inoculum leaches to a depth beyond that inhabited by the host, infection will not occur. Storey and Gardner (1987) reported that conidia applied to the surface of clay soil penetrates to a depth of 5 cm. Vertical movement in sandy soils low in organic matter can exceed 5 cm, particularly in sand saturated with water. Harrison (unpublished data) showed that conidial density per volume of sandy soil produced via fungal mycelia or a dead host decreases significantly at soil densities greater than 1.5 g/cm^3. The optimal soil temperature for fungal persistence and survival varies widely according to fungal strain, soil type, soil moisture and natural antagonists. Cammon and Rath (1994) identified strains of *Metarhizium anisopliae* that are infectious at 5°C and persist for two years. Studdert and Kaya (1990) have shown that conidial persistence of *Beauveria bassiana* increased with a decrease in temperature and also decreased as soil moisture approached saturation. Persistence declines rapidly at temperatures above 30°C, and 50°C is lethal. This relationship between soil, water potential, temperature and persistence of conidia in different soils has been demonstrated by Studdert and Kaya (1990).

NATURAL ANTAGONISTS

An estimated 464 species of fungi and filamentous yeasts have been found in soils (Kendrick and Parkinson 1990). In addition, bacteria and actinomycetes occur and play functional roles determined by the physiological and morphological characters in different soils. These naturally occurring microbes and their secondary metabolites appear to function as antagonists under certain soil conditions, influencing both infectivity and survival of entomopathogenic fungi.

This phenomenon has been demonstrated most clearly in numerous published comparative studies with natural and pasteurized soil or soil extracts. Studies with *Beauveria bassiana* suggest that inhibition in natural soils is caused by fungistasis rather than a result of microbial competition. Shields *et al.* (1981) identified patulin, a water-soluble metabolite of *Penicillium uriticae,* which was inhibitory to conidial germination and growth of *Beauveria bassiana.* Other research reports suggest that the level of inhibition varies between fungal species and also can be a function of soil moisture, temperature and type (Daoust and Pereira, 1986; Keller and Zimmermann, 1989; Studdert and Kaya, 1990). The addition of a carbon and nitrogen source as nutrient supplement to the soil to promote the biological activity of naturally and artificially applied entomopathogenic fungi in soil has been suggested as a feasible biological control strategy (Lingg and Donaldson, 1981). Lockwood (1964) and Bumbieris and Lloyd (1967) showed that fungal hyphae lyse faster in natural soil supplemented with glucose, peptone or bone meal. Actinomycetes are more likely to increase and lyse existing fungal hyphae in soil when peptone is a supplement compared to glucose. Very little research on soil supplementation with nutrients has been done with entomopathogenic fungi. Studies have shown that conidia of common soil-inhabiting species, such as *Metharizium, Paecilomyces* and *Beauveria,* require a carbon or nitrogen source for germination but will not germinate outside a host in soil (McCoy *et al.,* 1988). Lingg and Donaldson (1981) working with *Beauveria bassiana* in one soil type showed that the addition of both carbon and/or nitrogen to natural soil is not a good idea in view of fungistasis.

ARTIFICIAL ANTAGONISTS

Entomogenous fungi tolerate a wide range of hydrogen ion concentrations ranging from a pH of 5 to 10 with an optimum at pH 7. The chemistry of representative soil types can influence survival and susceptibility of the host to infection. Reports of both direct and indirect effects of organic fungicides, insecticides, acaricides, herbicides, and growth regulators on entomopathogenic fungal survival have been made during the past two decades (Gardner and Storey, 1985; McCoy *et al.,* 1988). These compounds interfere with fungal epizootics by reducing the available host population or inhibit spore germination and vegetative development of the fungi. In some cases, these artificial antagonists are innocuous to the fungal species or strain; sublethal doses of insecticides actually will enhance the biological activity of the fungus by stressing the host or changing its behaviour (Quintela and McCoy, 1997).

LARGE-SCALE PRODUCTION OF ENTOMOGENOUS FUNGI

Several techniques for the mass production of entomopathogenic fungi are available, mostly designed to yield infective conidia; the conidia are harvested and formulated for storage and field use. The entomogenous hyphomycetous fungi especially *Beauveria, Metarhizium* and *Nomuraea* etc. are most used for experimentation and mass production (Ferron, 1978, 1981). Products containing infective particles of these fungi are available (Soper, 1985) for use against insect-pests of field crops. For large-scale field experimentation and commercialization of the fungi, development of efficient mass production techniques is necessary. Prerequisites for the development of entomopathogenic fungi into successful commercial bioinsecticide are:

(*i*) large-scale production of fungal spores under controlled and standardized conditions, (*ii*) formulation of fungal spores into a product with an adequate shelf-life necessary for use in agriculture, horticulture and home and garden (*iii*) performance and (*iv*) the price of the final product in a competitive market. Fungi with a saprophytic life cycle can be grown on artificial nutrient media. In principle there are two different procedures: (*a*) Solid-state fermentation and (*b*) Liquid fermentation.

Solid-State Fermentation

In recent years, solid-substrate fermentation has gained significance for production of conidiospores and fungal biomass (Lisasky and Hall, 1983; Desgranges *et al.*, 1993; Antia-Londono *et al.*, 1992; Auld, 1992). One of the major advantages of solid fermentation is the use of cheaper, easily available, agricultural based and biodegradable substrate. This allows the cheaper and safe exploitation of this fungus in insect-pest management. Vyas *et al.* (1991) used various cereal grains and agricultural wastes for mass production of *B. brongniartii*. Maize grains were found most suitable substrate for mass culturing and sporulation of the fungus, which yielded maximum 4.8×10^8 conidia/g of crushed grain in the laboratory. This production system makes use of sterilized cereal seed like rice, sorghum, and wheat, which are inoculated with the fungal spores. High-density polyethylene bags and shallow galvanized iron trays were found very useful in the laboratory. After a few weeks of fermentation / incubation in sterilized bottles or bags, the kernels are overgrown by sporulating mycelium. The spores can be separated from the kernels and formulated into a fungal biopesticide. The mass produced spore dust can be used for testing the fungus efficacy against target pests in the laboratory as well as field conditions. In Brazil, *Metarhizium anisopliae* was mass cultured on sterilized rice and has been found to control effectively the population of sap sucking bug. This technique is used in many small local production facilities eg., in Brazil, Columbia and China. In China, more than 0.4 million ha. of maize infested by stemborer caterpillar was treated with *Beauveria bassiana* cultured on steamed rice, bran or corn stalks (Prior, 1989). Only recently this technology, simple in principle, has been improved so far that large quantities of non-contaminated fungal spores of good quality can be produced (Hassan *et al.*, 2002).

Liquid Fermentation

However, for less expensive mass production a single-phase liquid fermentation process is desirable. In liquid fermentation, biomass production can be achieved under controlled conditions (Jenkins and Prior, 1993; Jenkins and Lomer, 1994; Andersch *et al.*, 1990; Andersch *et al.*, 1992; Hirte *et al.*, 1989). The nutrient broths are required for efficient multiplication and growth varies between different entomopathogenic fungi. Besides oxygen and water, an entomopathogenic fungus requires a number of nutrients and minerals, which provides efficient multiplication and growth in liquid fermentation. As organic carbon sources, mainly glucose and polysaccharides and lipids are added to the broth. Nitrogen can be provided as nitrate, ammonium, amino-acids or proteins. Other nutrients are phosphates, potassium, magnesium and sulphur (as sulphate or in the form of cysteine or methionine). As microelements calcium, magnesium, iron, copper, molybdenum, zinc and vitamins like biotin and thaimin can be used.

Under liquid fermentation, production of mycelium, yeast-like cells (blastospores) and conidia is possible. Blastospores of Deuteromycetes, conidia of *Beauveria bassiana, Metarhizium anisopliae* (Jenkins and Thomas, 1996) and *Hirsutella thompsonii* and resting spores of *Conidiobolus obscurous, Conidiobolus thronboides* and *Lagnedium gigantium* has been produced in fermentation tanks. The composition of the nutrient broth together with other parameters like pH, partial oxygen, pressure, temperature, stirring velocity, are criteria determining biomass production as well as mycelium or blastospore formation. From the economic standpoint the evaluation of a cheap composition of the broth is of great importance. For example, Stephan *et al.* (1995) used a series of cheap media components for the fermentation of *M. flavoviride*. As nitrogen source, blood meal, horn meal or compost of chicken excrement was used in the fermentation broth and as carbon sources, glucose, molasses or sugar-beet syrup. Media compositions containing these ingredients were very effective in *M. flavoviride* blastospore formation. Best results were obtained with media containing compost of chicken excrement and sugar-beet syrup.

The production of multicellular mycelium is possible by liquid fermentation. One example is fermentation of stable pellets which consists of a tissue-like network of mycelium from *M. anisopliae* (Andersch *et al.*, 1990 and Andersch *et al.*, 1992). With this technology a product containing dry blastospores of *Verticillium lecanii* was produced. Techniques were also developed to produce *Beauveria brongniartii* dry blastospores (Blachere *et al.*, 1973) and *B. bassiana*; conidia are also produced by submerged fermentation (Thomas *et al.*, 1987 and Rombach, 1988). Submerged production of conidia is the most desirable single-phase process. In general, entomogenous hyphomycetes do not produce conidia in liquid culture, except for *Hirsutella thompsonii, Paecilomyces lilacinus* and *B. bassiana*. Both *M. anisopliae* (Adamek, 1963) and *B. bassiana* (Samsinakova, 1966) produce hyphal bodies. Blastospores in liquid culture. *B. bassiana* hyphal bodies have been stabilized, formulated and applied against insects in the field (Fargues *et al.*, 1980). However, formulation of the hyphal bodies is complicated (Blachere *et al.*, 1973) and their field stability and virulence is low compared to aerially produced conidia. Ideally, the fungi should produce the infective conidia in the liquid fermentation broth and conidia, processed directly from the fermenter slurry (Rombach, 1989).

Formulation

An important prerequisite for commercial use of microorganisms as a plant protection agent is the formulation of the spores, thus providing homogenous distribution of the spores, guaranteeing exact dosage and application. There are a number of possibilities of formulating fungal mycelium and spores, ranging from dry mycelium preparation, mycelium granules and spores formulation based on oil, pure oil and dry spores powders or suspendible spore granules. The most important prerequisite, of course, is a reproducible bioactivity of the product. A number of formulation additives used successfully in the formulation of chemical pesticides are not useful for the formulation of microorganisms. However, in case of hydrophobic conidia of *M. anisopliae* eg., emulators have to be added to the formulation, providing suspension and good distribution of the conidia in the spray solution. Furthermore, such additives can contribute to the biological activity and elevate the shelf-life of the microorganisms in the product. It is important to mention that formulations also can provide safety to the user. For example there are possibilities of formulating dusty conidia powders into user-friendly, less dusty wettable powders or water-soluble granules. Due to the nature of the cells eg. blastospores or conidia and the fungal species, different additives may be required to prevent cell damage and negative effects on viability or infectivity. Such negative effects of formulation additives are reasons for the limited number of additives which can be used for the formulation of microorganisms.

A very important point in the development of bioinsecticides based on entomopathogenic fungi is the shelf-life of the fungus in the final formulation. The fungus must keep its biological activity during production, formulation, packing, distribution and storage till the application. After production or during the process of formulation the conidia or blastospores have to be dried in most cases carefully dawn to approximately 5% – 8% relative humidity. This can be achieved by freeze-drying, spray drying, fluid-bed drying etc. and has to be adapted to the type of cells to be dried. In general, after drying and formulation a minimum shelf-life of the fungus of half a year, better one year or more at room temperature (20°C) are required in the product. However, for specific market segment or distribution channels, storage of the product at 4°C might be feasible.

APPLICATION STRATEGIES

Depending on the type of formulation (wettable and suspendable powders and granules, oil formulations, non-soluble granules) different application forms are possible. This includes soil treatments with granules or powders, drench applications and normal and ultra low volume spray applications. The entomopathogenic fungi produced in large scales and formulated can be applied in a variety of ways:

Introduction: An exotic fungus is introduced into and established in a new area or insect population and maintains their density below the economic threshold.

Augmentation: This is the strategy of increasing the density of a native entomopathogenic fungus within an insect population to maintain the pest density at an acceptable level.

Inundation: This is an insecticidal approach. The fungal pathogens are propagated in the laboratory and applied in the same manner as an insecticide to pests at critical periods of the life cycle of the pest for short-term suppression of pest numbers.

Integrated Pest Management: This strategy applies the simultaneous and sequential use of several control methods in managing insect populations. This includes joint uses of entomopathogenic fungus, low doses of insecticides, cultural methods, other parasites and predators etc.

REGISTRATION

The registration requirements for biopesticides are slightly different from country to country. In principle, biopesticides must be sufficiently effective against the target pests, must have no harmful effects on humans and animals and must not have any other harmful effects on the environment when used in a right manner and for the intended purpose. These requirements are very important in the German Plant Protection Act. The specific data requirements for the registration of a biopesticide are a product description which include strain identity, production, formulants and contaminants, an analytical method for the detection of microorganisms mainly used for quality control, data concerning phytotoxicity and quality of crops and processing of crops. A limited residual analysis study on the fate of the microorganisms in the food chain or in the environment might be requested. Toxicology studies should include toxicology data on acute toxicity, inhalation toxicology, skin and eye irritation and allergenic potential of the microorganisms. In the case of fungi as biopesticides, there might be a 90-day feeding study with fungi requested. Furthermore, mutagenicity tests eg. an Ames test and an *in vitro* test on sensitivity of microorganisms to antibiotics could be requested. Ecotoxicology data have to be provided on the effects of microorganisms on the microflora, on soil fauna and aquatic organisms, on animals and birds as well as effects on honeybees and other beneficials. The requirements for registering microbial pest control agent in the United States are included in 'Subdivision M of the pesticide testing guidelines: Microbial and biochemical and pest control agents.' EPA updated these guidelines in February 1996. The updated guidelines are available from various sources, including the government printing office and EPA public access gopher.

PRACTICAL CONSIDERATIONS

From the farmers' point of view the ideal biocontrol agent would be an inexpensive inoculant that could be applied once or infrequently and sustain itself in the agro ecosystem. The fungus should have a narrow host range to minimize impact on the system and should not infect beneficial insects. Agricultural practices could be modified to increase the effectiveness of the agent. Variations in disking, tilling, cover cropping, and intercropping and irrigation practices could sustain endozootic populations or could induce epizootic populations of the biocontrol microbes. The biocontrol agent could either be endemic or an introduced strain. Fungi provide a needed control of insects with sucking mouthparts. Viruses and bacteria must be ingested to cause disease, so fungi provide the only satisfactory microbial means of biocontrol for plant sucking insects. From a marketing point of view the best microbial agent would inundate the agro ecosystem and temporarily reach epizootic proportions. The product would be sold as a mycoinsecticide, which the farmer would purchase and apply when the pest population reaches the economic threshold.

The fungus must not be a human pathogen. The only entomopathogen, which has also induced disease symptoms in a human, is *Basidiobolus haptosporus*. The biocontrol agent should also be tested for allergencity. A number of fungi could become human health problems as our immune systems become increasingly compromised. *Coprinus* has been a human pathogen. There is a reported case where *Coprinus* spores fell from a hospital ceiling and infected an open-heart surgery patient. According to Hall and Papierok, five types of risks have been identified by researchers: (*i*) growth of the pathogen inside tissues, which is best assessed by the absence of sporulation and growth above 35°C; (*ii*) toxicity of the fungal entomopathogen or of its metabolites; (*iii*) allergencity, which for now appears to be comparable to that of other fungi produced by industry; (*iv*) irritability, which can be estimated by Draize's technique; and finally, (*v*) safety towards beneficial fauna. The lack of official norms is a real impediment to the registration of entomopathogens. It is imperative that such norms should be officially drawn up, approved and implemented, preferably on a worldwide-standardized basis.

CONCLUSIONS

Most of the fungi with reputed entomopathogenicity have been found in the agro ecosystem. This is where there are high populations of insects. A plethora of insect-fungus relations await discovery in the rain forests of the world. Many of these fungi could be valuable biocontrol agents or sources of otherwise useful metabolites. Knowledge of insect-fungus relationships in primary ecosystems could also help us create more stable and diverse agro ecosystems that would be less vulnerable to pest-insect epizootics. At the present rate of rain forest habitat destruction we are already losing uncharted insect-fungus relationships. It could prove invaluable to record these data and to archive fungal germplasm. If at some time in the future we attempt to recreate our rain forests this information would be a crucial part of the blueprint for design.

Generalizations on the environmental persistence of all entomopathogenic fungi are difficult because of the extreme diversity of the group. However, the following are correct for most species. Natural sunlight in the range of 290 to 400 nm will affect persistence on foliage and to a lesser extent on other substrates. The half-life of spores is one to four hours in full sun, less for mycelia. Ambient temperatures higher than 35°C reduce persistence; freezing temperatures have little effect. Ambient relative humidities higher than 90 per cent are required for entomophthorales, while hyphomyceteous species tolerate lower humidities as both, spores and mycelia. In the soil environment, soil temperatures higher than 30°C, soil moisture near saturation and the presence of antagonists, all reduce fungal persistence.

Under practical conditions the application of entomopathogenic fungi and generalization about the compatibility among entomopathogenic fungi and pesticides is difficult. Since penetration of the insect by the fungus takes some days, no immediate effects after the application can be observed. In the case of the application of mycelium or myceliar granules in soil, infectious conidia have to be developed. These delays in activity may lead to further damage to the plants. A crucial factor is the relatively slow activity compared to chemical insecticides. Some insecticides and fungicides are detrimental to fungal infections of insects, and even single applications of fungicides can weaken the pathogenicity of the fungal parasites. Furthermore, insect pathogenic fungi are sensitive to UV irradiation and extreme temperature conditions, which influence their viability and activity negatively. Little research has been conducted on the use of both compatible pesticides and entomopathogenic fungi for pest control. These inherent problems associated with living organism lead to possible restrictions in their efficient use as pesticides.

Major obstacles in the development of entomopathogenic fungi after years of long academic and industrial research still to some extent are: (*i*) economical mass production (*ii*) reproducible bioactivity (*iii*) formulation with a shelf-life (*iv*) slow activity and (*v*) competition with cheaper chemical insecticide. To date these hurdles are in opinion still very high so that in terms of economics no real success story has been made out of a product based on entomopathogenic fungi, although there are number of products in the market. But mass production and formulation have made some big step forward and some new product developments have been registered or tested under an experimental use. Furthermore, until recently there has been little application of molecular methods and gene technology to entomopathogenic fungi, since there was little information on the genes controlling pathogenicity or specificity of the strains. The lack of pesticidal genes that enhance pathogenicity has been a major constraint in the implementation of biotechnology in crop protection. However, there are more and more new physiological and biochemical data available, which allow the isolation of the genes that might open new possibilities for mycoinsecticide improvement by molecular methods and genetic engineering. There are few reports currently available of attempts to manipulate entomopathogenic fungi genetically, but clearly, the potential exists to overcome present limitations. Therefore the situation may change in future. With continuing increase in application and formulation technology, and the development of specific models based on increased ecological data and with a firm handle on molecular biology of entomopathogenic fungi and exploitation of their biotechnological potential, it may be possible to control pests by using fungi.

REFERENCES

Adamek, L (1963) *Folia Microbiologica.* **10**: 255-267.

Agarwal, G. P., Rajak, R. C., Sandhu, S. S. & Khan, A. R. (1990) *J. Trop. Forestry.* **6**: 172-174.

Alexopoulos, C. J. (1979) *Introductory Mycology*, 3rd edition. John Wiley and Sons, Inc., N.Y.

Andersch, W., Hartwig, J. & Reinecke, P. (1990) In *V^th^ International Colloquium Invert Pathol, Adelaide*, Australia, pp. 2-5

Andersch, W. J. Hartwig. & Reinecke, P. (1992) *Pflanzenschutz-Nachrichten* **45**(63): 129-142.

Antia-Londono, O. P., Posada-Florez, F. & Bustillo-Pardey, A. E. (1992) *Cenicafe.* **182**: 1-12.

Auld, B. A. (1992) *Biological control of locusts and grasshoppers*, (Eds. Lomer, C.J. & C. Prior), Waalinford, CAB International, pp. 219-229

Bartlet, M. & Jaronski, C. (1988*) Fungi in biological control systems.* (Eds. R.N. Burges) Manchester University Press, Manchester, pp. 61-85.

Bidochka, M. J. & Khachatourians, G. G. (1988) *Exp. Mycol.* **12**: 161-168.

Blachere, H., Calvez, J., Ferron, P., Corrieu, G. & Peringer, P. (1973) *Ann. Zool. Ecol. Ani.* **5**: 69-79.

Brokyn, P. J., Wilding, N. & Dark, S. J. (1985*) Ann. Appl. Biol.* **107**: 365-376.

Bumbieris, M & Lloyd, A. B. (1967) *Aust. J. Biol. Sci.* **20**: 1169-1172.

Burges, H. D. (1981) *Microbial control of pests and plant diseases.* 1970-1980. Academic Press, London.

Callaghan, A. A. (1969) *Trans. Br. Mycol. Soc.* **53**: 87-97.

Cammon, S. A.& Rath, A. C. (1994) *Mycol. Res.* **98**: 1253-1257.

Carruthers, R. I. & Hural, K. (1990) In *New Directions in Biological Control* (Eds. R.R. Baker and PE Dunn), Alan R Liss, New York, pp. 115-138.

Carruthers, R. I., Feng, Z., Ramos, M. E. & Soper, R. S. (1988) *J Invert Pathol.* **52**: 154-162.

Daoust, R. A. & Pereira, R. M (1986) *Environ. Entomol.* **15**: 1237-1243.

Desgranges, C., Vergoignam, C. & Lereec, A. (1993) *Biotech. Advances* **11**: 577-587.

Fargues, J. (1981) These Doctorat es Sci., University of Paris VI.

Fargues, J., Cugier, J. P. & Van de Wegle, P. (1980) *Acta Oceologica / oceol Appl.* **1**: 49-61.

Farques, P., Rougier, M. , Goujet, R. & Itier, R. (1988) *Entomophaga.* **33**: 357-370.

Ferron, P. (1978) *Ann. Rev. Entomol.* **23** : 409-423.

Ferron, P. (1981) In *Microbial control of Pests and Plant diseases* (Ed. H. D. Burges) Academic Press, London, pp. 465-485.

Fuxa, J. R. (1987) *Annu. Rev. Entomol.* **32**: 225-251.

Gardner, W. A. & Storey, G. K. (1985) *J. Econ. Entomol.* **78**: 1275-1279.

Gottwald, T. R. & Tedders, W. L. (1984) *Environ. Entomol.* **13**: 557-560.

Gunnarson, S. G. S. (1988) *J. Invert. Pathol.* **52**: 9-7.

Hajek, A. E. & Leger, R. J. S. (1994) *Annu. Rev. Entomol.* **39**: 293-322.

Hajek, A. E. & Soper, R. S. (1992) *Environ. Entomol.* **21**: 129-135.

Hassan, A. E. M., Dillon, R. J. & Charnley, A. K. (1989) *J. Invert. Pathol.* **54**: 277-279.
Hassan, S, Bhamra, A. K., Sil, K., Rajak, R. C. & SS Sandhu (2002) *J. Ind. Bot. Soc.* **8**: 85-88.
Hirte, W., Walter, C. & Grunberg, M. (1989) *Zentralbl Mikrobiol.* **144**: 405-420.
Ignoffo, C. M. & Garcia, C. (1992) *Environ. Entomol.* **21**: 913-917.
Ignoffo, C. M., Garcia, C. & Gardner, W. A. (1985) *Environ. Entomol.* **14**: 87-91.
Ignoffo, C. M., Hostetter, D. L., Sikorowski, P. P., Sutter, G. & Brooks, W. W. (1977) *Environ. Entomol.* **6**: 411-415.
Jaronski, S. T. & Axtell, R. C. (1982) *J. Med. Entomol.* **19**: 255-262.
Jenkins, N. E. & Prior, C. (1993) *Mycol. Res.* **97**: 1489-1492.
Jenkins, N. E. & Lomer, C. J. (1994) In *Microbial Control of Pests*, (Ed. P.H. Smits) IOBC wprs Bulletin, Bulletin ,OILB crop **17**(3): 181-184.
Jenkins, N. E. & Thomas, M. B. (1996) *Pesticide Science* **46**: 299-306.
Keller, S and G Zimmermann (1989) *In Insect-Fungus Interactions,* (Eds. Wilding, N, N Collins, Hammond, N. M., Webber, P. M. & J. F. Webber), Academic Press, London, pp. 240-269.
Kendrick, W. B. & Parkinson, D. (1990) *In Soil Biology Guide* (Ed. D.L. Dindal), John Wiley and Sons, N.Y., pp.49-68.
Khachatourians, G. G. (1991) In *Physiology and genetics of entomopathogenic fungi.* (Eds. Arora D. K., Ajello, L., Mukherji K.G.) *Handbook of applied mycology*, Vol. 2: Humans, animals and insects. Marcel Dekker, New York, pp. 613-661.
Lacey, L. A., Frutos, R., Kaya, H. K. & Vails, P. (2001) *Biological Control* **21**: 230-248.
Lacey, Y. & Kaya, H. K. (2000) *Field manual of techniques in invertebrate pathology: Application and evaluation of pathogens for control of insects and other invertebrate pests.* Kluwer Academic, Dordrecht.
Latge, J. P. & Boucias, D. G. (1984) *J. Gen. Appl. Microbiol.* **30**: 135-150.
Latge, J.P., Sampedro, L., Brev, P. & Diaquin, M. (1987) *J Gen Microbiol.* **133**: 1987-1997.
Li, D. P. & Holdum, D. G. (1993) *J. Invertebr. Pathol.* **62**: 273-277.
Lingg, A. J. & Donaldson, M.D. (1981) *J. Invertebr. Pathol.* **38**: 191-200.
Lisasky, S. G. & Hall, H. A. (1983) *The Filamentous fungi.* Vol. IV Edward Arnold, London.
Lockwood, J. L. (1964) *Ann. Rev. Phytopathol.* **22**: 341-362.
Lord, J. C. & Roberts, D. W. (1985) *J. Invertebr. Pathol.* **45**: 331-338.
Matewele, P, Trinci, A. P. J. & Gillespie, A. T. (1994) *Mycol. Res.* **98**: 1329-1333.
McCleod, D. M. (1963) In *Insect Pathology*, (Ed. E.A. Steinhaus) Vol. 2, Academic Press, New York pp. 189-231.
McCoy, C. W. (1990) *In New Directions in Biological Control,* (Eds. R.R. Baker & P.E. Dunn), Alan R Liss, New York, pp. 139-159.
McCoy, C. W., Samson, R. A. & Boucias, D. G. (1988) *In Handbook of Natural Pesticides* (Eds. CM Ignoffo and NB Mandava), Vol.V. *Microbial Pesticides.* Part A. *Entomogenous Protozoa and Fungi.* CRC Press, Inc., Boca Raton, FL, pp. 151-236.
Millstein, J. A. , Brown, G. C. & Nordin, G. L. (1983) *Environ. Entomol.* **12**: 1344-1349.
Pell, J. K. & Wilding, N. (1992) *Entomophaga* **37**: 649-654.
Pendland, J. C. (1982) *Can. J. Bot.* **60**: 1569-1576.
Pendland, J. C. & Boucias, D. G. (1986) *Mycologia.* 78: 818-824.
Prior, C (1989) *Biological pesticides for low external input agriculture.* CAB International Publication, Wallingford.
Quintela, E. D. & Mc Coy, C. W. (1997) *Environ. Entomol.* **26**: 1173-1182.
Rajak, R. C., Sandhu, S. S., Mukherjee, S., Kekre, S. & Gupta, A. (1991). *J. Biol. Control* **5**: 123-124.
Roberts, D. W. (1966) *J. Invert. Pathol.* **8**: 212-221.
Roberts, D. W. & Campbell, A. S. (1977) *Miscell. Pub. of the Entomol. Soc. Am.* **10**: 19-76.
Rombach, M. C. (1988) *Entomophaga.***33**: 315-324.
Rombach, M. C. (1989) *Entomophaga.* **34**: 45-52.
Sakamoto, M., Inove, Y. & Aoki, I. (1988) *Trans. Mycol. Soc. Japan* **26**: 499-508.
Samsinakova, A. (1966) *J. Invert. Pathol.* **8**: 394-400.

Samson, R. A. (1982) In *Invertebrate Pathology and Microbial Control,* Proc. XV Ann Meet Soc Invert Pathol, and IIIrd Int Colloq Invert Pathol, Brighton, U.K. September 6-10 pp. 182-187.

Sandhu, S. S. (1993) *Nat. Acad. Sci. Ind.* **16**: 133-135.

Sandhu, S. S. (1994) *Biotech. Ind.* pp.165-176.

Sandhu, S. S. (1995) *Nat. Acad. Sci. Letters* **18**(1&2): 1-5.

Sandhu, S. S. & Vikrant, P. (2004) *In Microbial Diversity: Challenges, Opportunities and Relevance in New Millennium,* (Eds. S. P. Gautam, A. Sharma, S.S. Sandhu & A.K. Pandey), Indica Publishers, New Delhi, India, pp. 373-386.

Sandhu, S. S., Rajak, R. C. & Agarwal, G. P. (1993a) *Annl. Forestry* **1**: 136-140.

Sandhu, S.S., Rajak, R.C.& Agarwal, G. P. (1993b). *Biocont Sci Tech (U.K.)* **3**: 47-53.

Sandhu, S.S., Unkles, S.E., Rajak, R. C. & Kinghorn, J. R. (2001) *Biocont. Sci. Technol.* **11**: 253-258.

Shields, M.S., Lingg, A. J. & Heimsch, R. C. (1981) *J. Invertebr. Pathol.* **38**: 374-377.

Smith, R. Pekrul, S. & Grula, E. A. (1981) *J. Invert. Pathol.* **38**: 335-344.

Soper, R. S. (1985) *Leafhoppers and plant hoppers*, (Eds. L.R. Nault, J.G. Rodriguez), John Wiley, New York, pp. 469-480.

St. Leger, R. J., Goettel, M., Roberts, D. W. & Staples, R. C. (1991) *J. Invert. Pathol.* **58**: 168-179.

St. Leger, R. J., Durrands, P. K., Cooper, R. M. & Charnley, A. K. (1988) *Arch. Microbiol.* **150**: 413-416.

Steinhaus, E. A. (1956) *Hilgardia.* **26**: 107-160.

Steinhaus, E. A. (1960) *J. Insect. Pathol.* **22**: 225-229.

Steinkraus, D. C. & Slaymaker, P. H. (1994) *J. Invertebr. Pathol.* **64**: 130-137.

Stephan, D., Welling, M. & Zimmermann, G. (1995) In *Proceedings Int Conf on Strategies in locust control. Bamako, Mali.*

Storey, G. K. & Gardner, W. A. (1987) *Environ. Entomol.* **16**: 178-181.

Studdert, J. P. & Kaya, H. K. (1990) *J. Invertebr. Pathol.* **56**:380- 386.

Sweeney, A. J. (1985) In *Integrated Mosquito Methodologies,* (Eds. M Laird, J.W. Miles), Academic Press, New York, pp. 269-284.

Tanada, Y. & Kaya, H. K. (1993) *Insect Pathology, Academic Press, Inc., San Diego.*

Tanada, Y. & Kaya, H. K. (1993) *Insect pathology*, Academic Press, New York.

Thakur, R., Sandhu, S. S. & Rajak, R. C. (2005) *Biocontrol Sci. Technol.* **15**: 733-744.

Thomas, K. C., Khachatourians, G. G. & Ingledew, W. M. (1987) *Can. J. Microbiol.* **33**: 12-20.

Tuveson, R. W. & McCoy, C. W. (1982) *Ann. Appl. Biol.* **l01**: 13-18.

Urbanczyk, M. J., Zabza, A., Balazy, S. & Cozh, B. W. (1992) *J. Invert. Pathol.* **59**: 250-257.

Vikrant, P. & Sandhu, S.S. (2005) In *Fungi: Diversity and Biotechnology,* (Eds. M.K. Rai and S.K. Deshmukh), Scientific Publishers, Jodhpur, India, pp. 391-402.

Vyas R. V., Yadav, D. N. & Patel, R. J. (1991) *Ind. J. Exp. Biol.* **21**: 795-797.

10

Application of Fusaria in Agricultural and Industrial Biotechnology

Piyush Pandey[1], Meenu Saraf[2], R. C. Dubey[3] and D. K. Maheshwari[3]

[1]*S.B.S.P.G. Institute of Biomedical Sciences and Research, Balawala, Dehradun- 248161 (India)*
[2]*Department of Microbiology, Gujarat University, Ahmedabad-380009 (India)*
[3]*Department of Botany and Microbiology, Gurukul Kangri University, Haridwar-249404 (India)*

Fusaria have long been considered either as organisms of academic curiosity or as organisms of nuisance value. Pioneering work done in the last decade has raised the status of this fungus to a level where it has been looked up biotechnologically. However, all is not well because its various species cause economic loss in crop yield. Some *Fusarium* species are important pathogens invading seeds, seedlings and older plants of almost all kinds of vegetables, flowers and cereals as well as many fruit and forest trees, causing manifold losses in the field of agriculture, horticulture and forestry. The mechanism of pathogenesis has been studied in detail, especially the major fusarial species and elicitors from the fungal cell wall used to induce defense mechanism in different ways in plants. The era of modern molecular biology has opened up new frontiers of investigation to study the genetic diversity of *Fusarium*. Besides, some species also produce potent mycotoxins like fumonisins, fusaric acid, fusaproliferin, etc.

Modern biotechnology has given some expectations from *Fusarium* as never before. They have been successful in providing the human beings some important secondary metabolites like gibberllin, bioactive compounds like cyclobutene-3, 4-dione herbicides, zearalenone and steroid transformations. In addition, much-prized mycoproteins, such as Quorn, are also being produced by *F. graminearum*. *F. solani* is used as weedicide to control emerging seedling of a prominent weed—*Cucurbita texana*. Induction of soil suppressiveness to soil-borne pathogens has found potential applications using non-pathogenic strains of *Fusarium*. This article reviews the diversity of *Fusarium*, their recent applications and new developments that are diversifying the directions for commercial applications.

INTRODUCTION

Hundreds of species of *Fusarium* (Fig. 1) are known which play multiplicity of role in the environment. They live in soil as saprophytes but also have facility to behave as pathogens. The saprophytic species of fusaria are the natural inhabitant of various soil types. *Fusarium* species are important pathogens invading seeds, seedlings and older plants of almost all kinds of vegetables, flowers, and cereals as well as many fruit and forest trees. Much of the interest in this fungus arises because of its ability to cause diseases to economically important plant hosts, but its near ubiquity in soils worldwide and its ecological activities indicate a much more diverse role in nature (Alves-Santos *et al.*, 1999). Hence, plant diseases caused by fusaria have attracted significant interest and are well studied, supported by enormous amount of available literature. Study of pathogenesis is relevant in agricultural biotechnology for the development of effective control measures. Various species of fusaria are exploited by biotechnological processes and utilized for several beneficial purposes. Use of fusaria in industrial fermentation of antibiotics and biochemicals are well known. Also, they are currently being widely utilized for various uses in agricultural, environmental and medical biotechnology. Many of the industrially important secondary metabolites like gibberllin, other bioactive compounds like cyclobutene-3, 4-dione are being produced by *Fusarium* spp. Recently, biocontrol of soil-borne pathogens using non-pathogenic strains of *Fusarium* has found potential applications. Our aim here is to emphasize the impact of *Fusarium* spp. (*Gibberella*) in modern biotechnology.

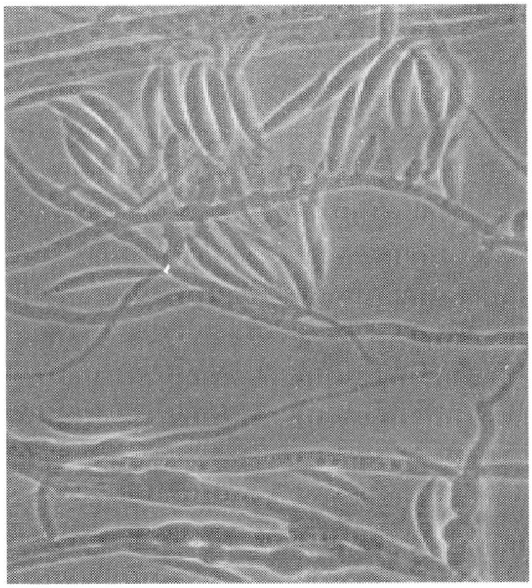

Fig. 1: Conidia of *Fusarium oxysporum* in phase contrast microscopy (X500).

PATHOGENESIS OF *FUSARIUM* SPP.

The *formae speciales* of *Fusarium oxysporum* can cause wilt diseases in more than 100 plant species (Armstrong and Armstrong, 1981). Similarly, Kolattukudy and Gamble (1995) have reported that at least 111 plant species belonging to 87 different genera of agronomic, horticultural, or of forestry importance were susceptible to *F. solani* attack. Ocamb and Juzwik (1995) found at least 7 species of *Fusarium* associated with seedling rot of eastern white pine (*Pinus strobes*).

Initiation of Infection

Fusarium sp. come in contact with host surface and recognizes the host. Macroconidia of *F. solani* release an extra cellular material from their tip which is involved in adhesion (Schuerger and Mitchell, 1993). Plant recognizes pathogen when physical contact occurs between them. Some of the cell-wall components act as elicitors in their recognition by host. *F. moniliforme* elicit chitinase activity in rice cells (Ren and West, 1992). A proteinaceous elicitor has also been isolated from *F. oxysporum* f. sp. *pisi*. (Dean *et al.*, 1989) identified as endoxylanase (Dean and Anderson, 1990). Elicitors from fungal cell wall signal induction of defense mechanism in different ways in plants. Application of chitosan induced cell-wall appositions such as papillae

(Benhamou and Theriault, 1992) and occlusion xylem with a coating material in tomato roots inoculated with *F. oxysporum* f.sp. *radicis-lycopersici*. Nichols *et al.*, (1980) reported that β-1, 3-glucanase and chitinase activity increased in pea-pod inoculated with *F. solani* f.sp. *pisi*.

Role of Hydrolytic Enzymes in Host Tissue Penetration

Fusarium spp. penetrate the cell wall of host by producing several hydrolytic enzymes. Production of cutinase is supposed to be of major importance. Cutinase has been isolated and purified from *F. roseum* (Lin and Kollatukudy, 1980a) and *F. solani* f.sp. *pisi* (Purdy and Kolattukudy, 1973). Two isozymes of cutinase have been also reported from *F. oxysporum* f.sp. *pisi*. Production of cutinase of *Fusarium* spp. is induced by the presence of cutin (Woloshuk and Kolattukuddy, 1986), however Koller *et al.* (1982) demonstrated that germinating spores of *F. solani* f.sp. *pisi* released a small quantity of cutinase. This small amount of cutinase generates soluble cutin monomers that may signal the bulk production of cutinase. Ultrastructural evidence for the involvement in cuticular penetration has been presented in host-pathogen interactions. *F. solani* spores were placed on pea stem surface and scanned with the electron microscope to determine the progress of penetration of the germinating spores. Treatment of the penetration area with ferritin-conjugated rabbit antibodies prepared against cutinase showed that the penetrating fungus secreted chitinase (Shaykh *et al.*, 1977). *F. oxysporum* is known to produce extracellular polygalacturonase and/or pectate lyase, the pectin enzymes, to breach pectinaceous barrier (Holz and Knox-Davies, 1985; Peres-Artes and Tena, 1989). *F. solani* f.sp. *pisi* also produce three pectin degrading enzymes (Crawford and Kolattukudy, 1987). Fusaria are also known to produce hemicellulases to break cell wall barrier of host cells. *F. oxysporum* f.sp. *lycopersici* produce high levels of arabinases, xylanases and galactanases *in vitro* (Cooper *et al.*, 1978). *F. roseum* (Mullen and Bateman, 1975) produces xylanases in high amount. Moreover, *F. culmorum*, a pathogen of monocots, is known to adapt to the nature of the cell walls of the host and produce suitable cell wall - degrading enzymes. When *F. culmorum* was grown on wheat seedling cell walls as sole carbon source, it produced early and high quantity of arabinase, followed by xylanase and laminariase, while low levels of pectinases and cellulases. This was in accordance to substrate concentrations in cell wall of host cells (Cooper *et al.*, 1988). *F. culmorum* produce β-1, 3-glucanase to degrade callose the main component of papilliae formed by host tissues in response to infection (Cooper *et al.*,1988).

Role of Toxin in Wilting

F. amygdali, the pathogen of peach and almond produces 'fusicoccin' that stimulates the activity of plasmalemma localized H^+ ATPase (Marre *et al.*, 1993). This opens stomata and enhances K^+ influx. Fusicoccin leads to a two-fold activation of ATP hydrolytic activity in the microsomal membrane (Schulz *et al.*, 1990). There was a stimulation of H^+ pumping by 15-130% and stimulation of ATP hydrolysis by 20-45% after incubation of microsomal membrane fractions from radish with fusicoccin (Rasi-Caldogno and Pagliarello, 1985). By this process the toxin increases stomatal transpiration. The uncontrolled water loss due to irreversible stomatal opening may result in rapid reduction of available water in intracellular spaces and in a subsequent loss of turgor of the mesophyll cells. As a result, large area of leaf blade wilts and withers (Turner and Graniti, 1969).

GENETIC DIVERSITY IN PATHOGENIC *FUSARIUM* SPP.

Phylogenetic relationships among groups of organisms are often determined by comparison of the nuclear rDNA regions. Partial phylogenetic tree for *Fusarium* spp. and fungi from related genera have been constructed by comparing nucleotide sequences in 5′ end of the 28S- like rRNA subunit (Guadet *et al.*,1989; Bruns *et al.*,1991; O'Donell, 1993). This approach grouped the species according to their teleomorph genus (when known) and indicated relationships within these groups. However, rDNA sequencing had been the main technique for resolving the taxonomy of *Fusarium* (Manicom *et al.*, 1990). Bateman *et al.* (1996) used indirect comparisons of sequences by restriction fragment length polymorphisms (RFLPs). They studied a region of the rDNA from 3′ end of the small ribosomal sub-unit to the 5′ end of the large ribosomal subunit and included the 5.8S coding sequence and the two internal transcribed spacer regions and observed at least two distinct genetic types among isolates (Fig. 2). Genetic diversity of *F. oxysporum* strains had also been studied by RFLP technique (Appel and Gordon, 1995; Alves-Santos *et al.*, 1999; Fiely *et al.*, 1995). Abdel-Satar *et al.* (2003) studied the molecular phylogeny of *Fusarium* spp. isolated from diseased roots of cotton seedlings, by amplified fragment length polymorphisms (AFLP). The UPGMA analyses indicated five distinct clusters corresponding to five taxa. AFLP technique has been utilized frequently for the analysis of *Fusarium* populations (Abd-Elsalam *et al.*, 2002; Klprop *et al.*, 2002; Sivaramakrishnan *et al.*, 2002) because of its utility, repeatability and efficiency.

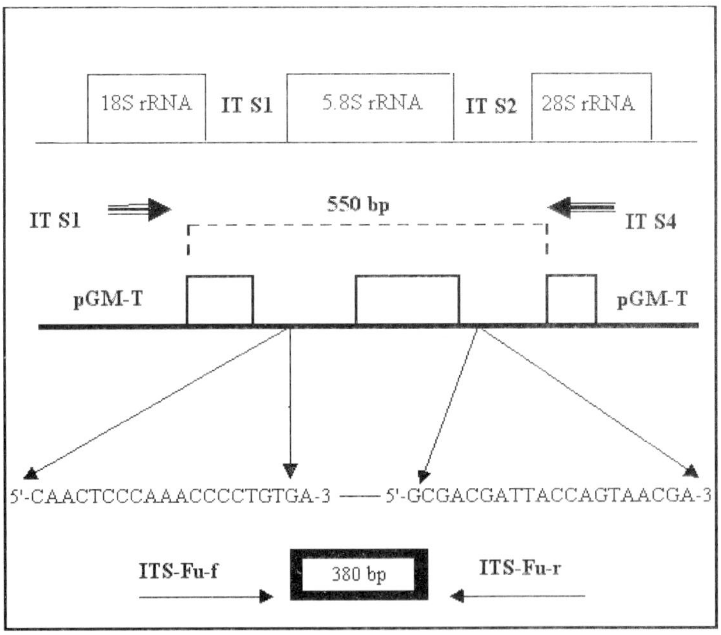

Fig. 2: Oligonucleotide primers, ITS-Fu-f and ITS-Fu-r, specific for the *Fusarium* genus, were constructed from the internal transcribed spacer (ITS) region.

Interestingly, *Fusarium* genome has been exploited for diagnostic PCR (Fernandez *et al.*, 1998; Plyler *et al.*, 1999; Woudt 1995; Schilling *et al.*, 1996). Abd-Elsalam *et al.* (2003) developed two 'taxon-selective' primers for quick identification of the *Fusarium* genus. These primers, ITS-Fu-f and ITS-Fu-r were designed by comparing the aligned sequences of internal transcribed spacer regions (ITS) of a range of *Fusarium* species. The primers showed good specificity for the genus *Fusarium* and the approximately 389-bp product was amplified exclusively.

BIOLOGICAL CONTROL OF *FUSARIAL* INFECTION

Pesticides related hazards have forced to adopt biological methods for the control of plant diseases. Application of biotechnological approach of biocontrol has emerged as potential alternative to chemical control. Considering the impact of fusarial diseases, it is worth discussing their biocontrol methods. Studies on biological control of fusarium wilts have a long history (Alabouvette *et al.*, 1998). The disease suppressive mechanisms of biocontrol agents include siderophore mediated competition for iron (Bakker *et al.*, 1988; Raaijmakers *et al.*, 1995), competition for substrate (Couteaudier and Alabouvette, 1990), induction of systemic resistance (Fuchs *et al.*, 1997; Leeman *et al.*, 1995; Van Loon *et al.*, 1998; Van Peer *et al.*, 1991; Van Wees *et al.*, 1997), and production of antibiotics (Chin-A-Woeng *et al.*, 1998). Rose *et al.* (2003) compared the biocontrol efficacy of *Pseudomonas chlororaphis, Trichoderma harzianum, T. virens,* with chemicals such as thiram and benomyl against *Fusarium oxysporum* f.sp. *radicis-cucumerinum* which causes root and stem rot in cucumber.

Role of Enzymes in Biocontrol

Production of chitinolytic enzymes had been hypothesized to be responsible for suppression of *Fusarium* wilt in cucumber by *Streptomyces* sp. and *Paenibacillus* sp. (Singh *et al.*, 1999). Chan *et al.*, (2002) reported antagonistic activity of *Bacillus subtilis* against *F. graminearum,* which causes ear rot of maize and scab of wheat and barley. Biological control of *Fusarium* wilt of pigeon pea (*Cajanus cajan*) had been reported with chitinolytic activity of *Alcaligenes xylosoxydans* (Vaidya *et al.*, 2003). Superior biocontrol activity by chitinase hyperproducing *Pseudomonas stutzeri* YPL-M26 against *F. solani* has been reported (Lim *et al.*, 1991). Similarly chitinase producing *P. fluorescens* was effective in reducing *F. moniliforme* infection in five different cultivars in sorghum (Raju *et al.*, 1999).

Role of Siderophores in Biocontrol

Role of siderophores of *P. fluorescens* in biocontrol of *F. solani* had been established by Ho-Seong and Kim (1997). Recently, we reported a *P. aeruginosa* strain which effectively inhibited growth of *F. oxysporum* (Gupta *et al.*, 1999) and promotes the growth of Indian mustard because of its siderophore producing ability (Pandey *et al.*, 2005).

Biocontrol of Fusarial Infection by Other Methods

Bacillus brevis inhibited the growth of pigeon pea pathogen-*F. oxysporum* f.sp. *udum* because of the production of an unknown antibiotic substance (Bapat and Shah, 2000). We isolated *Bacillus* sp. B1 from the rhizosphere of *Cajanus cajan,* which showed excellent biocontrol activity

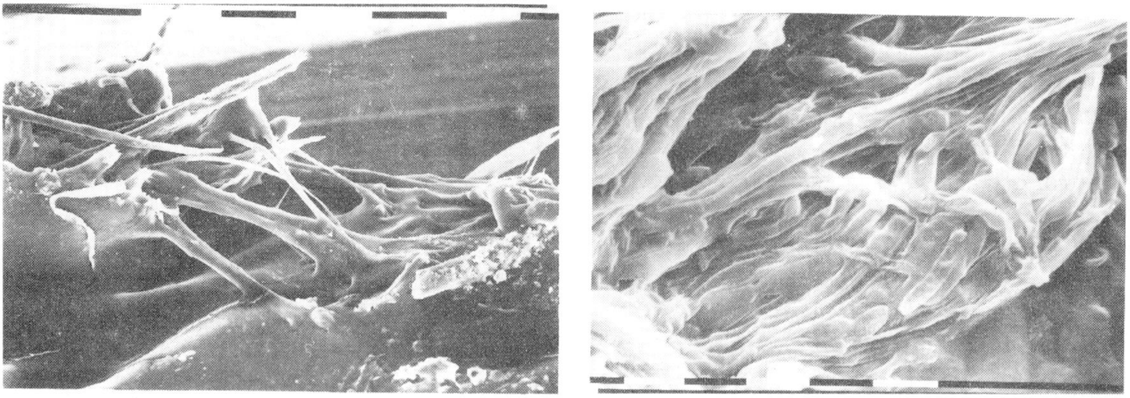

(a) (b)

Fig. 3: Scanning electron micrographs of F. oxysporum mycelial growth in (A) absence and (B) presence of biocrontrol agent Bacillus sp. B1 (bars = 10 ìm). Degenerated, leakage of cytoplasm and lysed hyphal biomass is evident (B) in contrast to healthy mycelial growth (A) (unpublished results).

against *F. oxysporum* (Fig. 3), because of the production of an unknown antibiotic. Another bacterial isolate from root nodules of *Mimosa pudica*, namely, *Burkholderia* sp. MSSP (Pandey *et al.*, 2005) produces 2-hydroxymethyl-chroman-4-one because of which it shows antifungal properties against many phytopathogens (Kang *et al.*, 2004), including *F. udum* (Fig. 4). Bello *et al.* (2002) isolated 25 bacterial and 10 fungal isolates that were suppressive for *F. graminearum*, among which *Stenotrophomonas maltophilia* was most efficient in pot trials. Biocontrol of *F. graminearum*, the causal agent of cereal damping-off had been reported by use of microbes (Harman *et al.*, 1989; Lazzaretti *et al.*, 1994; Hedegus and Farkas, 1997). Sindhu *et al.* (1999) reported a wide range antifungal activities of *Pseudomonas* sp. and *Bradyrhizobium* sp. against several phytopathogens including *F. oxysporum*. De Boer *et al.* (2003) observed that biocontrol

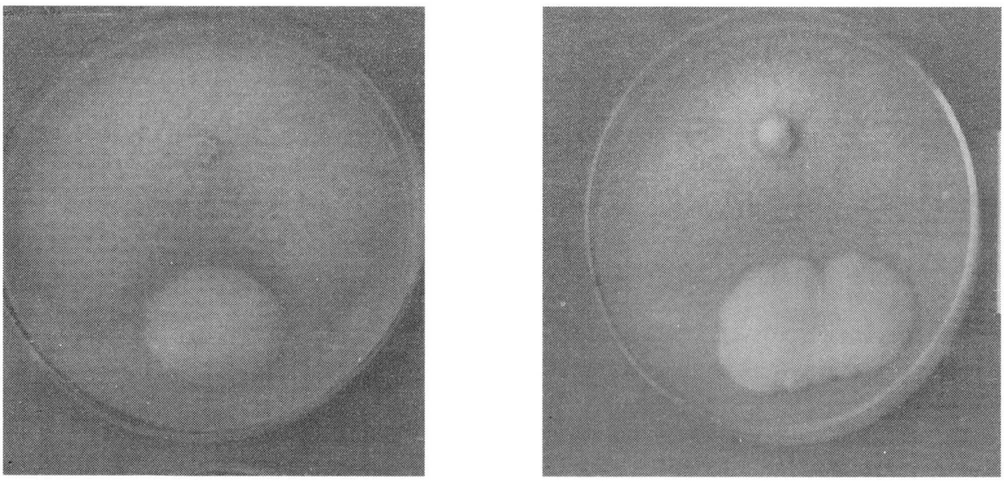

(a) (b)

Fig. 4: Dual culture showing colony interaction Burkholderia sp. MSSP against A) *F. udum* and B) *F. solani*. The zone of inhibition is relatively more prominent in A, compared to B (unpublished data).

of *Fusarium* wilt of radish increased significantly when treated by combining two different strains of *Pseudomonas putida* having different disease suppressive mechanisms. Biocontrol of fusarium root rot in *Phaseolus vulgaris* by using symbiotic *Glomus mosseae* and *Rhizobium leguminosarum* was reported by Dar *et al.*, (1997). In another study, Hwang *et al.* (1992) reported effect of vesicular arbuscular mycorrhiza on fusarium wilts of alfa alfa. Role of rhizobia in biocontrol of *Fusarium* has been suitably compiled (Deshwal *et al.*, 2003).

Role of Non-pathogenic *Fusarium* spp. in Suppression of Pathogenic Fusaria

Antagonistic non-pathogenic *Fusarium* spp. had been proved to reduce the incidence of fusarium wilt by suppressing the growth of pathogenic forms of *F. oxysporum* (Alabouvette and Couteaudier, 1992; Larkin *et al.*, 1996; Paulitz *et al.*, 1987). The main mechanisms of biocontrol of fusarium wilt by non-pathogenic fusaria are competition (for nutrients and/or infection site) and induced systemic resistance (ISR) (Larkin and Fravel, 1999). The phenomenon of ISR has been observed with other *F. oxysporum* strains (Fuchs *et al.*, 1997; Kroon *et al.*, 1991; Mandeel and Baker, 1991; Matta, 1989; Ogawa and Komada, 1986) as well as due to certain bacteria (Kloepper *et al.*, 1996; Leeman *et al.*, 1995; Liu *et al.*, 1995; Van Peer *et al.*, 1991) in several plants. *Fusarium oxysporum* f.sp. *radici-lycopersici* is known to induce systemic resistance in barley against powdery mildew (Nelson, 2005).

INDUSTRIAL APPLICATIONS OF FUSARIA

Secondary Metabolites for Industrial Use

Species of fusaria produce several kinds of secondary metabolites such as gibberellins. Gibberellins also occur as endogenous hormones in plants. They are of biotechnological interest because of their growth promoting properties. *Fusarium moniliforme* (perfect stage *Gibberella fujikuroi*) used for commercial production of gibberellins. It also produces an antibiotic Bikaverin simultaneously and only one gibberellin called gibberellic acid or gibberellin A3. *G. fujikuroi* species complex is composed of at least nine different biological species termed as mating populations (MP), MP-A to MP-I. All are known to produce secondary metabolites. MP-C is the gibberllic acid producer. Malonek *et al.*, (2005) analyzed the gibberellic acid gene cluster and the gibbrellic acid pathway of related species. The complex *G. fujikuroi* is composed of an increasingly large number of morphological, biological and phylogenetic species. Almost all of them can be isolated from agricultural ecosystems. Zeller *et al.*, (2003) isolated *Fusarium konzum* from prairie grass (a new species in the *Gibberella fujikuroi* complex).

Low nitrogen content and a mixture of carbon sources are used for the production of gibberellin. After 120 –150 h yields in production scale fermenters are 1-2 g/l of total gibberellin at 25°C (Crueger and Crueger, 2000). Because of its ability to increase yields of amylase, this compound is applied during malting to germinate barley. Supplementation of glycerol increases yields when added to fermentation medium (Wainwright, 1992). Sometimes, coffee husk can also be used as carbon source or for secondary metabolite production on solid state fermentation using wheat bran, corn cob, cassava flour, sugarcane, baggase. The complete process of production, extraction recovery and performance can be obtained from Pandey *et al.* (2001) whereas the use of immobilization techniques in the process can be obtained from Saucero *et al.*, (1989).

Production of Pectinases

A large number of pectic enzymes are produced by *Fusarium* spp. *F. moniliforme* secretes macerating enzymes in liquid media and in presence of pectic substance. Another enzyme endopolymethyl galacturonate lyase, a pectinolytic enzyme is obtained from *F. solani* and is different from other pectinolytic enzymes in having optimum pH 8.6 (Fogarty, 1983). Miller and Mac Millian (1971) analyzed pectin esterase having a molecular weight of 35000 Da, 203 specific activity units and pH optimum 7.0. The activity occurs at reducing end of highly esterified pectin chains and the remaining activity at some other locus along with protein module. Pectic enzyme patterns from chick-pea have been studied by Zamani *et al.* (2000). They compared polygalacturonase activity from different isolates of *F. oxysporum*. *F. oxysporum* can directly ferment cereal straw to ethanol. However, ball-milled straw is required for this purpose. *F. oxysporum* can also be cultivated on sugar cane bagasse for the production of ethanol (Pandey *et al.*, 2001).

Application in Food Industry

One of the positive sides of *Fusarium graminearum* is the production of mycoprotein. The successful introduction of this fungal food product in the market is an important landmark. One of the advantages that mycoprotein has over soya yeast and bacterial proteins is its filamentous nature which gives it a texture and bite similar to that of meat. This fungus grows on a medium containing only glucose and inorganic salts. It can convert each kilogram of food to 136g of cell protein and doubles in every 6h. This has been approved for human consumptions all over the world under the trade name Quorn (Trinci, 1991). Quorn is produced by growing *F. graminearum* in a 1300 l fermenter by continuous fermentation process. The organism is grown at 30°C at pH 6 on a sterilized medium containing glucose syrup, mineral salts, trace metals, choline and biotin. Oxygen is supplied as compressed air to maintain dissolved oxygen concentrations. Ammonia is used to control the pH and supply nitrogen. Mycoproteins typically contain 44.3% protein, 18.3% dietary fibre and only 13% fat. It has an ideal blend of amino-acids and low sodium content (Glazer and Nikaido, 1998).

Important secondary metabolite, which is used as anabolic growth stimulating compound in swine and sheep is zearalenone. It is produced from *F. graminearum*, (imperfect stage *Gibberella zeae*) in submerged process at 32°C for a period of three weeks of incubation period to obtain 30g/l (Crueger and Crueger, 2000).

Detoxification or degradation of aflatoxins by *Fusarium* sp. has also been studied. Cuero *et al.* (1988) have shown about 40% reduction in aflatoxin concentration when *A. flavus* (toxigenic strain) was co inoculated on maize with *F. graminearum*.

Biotransformations

Many species of *Fusarium* bring about a variety of steroid transformations. Dehydrogenation of steroids in 1, 2 positions is carried out by *F. solani*. Steroid transformations are achieved by growing the fungus in the minimum amount of nutrients, because it reduces subsequent extraction problems (Sebek, 1983). Often steroid transformations are achieved using fungal spores rather than mycelium. Commercially, progesterone is converted to 1, 4 – and nostadiene – 3, 17 dione in a medium containing peptone, corn steep liquor and glucose at pH 6.0.

Progesterone (0.25 g) is added after 48 h at 25°C and 85% conversion is obtained. Some other species of *Fusarium* have other diverse qualities and are being investigated. A number of species of *F. avenaceum, F. couglutinans* and *F. semitectum* are examples of commercially useful bioconversions like penicillin V (phenoxy methyl penicillin) to 6 amino penicillanic acid (6-APA).

Applications in Environment

Recently attempts have been made to treat industrial effluents on a more rational basis with the aim of producing valuable biomass protein. *F. moniliforme* has been used to treat the effluent from a food processing plant. Starch and amino-acids are utilized while the biomass produced potentially valuable steroids. Biologically active secondary metabolites produced by fungi are being evaluated as potential pesticides, particularly for controlling growth of weeds or herbs. Moniliformin, initially isolated from *F. moniliforme,* became the template for the production of cyclobutene-3, 4-dione herbicides, which are post-emergence herbicides. It was later granted a Swiss patent (Wainwright, 1992).

The use of fungal enzymes, either in free form or immobilized in a suitable support provides an alternative to use mycelium or yeast cells to treat pollutants (Munnecke, 1978). For example, the enzyme acylamidase from *F. solani* is immobilized on glass beads, which are then packed into a column into which liquid effluent in trickled. The enzyme acts on the pollutant substrate resulting in reduction of toxicants BOD and COD. The system may be re-used as long as enzymes remain active. Here *F. solani* is used to cleave herbicide propanil (3, 4 - dichloroprop-ionilide) to form 3, 4 - dichloroaniline.

Other Applications in Biotechnology

F. solani has been found to be a potential biocontrol agent of weeds. Its conidia have been used as immobilized or encapsulated granules or pellets to control emerging seedling in the control of *Cucurbita texana* (Weidemann and Templeton, 1988).

The latest innovation in the field of biotechnology is nanoparticle synthesis using *F. oxysporum.* The fungus produces nanoparticles of high redox potential elements like platinum, palladium, semiconductor oxides and quantum dots (Pbs, ViS, MnS, ZnS) as well as bimetallic Ag/Au alloy nanoparticles. The fungus when suspended in water, releases a number of enzymes and proteins, which are involved in the reduction of the metal salts and capping of these reduced metal and metal sulphides at nanoscale (Ahmad *et al.,* 2005).

Some other species of fusaria can degrade the complex phenolic compounds found in wood. Conidia of non-wood decomposing species like *F. flocciferum* are used to degrade phenol following immobilization on diatomite particles (Anselmo *et al.,* 1989). Cell loading up to 50% dry weight have been obtained and the immobilized system can efficiently degrade high concentration of substrate being active over a period of at least two months. Production of many potential products like beavericin, moniliformin, fusaproliferin and fumonisins B_1, B_2, B_3 by different species of *Fusarium* have been studied (Fotso *et al.,* 2002).

Products of some of the species of fusaria bear anti-tumour properties. *Fusarium* sp. elaborates a compound called augidine that has been found active against transplantable mouse tumour lymphatic leukaemia (Wainwright, 1992). Some amino acids like tryptophan produced by *Fusarium*. *F. heterosporum* has been found to produce 'equistein', a natural product possessing anti-HIV activity. This inhibits the 3′ end processing and strand transfer at an IC-50 of appoximately 10 µm. It also inhibits strand transfer reactions catalyzed by pre-integration complexes isolated from HIV infected cells (Singh *et al.*, 2005)

Mycotoxins of Fusaria

Mycotoxin contamination associated with head blight of wheat and other grains caused by *Fusarium culmorum* and *F. graminearum* is a chronic threat to crop, human and animal health throughout the world. One of the most important toxins in terms of human exposure is deoxynivalenol (DON) (formerly called vomitoxin), an inhibitor of protein synthesis with a broad spectrum of toxigenicity against animals (Gilbert and Tekauz, 2000; Walker *et al.*, 2001). Certain *Fusarium* toxins have additional antimicrobial activity, and the phytotoxin fusaric acid has recently been shown to modulate fungus-bacterium interactions that affect plant health (Duffy and De´fago, 1997).

Fusarium moniliforme, *F. proliferatum* and *F. subglutinans* commonly occur on maize (Nagaraj *et al.*, 1994). These species can produce potent mycotoxins, including fumonisins (Leslie *et al.*, 1992), fusaric acid and moniliformin (Marasas, 1984). Fusaproliferin is a toxic metabolite originally isolated from cultures of *F. proliferatum* (Manetti *et al.*, 1995; Ritieni *et al.*, 1995; Santini *et al.*, 1996). It is a sesterterpene compound that has toxic activity against brine shrimp (*Artemia salina* L.), insect cells, and human B-lymphocytes and has teratogenic effects on chicken embryos (Logrieco *et al.*, 1996; Ritieni *et al.*, 1995). Logrieco *et al.* (1996) showed that fusaproliferin could also be produced by *F. subglutinans*. Beauvericin is a cyclodepsipeptide compound known to have insecticidal properties (Grove and Pople, 1980). It has shown to be highly toxic to human cell lines inducing apoptosis (Macchia *et al.*, 1996, Ojcius *et al.*, 1991). It can be produced by strains of *F. proliferatum* (Moretti *et al.*, 1994), *F. semitectum*, and *F. subglutinans* (Gupta *et al.*, 1991).

Expression of *Trichoderma* chitinase genes, *ech42* and *nag1*, which contribute to biocontrol activity, was monitored *in vitro* and on crop residues of two maize cultivars by using *goxA* reporter gene fusions by Lutz *et al.*, (2003). They found that DON-producing *F. culmorum* and *F. graminearum* strains repressed expression of *nag1-gox*. Other mycotoxins of fusaria include Zearalenone (benzoic acid-u-lectone, Nivalenol, Diacetoxyscirpenol or anguidin), Fusarenon-X, Fusarochromanone and Neosolaniol. Details on *Fusarium* toxins have been excellently compiled (Chauhan and Agarwal, 2001; Sumbali, 2001)

CONCLUSION

Various species of fusaria possess a treasure of unique qualities that needs to be explored to realize its full glory. Recent advances made in genomics, proteomics and combinatorial chemistry show a formidable pool of bioactive compounds and functions present in *Fusarium*, several of which are valuable pharmaceuticals, agrochemicals and industrial products. This brief review into the world of *Fusarium* provides a fascinating and almost endless source of diversity that can be exploited commercially. Development of new culture techniques and more innovative tools in molecular biology related to genomic library construction in culturable members of *Fusarium* is required which can be expected to revolutionize the field of biocatalysis and drug discovery. Long-term benefits, which may accrue from such studies includes clues to the emergence of new pathogens and improved understanding of pathogenicity *per se*. The extensive and intensive efforts world over in understanding the various faces of *Fusarium* has indicated that what we know today is just the tip of the iceberg. Only sporadic attempts have been made to isolate extremophilic species. Extensive funding and concerted and collaborative efforts are needed for understanding the diversity of these fungi in extreme environments and their exploitation.

REFERENCES

Abd-Elsalam, K.A., Aly, I.N., Abdel-Satar, M.A., Khalil, M.S. & Verreet, J.A. (2003) *African J. Biotechnol.* **2**: 82-85.

Abd-Elsalam, K.A., Khalil, M.S., Aly, A.A. & Asran-Amal, A. (2002) *Phytopathol. Mediterr.* **41**: 1-7.

Abdel-Satar, M.A., Khalil, M.S., Mohmed, I.N., Abd-Elsalam, K.A. & Verreet, J.A. (2003) *African. J. Biotechnol.* **2**: 51-55.

Ahmad, A., Khan, M.I., Kuma, R. & Sastry, M. (2005) International conference on *microbial diversity: Current perspectives and Potential applications,* Deptt. of Microbiology, Delhi Univ. April 16-18, New Delhi.

Alabouvette, C. & Couteaudier, Y. (1992) In *Biological control of plant diseases,* (Eds. E.C. Tjamos, G.C. Papavizas & R.J. Cook), New York. Plenum Press, pp. 415-426

Alabouvette, C., Schippers, B., Lemanceau, P. & Bakker, P.A. (1998) In *Plant microbe interactions and biological control,* (Eds. G.J. Boland & L.D. Kuykendall), Marcel Dekker Inc., New york, pp. 15-36

Alves-Santos, F.M., Benito, E.P., Eslawa, A.P. & Diaz-Minguez, J.M. (1999) *Appl. Environ. Microbiol.* **65**: 3335-3340.

Anselmo, A.M., Cabral, J.M.S. & Novais, J.M. (1989) *Appl. Microbiol. Biotechnol.* **31**: 200-203.

Appel, D.J. & Gordon, T.R. (1995) *Exp. Mycol.* **19**: 120-128.

Armstrong, G.M. & Armstrong, J.K. (1981) In *Fusarium: Disease, biology and taxonomy*, (Eds. P.E. Nelson, T.A. Toussoun & R.J. Cook, University Park. Pennsylvania State University Press. pp 391-399.

Bakker, P.A.H.M., Weisbeck, B.J. & Schippers, B. (1988) *Plant Nutr.* **11**: 925-933.

Bapat, S. & A.K. Shah (2000) *Can. J. Microbiol.* **46**: 125-132.

Bateman, G.L., Kwasna, H. & Ward, E. (1996) *Can. J. Microbiol.* **42**: 1232-1240.

Bello, G.M.D., Monaco, C.I. & Simon, M.R. (2002) *World J. Microbiol. Biotechnol.* **18**: 627-636.

Benhamou, N. & Theriault, G. (1992) *Physiol. Mol. Plant Pathol.* **41**: 33-52.

Bruns, T.D., White, T.J. & Taylor, J.W. (1991) *Annu. Rev. Ecol. Syst.* **22**: 525-564.

Chan, Y.K., Mc Cormick, W.A. & Seifert, K.A. (2002) *Can. J. Microbiol.* **49**: 253-262.

Chauhan, S. & Agarwal, S. (2001) In: *Innovative Approaches in Microbiology* (Eds. Maheshwari, D.K. & Dubey, R.C.), Bishen Singh Mahendra Pal Singh Publishers, Dehradun, India, pp. 191-211.

Chin, A., Woeng, T.F.C., Bloemberg, G.V., Van der Bij, A.J., Van Der Drift, K.M.G.F., Schripsema, J., Kroon, B., Scheffer, R.J., Keel, C., Bakker, P.A.H.M., Tichy, H.V., De Bruijn, F. J., Thomas-Oates, J.E. & Lugtenberg, B.J.J. (1998) *Mol. Plant Microbe Interct.* **11**: 1069-1077.

Cooper, R.M., Longman, D., Campbell, A., Henry, M. & Lees, P.E. (1988) *Physiol. Mol. Plant Physiol.* **32**: 33-47.

Cooper, R.M., Rankin, B. &Wood, R.K.S. (1978) *Physiol. Plant Pathol.* **13**: 101-134.

Couteaudier, Y. & Alabouvette, C. (1990) *FEMS Microbiol. Ecol.* **74**: 261-268.

Crawford, M.S. & Kolattukudy, P.E. (1987) *Arch. Biochem. Biophysics.* **258**:196-205.

Crueger, W. & Cruegar, A. (2000) In *A textbook of Industrial Microbiology,* Panima Publishing Corporation, New Delhi.

Cuero, R., Smith, J.E. & Lacey, J. (1988) *J. Food Protect.* **5**: 452-456.

Dar, G.H., Zargar, M.Y. & G.M. Beigh (1997) *Microb. Ecol.* **34**: 74-80.

De Boer, M., Bom, P., Kindt, F., Keurentjes, J.J.B., Van der Sluis, I., Van Loon, L.C., & Bakker, P.A.H.M. (2003) *Phytopathol.* **93**: 626-632.

Dean, J.F.D. & Anderson, J.D. (1990) *Plant Physiol.* **94**: 1849-1854.
Dean, J.F.D., Gamble, H.R. & Anderson, A.J. (1989) *Phytopathol.* **79**: 1071-1078.
Deshwal, V.K., Pandey, P., Kang, S.C. & Maheshwari, D.K. (2003) *Ind. J. Exp. Biol.* **41**: 1160-1164.
Duffy, B.K. & Defago, G. (1997) *Phytopathology.* **87**: 1250–1257.
Fernandez, D., Ouinten, M., Tantaoui, A., Geiger, J.P., Daboussi, M. J. & Langin, T. (1998) *Appl. Environ. Microbiol.* **64**: 633-636.
Fiely, M.B., Correll, J.C. & Morelock, T.E. (1995) *Plant Dis.* **79**: 990-993.
Fogarty, W.M. (1983) In *Microbial Enzymes and Biotechnology*, Applied Science Publishers.
Fotso, J., Leslie, J.F. & Smith, J.S. (2002) *Appl. Environ. Microbiol.* **68**: 5195–5197.
Fuchs, J.G., Moenne-Loccoz, Y. & Defago, G. (1997) *Plant Dis.* **81**: 492-496.
Gilbert, J. &. Tekauz, A. (2000) *Can. J. Plant Pathol.* **22**: 1–8.
Glazer, A.N. & Nikaido, N. (1998) *Microbiol. Biotechnol.* **1**: 43-42.
Grove, J.F. & Pople, M. (1980) *Mycopathologia.* **70**: 103–105.
Guadet, J., Julien, J., Lafay, J.F. &. Brygoo, Y. (1989) *Mol. Bio. Evol.* **6**: 227-242.
Gupta, C.P., Dubey, R.C. & Maheshwari, D.K. (1999) *Cytobios.* **99**: 183-189.
Gupta, S., Krasnoff, S.B., Underwood, N.L., Renwick, J.A.A. & Roberts, D.W. (1991) *Mycopathol.* **115**: 185–189.
Harman, A., Taylor, A. & Stasz, T. (1989) *Plant Dis.* **73**: 631-637.
Hedegus, A. & Farkas, S. (1997) *Cereal. Res. Comm.* **25**: 749-751.
Holz, G. & Knox-Davies, P.S. (1985) *Phytopathol.* **112**: 69-80.
Ho-Seong, L. & Kim, S. D. (1997) *J. Microbiol. Biotechnol.* **7**: 13-20.
Hwang, S.F., Chang, K.F. & Chakravaty, P. (1992) *Plant Dis.* **76**: 239-243.
Kang, J.G., Shin, S. Y., Kim, M. J., Bajpai, V., Maheshwari, D. K. & Kang, S. C. (2004) *J. Antibiotics* **57**: 726-731.
Kloepper, J.W., Zehnder, G.W., Tuzun, S., Murphy, J.F., Wei, G., Yao, C. & Raupach, G. (1996) In *Advances in biological control of plant diseases* (Eds. T. Wenhua, R.J. Cook & A. Rovira), China Agricultural University Press. Beijing, pp 165-174.
Klprop, E.K., Baudoin, J.P., Myang'ombe, A.W., Kimani, P.M., Mergeai, G. & Maquet, A. (2002) *J. Phytopathol.* **150**: 517-527.
Kolattukudy, P.E. & Gamble, D.L. (1995) *Pathogenesis and host specificity in plant disease. Vol II- Eukaryotes*, Oxford. U. K: Pergamon, Elservier Science Ltd.
Koller, W. & D.M. Parker (1989) *Phytopathol.* **79**: 278-283.
Koller, W. & Kolattukudy, P.E. (1982) *Biochemistry.* **21**: 3083-3090.
Koller, W., Allan, C.R. & Kolattukudy, P.E. (1982) *Physiol. Plant Pathol.* **20**: 47-60.
Kroon, B.A.M., Scheffer, R.J. & Elgersma, D.M. (1991) *J. Plant Pathol.* **97**: 401-408.
Larkin, R.P. & Fravel, R. (1999) *Phytopathol.* **89**: 1152-1161.
Larkin, R.P., Hopkins, D.L. & Martin, F.N. (1996) *Phytopathol.* **86**: 812-819.
Lazzaretti, E., Menten, J. & Bettiol, W. (1994) *Fitopatologia Venezolana.* **7**: 42-46.
Leeman, M., Van Pelt, J.A., Den Ouden, F M., Heinsbroek, M., Bakker, P.A.H.M. & Schippers, B. (1995) *Eur. J. Plant Pathol.* **101**: 655-664.
Leslie, J.F., Plattner, R.D., Desjardins, A.E. & Klittich, C.J.R. (1992) *Phytopathol.* **82**: 341–343.
Lim, H.S., Kim, Y.S. & Kim, S.D. (1991) *Appl. Environ. Microbiol.* **57**: 510-515.
Lin, T.S. & Kolattukudy, P.E. (1980) *Eur. J. Biochem.* **106**: 341-351.
Liu, L., Kloepper, J.W. & Tuzun, S. (1995) *Phytopathol.* **85**: 695-698.
Logrieco, A., Moretti, A., Fornelli, F., Fogliano, V., Ritieni, A., Caiaffa, M.F., Randazzo, G., Bottalico, A. & Macchia, L. (1996) *Appl. Environ. Microbiol.* **62**: 3378–3384.
Lutz, M.P., Feichtinger, G., De'fago, G. & Duffy, B. (2003) *Appl. Environ. Microbiol.* **69**: 3077-3084.
Macchia, L., Di Paola, R., Nenna, S., Fornelli, F., Moretti, A., Napoletano, R., Logrieco, A., Tursi, A., Bottalico, A. & Caiaffa, M.F. (1996) In *XI Congresso della Societa' Italiana di Tossicologia*, pp. 32.
Malonek, S., Rojas, M.C., Hedden, P., Gaskin, P., Hopkin, P. & Tudzynski¸ B. (2005) *Appl Environ Microbiol.* **71**: 1462-1472.

Mandeel, Q. & Baker, R. (1991) *Phytopathol.* **81**: 462-469.

Manetti, C., Fogliano, V., Ritieni, A., Santini, A., Randazzo, G., Logrieco, A., Mannina, L. & Segre, A. L. (1995) *Struct. Chem.* **6**: 183–189.

Manicom, B.Q., Bar-Joseph, M. & Kotze, J.M. (1990) *Phytophylactica.* **22**: 233-239.

Marasas, W.F.O., Nelson P.E. & Toussoun, T.A. (1984) In *Toxigenic Fusarium species: identity and mycotoxicology.* Pennsylvania State University Press, University Park, Pa.

Marre, M.T., Vergani, P. & Albergoni, F.G. (1993) *Physiol. Mol. Plant Pathol.* **42**: 141-157.

Martin, J.T. (1964) *Ann. Rev. Phytopathol.* **2**: 81-100.

Matta, A. (1989) In *Vascular wilt diseases of plants.* (Eds. E.C. Tjamos & C.H. Beckman) Springer-Verlag, New York. pp 175-195.

Miller, L. & Mac Millian, J.D. (1971) *Biochem.***10**: 570-576.

Moretti, A., Logrieco, A., Bottalico, A., Ritieni, A. & Randazzo, G. (1994) *Mycotoxin Res.* **10**: 73–78.

Mullen, J. M. & Bateman, D.F. (1975) *Physiol. Plant Pathol.* **6**: 233-246.

Munnecke, D.M (1978) *Appl. Env. Microbiol.* **54**: 1719-1723.

Nagaraj, R.Y., Wu, W.D. & Vesonder, R.F. (1994) *Poult. Sci.* **73**: 617–626.

Nelson, H.E. (2005) *J. Phytopathol.* **153**: 366-372.

Nichols, E.J., Beckman, J.M. & Hadwiger, L.A. (1980) *Plant Physiol.* **66**: 199-204.

O' Donnell, K. (1993) In *The fungal holomorph: mitotic, meiotic and pleomorphic speciation in fungal systematics,* (Eds. D.R. Reynilds & J.W. Taylor), CAB International, Wallingford. U. K., pp. 225-233

Ocamb, C.M. & Juzwik, J. (1995) *Can. J. Plant Pathol.* **17**: 325-330.

Ogawa, K. & Komada, H. (1986) *Ann. Phytopathol. Soc. Jpn.* **52**: 15-21.

Ojcius, D.M., Zychilinsky, A., Zheng, L.M. & Young, J.D.E. (1991) *Exp. Cell Res.* **197**: 43–49.

Pandey, A., Soccol, R., Leon – Rodriguez, J.A. & Nigam, P. (2001) In *Solid State Fermentation in Biotechnology: Fundamentals and Applications,* Asiatech Publishers, New Delhi.

Pandey, P., Kang, S. C & Maheshwari, D. K. (2005) *Curr. Sci.* **89**: 177-180.

Pandey, P., Kang, S.C., Gupta, C.P. & Maheshwari, D.K. (2005) *Curr. Microbiol.* 51. (In Press).

Paulitz, T.C., Park, C.S. & Baker, R. (1987) *Can. J. Microbiol.* **33**: 349-353.

Peres-Artes, E. & Tena, M. (1989) *J. Phytopathol.* **124**: 39-51.

Plyler, T.R., Simone, G.W., Fernandez, D. & Kistler, H.C. (1999) *Phytopathol.* **89**: 407-413.

Purdy, R.E. & Kolattukudy, P.E. (1973) *Arch. Biochem. Biophys.* **159**: 61-69.

Raaijmakers, J.M., Leeman, M.,Van Oorchot, M.M.P., Van der Sluis, I., Schippers, B. & Bakker, P.A.H.M. (1995) *Phytopathol.* **85**: 1075-1081.

Raju, N.J., Niranjana, S.R., Janardhana, G.R., Prakash, H.S., Shetty, H.S. & Mathur, S.B. (1999) *J. Sci. Food Agri.* **79**: 206-212.

Rasi-Caldogno, F. & Pugliarello, M.C. (1985) *Biochem. Biophys. Res. Commun.* **133**: 280-285.

Ren, Y.Y. & West, C.A. (1992) *Plant Physiol.* **99**: 1169-1178.

Ritieni, A., Fogliano, V., Randazzo, G., Scarallo, A., Logrieco, A., Moretti, A., Mannina, L. & Bottalico, A. (1995) *Nat. Toxins* **3**: 17–20.

Rose, S., Parker, M. & Punja, Z.K. (2003) *Plant Dis.* 87: 1462-1470.

Santini, A., Ritieni A., Fogliano, V., Randazzo, G., Mannina, L., Logrieco, A. & Benedetti, E. (1996) *J. Nat. Prod.* **59**: 109–112.

Saucero, O.N.J.E., Barbotin, J.N. & Thomas, D. (1989) *Appl. Environ. Microbiol.* **55**: 226–233.

Schilling, A.G., Moller, E.M. & Geiger, H.H. (1996) *Phytopathol.* **86**: 515-522.

Schuerger, A.C. & Mitchell, D.J. (1993) *Phytopathol.* **83**: 1162-1170.

Schulz, S., Oelgemoller, E. & Weiler, E.W. (1990) *Planta* **183**: 83-91.

Sebek, D.K. (1983) *Mycologia.* **7**: 383-394.

Shaykh, M., Soliday, C.L. & Kolattukudy, P.E. (1977) *Plant Physiol.* **60**: 170-172.

Sindhu, S.S., Gupta, S.K. & Dadarwal, K.R. (1999) *Biol. Fertil. Soils.* **29**: 62-68.

Singh, I.P., Bharate, S.B. & Bhutani, K.K. (2005) *Curr. Sci.* **89**: 269-290.

Singh, P.P., Shin, Y.C., Park, C.S. & Chung, Y.R. (1999) *Phytopathol.* **89**: 92-99.

Sivaramakrishnan, S., Kannan, S. & Singh, S.D. (2002) *Mycopathol.* **155**: 171-178.

Sumbali, G (2001) In *Innovative Approaches in Microbiology* (Eds. Maheshwari, D.K. & Dubey, R.C.) Bishen Singh Mahendra Pal Singh Publishers, Dehradun, India, pp. 213-227.

Trinci, A.P.J. (1991) *Mycologist* **5**: 106-109.

Turner, N.C. & Graniti, A. (1969) *Nature* **223**: 1070-1071.

Vaidya, R.J., Macmil, S.L.A., Vys, P.R., Ghetiya, L.V., Thakor, K.J. & Chhatpar, H.S. (2003) *Ind. J. Exp. Biol.* **41**: 1469-1472.

Van Loon, L.C., Bakker, P.A.H.M. & Pieterse, C.M.J. (1998) *Annu. Rev. Phytopathol.* **36**: 453-483.

Van Peer, R., Niemann, G.J. & Schippers, B. (1991) *Phytopathol.* **81**: 728-734.

Van Wees, S.C.M., Pieterse, C.M.J., Trijssenaar, A., Vant Westende, Y.A.M., Hartog, F. & Van Loon, L.C. (1997) *Mol. Plant Microbe Interact.* 10: 716-724.

Wainwright, M. (1992) In *An Introduction to fungal Biotechnology.* John Wiley and Sons.

Walker, S.L., Leath, S., Hagler, W.M. & Murphy, J.P. (2001) *Plant Dis.* **85**: 404–410.

Weidemann, G.J. & Templeton, G.E. (1988) *Plant Diseases.* **72**: 36-38.

Woloshuk, C.P. & Kolattukudy, P.E. (1986) *Proc. Natl. Acad. Sci. USA.* **83**: 1704-1708.

Woudt, L.P. (1995) *Bull. OEPP.* **25**: 109-112.

Zamani M.R., Motallebi, M. & Arefoour, M.A. (2000) *Iran. J. of Agri. Sci.* **31**: 293-302.

Zeller, K.A., Summerella, B.A., Bullock, S. & Leslie, J.F. (2003) *Mycologia.* **95**: 943-954.

11

Microbial Production and Applications of 1, 3-Propanediol

R. K. Saxena, Jasmine Isar, Saurabh Saran and L. Agarwal
Department of Microbiology, University of Delhi South Campus Benito Juarez Road, New Delhi 110021, (India)

1, 3-propanediol, also known as trimethylene glycol, has traditionally been considered a 'specialty chemical'. It is one of the oldest known fermentation products and is a versatile intermediate compound. It is considered a good starting chemical for the production of polymers such as polyesters, polyurethanes, etc and other organic chemicals. This diol can also be used in fine chemistry to produce biodegradable plastics. In addition to all these use, it can advantageously be used in cosmetic industries and as a solvent and additive for lubricants.

Traditionally, 1, 3-propanediol is obtained by chemical route from non-renewable resources however, now efforts are on to produce this chemical from renewable resources through microbes. For the successful commercial production of this diol, it is necessary to establish the metabolic pathways and production hosts (microorganisms), which are capable of delivering final product at high yields. By virtue of being a natural product, relevant biochemical pathways exist that can be harnessed into fermentation processes to produce 1,3-propanediol. Furthermore, genetic and metabolic engineering have the potential to significantly improve product yields, titer, and productivity and overcome the limitations of fermentation technology.

This article presents an exhaustive review of different strategies, which have been employed for enhancing the microbial production of 1,3-propanediol and its applications. It encompasses around these lines and highlights the studies on potential microorganisms involved in the production, biochemical pathways, product yields, and metabolic and genetic engineering since these are excellent tools to help science to produce this bulk chemical from renewable resources using microbes as the production entities.

INTRODUCTION

Fossil fuels served as the main resource for the production of large volume chemicals for ages. However, in due course of time it was realized that these resources could not be used for a long time as they are non-renewable and cause pollution. Driven by the environmental concerns and concept of sustainability, the chemical industries seriously began exploring bio-based or renewable feedstocks for producing bulk chemicals. Initially, this transition to renewable feedstocks was impossible due to the lack of appropriate technologies and infrastructures (Zeng and Bielb, 2002).

However, in course of time, science touched all edges in various spheres of life and successfully developed techniques and processes to fulfil the demands of the consumers in terms of 'natural' or 'green' products. The driving forces for developing environmentally benign technology or green technology are the current need to decrease pollution caused by petro-chemical processing and the future needs to replace the dwindling hydrocarbon economy with a renewable, environmentally sound carbohydrate economy. Microbial fermentation technologies are particularly attractive in this regard as they (a) involve renewable feed stock and (b) do not produce toxic byproducts. The development of this technology has given the industries new ways to make valuable chemicals and as a result there is a strategic shift of major chemical companies into this area.

Today, there is a paramount interest in producing bulk chemicals 1, 3-propanediol etc. using microorganisms by fermentation focusing on the environmental aspects, industrial safety and renewable nature of this mode of production. Earlier, propanediol had negligible market value but with the development of new biological processes, novel strains, metabolic and genetic engineering, the biologically produced propanediols can compete with the propanediols produced from the petrochemicals. Microbiologically produced propanediol are pure stereoisomers and find wide applications in pharmaceutical industries. Propanediol are 3 carbon diol and exist in two forms–1, 2-propanediols and 1, 3 propnadiol (Fig.1). They also find use in polyester industries and resin manufacture.

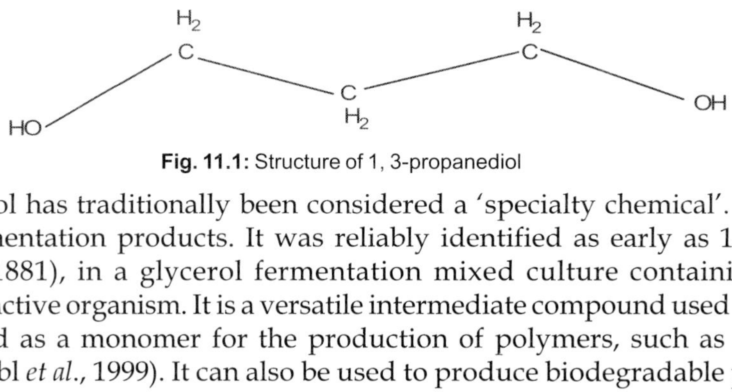

Fig. 11.1: Structure of 1, 3-propanediol

1, 3-propanediol has traditionally been considered a 'specialty chemical'. It is one of the oldest known fermentation products. It was reliably identified as early as 1881 by August Fruand (Fruand, 1881), in a glycerol fermentation mixed culture containing *Clostridium pasteurianum* as an active organism. It is a versatile intermediate compound used in the synthesis of heterocycles and as a monomer for the production of polymers, such as polyesters and polyurethanes (Biebl *et al.*, 1999). It can also be used to produce biodegradable plastics (Witt *et al.*, 1994). Other specific uses are in fine chemistry and cosmetic industry. In addition to all these use, it can advantageously be used as a solvent and additive for lubricants (Zhu *et al.*, 2002).

1, 3-propanediol was earlier manufactured either by the hydration of acrolein (Lawrence and Sullivan, 1972) or by the hydroformylation of ethylene (Colin *et al.*, 2000). Currently, microbial pathways are exploited to manufacture 1, 3-propanediol. It is a typical product of glycerol fermentation and cannot be produced by the anaerobic conversion of any other organic substrate. In earlier times, high price of glycerol was prohibitive for the production of 1, 3-propanediol through biological means. However, recently the development of new biocatalysts, molecular genetic tools, new process configurations has provided a flexible and sustainable way for the use of cheap and abundant renewable resources such as glucose, starch and sucrose for the production of propanediol.

This review has been presented in the following subsections:

- Microorganisms producing 1, 3-propanediol.
- Biochemistry and enzymology of propanediol production.
- Genetic engineering and metabolic engineering of propanediols production.
- Economic importance and application of 1, 3-propanediol production.
- Future prospects and outlook.

Microorganisms Producing 1, 3-Propanediol

Lately, considerable attention is being paid to the production of bulk chemicals from microorganisms. However,the production of 1,3-propanediol has only been known from bacteria (Biebl *et al.*, 1999). In this section, we will discuss the microorganisms involved in the production of 1,3-propanediol.

Some natural microorganisms can ferment sugars to glycerol but they cannot convert glycerol to 1, 3-propanediol. Others may have the ability to ferment glycerol to 1, 3-propanediol and still others that can ferment mixtures of glycerol and sugars to 1, 3- propanediol. However, none can ferment sugars directly to 1, 3-propanediol (Cameron *et al.*, 1998) With recent advances in biotechnology, particularly the development of metabolic engineering, it is possible to design and construct totally new metabolic pathways in cell by combining the ability to ferment sugar to glycerol and glycerol to 1, 3-propanediol (Cameron *et al.*, 1998). Bacteria, commonly producing 1, 3-propanediol are of genera *Klebisiella* (Homann *et al.*, 1990; Biebl *et al.*, 1998; Huang *et al.*, 2002; Nemeth *et al.*, 2003), *Enterobacter* (Barbirato *et al.*, 1995; Zhu *et al.*, 2002) and *Citrobacter* (Homann *et al.*, 1995), *Lactobacilli* (Schutz and Radler, 1984.) and *Clostridia* (Raynaud *et al.*, 2003; Forsberg, 1987; Bielb *et al.*, 1992; Dabrock *et al.*, 1991; Luers *et al.*, 1997; Malaoui and Marczak, 2001).

Biochemistry and Enzymology of 1, 3-Propanediol Production

Enzyme kinetics of propanediol formation and studies on their metabolic pathways have been carried out over the past two decades. On the basis of these studies, the biochemistry of propanediol production has been elucidated in detail.

For large-scale biological production of any bulk chemical, it is desirable to use cheaper and more abundant technical substrates such as starch or glucose. The same should be kept in mind for the production of 1, 3-propanediol. However, 1, 3-propanediol is a typical product of glycerol fermentation. It is not formed in anaerobic conversion of any other organic substrate. Therefore, the strategy adopted is to convert glucose to glycerol than convert glycerol to 1,3-propanediol. Glycerol or 1, 2, 3-propanetriol is a simple alcohol and can be obtained by hydrolysis of natural fats and oil. This can also be obtained from glucose with the help of microorganisms (Dechwer, 1995).

Different Strategies for the Production of 1, 3-Propanediol

Conversion of glucose to glycerol: Osmotolerant yeasts are best for glycerol production and the strains *Saccharomyces rouxii, Torulopsis magnoliae* and *Pichia farinosa* are commonly used.

An interesting process for glycerol production has been developed in recent years in Israel, using the halophilic alga *Dunaliella salins.* Most mesophillic *Bacilli* can grow anaerobically at the expense of sugars. They can carry out a distinctive fermentation, in which 2,3-butanediol, glycerol, and CO_2 are the major end products accompanied by small amounts of lactate and ethanol. The fermentation can be approximately represented as:

$$\text{3-glucose} \longrightarrow \text{2 butanediol} + \text{2 glycerol} + 4CO_2$$

Glucose is initially dissimilated through the Embden-Meyerhof pathway to the level of triose phosphate, a point after which a metabolic divergence occurs (Fig. 2). Part of the triose phosphate is converted to pyruvate, from which Butanediol and CO_2 are produced; the rest is reduced to glycerol.

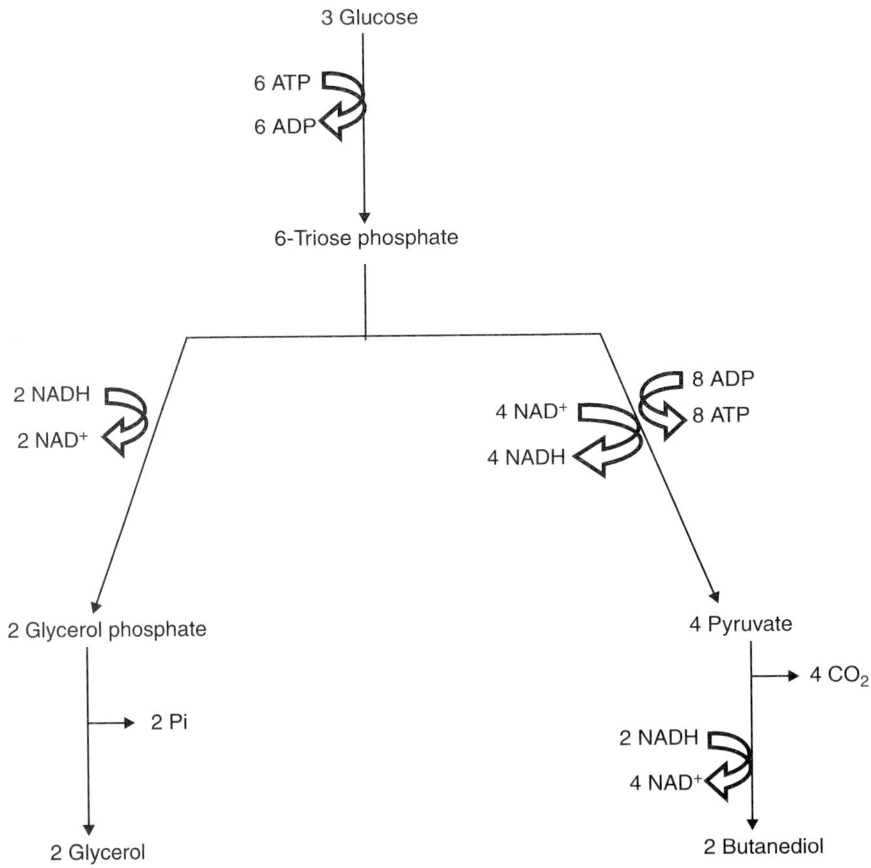

Fig. 11.2. Production of glycerol from glucose

Generally, the amount of glycerol formed is quite small, but by modifying the fermentation balance the yield can be greatly increased. This can be done by the addition of sodium sulphite which forms an acetaldehyde sulphite complex. Normally, the $NADH_2$ formed during the first part of glycolysis is reoxidized by acetaldehyde to form ethanol, but if the acetaldehyde is removed in this sulphite complex, the $NADH_2$ is available for the reduction of dihydroxyacetone phosphate (also formed during glycolysis) via the enzyme glyceraldehyde phosphate dehydrogenase to glycerol phosphate (which is dephosphorylated to glycerol). Through the dismutation of the two-acetaldehyde molecules into acetic acid and ethanol, some glycerol enrichment is also attained by adding Na_2HPO_4. Owing to the formation of additional by products, yields never exceed more than 20-30% of the sugar used.

Conversion of glycerol to 1,3-Propanediol: From the chemical point of view, the microbial conversion of glycerol to propanediol is disproportionate. On one hand, glycerol is dehydrated to 3-hydroxyl propionaldeyde followed by hydrogenation to propanediol by $NADH_2$. On the other hand, glycerol is converted to pyruvate, which in turn produces various by- products (acetate, lactate, butyrate, ethanol, H_2, CO_2) and $NADH_2$ (Gunzel *et al*., 1991).

Thus, a part of the glycerol is transformed via dihydroxyacetone to glyceraldehyde -3-phosphate and further to pyruvate. The reduction equivalents ($NADH_2$) generated along this route may be transferred to an external electron acceptor or to the dehydration product of glycerol namely, 3-hydroxypropionaldehyde. Scheme of major pathway involved in the overall disproportion as proposed for *Klebsiella pneumoniae* is given in the Fig. 3 (Dechwer, 1995).

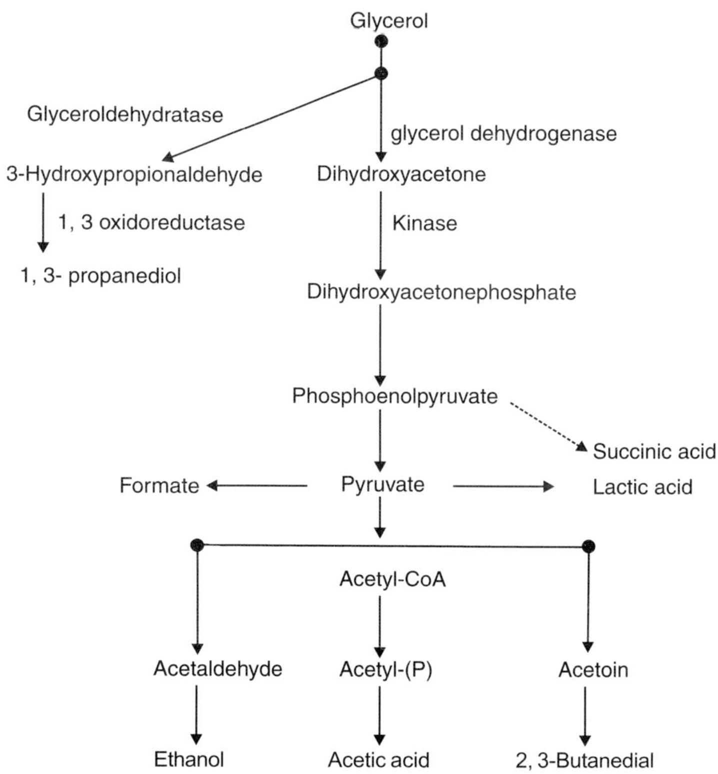

Fig. 11.3. Glycerol metabolism in *K. pneumoniae* showing the production of 1, 3-propanediol(Dechwer, 1995)

The highest propanediol yield is obtained if only acetate is formed as a by-product. In this case, two molecules of $NADH_2$ are formed and therefore, the yield amounts to 2/3 mol of propanediol /mol of glycerol consumed (without consideration of biomass formation). Generally, more the availability of $NADH_2$ more is the production of propanediol (Gunzel *et al.*, 1991). This 1,3-propanediol is not metabolized further and, as a result accumulates in the media. The physiological role of 1,3-propanediol formation is to regenerate the reducing equivalents that are released from the formation of dihydroxyacetone (DHA) and during the further oxidation of dihydroxyacetonephosphate (DHAP), (Zhu *et al.*, 2002).Under the conditions of glycerol limitation, hydrogen production is found to be higher than the maximum amount that can be generated from the cleavage of pyruate to acetyl CoA. Consequently, there is a lower recovery reducing equivalent in the reduced product—propanediol. However, under condition of glycerol excess the formation of hydrogen is significantly reduced and a surplus of $NADH_2$ was generated for the formation of propanediol. These findings indicate the existence of enzymes in *Klebsiella pneumoniae* that transfer reducing equivalents from $NADH_2$ to hydrogen and propanediol flexibly Fig. 4 (Dechwer, 1995).

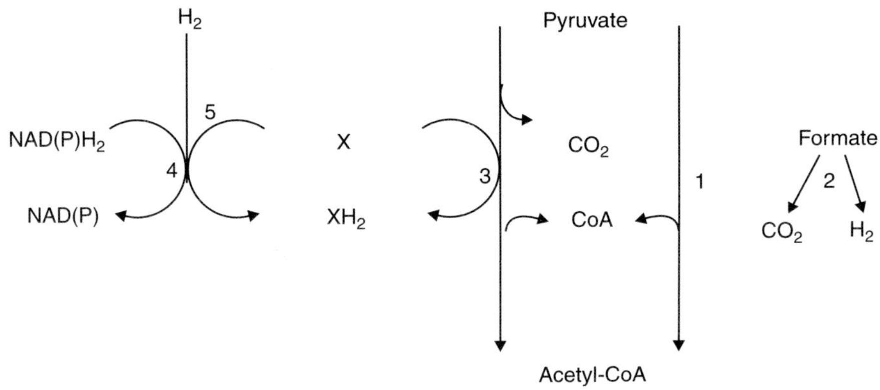

Fig. 11.4. Flow scheme of reducing equivalents suggested for anaerobic glycerol metabolism of *K. pneumoniae* [Zeng *et al.*, (1993)]

1. Pyruvate lyase
2. Formate lyase
3. Pyruvate : ferredoxin (flavodoxin) oxidoreductase
4. NAD (P) H : feredoxin (flavodoxin) oxidoreductase
5. Hydrogenase.

X. Low potential redox carrier such as ferredoxin or flavodoxin.

Two-stage fermentation procedure: To produce 1, 3-propanediol directly from glucose, two approaches can be used, one is the 'Process Approach' and the other is the 'Genetic Approach' (Hartlep *et al.*, 2002). In process approach, the two fermentation processes i.e. fermentation of glucose to glycerol and fermentation of glycerol to 1, 3-propanediol can be carried out as a two stage process. Thus, a glucose to 1, 3-propanediol conversion process can be achieved by mixed culture or a two stage process with yeast and a member of *Enterobacteriaceae* in two consecutive stages (Haynie and Wagner, 1996). Two-stage process could have the following advantages: (1) better control of cultivation conditions for the individual stages, (2) flexibility of the process to use different substrates (glucose or glycerol) and to produce different products (glycerol or 1, 3-propanediol) depending on the market situation and substrate prices (Hartlep *et al.*, 2002).

Co-fermentation: Co-fermentation of glycerol with sugars is desirable for the commercial production of 1, 3-propanediol because sugars such as glucose are significantly less expensive than glycerol. Co-fermentation is typically not possible with native 1, 3 -propanediol producers because glucose represses the 1, 3-propanediol pathway. Native *E. coli* does not ferment glycerol to 1, 3-propanediol (Cameron *et al.*, 1998). In one of the reports by Tong and Cameron (1992), it was reported that in *E.coli* cells transformed with *K. pneumoniae* dha regulon genes is able to co-ferment glycerol with glucose. 1,3-propanediol yield may be enhanced by co-fermentation of glycerol with sugars provided that kinetic constraints are overcome. The yield of 1, 3-propanediol from glycerol was improved from 0.46 mol/mol with glycerol alone to 0.63 mol/mol with glucose co-fermentation and 0.55 mol/mol with xylose co-fermentation (Tong and Cameron,1992).

Abbad- Andaloussi *et al.* (1998) also studied the effect of glucose on glycerol metabolism in the case of *Clostridium butyricum* DSM 5431. They found that with glycerol alone, 43% of glycerol is oxidized to organic acids to obtain energy for growth and 57% to produce 1, 3-propanediol. With glucose–glycerol mixture, glucose catabolism is used by the cells to produce energy whereas glycerol is used chiefly in the utilization of the reducing power.

Enzymes Involved in 1, 3-Propanediol Production

Glycerol dissimilation involves two parallel pathways: oxidative and reductive. Through the oxidative pathway glycerol is dehydrogenated by a nicotinamide–adenine dinucleotide (NAD^+)-linked enzyme, glycerol dehydrogenase, to dihydroxyacetone (DHA), and then to dihydroxyacetonephosphate (DHAP) by an adenosine triphosphate (ATP)-dependent DHA kinase. On the other hand, through the reductive pathway, a B_{12}-dependent glycerol dehydratase removes a water molecule from glycerol to form 3-hydroxpropionaldehyde (3-HPA) which is then reduced to 1, 3-propanediol by an NADH-linked oxidoreductase (Forage and Lin, 1982).

The key enzyme for the conversion of sugars to glycerol is glycerol phosphatase. This enzyme has been purified from *Bacillus licheniformis*, a glycerol producing bacteria. The enzyme was found to have a unique activity which is specific for D-glycerol-1-phosphate. The other enzymes of glycerol metabolism are glycerol dehydrate (encoded by *dha B, C, & E*), 1, 3-propanediol oxidoreductase (encoded by *dhaT*), glycerol dehydrogenase (encoded by *dhaD*), dihydroxyacetone phosphate kinase (encoded by *dha K*), (Biebl *et al.*, 1999).

The genes for these enzymes have been cloned and sequenced from *K. pneumoniae* and *C. freundii.* The enzyme dehydratase is encoded by three structural genes *gld A, gld B,* and *gld C* in the *K. pneumoniae* (Tobimatsu *et al.*, 1996) or *dha B, dha C,* and *dha E* in the case of *C. freundii* (Seyfried *et al.*, 1996).

Klebsiella pneumoniae is one of the several organisms that naturally ferment glycerol to 1, 3-propanediol and has been studied in details. The conversion is carried out in two-enzymatic steps .The first enzyme glycerol dehydratase (E.C. 4.2.1.3.0.), which requires adenoslycobalamin (co enzyme B12), removes a water molecule from glycerol to form 3-hydroxypropionaldehyde (3-HPA). The second enzyme, 1, 3-propanediol oxidoreductase (E.C. 1.1.1.202) transfers a reducing equivalent from NADH to 3-HPA, yielding 1, 3-propanediol. The genes encoding glycerol dehydratase are designated *dhaB* and *dha T*, respectively (Skraly *et al.*, 1998). Nemeth *et al.* (2003) had reported *Klebsiella pneumoniae* DSM 2026 fermentation procedure for the efficient production of this key enzyme.

E. coli, when grown in a vitamin B12 free medium with glycerol as carbon source, produce a new type of glycerol dehydratase and its protein has been found (operon composed of three genes *dha B1, dhaB2* and *dha T*). The deduced proteins show no homology with the known glycerol dehydratase but show significant similarity with the pyruvate formate lyase. The discovery of glycerol dehydratase, a co-enzyme independent enzyme should significantly influence the development of an economical Vitamin B12 free biological process for the production of 1, 3-propanediol from renewable resources (Raynaud *et al.*, 2003).

1, 3-propanediol dehydrogenase, the other important enzyme involved in the 1, 3-propanediol production has been purified to homogeneity from *Clostridium butyricum* E5 grown anaerobically on glycerol in continuous culture. The native enzyme has a molecular weight of 384200 +/- 31100 Da. It exists as an octamer or a decamer of identical molecular weight sub-units. The apparent K_m value of the enzyme for 3-HPA and NADH are 0.17 mM and 0.06 mM respectively. The enzyme requires only Mn^{2+} for full activity and has similar properties to *K. pneumoniae*. However, when *C. butyricum* E5 wild type and mutant E5 MD were cultivated in a chemostat culture on glycerol, it was found that these two enzymes display similar properties such as specificity towards physiological substrates, pH optima and dependence on Mn^{2+}. These properties make *C. butyricum* mutant D strain the best candidate so far to be used as a biotechnological agent for the bioconversion of glycerol to 1, 3-propanediol (Malaoui and Marczak, 2000).

1, 3-propanediol dehydrogenase of *Enterobacter agglomerans* CNCM 1210 has also been purified to homogeneity. The enzyme activity was optimal at pH 7 to 8 and the enzyme is competitively inhibited by NAD^+ (Ki of 0.29 mn), (Barbirato *et al.*, 1995).

1, 3-propanediol NAD^+ oxidoreductases have been purified from *Lactobacillus brevis* and *L. buchneri*. This enzyme is produced during the co-fermentation of glycerol with sugar. It provides an additional method of NADH disposal. The enzyme requires Mn^{2+} and is an octamer with a molecular mass of 350 kDa. The enzymes isolated from both these strains show properties very similar to those reported for *K. pneumoniae*. However, these properties contrast sharply with those of a protein isolated from another *Lactobacillus* sp. (*L . reuteri*) that ferment glycerol with glucose (Veiga da Cunha and Foster, 1992).

Genetic and Metabolic Engineering of Propanediol Production

Although there is no simple solution to the limitations of fermentation processes like low yields\productivities\titers, recovery and purification difficulties however, these problems can be significantly reduced through the application of metabolic and genetic engineering procedures (Cameron *et al.*, 1998). Metabolic engineering has the potential to considerably improve process productivity by manipulating the throughput of certain pathways.

However, despite its powerful and precise nature, the current status of metabolic engineering is still hindered by the lack of full understanding of cellular metabolism. The aspects of integrated dynamics and overall control structure are the common obstacles for the optimal design of pathways to achieve a desired goal. In many cases, when a natural fermentor (microorganism) of sugars to a desirable metabolite is discovered, there is very limited knowledge about the organism. In this case, the production of the desirable metabolite can be achieved by random mutagenesis, bioprocess improvement and media optimization. With the advances of sequencing and other genetic engineering methods, it is becoming much easier to apply metabolic engineering to natural fermentors and to improve productivity (Altaras *et al.*, 2001). Applications of genetic engineering or metabolic engineering have increased in both, academic and industrial institutions, and the area has been profoundly reviewed (Stephanopoulos and Vallino, 1990; Bailey, 1991; Cameron and Chaplen, 1997; Stephanopoulos *et al.*, 1998; Lee and Papoutsakis, 1999). Most current metabolic engineering studies have mainly focused on manipulating enzyme levels through the amplification, addition, or deletion of a particular. Furthermore, these studies also provide an additional means to study cellular metabolism, in particular the interplay between cofactor levels/fluxes and metabolic fluxes.

The story of 1, 3-propanediol represents a benchmark in metabolic engineering. Expanding the glycerol based natural process to more efficient process based on lower cost carbon feed-stock generally involves two approaches. One is to introduce the gene that allows formation of glycerol from sugars or intermediates of glycolysis into a bacterial strain producing 1, 3-propanediol from glycerol. Another is to introduce the genes that allow conversion of glycerol to 1, 3-propanediol into a bacterium producing glycerol from sugars. The combination of natural biological pathways for the production of glycerol from glucose and the production of 1, 3-propanediol from glycerol into a single host should yield a superior process for the production of 1, 3-propanediol. The conversion of glucose to 1, 3-propanediol requires two reducing equivalents per 1, 3-propanediol molecule. Their results showed that the classical anaerobic fermentation process to produce 1, 3-propanediol from glycerol is limited by the amount of reducing equivalents, which can be generated from the production of organic acid such as acetate. The single organism approach (under development by comparison like Genencor International and Du Pont) overcomes this problem by allowing some of the glucose to be converted all the way to CO_2 providing more reducing power per glucose molecule the ability to control both, carbon and energy, in the single organism allow more efficient use of the in-put carbon source.

For construction of a single organism, the first strategy was applied by the expression of the glycerol –3-phosphatase gene. Glycerol –3- phosphate is converted to glycerol by the enzyme glycerol –3-phosphatase. The genes named *GPP*1/2 which codes two isoenzymes of glycerol 3-phosphatase have been cloned from *Saccharomyces cerevisiae* (Norbeck *et al.*, 1996). They further reported that the glycerol-3-phosphatase can be expressed in 1, 3- propanediol producers or in organisms already transformed with glycerol dehydratase (*dhaB*) and 1, 3- propanediols oxidoreductase (*dhaT*), (Fig. 5). Over expression of this enzyme in the other organism would link glucose metabolism and 1, 3–propanediol formation because the substrate glycerol–3-phosphate is an intermediate in lipid synthesis in all microorganisms (Hartlep *et al.*, 2002). Nakamura *et al.* (2000) reported that 1, 3–propanediol can be produced by recombinant microorganisms by applying this strategy. They reported that various sugars such as glucose, fructose, lactose, sucrose, maltose and mannose can be converted to 1, 3 –propanediol by cultivation of recombinant *K.pneumoniae* expressing glycerol- 3- phosphatase genes of *S. cerevisiae,* resulting in the production of 10.9 gL^{-1} of the 1,3-propanediol.

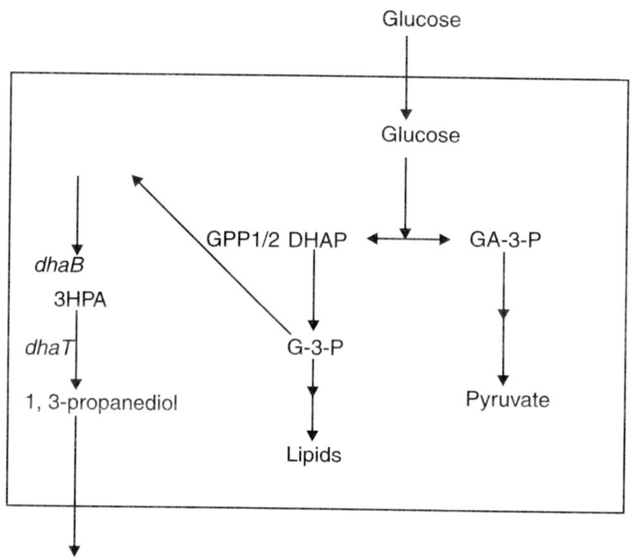

Fig. 11.5. Conversion of glucose/glycerol to 1, 3-propanediol by the expression of glycerol-3-phosphatase in organisms already transformed with *dhaB* and *dhaT* [Norbeck *et al.*, (1996)]

DHAP = Dihydroxy Acetone Phosphate,
G-3-P = Glycerol 3-phosphate
GA-3-P = glyceral dehyde 3-phosphate,
3HPA = hydroxypropionaldehyde,
1, 3-PD, *GPP 1/2* = Glycerol-3-phosphatease gene,
dha B = gene of glycerol dehydratase,
dha T = gene of 1, 3-propanediol oxidoreductase.

Based on this, Du Pont (Wilminjgton, DE, USA) is currently commercializing the process for the fermentative production of 1, 3-propanediol.

The second strategy was demonstrated by over-expressing glycerol dehydratase and 1, 3-propanediol oxidoreductase genes in an organism which produces glycerol naturally such as *S. cerevisiae* (Nevoigt and Stahn, 1997). For expression in yeast, bacterial genes were expressed under the control of eucaryotic promoters. However, only a small amount of 1, 3 propanediol (< 0.1 gL^{-1}) was produced (27, 10). To overcome this problem, recombinant *S. cerevisiae* has recently been developed by integrating glycerol dehydratase and 1, 3 propanediol oxidoreductase gene of *K. pneumoniae* into the chromosome (Nakamura *et al.*, 2000). In this way, 1,3-propanediol concentration of 0.53 gL^{-1} was achieved from glucose.

Although construction of a single organism to produce 1, 3- propanediol from glucose by inserting the glycerol pathway into a natural 1, 3-propanediol producer or the 1, 3 –propanediol pathway into a glycerol producer seems simple and direct, there are certain problems involving natural regulation of the pathways. Therefore, it was decided to use another approach to build both pathways into a host that could do neither step. In this respect, Chottani *et al.* (2000) constructed a strain of *E. coli* containing the genes from *Saccharomyces* for glycerol production and the genes from *K. pneumoniae* for 1, 3-propanediol production. *E.coli* provides several advantages as a study system, since, this provides a rich set of genetic tools sequenced genome, vectors, promoters etc. In addition to this, *E. coli's* metabolism and physiology are well characterized and a large number of metabolic mutants have been constructed and analyzed. Besides this, *E.coli* is closely related to the natural producer of 1,3-propanediol. Since *E.coli* does not naturally produce glycerol or 1, 3–propanediol, there is a need to overcome this through the construction of artificial operons for the optimized expression of the genes for the 1, 3- propanediol pathway. As reviewed above, Chottani *et al.* (2000) have built an *E.coli* strain which can produce 1, 3-propanediol from glucose.

The final strategy was demonstrated by expressing the genes of 1,3-propanediol pathway heterologously. The gen codinges for the enzymes of glycerol metabolism, i.e. glycerol dehydratase (*dha B,C,E*), 1, 3-propanediol oxidoreductase (*dha T*), glycerol dehydrogenase kinase (*dha K*) and a putative regulatory gene (*dha R*) belong to one and the same region named *dha* (Forage and Lin, 1982). The *dha* regulon in *Klebseilla pneumoniae* enables the organism to grow anaerobically on glycerol and produce 1, 3-propanediol. *E.coli*, which does not have a *dha* system, is unable to grow anaerobically on glycerol without an exogenous electron acceptor and does not produce 1,3-propanediol. A genomic library of *Klebseilla pneumoniae* ATCC 25955 has been constructed in *E. coli* AGI, by which, it grows anaerobically on glycerol and dihydroxyacetone. The cosmid pTCI (42.5 kb, total with an 18.2 kb major insert) has been isolated from a 1, 3-propanediol producing strain of *E. coli* and has been found to possess enzymatic activities associated with four genes of *dha* regulon (*dha B, dha T, dha D, dha K*). The activities of all the four genes are inducible by the presence of glycerol. When this *E.coli* AGI/ pTC1 was grown on complex medium plus glycerol, the yield of 1,3-propanediol from glycerol was 0.46 mol/mol. (Tong and Cameron, 1992). Zhu *et al.* (2002) has reported a plasmid, pTC49 of 15.9 Kb containing the *dha* regulon from *K. pneumonia*. It can be used as a genetic source for production of 1, 3-propanediol in *E.coli*. Similarly, Skarly *et al.* (1998) have constructed a variety of synthetic operons with *dha B* and *dha T* under the control of a single *trc*-promoter, which are adaptable for expression in different hosts. Glycerol/glucose fed-batch co-fermentation of *E. coli* AGI harbouring this synthetic operon has resulted in the final concentration of 6.33 gL^{-1} of 1, 3-propanediol.

There are some more strategies of developing a single organism which has been reviewed by Nakamura *et al.* (2003) eg. Larsson *et al.* (1993) and Norbeck *et al.* (1996) reported that the carbon pathway utilizes glycerol 3-phosphate dehydrogenase (DAR1) and glycerol 3-phosphate phosphatase (GPP2) genes, obtained from *Saccharomyces cerevisiae,* to provide glycerol. A couple of patents (Nair *et al.*, 1999; Diaz-Torres *et al.*, 2003) showed that glycerol dehydratase (*dhaB1, dhaB2, dhaB3*) and its reactivating factors (*dha* BX, *orf*X), obtained from *K.pneumoniae,* enable the conversion of glycerol to 3- hydroxypropanaldehyde. Empatage *et al.* (2001) reported that an oxidoreductase endogenous to *E.coli* (*yqh D*) completes the pathway.

Nair *et al.* (1999) further reported that D-glucose-utilizing strain incorporates gene deletions that eliminate non-productive reactions. For example, once formed, glycerol is prevented from re-entering central carbon metabolism by deletion of the genes encoding glycerol kinase (*glp K*) and glycerol dehydrogenase (*gldA*). Empatage *et al.* (2001) suggested that a triosephosphate isomerase (*tpi*) deletion forces a 1:1 split of the carbon flux at the fructose biphosphate aldolase step of the Embeden – Meyerhof Paranas pathway, one branch being forced towards 1, 3–

propanediol formation and another towards the tricarboxylic acid cycle. With the incorporation of this deletion the yield increase up to 40%.

Nakamura and Whited, (2003) recently reported that the two most fundamental changes imposed on the natural reactions of the *E. coli*-based strain are the elimination of D-glucose transport by the phosphotransferase system (PTS) are down regulation of glyceraldehydes 3- phosphate dehydrogenase (*gap*). The former modification brings about the limitations that D-glucose phosphorhylation in *E. coli* is largely PEP-dependent. Glucose phosphorhylation is energetically more efficient, if the PEP-dependent system is replaced with ATP-dependent phophorhylation. As a result, the PTS system is replaced with a synthetic system comprising galactose permease (*galP*) and glucokinase (*glk*), both genes are endogenous to *E. coli*. The *gap* modification is motivated by the requirement to remove the critical ceiling imposed by the *tpi* deletion. Down regulation of *gap*, concomitant with reinstatement of *tpi*, provides an improved flux control point (Fig. 6).

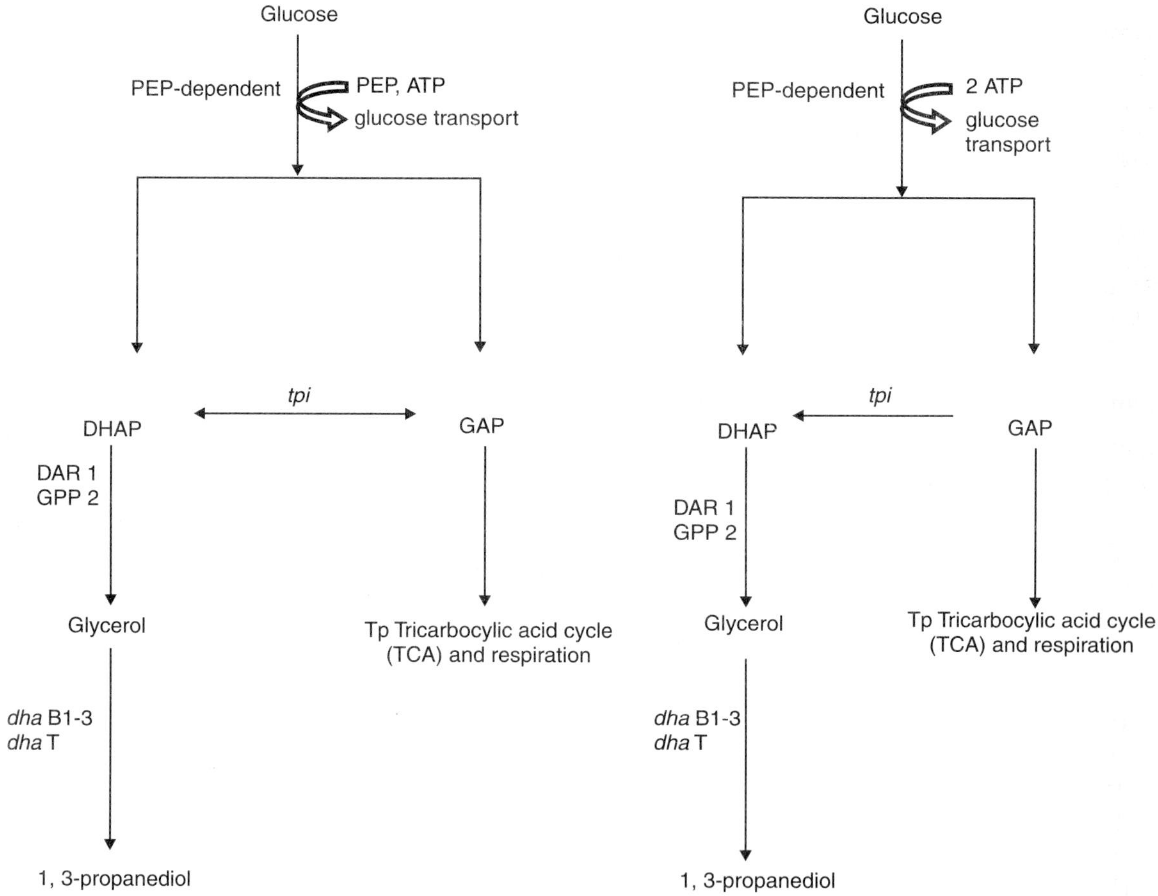

Fig. 11.6. Metabolic engineering for the production of 1, 3-propanediol from D-glucose [Nakamura and Whited, (2000)]

The yields and the productivity of 1, 3-propanediol achievable with metabolically engineered microorganisms, should be carefully evaluated and require detailed analysis.

Important aspects in this regard include: (*i*) the balance of reducing equivalent, (*ii*) the toxicity of 1, 3-propanediol and other by-products, and (*iii*) the regulation of gene expression.

***Protein stability in the recombinant microorganisms*:** A brief account of the reported values for the production of propanediols from various organisms and engineered strains is presented in Table 1.

Table 1: Production values of propanediol (PD) from different organisms and engineered strains

Organism	Plasmid or	Substrate recombinant genes added	Yield (g/L)	Propanediol (PD)	References
Clostridium butyricum DSM 5431	—	glycerol	58.0	1, 3-PD	Gunzel *et al.* (1991)
Clostridium butyricum ATTCC 25955	—	glycerol	25.0	1, 3-PD	Tong (1992)
Klebsiella pneumoniae	—	glycerol	56.0	1, 3-PD	Deckwer (1995)
C. butyricum DSM 5431	—	glycerol	56.0	1, 3-PD	Deckwer (1995)
C. freundii	—	glycerol	25.0	1, 3-PD	Held (1996)
Clostridum butyricum	—	glycerol	83.76	1, 3-PD	Colin *et al.* (2000)
E. coli	K10/ptc9	glycerol+glucose	1.21	1, 3-PD	Zhu *et al.* (2002)
E. coli	Lim61/pTC9	glycerol+glucose		1, 3-PD	Zhu *et al.* (2002)
Klebsiella pneumoniae	—	glycerol without acetate	14.1	1, 3-PD	Hartlep *et al.* (2002)
E. coli	—	glycerol	130.0	1, 3-PD	Nakamura & Whited (2003)
Klebsiella pneumoniae DSM 2026	—	glycerol	78.0	1, 3-PD	Nakamura & Whited (2003)
Klebsiella pneumoniae	—	glycerol	70.0	1, 3-PD	Cheng et al. (2004)

Economic Importance and Applications of 1, 3-Propanediol

Further improvements of microbial and fermentation/recovery processes will permit the cost effective production of propanediols from renewable resources, and hence will be made available for use in a wide range of applications.

On the other hand, the market for 1, 3-propanediol was quite small in the beginning. However, it has now become a large volume commodity chemical. Recent application of 1, 3-propanediol as a monomer for novel polyester and biodegradable plastics has led to strong market need for production of this bulk chemical preferably by biological route. The main advantage of this biodegradable polyester compared to polyhydroxy alkanoate or polylactides (two biodegradable polyesters from renewable sources) is the possibility that the physical properties of the polyester can be easily modified to meet the manufacturing specifications for the final article made from the plastic (Bio. Mat. Net. Item, 1912).

Shell Chemical Company (1995) developed a new chemical route for 1, 3-propanediol by the reaction of ethylene oxide with CO and H_2 (Stinson, 1995; Sullivan, 1993). In 1995, the market situation for 1, 3-propanediol changed significantly as this company announced the commercialization of new polyester, 'corterra' based on terephthalic acid and 1, 3–propanediol. This polymer (Trimethylene terephthalate) is particularly appropriate for fibre and textile applications and combines excellent properties (good resilience, inherent stain resistance, low static generation) with an eco-friendly manufacturing process (Chuah *et al.*, 1995).

1, 3-propanediol is a key ingredient in Sorona ^{Tm} -Du Pont's newest advanced polymer platform. In addition to benefits of softness, stretch recovery, vibrant colour and strain resistance, the textiles made with Sorona ^{Tm} will have the dimension of a naturally sourced material. Du Pont and its collaborator, Genencor, have undertaken an extensive effort over the past seven years to develop a biocatalyst which would enable a fermentation based process for 1,3-propanediol. Since the year 2000, Du Pont and Tate- Lyle have been jointly developing the commercial manufacturing process for fermentation-based 1,3-propanediol. The partnership meshes Du Pont's expertise in biocatalyst engineering and polymer processes with Tata and Lyle's expertise in commercial fermentation and carbohydrate processing (Genencor International and DuPont Expand R & D Collaboration, 2001).

This propanediol may also be used as minor tranquilizer. Merprobamate, which was synthesized in 1951, is a prototype of propanediol. It was the first known propanediol tranquilizer. Propanediol may also be used as biocides for example-PCT 3015; this is a low-cost high performance biocide. Besides this, 2-bromo-2 –nitro-1, 3-propanediol is used as an industrial biocide for the prevention of bio-fouling in areas such as recirculating water in cooling tower and evaporation condenser, air conditioners and humidifier systems. It is also used in the preservation of consumer, household and institutional products (Federal register, 1996).

Future Prospects and Outlook

Owing to the high cost of production procedures, propanediol produced by chemical route was expensive (eg. 30 US dollars per kilogram of 1, 3 –propanediol) and could not compete with petrochemically available diols like 1, 4-butanediol, 1, 2-ethanediol etc. Therefore, several attempts were made and studies were carried out to produce propanediols from microbes. The biggest constraint was no microbial wild type strain ferments sugars directly to 1, 3-propanediol. Although, sugars and glycerol are renewable resources, sugars such as glucose are much less expensive than glycerol ($0.10 per lb >$0.60 per lb). Therefore, efforts were concentrated to metabolically or genetically modify organisms to produce 1, 3-propanediol. The main objective was to extend the substrate spectrum of microbial propanediol to cheaper and more abundant renewable resources such as glucose, starch, sucrose etc. Success has now been achieved in opening new ways for flexible utilization of renewable feed-stock for the production of bulk chemicals.

Even today, both these diols are produced by synthetic processes starting with petrochemical feed-stock and catalysis and involving a variety chemical intermediates. However, it is suggested that microbiological catalysis can potentially provide route to 1, 3-propandiol from renewable feedstocks by processes that would involve benign intermediates and by products. The use of biotechnology combined with chemical conversions is one means to achieve integrated cost effective processes for the conversion of biomass to chemicals and fuels. This approach is currently being pursued at Cargill Dow, LLC in United States to produce chemicals other than propanediols. These processes combine fermentation with chemical process steps to achieve economics that give the products at commodity prices. However, efforts are currently being made worldwide to minimize the chemical process steps and stick to purely biological routes for the production of bulk chemicals. Recently, a joint research effort between DuPont and Genencor has reported the production of 1, 3 – propanediol from sugar by a variety of microorganisms.

Some of the technology developments that are necessary for the successful commercialization of products or chemicals, are the integration of metabolic pathway engineering and fermentation production technology. The toolbox to do so is ever expanding to maximize productivity, improve product purity, expand product line and broaden markets. Knowledge of biochemical pathways and fermentation processes will facilitate metabolic engineering, a rapidly developing technology with great potential to impact dramatically the development of the bio-based economics.

Attempts are being made to bring together, the appropriate genes from yeasts and bacteria, which would offer the possibility to produce the diol directly from a cheaper substrate. However, the yields of such metabolically tailored pathways are still inefficient. It remains a challenge for the scientists to develop improved metabolic processes. This challenge is indicative of the advances that lie ahead in this field.

Future improvements of microbial strains, fermentation/recovery process, metabolic and genetic engineering will pave way for the production of cost effective propanediol on a large scale.

It is envisaged that success in this direction will certainly result in a big 'industrial revolution' and environmentally benign technology.

ACKNOWLEDGEMENTS

Authors wish to acknowledge with thanks the help provided by Anoop, Gautam, Pritesh, Rekha and Kakoli for constructive suggestions and critically evaluating the manuscript.

REFERENCES

zAbbad- Andaloussi., Amne, J., Ferard, P. & Petitdemange (1998) *J. Appl. Microbiol.* **84**: 515-522

Altaras, N. E., Etzel, M. R. & Cameron, D.C. (2001) *Biotechnol. Prog.* **17**: 52-56.

Bailey, J. E. (1991) *Sci* **252**: 1668-1675

Barbirato, F., Camarasa, C., Claret Grivet, J. P. & Bories, A. (1995) *Appl. Microbiol. Biotechnol.* **43**: 786-793

Biebl, H., Menzel, K., Zeng, A.P. & Deckwer, W.D. (1999) *Appl. Microbiol. Biotechnol* **52**: 289-297

Biebl, H., Zeng, A.P., Menzel, K. & Deckwer, W.D. (1998) *Appl. Microbiol. Biotechnol.* **50**: 453-57

Bielb, H., Marten, S., Hippe, H. & Deckwer, W. D. (1992) *Appl. Microbiol. Biotechnol.* **36:** 542-547

Bio. Mat. Net. Item: FAIR-CT-96 (1912) Biodegradable polyester from 1,3-propanediol and succinate produced by fermentation of regrowing resources.

Cameron, D. C. & Chaplen, F.W.R. (1997) *Curr. Opin. Biotechnol.* **8:** 175-180

Cameron, D. C., Altaras, N. E., Hoffman, M.L. & Straw, A. J. (1998) *Biotechnol. Prog.* **14**: 116-125

Chotani, G., Dogde, T., Hsu, A., Kumar, M., La Duca, R., Trimbur, D., Weler, W. & Sanford, K. (2000) *Biochem. Biophys. Acta.* **1543:** 434-455

Chuah, H., Brown, H.S. & Dalton, P.A. (1995) *Int Fiber. J.* Oct. 1995

Colin, T., Bories, A. & Moulin, G. (2000) *Appl. Microbiol. Biotechnol.* **54**: 201-205

Dabrock, D., Bahl, H. & Gottschak, G. (1992) *Appl. Environ. Microbiol* **58**: 1253-1239

Dechwer, W.D. (1995) *FEMS Microbiol. Rev.* **16**: 143-149

Diaz-Torres, M., Dunn-Coleman, N. S., Chase, M. W. & Trimbur, D. (2003) *United States Patent No.* 6, 514, 733

Empatage, M., Haynie, S. L., Laffend, L.A., Pucci, J. & Whited, G. M. (2001) *Patent No* 01/12833 (El Du Pont de Nemours and Genencor International)

Federal Register (1996) *Fed. Regist.* **61:** 19542-19544.

Forage, R. & Lin, E C. C. (1982) *J. Bacteriol.* **151**: 591- 599.

Forsberg, C. W. (1987) *Appl. Environ. Microbiol.* **53**: 639-643.

Freund, A. (1881) Uber die Bildung und Darstellung von Trimethylenalkohol aus Glycerin, *Monatsh. Chem.* **2:** 636-641.

Gunzel, B., Yonsel, S. & Deckwer, W.D. (1991) *Appl. Microbiol. Biotechnol.* **36**: 289-294

Hartlep, M., Hussman, W., Prayitno, N., Meynial –Salles, I. & Zeng, A.P. (2002) *Appl. Microbio. Biotechnol.* **60**: 60-66.

Haynie, S.L & Wagner, L. W. (1996) **Patent** WO 35.799 (El Du Pont de Nemours).

Homann, T., Tag, C., Bielb, H., Deckwer, W.D. & Schink , B. (1990) *Appl. Microbiol. Biotechnol* **33**: 121-126.

Huang, K., Gong, C.S. & Tsao, G. T. (2002) *Appl. Biochem. Biotechnol.***98**: 687-98.

Laffend, L.A., Nagarajan, V. & Nakamura, C. E. (1996) xWO 96/53.796. (El Du Pont de Nemours).

Larsson, K., Ansell, R., Eriksson, P. & Aldler, L. (1993) *Mol. Microbiol.* **10**: 1101-1111.

Lawrence, F. R. & Sullivan, R. H. (1972) *US Patent* **3:** 87, 981.

Lee, S. Y. & Papoutsakis, E.T. (1999) In: *Metabolic Engineering*, Dekker, New York.

Luers, F., Seyfried, M., Daneil, R. & Gottaschalk, G. (1997) *Microbiol Lett.* **154**: 3337-3345.

Malaoui, H. & Marczak, R. (2000) *J. Appl. Microbiol.* **27**: 399-405.

Malaoui, H. & Marczak, R. (2001) *Appl. Microbiol. Biotechnol* .**55**: 226-233.

Nair, R. V., Payne, M.S., Trimbur, D. F. & Valle, F. (1999) PCT Int. Application, Publication No. WO 99/28480.

Nakamura, C. E. & Whited G. M. (2003) *Curr. Opi. Biotechnol* .**14**: 454-459

Nakamura, C. E., Gatenby, A. A., Hsu, A. K., Reau, R.D., Haynie, S. L., Diaz-Torres, M., Trimbur, D. E., Whited, G. M., Nagaarajan, V., Payne, M. S., Picapaggio, S. K. & Nair, R.V. (2000) *US Patent* No. 6, 013, 494.

Nakas, J.P., Schaedle, M., Parkinson, C.M., Coonley, C.E. & Tannenbaum, S.W. (1983) *Appl. Environ. Microbiol.* **46**: 1017-1023.

Nemeth, A., Kupesulik, B., & Sevetla B. (2003) *W. J. Microbiol. Biotechnol.* **19**: 659-663.

Nevoigt, E. & Stahn, U. (1997) *FEMS Microbiol. Rev.* **21**: 231-241.

Norbeck, J., Pahlman, AK., Akhtar, N., Blomberg, A. & Adler, L. (1996) *J. Biol. Chem.* **271**: 13875-13881.

Raynaud, C., Sarcabal, P., Meynial , Salles, I., Croux, C. & Soucaille P. (2003) *Proc. Natl. Acad. Sci. USA.* **100**: 5010-5015.

Schutz, H. & Radler, F. (1984) *Syst. Appl. Microbiol.* **5**: 169-178.

Seyfried, M., Daniel. R., & Gottschalk, G. (1996) *J. Bacteriol.* **178**: 5793-5796

Shell Chemical Company, (1995) Shell Chemical Company announces commercialization of new polymer (Press release)

Skraly, F. A., Lytle, B.L. & Cameron, D.C. (1998) *Appl. Environ. Microbiol.* **64**: 98-105

Stephanopoulos, G. N. & Vallino, J.J. (1990) *Sci.* **252:** 1675-1681.

Stephanopoulos, G. N., Aristidou, A. A. & Nielson, J. (1998) In: *Metabolic Engineering*, Academic Press, San Diego.

Stinson, S.C. (1995) *Chem. Eng.* **17**: 10-14.

Sullivan, J.C. (1993) *Ull mann'Encyklopedia of Indus. Chem.* VCH Press, Weinheim, vol. A-22 .

Tobimatsu, T., Azuma, M., Matsubara, H., Takatori, H., Niida, T., Nishimoto, K., Satoh, H., Hayashi, R. &Toraya, T. (1996) *J. Biol. Chem.* **271**: 22352-22357.

Tong, L.T. & Cameron, D.C. (1992) *Appl. Biochem. Biotechnol.* **34\35**: 149-159.

Veiga da Cunha, M. & Foster, M. A. (1992) *J. Bacteriol.* **174**: 1013-1019.

Witt, U., Muller, R. J., Augusta, J., Widdecke, H. & Deckwer, W.D. (1994) *Makromol. Chem. Phys.* **195**: 793-802.

Zeng, A.P. & Bielb, H. (2002) In: *Advances in Biochem Eng/Biotechnol* (Eds. Zong, A.) Schngrel Press, 74, pp. 239-259.

Zeng, A.P., Bielb, H., Schlier, H. & Deckwer, W. D. (1993) *Enzyme. Microb. Technol.* **15**: 770-779.

Zhu, M. M., Lawman, P. D. & Cameron, D. C. (2002) *Biotechnol. Prog.* **18**: 694-699.

12

Biochemical Methods for the Production of Aldehydes and Organic Acids

Kimiyasu Isobe
Department of Agro-bioscience,
Iwate University, 3-18-8 Ueda, Morioka-020-8550 (Japan)

Aldehydes and organic acids are currently manufactured by chemical and biochemical methods, but the former produce undesirable by-products due to the low specificity and complexity of the reactions. Biochemical methods, on the other hand, exhibit high specificity and are generally carried out under mild conditions, thus giving them advantages in the specific production of aldehydes and organic acids. This review focuses on the characteristics of biochemical methods for the production of aldehydes and organic acids.

Among biochemical methods, fermentation methods have proven effective for the production of some organic acids, but not useful for a variety of organic acids and aldehydes, since fermentation methods require cell growth with a specific metabolic pathway for the production of organic acids. On the other hand, enzymatic methods including resting cell methods are applicable to the production of a wide variety of aldehydes, organic acids, and their derivatives, since they are simple and do not involve cell growth. Among enzymatic methods, those using transferases and dehydrogenases have drawbacks such as the requirement of a second substrate and cofactor regeneration system, which is a disadvantage that methods using oxidases do not share. Thus, methods using oxidases have been the most effective in the production of a wide variety of aldehydes, organic acids and their derivatives.

INTRODUCTION

Many aldehydes and organic acids are widely used in food, pharmaceutical, textile, and other chemical industries, and are currently manufactured by chemical as well as biochemical methods. Although, chemical methods have the advantage that products are obtained at high concentrations, they also suffer the disadvantage that undesirable by-products are produced in the same reaction due to their low specificity and complex reactions under high temperature. Recently, new chemical methods using heterogeneous catalysts have been investigated for the selective production of aldehydes and organic acids (Creca *et al.*, 2001; Biella *et al.*, 2002; Yoneda *et al.*, 2001; Vedrine *et al.*, 2003; Yokoyama and Yamogata, 2001).

On the other hand, biochemical methods have the advantage of being carried out under mild conditions with high specificity in comparison with chemical methods, although they yield product concentrations lower than those by chemical methods. Recently, new biochemical methods with high yield have been extensively investigated. Biochemical methods are generally divided into two main types: the microbial fermentation method and the enzymatic method. Methods using microbial fermentation are characterized by their ability to yield final products by culturing the microorganisms with a specific metabolic pathway under the conditions suitable for the production of such products. Enzymatic methods, on the other hand, generally utilize a simple reaction of isolated enzymes or resting cells, and do not require cell growth. Here, we review the characteristics of and recent investigations into the biochemical methods for the production of aldehydes and organic acids.

PRODUCTION OF ORGANIC ACIDS

Since organic acids such as acetic, lactic, citric, succinic and other organic acids are widely used both in the food industry and as chemical feed-stock, the utilization of inexpensive and readily available raw materials is essential to their production on an industrial scale. Organic acids used in the food industries are mainly manufactured by biochemical methods, while those in the chemical industries are produced by chemical methods.

Acetic Acid

Acetic acid was originally produced from ethanol by the aerobic fermentation of acetic acid bacteria such as *Acetobacter aceti* and *Gluconobacter suboxydans* (Macauley *et al.*, 2001; Sun and Furusaki, 1990). This acetic acid fermentation was catalyzed by alcohol dehydrogenase and aldehyde dehydrogenase bound on a membrane of acetic acid bacteria according to scheme 1 (Matsushita *et al.*, 1994). These microbial fermentation methods are still the most widely used in production of vinegar. On the other hand, acetic acid used as a raw material for vinyl acetate monomer and acetic anhydride synthesis is produced by chemical methods. The synthetic production of acetic acid is based on the oxidation of acetaldehyde with a manganese acetate catalyst. Methods using methanol carbonylation with an Rh catalyst or using direct oxidation of ethylene and ethane were subsequently developed (Yoneda *et al.*, 2001; Smejkal *et al.*, 2005).

$$CH_3CH_2OH \xrightarrow[\text{dehydrogenase}]{\text{Alcohol}} CH_3CHO \xrightarrow[\text{dehydrogenase}]{\text{Aldehyde}} CH_3COOH$$

Scheme 1: Production of acetic acid from ethanol by acetic acid bacteria

Lactic Acid

Lactic acid and its derivatives are widely used in the food, pharmaceutical, leather and textile industries. Recently, it has also attracted much attention as a feed-stock for the manufacture of polylactic acid, a biodegradable polymer used as an environmentally friendly biodegradable plastic. At present, lactic acid is commercially produced by either microbial fermentation or chemical synthesis. Approximately 90% of the total lactic acid produced worldwide is made by the microbial fermentation method, and the remainder by hydrolysis of lactonitrile (Ross *et al.*, 2002). Fermentative production offers advantages in the production of optically pure L- or D-lactic acid depending on the strain selected, although many lactic acid bacteria are hetero-fermentative (Scheme 2). *Lactobacillus delbrueckii* NCIM 2365, *Enterococcus faecalis* RKY1, and *Bacillus* sp. have been reported to be excellent lactic acid bacteria for the high yield production of optically pure L-lactic acid (Ohara and Yahata, 1996; Kadam *et al.*, 2005; Yun *et al.*, 2003), while *L. delbrueckii* IFO 3202 and *L. bulgaricus* have been found most useful in producing optically pure D-lactic acid with a high yield (Benthin and Villadsen, 1995; Tanaka *et al.*, 2005). Although those lactic acid fermentations are carried out under anaerobic conditions, strains of the genus *Rhizopus* produce L-lactic acid under the aerobic conditions (Bai *et al.*, 2004). A major disadvantage of the chemical synthesis of lactic acid is that it always results in a racemic mixture of lactic acid. Enzymatic methods have not yet been used in industrial production.

(A) Homo-fermentation

$$C_6H_{12}O_6 \longrightarrow 2\ CH_3CH\ (OH)\ COOH$$

(B) Hetero-fermentation

$$C_6H_{12}O_6 \longrightarrow CH_3CH\ (OH)\ COOH + C_2H_5OH + CO_2$$

Scheme 2 : Production of lactic acid from glucose by microbial fermentation methods with lactic acid bacteria

Citric acid

Citric acid is the principal organic acid found in citrus fruits, and is widely used in the food industry. Since it is a key compound of the TCA cycle, it is mainly produced by microbial fermentation methods using *Aspergillus niger* and *Candida lipolytica* (Grewal and Kalra, 1995). In the case of *A. niger*, 71.4 and 92.9 g/l citric acid were produced from 200 g/l corn and potato starch, respectively (Haq *et al.*, 2003). For cost-effectiveness of the industrial production of citric acid, the surface, submerged, and solid-state fermentation methods, and the utilization of inexpensive raw materials were also investigated (Usami and Fukutoni, 1977). Chemical methods and enzymatic methods have yet to be used in the industrial production of citric acid (Scheme 3).

$$C_6H_{12}O_6 + 3/2\ O_2 \longrightarrow HOOCCH_2C(OH)(COOH)CH_2COOH + 2\ H_2O$$

Scheme 3: Production of citric acid from glucose by microbial fermentation methods

Succinic acid

Succinic acid is an intermediary compound of the TCA cycle, and serves as an important raw material for producing a wide variety of industrially important chemicals such as food additives, synthetic resins, biodegradable polymers and paints. Although it is commercially manufactured by hydrogenation of fumaric or maleic acids, microbial fermentation methods have also been investigated, since succinic acid is the main taste component produced by yeast during sake (Japanese rice wine) fermentation. Thus, the production of succinic acid from glucose, n-paraffin, and organic acids has been investigated using *Saccharomyces cerevisiae, Candida brumptii, Brevibacterium flavum* or other microorganisms (Arikawa *et al.*, 1999; Sato *et al.*, 1972). Recently, *Anaerobiospirillum succiniciproducens* and *Mannheimia succiniciproducens* were also isolated as producers of succinic acid, and their fermentation conditions were investigated (Lee *et al.*, 1999; Kim *et al.*, 2004). However, yields of succinic acid by these microbial fermentation methods were low, since these microorganisms also produced other organic acids (Scheme 4).

(A) Enzymatic method with succinate dehydrogenase

$$HOOCCH = CHCOOH + H_2 \longrightarrow HOOCCH_2CH_2COOH$$

(B) Microbial fermentation method

$$C_6H_{12}O_6 + CO_2 \longrightarrow HOOCCH_2CH_2COOH + CH_3COOH + HCOOH$$

$$C_6H_{12}O_6 + 2CO_2 + 2H_2 \longrightarrow 2\ HOOCCH_2CH_2COOH + 2H_2O$$

Scheme 4: Production of succinic acid by enzymatic and microbial fermentation methods

Oxalic Acid

Oxalic acid is chemically produced from sodium formate by heating followed by H_2SO_4 treatment or from carbohydrates by oxidation with HNO_3. Although microbial fermentation methods have also been investigated using *Aspergillus niger, Sclerotinia sclerotiorum* and other microorganisms, yields of oxalic acid by these microorganisms were low (Marciano *et al.*, 1989; Mandal and Banerjee, 2005; Shimada *et al.*, 1977). No enzymatic method has yet been developed.

L-Malic Acid

L-Malic acid is extensively used in the pharmaceutical industry and as a food additive. It is now produced on an industrial scale from fumaric acid by the enzymatic method using immobilized cells of *Brevibacterium flavum* (Chibata *et al.*, 1983). Direct production from glucose has been achieved by microbial fermentation using *Aspergillus flavus, A. oryzae,* and combinations of *Rhizopus chinensis* with *Pichia membranaefaciens,* or of *R. arrhizus* with *Proteus vulgaris* (Takao and Holta, 1977). Direct production from n-paraffin and acetic acid has been investigated using the microbial fermentation method with *Candida brumptii* and *Paecilomyces varioti,* respectively (Sato *et al.*, 1977; Takao *et al.*, 1977). A method using bubble-column fermentors has also been investigated as a novel microbial fermentation method using *Schizophyllum commune* (Kawagoe *et al.*, 1977). However, since such methods were inferior to the enzymatic method using immobilized cells, they have not been used for industrial production (Scheme 5).

(A) Enzymatic method with fumarase

$$HOOCCH{=}CHCOOH + H_2O \longrightarrow HOOCCH(OH)CH_2COOH$$

(B) Microbial fermentation method

$$C_6H_{12}O_6 \longrightarrow 2\ CH_3COCOOH \longrightarrow 2\ HOOCCOCH_2COOH$$

$$\longrightarrow HOOCCH(OH)CH_2COOH$$

Scheme 5: Production of L-malic acid by enzymatic and microbial fermentation methods

Fumaric Acid

Fumaric acid is widely used as an acidulant in food industry and also as an intermediary in chemical synthesis. Although most of the fumaric acid is now produced by chemical method through the catalytic oxidation of benzene, several microbial fermentation methods have been also investigated for the food and pharmaceutical industries. Among microbial fermentation methods, the production of fumaric acid from glucose was investigated using strains of the genus *Rhizopus* such as *R. nigricans* and *R. arrhizus* (Petruccioli *et al.*, 1996). The production from n-paraffin was investigated using *Candida hydrocarbofumarica* (Yamada *et al.*, 1970). Other microbial fermentation methods using immobilized cells or renewable resources were also investigated to reduce the production cost of fumaric acid (Carta *et al.*, 1999). An enzymatic method has been investigated by incubating maleic acid with cells from *Pseudomonas alcaligenes*. The yield of fumaric acid from that method increased using cells in which fumarase activity was suppressed by heating with $CaCl_2$ (Ichikawa *et al.*, 2003).

Gluconic Acid

Gluconic acid is widely used in the food, pharmaceutical, and allied industries. Although the chemical production of gluconic acid has been developed using Pd group metals, Bi-promoted catalysis and Au-catalysis, fermentation and enzymatic methods have also been developed. As for microbial fermentation methods, gluconic acid was produced from glucose or starch by culturing *Aspergillus niger* or acetic acid bacteria such as *Gluconobacter oxydans* and *Acetobacter diazotrophicus* (Singh *et al.*, 2003; Pronk *et al.*, 1989; Attwood, 1991). Two enzymatic methods have also been developed, since it was discovered that the production of gluconic acid with *A. niger* was catalyzed by glucose oxidase, and that those with *G. oxydans* and *A. diazotrophicus* were catalyzed by glucose dehydrogenase. The former used an immobilized glucose oxidase or mycelia of *A. niger* (Beltrame *et al.*, 2004; Nakao *et al.*, 1997), while the later used glucose dehydrogenase or cells from *G. oxydans* and *A. diazotrophicus* (Shiraishi *et al.*, 1989). The enzymatic methods with glucose oxidase or mycelia of *A. niger* have the advantage of not requiring the addition of a cofactor or cofactor regeneration system, which is required by methods using isolated glucose dehydrogenase.

5-Keto-D-gluconic Acid

5-Keto-D-gluconic acid is useful as a precursor of L-tartaric acid production. Some strains belonging to *Pseudomona, Klebsiella* and acetic acid bacteria oxidatively convert glucose into 2-keto- or 5-ketogluconic acid via gluconic acid. Although most of these microorganisms produce 2-keto-D-gluconic acids as a main product, a few strains such as *Gluconobacter gluconicus, G. oxydans* and *G. suboxydans* selectively produce 5-keto-D-gluconic acids. Microbial fermentation methods have been developed using *G. oxydans* or recombinant *G. oxydans* (Elfari *et al.*, 2005;

Herrmann *et al.*, 2004), while enzymatic methods have been developed using pyrroloquinoline quinone-dependent membrane-bound D-gluconate dehydrogenase from *G. suboxydans* (Shinagawa *et al.*, 1999).

Glyoxylic Acid

Glyoxylic acid is used as a raw material for chemical synthesis of fine chemicals such as pharmaceuticals or perfumes. Although it has been chemically produced by either the oxidation of glyoxal using nitric acid or a three-step reaction involving the ozonolysis of dimethylmaleate, a new chemical method using trimetallic catalysts has recently been reported (Defferenez *et al.*, 2005). Biochemical methods from glycolic acid were first developed using glycolic acid oxidase from spinach (Seip *et al.*, 1993). They improved this enzymatic method by immobilizing glycolic acid oxidase and catalase (Seip *et al.*, 1994), and by preparing a new recombinant strain of *Aspergillus* sp. containing fungal catalase and glycolic acid oxidase from spinach (Gavagan *et al.*, 1995). A chemoenzymatic method was also reported using NaOH and glycolic acid oxidase from spinach (Yadav and Gupta 2000) as shown in scheme 6. This method has the advantage that glyoxal can be used as a starting material for the production of glyoxylic acid. However, such enzymatic methods have the disadvantage of utilizing glycolic acid oxidase from spinach, which weakly oxidizes glyoxylic acid as well. Isobe revealed that glycerol oxidase from *Aspergillus japonicus* oxidized glycolic acid to glyoxylic acid, and developed a new enzymatic method for the production of glyoxylic acid using glycerol oxidase (Isobe, 1995). However, his method did not prove applicable to industrial production due to the product inhibition by glyoxylic acid. Isobe and Nishise have further developed a novel

Table 1. Enzymatic production of glycolaldehyde, glyoxal and glyoxylic acid.

		Glyoxylic acid from glycolic acid		
	Glycolic acid (mM)	Time (h)	Glyoxylic acid (mM)	Conversion yield (%)
Glycolate oxidase[a]	750	77	750	99.8
Glycerol oxidase[b]	50	10	40.8	81.6
	500	24	120	24.0
Alcaligenes sp. GOX 373 [c]	1000	360	950	95.0
		Glycolaldehyde from ethylene glycol		
	Ethylene glycol (mM)	Time (h)	Glycolaldehyde (mM)	Conversion yield (%)
Alcohol oxidase (*Pichia*) [d]	1000	48	920	92.0
Glycerol oxidase [d]	1000	60	970	97.0
		Glyoxal from ethylene glycol		
	Ethylene glycol (m)	Time (h)	Glyoxal (mM)	Conversion yield (%)
Alcohol oxidase (*Pichia*)[e]	100	10	98.5	98.5
Alcohol oxidase (*Candida*)[e]	100	10	100	100
Glycerol oxidase[b]	100	5	96.8	96.8

a, Seip *et al.* (1993); b, Isobe (1995); c, Isobe & Nishise (1999); d, Isobe & Nishise (1995); e, Isobe & Nishise (1999).

enzymatic method using the resting cells of a new bacterial strain, *Alcaligenes* sp. GOX 373, which was isolated as a strain exhibiting strong activity for converting glycolic acid into glyoxylic acid without product inhibition (Isobe, 1999). Using this new strain, approximately 0.95 M of glyoxylic acid was produced from 1.0 M glycolic acid by incubating at 20°C and pH 6.5 for 15 days (Table 1). No microbial fermentation method has yet been developed.

$$HOOCCH_2OH + O_2 \longrightarrow HOOCCHO + H_2O_2$$

Scheme 6: Production of glyoxylic acid by enzymatic methods with glycolate oxidase or glycerol oxidase

L-α-Aminoadipic Acid

L-α-Aminoadipic acid (L-α-AAA) is a precursor of the biosynthesis of β-lactam antibiotics. Both, it and its related compounds, have been widely recognized as important raw materials for pharmaceuticals. Although chemical methods have been developed (Scott and Wilinson, 1980; Barton *et al.*, 1987), they proved to be complicated and their yield of L-α-AAA was low. Moreover, some alternative biochemical methods have been developed to overcome the methodological shortcomings of such chemical methods. For example, an enzymatic method has been reported by Matos and Wong using glutamate dehydrogenase and alcohol dehydrogenase in a water-organic solvent biphasic system (Matos and Wong, 1986). However, this method also proved unfit for the industrial production of L-α-AAA. Agematsu and Fujii have developed a new microbial method applicable to industrial-scale production using recombinant cells containing L-lysine 6-aminotransferase (EC 2.6.1.36) and Δ-1-piperideine-6-carboxylate dehydrogenase (Agematsu, 2002).

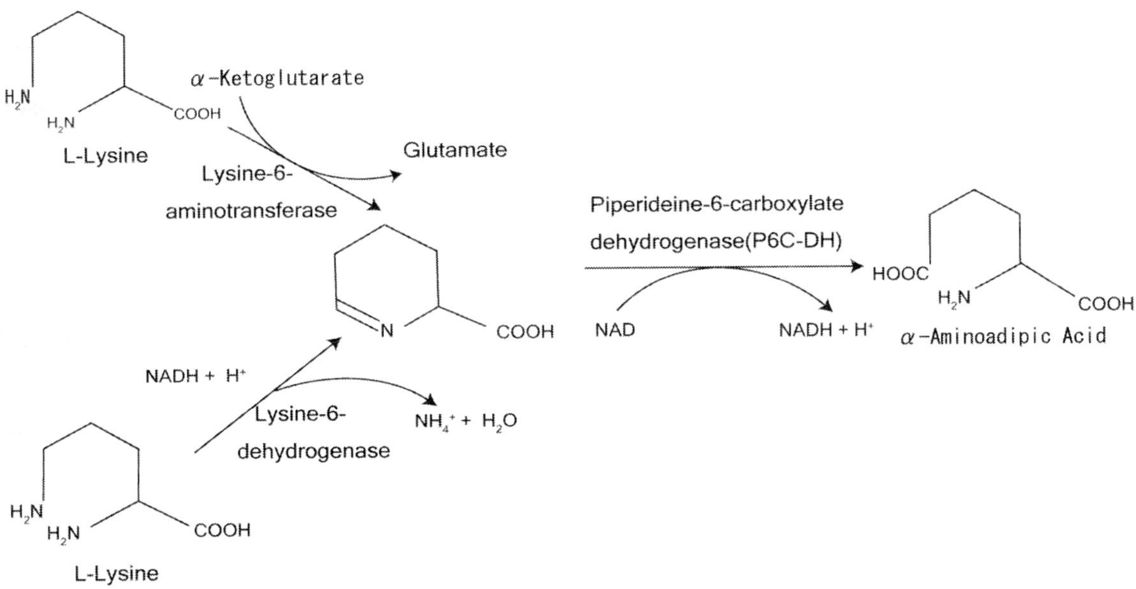

Scheme 7: Production of L-a-aminoadipic acid by enzymatic methods.

N^{α}-Benzyloxycarbonyl-L-aminoadipic Acid

Although chemical methods for the production of amino adipic acid derivatives have been reported, they were shown to be both, complex and low yielding (Baldwin el al., 1988; Ezuerra and Pedregal, 1995). Recently, a new biochemical method for the production of N^{α}-benzyloxycarbonyl-L-aminoadipic acid (N^{α}-Z-L-AAA) has been developed by Isobe *et al.* using mycelia from *A. niger* AKU 3302 (Isobe *et al.*, 2004). Since this enzymatic method utilizes the catalysis of two oxidases, amine oxidase and aldehyde oxidase, this method is simple. However, the formation speed of N^{α}-Z-L-AAA is slow and its yield of N^{α}-Z-L-AAA is poor, since the enzyme productivity of this strain is low, and the N^{α}-Z-L-AAA produced is gradually degraded. The authors also developed a new enzymatic method for the production of N^{α}-Z-L-AAA and N^{α}-Z-D-AAA using the resting cells from a new bacterial strain, *Rhodococcus* sp. AIU Z-35-1, which was isolated as a high producer of N^{α}-Z-L-AAA and N^{α}-Z-D-AAA by the same group (Isobe *et al.*, 2004) as stated in scheme 8. As a result, more than 40 mM N^{α}-Z-L-AAA was produced from 50 mM N^{α}-Z-L-lysine by incubating at pH 7.0 and 30°C for 6 days. A similar amount of N^{α}-Z-D-AAA was also obtained from N^{α}-Z-D-lysine using the same strain under the same reaction conditions (Table 2). At present, this enzymatic method is most effective for the production of N^{α}-Z-L-AAA and N^{α}-Z-D-AAA.

Table 2: Enzymatic production of N^{α}-Z-L-AAA, N^{α}-Z-D-AAA, N^{α}-Z-L-AASA and N^{α}-Z-D-AASA

	N^{α}-Z-L-AAA from N^{α}-Z-L-lysine			
	N^{α}-Z-L-lysine (mM)	Time (h)	N^{α}-Z-L-AAA (mM)	Conversion yield (%)
A. niger [a]	25	12	4.7	19
Rhodococcus sp. [b]	50	6	42	84
	N^{α}-Z-D-AAA from N^{α}-Z-D-lysine			
	N^{α}-Z-D-lysine (mM)	Time (h)	N^{α}-Z-D-AAA (mM)	Conversion yield (%)
A. niger [a]	25	12	<1	<1
Rhodococcus sp. [b]	50	6	42	84
	N^{α}-Z-L-AASA from N^{α}-Z-L-lysine			
	N^{α}-Z-L-lysine (mM)	Time (h)	N^{α}-Z-L-AASA (mM)	Conversion yield (%)
Amine oxidase [c]	50	4	50	100
Rhodococcus sp. [d]	100	1	97	97
	N^{α}-Z-D-AASA from N^{α}-Z-D-lysine			
	N^{α}-Z-D-lysine (mM)	Time (h)	N^{α}-Z-D-AASA (mM)	Conversion yield (%)
Amine oxidase [c]	50	4	50	100
Rhodococcus sp. [d]	100	1	97	97

a, Isobe *et al.* (2004a); b, Isobe *et al.* (2004b); c, Isobe *et al.* (2004c); d, Isobe *et al.* (2001)

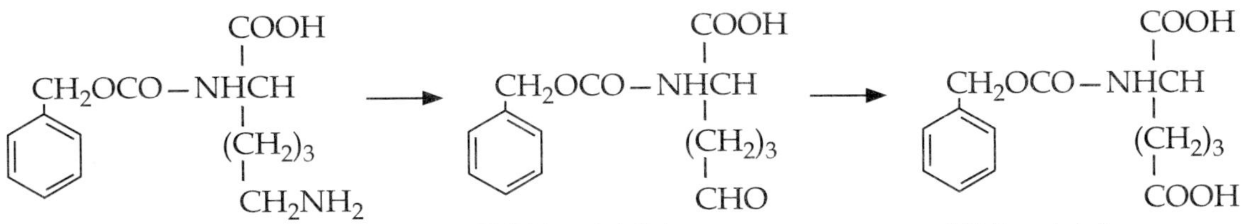

N^α-Z-L-lysine N^α-Z-D-lysine → N^α-Z-L-AASA N^α-Z-D-AASA → N^α-Z-L-AAA N^α-Z-D-AAA

Scheme 8: Production of N^α-Z-L-AAA from N^α-Z-L-lysine and N^α-Z-D-AAA from N^α-Z-D-lysine by *A. niger* AKU 3302 or *Rodococcus sp.* AIU Z-35-1.

Other Organic Acids

Propionic and butyric acids, which are widely used in food, pharmaceutical, and allied industries, are now manufactured by chemical methods. Recently, biochemical methods for the production of these two organic acids have been reported using microbial fermentation methods with *Propionibacterium acidipropionici* and *Clostridium tyrobutylicum*, respectively. However, the growth of these microorganisms was slow and their productivity level of organic acids was low. Although these disadvantages have been somewhat mitigated by the immobilized cell fermentation method, organic acid productivity has remained poor (Huang *et al.*, 2002).

In general, enzymes catalyze the conversion of synthetic compounds within the substrate specificity. Therefore, enzymatic methods using isolated enzyme or whole cells are applicable to the production of a wide variety of synthetic organic acids. Recently, the oxidation of racemic 2-phenyl-1-propanol, 2-chloropropanol, and 1,3-butandiol into their corresponding carboxylic acids has been achieved using whole cells of acetic acid bacteria (Romano *et al.*, 2002; Jongegan *et al.*, 2000).

PRODUCTION OF ALDEHYDES

Aldehydes are widely used as building blocks for the synthesis of pharmaceuticals, agrochemicals or perfumes, and various aldehydes are being produced on a small or medium scale by chemical and enzymatic methods. The chemical methods so far under development are the vapour-phase oxidation of alkyl aromatics, halogenation of alkyl aromatics followed by hydrolysis, Rosenmund reduction of acid halides, and oxidation or dehydrogenation of aromatic alcohols. However, some chemical methods had drawbacks such as low yields or the production of undesirable by-products. To counteract these disadvantages, methods for direct production of aldehydes have been investigated using a heterogeneous chemical catalyst (Yokoyama and Yomogata, 2001) or biochemical methods. Among the later, microbial fermentation methods were not always useful for the efficient production of aldehydes, whereas many such methods could be employed for the production of organic acids, since microorganisms produce enzymes for degrading aldehydes which often exhibit toxicity to microorganisms. On the other hand, enzymatic methods proved to be a useful tool in the production of aldehydes, since enzymes exhibit chemo-, regio- and stereo-specificity. Here, we compared certain enzymatic methods for their effectiveness in the production of aldehydes.

Short-chain Aliphatic Aldehydes

Formaldehyde, which is widely used as feed-stock for the production of resin and fine chemicals, is manufactured from methanol by chemical methods. Enzymatic methods have also been developed by Tani and his group using alcohol oxidase or whole cells of the methanol yeast, *Candida boidinii* (Sakai and Yani, 1988). Using this new enzymatic method, approximately one molar of formaldehyde was produced after a 10-hour reaction with cells.

Acetoaldehyde and acrolein are chemically produced mainly through the direct oxidation of propylene and ethylene, respectively. However, yields of these products were low due to the formation of by-products. Enzymatic production methods for acetoaldehyde have been developed using alcohol oxidases from *C. boidinii* or *Pichia pastrie,* since such oxidases rapidly oxidize short-chain aliphatic alcohols (Nelles *et al.*, 1990; Sakai and Tani, 1987). The alcohol oxidase from *C. boidinii* has also been used in the enzymatic production of acrolein (Sakai and Tani, 1987). The immobilized cells from *Saccharomyces cerevisiae* were applied to the biotransformation of short-chain alcohols into their respective aldehydes (Norouzian *et al.*, 2003).

$$R\text{-}CH_2OH + O_2H \longrightarrow R\text{-}CHO + H_2O_2$$

Scheme 9: Production of aldehydes from alcohols by alcohol oxidase.

Aromatic and Middle-chain Aliphatic Aldehydes

Phenylacetaldehyde is a valuable component of flavours, and its biochemical production would be welcome. An efficient method for its production has been developed by incubating the immobilized cells of *Acetobacter aceti* with 2-phenyl-1-ethanol under a two-liquid phase system comprising water and isooctane (Molinari *et al.*, 1999). Such a system was effective in the separation of this aldehyde before its further oxidation, and it was further oxidized to phenylacetic acid when the same reaction was carried out in water.

Enzymatic methods for the production of the corresponding aldehydes from 1-pentanol, 3-methyl-1-butanol, 1-hexanol, geraniol, 2-phenylthioethanol, cinnamyl alcohol and other alcohols have also been reported using the cells of *Acetobacter* sp. and *Gluconobacter oxydans* under the above two-liquid phase system (Villa *et al.*, 2002; Gandolfi *et al.*, 2001).

Enzymatic methods for the production of n-hexanal from n-hexanol and of benzaldehyde from benzylalcohol have also been reported using alcohol oxidase from *Pichia pastries,* which has exhibited high activity toward these alcohols (Shiraishi *et al.*, 1989; Singh *et al.*, 2003).

Glycolaldehyde

Although it has been reported that glycolaldehyde was chemically formed by the oxidation of diethylene glycol or by hydroformylation of formaldehyde, the fact remains that undesirable by-products were also formed (Turan-Ertas and Gurol, 2002; Marchionna and Garlaschelli, 1989). An enzymatic method for the production of glycolaldehyde was first developed by Isobe and Nishise using regiospecific oxidation of ethylene glycol by alcohol oxidases from *Candida* sp. and *Pichia pastoris,* and glycerol oxidase from *A. japonicus,* although it has been reported that alcohol oxidases and glycerol oxidase were specific to the primary alcohols and glycerol, respectively (Tani *et al.*, 1985; Couderc *et al.*, 1980; Vwajima et. al., 1980). Thus, they obtained new results showing that those oxidases also catalyzed oxidation of ethylene glycol into glyoxal via glycolaldehyde, and that the formation rate of glyoxal declined whereas that of glycolaldehyde rose by increasing the ethylene glycol concentration through kinetic analyses (Agematsu and Fujii, 2002; Isobe and Nishise, 1994). Using the new enzymatic method, 920 mM glycolaldehyde was produced from 1.0 M ethylene glycol after 48-hr incubation with alcohol oxidase. The concentration of glyoxal was approximately 8% using that method. In the case of glycerol oxidase, 1.0 M ethylene glycol was completely oxidized after 60 hrs of incubation, and 970 mM glycolaldehyde was formed with the ratio of glyoxal to glycolaldehyde concentration being less than 3% (Table 1) (Isobe and Nishise, 1995). The conversion yield and selectivity for glycolaldehyde by both enzymatic methods were thus higher than those of chemical methods. In addition, using the alcohol oxidase immobilized onto Chitopearl BCW 3501, 970 mM glycolaldehyde was formed from 1.0 M ethylene glycol with a ratio of glyoxal to glycolaldehyde concentration less than 1%. Since this immobilized enzyme made four repetitions possible, these enzyme methods appear to make possible continuous production of glycolaldehyde (Vkeda *et al.*, 1998).

$$HOH_2CCH_2OH + O_2 \longrightarrow OHCCH_2OH + H_2O_2$$

Scheme 10: Production of glycolaldehyde from ethylene glycol by alcohol oxidase and glycerol oxidase.

Glyoxal

Glyoxal is widely used in the production of pharmaceuticals and resins, as well as in the fields of cement, paper, leather, rubber, adhesives, photography and food deodorants. Although it is now chemically produced by the oxidation of acetaldehyde with nitric acid or by the dehydrogenation of ethylene glycol, the yield is low due to the formation of by-products such as glyoxylic and oxalic acids.

A new enzymatic method for the production of glyoxal has been developed using alcohol oxidases and glycerol oxidase, which oxidizes ethylene glycol to glyoxal via glycolaldehyde as mentioned above. When 100 mM ethylene glycol was incubated with alcohol oxidase and catalase, 98.5 mM glyoxal was produced after 10 hrs of incubation (Isobe and Nishise, 1994). In the case of glycerol oxidase, 100 mM ethylene glycol was also readily converted into glyoxal, and 96.8 mM glyoxal was produced after incubation at 20°C for 5 hrs (Table 1) (Isobe and Nishise, 1995). Although the concentration of glyoxal produced was less than that of glycolaldehyde, these results indicated the possibility of production of glyoxal from ethylene glycol with oxidases.

$$HOH_2CCH_2OH \longrightarrow OHCCH_2OH \longrightarrow OHCCHO$$

Scheme 11: Production of glyoxal from ethylene glycol by alcohol oxidase and glycerol oxidase

L-α-Aminoadipate-δ-semialdehyde

L-α-Aminoadipate-δ-semialdehyde (L-α-AASA) is a key intermediary in the biosynthesis of β-lactam antibiotics, and has been widely recognized as a raw material for the chemical synthesis of pharmaceuticals. Although different chemical methods have been developed, they have proven to be complicated, and their yield of L-α-AASA was low (Asper and Meister, 1962; Rumbero *et al.*, 1995). To overcome these methodological shortcomings, some alternative biological methods have been developed. For example, an enzymatic method for the production of L-α-AASA has been developed using L-lysine 6-dehydrogenase (EC 1.4.1.-) from *Candida albicans* (Hammer and Bode, 1992) or using L-α-AASA synthetase with lysine-ketoglutarate reductase and sacchropine dehydrogenase activities (Markovitz *et al.*, 1984). However, since these enzymatic methods have required a cofactor regeneration system, the development of alternative new biochemical methods is expected. No microbial fermentation method has yet been reported.

N^{α}-Benzyloxycarbonyl-L-aminoadipate-δ-semialdehyde

Derivatives of L-α-AASA such as N^{α}-benzyloxycarbonyl-L-aminoadipate-δ-semialdehyde (N^{α}-Z-L-AASA) are widely recognized as raw materials suitable for the chemical synthesis of new antibiotics or physiological peptides. Although chemical methods for the production of N^{α}-Z-L-AASA have been developed, they were complicated and their yield of N^{α}-Z-L-AASA was low (Asper and Meister, 1962). To overcome these methodological disadvantages, new biological methods have recently been developed by Isobe and his group. Their results have revealed that amine oxidase from *Aspergillus niger* AKU 3302, which has exhibited oxidase activity toward a wide variety of aliphatic and aromatic amines (Yamada *et al.*, 1965), has also oxidized the ε-amino groups of N^{α}-Z-L-lysine and N^{α}-Z-D-lysine, both of which were converted to N^{α}-Z-L-AASA and N^{α}-Z-D-AASA, respectively, according to scheme 12. Using this new enzymatic method, equimolar amounts of N^{α}-Z-L-AASA and N^{α}-Z-D-AASA were produced from 50 mM of N^{α}-Z-L-lysine and N^{α}-Z-D-lysine, respectively, by incubation at 25°C and pH 7.5 for 4 days (Table 2) (Isobe *et al.*, 2004). This enzymatic method is simple, and its conversion yields of N^{α}-Z-L-AASA and N^{α}-Z-D-AASA are higher than those of the chemical method, though it has the disadvantage of requiring the purified amine oxidase, since *A. niger* AKU 3302 also produced aldehyde oxidase that catalyzed a further conversion of N^{α}-Z-L-AASA into N^{α}-Z-L-AAA as mentioned above.

Isobe and his group further developed a new enzymatic method using resting cells of a new bacterial strain, *Rhodococcus* sp. AIU Z-35-1, which was isolated as an effective producer of N^{α}-Z-L-AAA and N^{α}-Z-D-AAA. Thus, this strain was also applicable to the efficient production of N^{α}-Z-L-AASA and N^{α}-Z-D-AASA by changing the pH and temperature of the reaction and the culturing time of the cells. Using cells harvested after one day of cultivation, more than 95 mM N^{α}-Z-L-AASA was produced from 100 mM N^{α}-Z-L-lysines by incubating at pH 5.0 and 30°C for one day or at pH 7.0 and 10°C for two days. A similar amount of N^{α}-Z-D-AAA was also produced from N^{α}-Z-D-lysine using the same strain under the same conditions (Table 2) (Isobe *et al.*, 2001). These enzymatic methods have proven highly effective for the production of N^{α}-Z-L-AASA and N^{α}-Z-D-AASA.

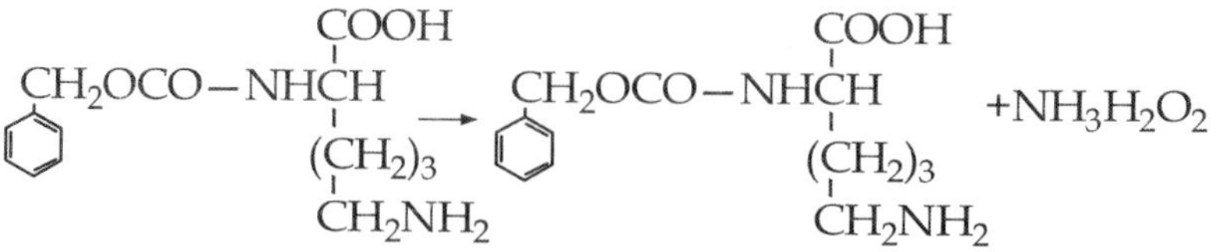

N^{α}-Z-L-Lysine or N^{α}-Z-D-Lysine N^{α}-Z-L-AASA or N^{α}-Z-D-AASA

Scheme 12: Oxidation of N^{α}-Z-L-lysine and N^{α}-Z-D-lysine by amine oxidase from *A. niger* AKU 3302 and by cells of *Rhodococcus* sp. AIU Z-35-1.

Other Aldehydes

As mentioned above, amine oxidase from *Aspergillus niger* AKU 3302 has exhibited a wide substrate specificity in aliphatic and aromatic amines. Therefore, it has also been applied to the production of vanillin from vanillylamine (Yoshida *et al.*, 1997).

The enzymatic oxidation of alcohols using alcohol dehydrogenase has played only a minor role until recently due to the lack of efficient recycling methods for oxidized nicotinamide cofactors. Recently, NADH- and NADPH-oxidases have been isolated from bacteria (Geueke *et al.*, 2003; Jiang and Bommarius *et al.*, 2004) and impressive progress in the enzymatic oxidation of alcohols has been achieved in enzyme-coupled alcohol oxidation using NADH- and NADPH-oxidases for cofactor regeneration. Therefore, investigations into enantioselective oxidations of primary and secondary alcohols have recently increased (Stigter *et al.*, 1997; Kroutil *et al.*, 2004).

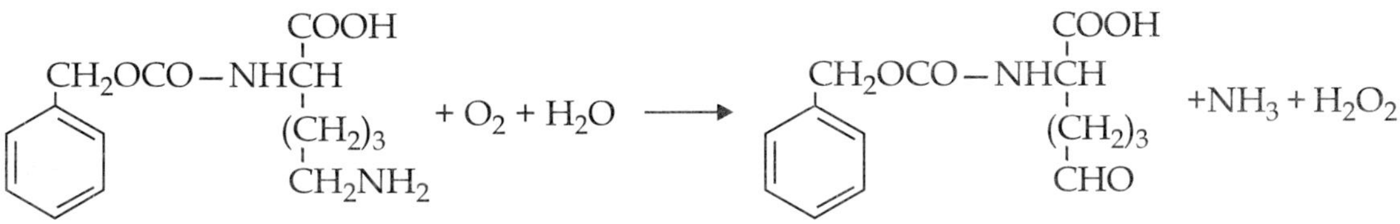

Scheme 13: Schematic presentation of regeneration of NAD using NADH oxidaase

CONCLUSION

At present, organic acids and aldehydes used in the pharmaceutical and chemical industries are manufactured by chemical methods, while those in food industries are mainly produced by biochemical methods. Recently, new biochemical methods have been extensively investigated, since they have a high possibility of selective production of organic acids and aldehydes, and are environmentally friendly, compared with chemical methods. This review mainly evaluated two biochemical methods, i.e., the microbial fermentation and enzymatic methods for the production of organic acids and aldehydes.

The microbial fermentation methods produced final products from inexpensive raw materials such as glucose, glycerol, or starch by culturing the microorganisms with a specific metabolic pathway under conditions suitable for the production of final products. For example, acetic, citric and other organic acids in the TCA cycle have been manufactured by microbial fermentation under aerobic conditions using acetic acid bacteria of the genera *Acetobacter* and *Gluconobacter* (Macauley *et al.*, 2001; Sun and Furusaki, 1990), *Aspergillus niger* (Grewal and Kalra, 1995; Marciano *et al.*, 1989), and *Rhizopus arrhizus* (Petruccioli *et al.*, 1996), respectively. L-lactic, propionic, and butyric acids have been produced under anaerobic conditions using lactic acid bacteria of such genera as *Lactobacillus, Lactococcus,* and *Enterococcus* (Yun *et al.*, 2003), as well as *Propionibacterium* sp. and *Clostridium* sp. (Huang *et al.*, 2002). Thus, in microbial fermentation methods, cell growth through a specific metabolic pathway together with a lack of the enzymes catalyzing the degradation of final products has proven essential for the efficient production of organic acids. Microbial fermentation methods were shown to be effective for the production of a limited number rather than a wide variety of organic acids. Moreover, they are probably not effective for the production of aldehydes, since there are few reports on the success of microbial fermentation in aldehyde production. Microorganisms probably produce enzymes catalyzing aldehyde degradation, since aldehydes often exhibit toxicity against microorganisms.

On the other hand, enzymatic methods have utilized the specific reaction of enzymes without requiring the cell growth, thus rendering them applicable to the production of a wide variety of aldehydes, organic acids, and their derivatives. For example, L-malic acid was produced from fumaric acid by fumarase (Chibata *et al.*, 1983), and alcohol oxidase from *Candida boidinii* was used in the production of formaldehyde, acetaldehyde, acrolein, glycolaldehyde and glyoxal (Sakai and Tani, 1987, 1988; Isobe and Nishise, 1994, 1995). Moreover, alcohol oxidase from *Pichia pastries* was used in the production of n-hexanal from n-hexanol and that of benzaldehyde from benzylalcohol (Murray and Duff, 1990; Karra-Chaabouni *et al.*, 2003). Amine oxidase from *A. niger* was also used to produce not only short-chain aldehydes but also vanillin, N^{α}-Z-L-AASA and N^{α}-Z-D-AASA (Yamada *et al.*, 1965; Isobe *et al.*, 2004; Yoshida *et al.*, 1997), and resting cells from *Pseudomonas alcaligenes* for the production of fumaric acid from maleic acid (Ichikawa *et al.*, 2003). Immobilized cells of *Saccharomyces cerevisiae* were also applied to the biotransformation of short-chain alcohols into their respective aldehydes, while those of *Acetobacter aceti* were effective in phenylacetaldehyde production. The cells of *Gluconobacter oxydans* were used to produce corresponding aldehydes from 1-pentanol, 3-methyl-1-butanol, 1-hexanol, geraniol, 2-phenylthioethanol, and cinnamyl alcohol. Resting cells of *Rhodococcus* sp. AIU Z-35-1 were applied to the production of not only N^{α}-Z-L-AAA and N^{α}-Z-D-AAA but also N^{α}-Z-L-AASA and N^{α}-Z-D-AASA. As the above examples show, enzymatic methods have proven both simple and readily applicable to the production of a wide variety of aldehydes and organic acids within the substrate specificity of the enzymes. In addition, reaction conditions were easily controlled, since these methods were carried out by incubating the substrates with isolated enzymes or whole cells containing one or more enzymes. Moreover, enzymatic methods also enjoy the advantage that their substrate concentrations are higher than those of fermentation methods, although their substrates are more expensive than the starting materials used in microbial fermentation methods. Chiral compounds have also been produced by enzymatic methods, since enzymes exhibit stereo- or regio-specificity.

Until now, enzymatic methods for the production of aldehydes and organic acids have been developed using hydrolases, transferases, dehydrogenases or oxidases. Enzymatic methods using transferases and dehydrogenases have involved drawbacks such as requiring either a cofactor regeneration system or a second substrate, whereas those using oxidases did not, since oxidases contain prosthetic groups. In addition, the hydrogen peroxide from the oxidase-reaction was recycled into oxygen by coupling it with catalase. Thus, methods using oxidases or oxidase-containing cells proved to be simpler and more effective than other enzymatic methods as well as both microbial fermentation and chemical methods in the production of a wide variety of aldehydes, organic acids, and their derivatives. Although these enzymatic methods using oxidases or oxidase-containing cells enjoy these important advantages, they are not currently utilized on an industrial scale. Thus, their conversion speed should be further enhanced by immobilizing the oxidases or microorganisms, or by constructing recombinant strains containing oxidase genes suitable for industrial use. Furthermore, the screening or construction of oxidases exhibiting a high resistance against strong concentrations of substrates and products will also be important for eventually allowing their use in the industrial production of aldehydes and organic acids.

REFERENCES

Agematsu, H. & Fujii, T. (2002) *Bio. Industry.* **19**: 40-47.

Arikawa, Y., Kuroyanagi, T., Shimosaka, M., Muratsubaki, H., Enomoto, K., Kodaira, R. & Okazaki, M. (1999) *J. Biosci. Bioeng.* **87**: 28-36.

Aspen, A. J. & Meister, A. (1962) *Biochem.* **1**: 600-605.

Attwood, M. M., Dijken, J. P. & Pronk. J. T. (1991) *J. Ferment. Technol.* **72**: 101-105.

Bai, D.M., Zhao, X.M., Li, X.G. & Xu, S.M. (2004) *Biochem. Eng. J.* **18**: 41-48.

Baldwin, J. E., Killin, S. J., Adlington, R. M. & Spiegel, U. (1988) *Tetrahedron.* **44**: 2633-2636.

Barton, D. H. R., Herve, Y., Potier, P. & Thierry, J. (1987) *Tetrahedron.* **43**: 4297-4308.

Beltrame, P., Comotti, M., Pina, C. D. & Rossi, M. (2004) *J. Catal.* **228**: 282-287.

Benthin, S. & Villadsen, J. (1995) *Appl. Microbiol. Biotechnol.* **42**: 426-429.

Biella, S., Prati, L. & Rossi, M. (2002) *J. Catal.* **206**: 242-247.

Carta, F. S., Soccol, C. R., Ramos, L. P. & Fontana, J. D. (1999) *Biores. Technol.* **68**: 23-28.

Chibata, I., Tosa, T. & Takata, I. (1983) *Trends Biotechnol.* **1**: 9-11.

Couderc, R. & Baratti, J. (1980) *Agric. Biol. Chem.* **44**: 2279-2289.

Creca, M. C., Moraes, C. & Sgadaes, A. M. (2001) *Appl. Catal. A: General* **216**: 267-276.

Deffernez, A., Hermans, S. & Devillers, M. (2005) *Appl. Catal. A: General* **282**: 303-313.

Elfari, M., Ha, S.W., Bremus, C., Merfort, M., Khodaverdi, V., Herrmann, U., Sahm, H. & Goerisch, H. (2005) *Appl. Microbiol. Biotechnol.* **66**: 668-674.

Ezquerra, J. & Pedregal, C. (1995) *Tetrahedron Lett.* **36**, 3247-3250.

Gandolfi, R., Ferrara, N. & Molinari, F. (2001) *Tetrahedron Lett.* **42**: 513-514.

Gavagan, J. E., Fager, S. K., Seip. J. E., Payne, M. S., Anton, D. L. & DiCosimo, R. (1995) *J. Org. Chem.* **60**: 3957-3963.

Geueke, B., Riebel, B. & Hummel, W. (2003) *Enzyme Microb. Technol.* **32**: 205-211.

Grewal, H. S. & Kalra, K. L. (1995) *Biotechnol. Advances* **13**: 209-234.

Hammer, T. & Bode, R. (1992) *Zentralbl. Microbiol.* **147**: 65-70.

Haq, I.U., Ali, S. & Iqbal, J. (2003) *Process Biochem.* **38**: 921-924.

Herrmann, U., Merfort, M., Jeude, M., Bringer-Meyer, S. and Sahm, H. (2004) *Appl. Microbiol. Biotechnol.* **64**: 86-90.

Huang, Y. L., Wu, Z., Zhang, L., Cheung, C. M. & Yang, S.T. (2002) *Bioresource Technol.* 82: 51-59.

Ichikawa, S., Iino, T., Sato, S., Nakahara, T. & Mukataka, S. (2003) *Biochem. Eng. J.* **13**: 7-13.

Isobe, K. & Nishise, H. (1994) *Biosci. Biotech. Biochem.* **58**: 170-173.

Isobe, K. & Nishise, H. (1995) *J. Mol. Cat. B: Enzym.* **1**: 37-43.

Isobe, K. & Nishise, H. (1999) *J. Biotechnol.* **75**: 265-271.

Isobe, K. (1995) *Biosci. Biotech. Biochem.* **59**: 576-581.

Isobe, K., Nagasawa, S., Tokuta, Y., Matsuura, A., Sakaguchi, T., & Wakao, N. (2001) *J. Biosci. Bioeng.* **100**: 288-291.

Isobe, K., Tokuta, K., Narita, Y., Matsuura, A., Sakaguchi, T. & Wakao, N. (2004a) *J. Mol. Catal. B, Enzym.* **30**: 13-18.

Isobe, K., Tokuta, K., Narita, Y., Matsuura, A., Sakaguchi, T. & Wakao, N. (2004b) *J. Mol. Catal. B, Enzym.* **32**: 27-32.

Isobe, K., Tokuta, K., Narita, Y., Matsuura, A., Sakaguchi, T. & Wakao, N. (2004c) *J. Mol. Catal. B, Enzym.* **30**: 119-123.

Jiang, R. & Bommarius, A. S. (2004) *Tetrahedron: Asymmetry* **15**: 2939-2944.

Jongejan, A., Machado, S. S. & Jongejan, J. A. (2000) *J. Mol. Catal. B: Enzym.* **8**: 121-163.
Kadam, S. R., Patil, S. S., Bastawde, K. B., Khire, J. M. & Gokhale, D. V. (2005) *Process Biochem.* In Press.
Karra-Chaabouni, M., Pulvin, S., Meziani, A., Thomas, D., Touraud, D. & Kunz, W. (2003) *Biotech. Bioeng.* **81**: 27-32.
Kawagoe, M., Hyakumura, K., Suye, S., Miki, K. & Naoe, K. (1997) *J. Ferment. Bioeng.* **84**: 333-336.
Kim. D. Y., Yim, S. C., Lee, P. C., Lee, W. G., Lee, S. Y. & Chang, H. N. (2004) *Enzyme Microb. Technol.* **35**: 648-653.
Kroutil, W., Mang, H., Edegger, K. & Faber K. (2004) *Current Opinion Chem. Biol.* **8**:120-126.
Lee, P. C., Lee, W. G., Kwon, S., Lee, S. Y. & Chang, H. N. (1999) *Enzyme Microb. Technol.* **24**: 549-554.
Macauley, S., McNeil, B. & Harvey, L. M. (2001) *Crit. Rev. Biotechnol.* **21**: 1-25.
Mandal, S. K. & Banerjee, P. C. (2005) *Process Biochem.* **40**: 1605-1610.
Marchionna, M. & Garlaschelli, L. (1989) *J. Mol. Catal.* **57**: 221-235.
Marciano, P., Magro, P. and Favaron, F. (1989) *FEMS Microbiol. Lett.* **61**: 57-59.
Markovitz, P. J., Chuang, D. T. & Cox, R. P. (1984) *J. Biol. Chem.* **259**: 11643-11646.
Matos, J. R. & Wong, C.H. (1986) *J. Org. Chem.* **51**: 2388-2389.
Matsushita, K., Toyama, H. & Adachi, O. (1994) *Advances in Microbial Physiology*: (Eds. Rose, A. H. & Tempest, D. H.), Academic Press, London, p. 247-301.
Molinari, F., Gandolfi, R., Aragozzini, F., Leon, R. & Prazeres, D. M. F. (1999) *Enzyme Microb. Technol.* **25**: 729-735.
Murray, W. D. & Duff, S. J. B. (1990) *Appl. Microbiol. Biotechnol.* **33**: 202-205.
Nakao, K., Kiefner, A., Furumoto, K. & Harada, T. (1997) *Chem. Eng. Sci.* **52**: 4127-4133.
Nelles, L. P., Arnold, J. A. & Willman, D. S. (1990) *Biotechnol. Bioeng.* **36**: 834-838.
Norouzian, D., Akbarzadeh, A., Inanlou, D. N., Farahmand, B., Sleh, M., S.-U.-Eslam, F. & Vaez, J. (2003) *Enzyme Microb. Technol.* **33**: 150-153.
Ohara, H. & Yahata, M. (1996) *J. Ferment. Technol.* **81**:272-274.
Petruccioli, M., Angiani, E. & Federici, F. (1996) *Process Biochem.* **31**: 463-469.
Pronk. J. T., Levering, P. R., Olijve, W. & Dijken, J. P. (1989) *Enzyme Microb. Technol.* **11**: 160-164.
Romano, A., Gandolfi, R., Nitti, P., Rollini, M. & Molinari, F. (2002) *J. Mol. Catal. B: Enzym.* **17**: 235-240.
Ross, R. P., Morgan, S. & Hill, C. (2002) *Int. J. Food Microbiol.* **79**: 3-16.
Rumbero, A., Martin, J. F., Lumbreras, M. A., Liras, P. & Esmahan, C. (1995) *Bioorg. Med. Chem.* **3**: 1237-1240.
Sakai, Y. & Tani, Y. (1987) *Agric. Biol. Chem.* **51**: 2617-2620.
Sakai, Y. & Tani, Y. (1988) *Appl. Environ. Microbiol.* **54**: 485-489.
Sato, M., Nakahara, T. & Yamada, K. (1972) *Agric. Biol. Chem.* **36**: 2025-2026.
Sato, S., Nakahara, T. & Minoda, Y. (1977) *Agric Biol. Chem.* **41**: 967-973.
Scott, A. I. & Wilkinson, T. J. (1980) *Synthetic Communications.* **10**: 127-131.
Seip. J. E., Fager, S. K., Gavagan, J. E., Anton, D. L. & DiCosimo, R. (1994) *Bioorg. Med. Chem.* **2**: 371-378.
Seip. J. E., Fager, S. K., Gavagan, J. E., Gosser, L. W., Anton, D. L. & DiCosimo, R. (1993) *J. Org. Chem.* **58**: 2253-2259.
Shimada, M., Akamatsu, Y., Tokimatsu, T., Mii, K. & Hattori, T. (1997) *J. Biotechnol.* **53**: 103-113.
Shinagawa, E., Matsushita, K., Toyama, H. & Adachi, O. (1999) *J. Mol. Catal. B: Enzym.* **6**: 341-350.
Shiraishi, F., Kawakami, K., Tamura, A., Tsurada, S. and Kusunoki, K. (1989) *Appl. Microbiol. Biotechnol.* **31**: 445-447.
Singh, O. V., Jain, R. K. & Singh, R. P. (2003) *J. Chem. Tech. Biotech.* **78**: 208-212.
Smejkal, Q., Linke, D. & Baerns, M. (2005) *Chem. Eng. Process* **44**: 421-428.
Stigter, E. C., Lugt, J. P. & Somers, W. A. C. (1997) *J. Mol. Catal. B: Enzym.* **2**: 291-297.
Sun, Y. & Furusaki, S. (1990) *J. Ferment. Technol.* **69**: 102-110.
Takao, S. & Hotta, K. (1977) *Agric Biol. Chem.* **41**: 945-950.
Takao, S., Tanida, M. and Kuwabara, H. (1977) *J. Ferment. Technol.* **55**: 196-199.
Tanaka, T., Hoshina, M., Tanabe, S., Sakai, K., Ohtsubo, S. & Taniguchi, M. (2005) *Bioresource. Technol.* In Press.

Tani, Y., Sakai, Y. & Yamada, H. (1985) *Agric. Biol. Chem.* **49**: 2699-2706.
Turan-Ertas, T. & Gurol, M. D. (2002) *Chemosphere* **47**: 293-301.
Ukeda, H., Ishii, T., Sawamura, M. & Isobe, K. (1998) *Biosci. Biotech. Biochem.* **62**: 1589-1591.
Usami, S. & Fukutomi, N. (1977) *Hakkoukougaku* **55**: 44-50.
Uwajima, T., Akita, H., Ito, K., Mihara, A., Aisaka, K. & Terada, O. (1980) *Agric. Biol. Chem.* **44**: 399-406.
Vedrine, J. C., Novakova, E. K. & Derouane, E. G. (2003) *Catal. Today* **81**: 247-262.
Villa, R., Romano, A., Gandolfi, R., Sinisterra, J. V. & Molinari, F. (2002) *Tetrahedron Lett.* **43**: 6059-6061.
Yadav, G. D. & Gupta, V. R. (2000) *Process Biochem.* **36**: 73-78.
Yamada, H., Adachi, O. & Ogata, K. (1965) *Agric. Biol. Chem.* **29**: 117-123.
Yamada, K., Furukawa, T. & Nakahara, T. (1970) *Agric Biol. Chem.* **34**: 670-675.
Yokoyama, T. & Yamagata, N. (2001) *Appl. Catal. A: General* **221**: 227-239.
Yoneda, N., Kusano, S., Yasui, M., Pujado, P. & Wilcher, S. (2001) *Appl. Catal. A: General* **221**: 253-265.
Yoshida, A., Takenaka, Y., Tamaki, H., Frebort, I., Adachi, O. & Kumagai, H, (1997) J. *Ferment. Bioeng.* **84**: 603-605.
Yun, J. S., Wee, Y.J. & Ryu, H.W. (2003) *Enzym. Micro. Technol.* **33**: 416-423.

13

GM in Sustainable Crop Production

Meenu Saraf
Department of Microbiology
Gujarat University, Ahmedabad (Gujarat)

The unprecedented rapid adaptation of transgenic crops during the initial five years (1996-2000) when genetically modified crops were first adopted reflects the significant multiple benefits realized by large and small farmers in industrial and developing countries that have grown transgenic crops commercially. Genes and genetic sequences from varied sources like bacteria, viruses and fungi can be transferred to plants, widening the scope of genetic exchange far beyond what sexual reproduction generally allows. A large number of insecticidal molecules are effective against insects and pests being deployed in crop plants. The efficiency of various categories of such proteins and molecules derived from microorganisms and their prospects shall be discussed. Significant losses due to bacterial and fungal attack are frequently reported and transgenic crops can basically be classified into two categories based on: (i) production of GM crops with antifungal molecules, (ii) generation of hypersensitive response through R genes following the availability of molecular information on viruses, initiatives have been taken in the development of transgenic virus resistance in important crops. While insects, viral and herbicide resistant GM crops are being field tested. The production of abiotic stress tolerant transgenics is a recent development. This is largely because of complex genetic mechanisms that govern it. Moreover the genes that have proven somewhat effective is providing stress tolerance using transgenic approach belong to both structural and regulatory gene categories. Plant transformation vectors and methodologies use a wide range of tool compounds used *Agrobacterium* T plasmid vectors and microprojectile bombardment. There is a need to increase productivity and enhance the nutritional value of pulse crops. The development of transgenic in major grain legumes like, Groundnut, Pegionpea, Chickpea and Mungbean remains dogged by poor regeneration and low frequency transformation protocols and this situation needs to be rectified. In this article the role of GM in crop production shall be reviewed focusing on the input of transgenes from the microbial world. Conflicting scientific public and farmer opinion on the benefits versus evils of GM crops continues, but the science behind the GM crops are not under dispute and probably may be the only solution to feed crores of us in the near future.

INTRODUCTION

Crop biotechnology broadly includes areas of development of transgenic crops, structural and functional genomic and marker assisted breeding could provide us with vital breakthrough to achieve improvements in both quality and quantity in a sustainable manner. With the advent utilizing host resistance genes and other genes responsible for adaptive host processes elicited in response to pathogen to obtain transgenic resistant to the virus.

Plant transformation vectors and methodologies have been improved to increase the efficiency of plant transformation and to achieve stable expression of modified genes in plants. Plant transformation is performed using a wide range of tools such as *Agrobacterium* Ti plasmid vectors, microporjectile bombardment, microinjection, polyethylene glycol (PEG) treatment of protoplasts and electroporation of protoplasts. Though all methods have advantages that are unique to each of them, transformation using *Agrobacterium* and micro projectile bombardment are currently the most extensively used methods (Dai, 2001).

GENES FROM BACTERIAL SOURCES

Insect-pest menace is the major factor that destabilizes crop productivity in agricultural ecosystems. A variety of insect-pest ranging from lepidopteron to orthopterans damaged crops and stored seed. Insect-pest management in an eco-friendly manner is no longer a dream (Grover and Pental, 2003). A large number of insecticidal molecules which are effective against insects only have been reported and advancements in molecular biology and genetic engineering can facilitate harnessing and deployment of these molecules in plants in a safe and sustainable fashion (Pathanayak and Kumar, 2000). *Bacillus thuriugenesis* is a Gram-positive, aerobic, sporulating bacterium which synthesizes *Cry*stalline proteins during sporulation. Those *Cry*stalline proteins are highly insecticidal even at low concentration (Staskawicz *et al.*, 1995) *Bt* strains and their insecticidal *Cry*stal proteins (ICPs) are being used effectively and rationally by incorporating them into crop plants. Bt strains and ICP's were first found to affect a range of lepidopteron insects, subsequently discovery of new strain expanded the host range to coleopterans, dipterans, lice, mice and nematodes (Kumar *et al.*, 1996). *B. thuringiensis* serotypes Kurstaki and Aizawai produced *Cry*stal proteins *Cry*1 and *Cry*2, which was responsible for increase in range of toxicity. Currently the *Cry*stal toxins are classified on the basis of amino-acid sequence homology. The ICPs fall under the forty different classes with some toxins exhibiting specificity to multiple insect orders, which have been studied and catalogued: (http://www.glfc.forestry ca/Bacukkys/Bt_ Home page/net intro99.htm). Extensive screeming programmes are in progress as Bt - ICP has high commercial value.The delivery of *Bt*-ICP through spray formulations, engineered *Bt* and other bacteria have certain limitations.

Table 3: Some important *Bt* - transgenic crops.

Crop	Gene	Target Pest	Reference
Cotton	*Cry*1Ab/*Cry*1 Ac	Boll Worms	Perlak (1990)
Corn	*Cry*1Ab	European Corn Borer	Koziel (1993)
Potato	*Cry* 3a	Colorado Potato beetle	Perlak (1993)
Rice	*Cry*1Ab/*Cry*1 Ac	Stem borers and leaf folders	Nayak (1997)
Tomato	*Cry*1 Ac	Fruit Borers	Mandaokar *et al.* (2000)
Potato	*Cry*1Ab	Tuber moth	Jansens (1995)
Eggplant	*Cry*1Ab/*Cry*1B	Shoot and fruit borer	Kumar *et al.* (1998)
Canola	*Cry*1 Ac	Diamondback moth	Stewart *et al.* (1996a)
Soyabean	*Cry*1 Ac	Soyabean borer	Stewart *et al.* (1996b)
Corn	*Cry*1H/*Cry* 9C	European Corn borer	Jansens *et al.* (1997)

The biopesticidal spray suffers from short half-life, physical removal (wind and rain) and inability to reach burrowing insects. Engineered bacteria very often proliferate at a rate and quantity not sufficient to kill the target insects. The first genetically modified crop using *Cry* genes were developed in 1987. The tobacco plants engineered with truncated genes encoding *Cry*1 Aa and *Cry*1 Ab toxins were found to be resistant to the tobacco horn worm (Jouanin *et al.*, 1998). As of now, more than thirty plant species have been transformed with *Bt Cry* genes (Table 3) (Schuler *et al.*, 1998). The commercialization of *Bt* crops started in 1996 with the introduction of boll worm resistant cotton (Bollgard) in the USA. Subsequently, potato and maize are also commercialized. In India, intensive efforts are underway to introduce *Cry* genes in crop plants such as rice, potato, cotton, sorghum and vegetable (Mandaokar *et al.*, 1998, Rao *et al.*, 1999).

Research efforts in the past five years have led to the discovery of novel insecticidal proteins *etal.*, which are produced during vegetative growth of *Bacillus thuringenesis* and are secreted into medium. These proteins are termed as vegetative insecticidal proteins (Vip) sequences, encoding for a Vip have been cloned, sequenced (Estruch *et al.*, 1996). This 88 kDa vegetative insecticidal protein has a putative bacillar secretory signal at the N- termical which is not processed during its secretion. It does not show any homology with the known *Cry*stalline insecticidal proteins.The range of novel insecticidal *Cry*stal proteins produced by different isolates of *B. thuringenesis* is large (Table 4).

Table 4: Range of insecticidal *Crystal* proteins in individual *Bacillus thuringensis* strains*.

***B. thuringiensis* subspecies and strains**	***Crystal* protein**
aizawai	*Cry*1 Aa, *Cry* 1Ab, *Cry*1 Ad,*Cry*1 Ca, *Cry* 1 Da, *Cry*1 Bb *Cry* 9Ea, *Cry*1 Fa, *Cry*39Aa, *Cry* 40 Aa
entomocidus	*Cry*1 Aa, *Cry*1 Ba, *Cry*1 Ca, *Cry*, 1 Ib
galleriae	*Cry*1Ab, *Cry*1Ac, *Cry*1 Da, *Cry*1 Cb, *Cry*7 Aa, *Cry* 8 Da *Cry* 9 Aa, *Cry* 9 Ba
israelensis	*Cry* 10 Aa, *Cry* 11 Aa
japonensis	*Cry*8Ca, *Cry* 9 Da
jegathesan	*Cry*11Ba, *Cry*19Aa, *Cry*24 Aa, *Cry*25 Aa
kenyae	*Cry*ZAa, *Cry*1 Ea
kumamotoensis	*Cry*7 Ab, *Cry*8 Aa, *Cry* 8 Ba
krstaki	*Cry*1 Aa, *Cry*1 Ab, *Cry*1 Ac, *Cry* 11a, *Cry* ZAa, *Cry*2 Ab
morrisoni	*Cry*1 Bc, *Cry*1 fb, *Cry*3 Aa
tenebrionis	*Cry* 3 Aa
tolworthi	*Cry* 3 Ba, *Cry* 9 Ca

*Source: Slater (2003)

In 1911, Bernard reported chitinases to be a thermo-sensitive and diffusible antifungal factor. Besides inability to attach fungal cell directly, chitinases release oligo–N-acetyl glucosamine that function as elicitors for the activation of defense related response in plant cell (Ren and West, 1992). This has aroused considerable interest for the development of mycolytic enzymes based formulation for the control of fungal plant pathogens. Chitinases are present in wide range of organisms including those that does not contain chitin, such as bacteria, viruses, higher plants and animals where they are known to play important physiological and ecological role (Connell *et al* 1998) . Chitinases producing organism is used as an effective biocontrol agent as chitin is present in the exoskeleton and gut linings of insects. *Aeromonas caviae* controlled infection caused by *Rhizoctonia solani* and *Fusarium oxysporum* in cotton and *Sclerotium rolfsi* in beans by the production of chitinases (Chet and Inbar ,1994). *Pantoea* sp. has been found to produce chitinase and protease when grown in chitin medium (Monreal and Reese, 1969). Also many other bacteria like *Serratia marcescens* (Monreal and Reese, 1969) and *Alcaligenes oylosoxydans* (Vaidya *et al.*, 2001) have been found to produce chitinases under laboratory conditions. In addition, it has been demonstrated that including chitinase protein together with insecticidal proteins of *B. thuringenesis* potentiates the effect of these toxin (Schuler, 1998).An exochitinase gene from *Serratia marcescens* has expressed in transgenic tobacco renders the host plants susceptible to *R. solani* (Howie *et al.*, 1992). The role of PGPR (Plant Growth Promoting Rhizobacteria) in sustainable crop production remains undisputed. Although they do not genetically modify the crop, but the presence of these bacteria in roots and rhizosphere is promotory to the growth of plant and inhibitory to the pathogens (Glick, 1995). In recent years, biological control of plant diseases by production of siderophores, antibiotics, hydrogen cyanide, indole acetic acid etc. (Glick and Pasternak, 1998) has been well documented. Aggressive root colonization of peanut, plant growth enhancement and suppression of *Macrophonia phaseolina* by fluorescent *Pseudomonas* have been studied (Gupta *et al.*, 2002). Promotion of root growth, one of the important factors linked to the synthesis of IAA produced by *Pseudomonas putida,* has been studied (Patten and Glick, 2002). *Achromobacter piechandii* has been identified as a bacterium that promotes growth in saline conditions (Mayak *et al.*, 2004). While a gene encoding an anti-freeze protein was cloned and expressed from *Pseudomonas putida,* which had anti-freezes and ice nucleation activities. (Muryei *et al.*, 2004). The plant hormone ethylene is an essential signalling molecule involved in many process from germination flower development, fruit ripening and responses to many environmental stimuli: modi-

Table 5: Plants with altered ethylene synthesis*.

Gene & Promoter	Plant Transformed	Consequences
ACCdeaminase from *P.Chloraphis* 6G5 with CaMV355	Tomato	No phenotypic change but at ethylene peak (3 days into ripening) Ethylene decreased by 90%
Ethylene forming enzyme from *P.syringae, Pv. phaseolicola Pk2*	Tobacco	Compressed phenotype savoller roots, excess root hairs with excess calls
ACCdeaminase from *E.cloacae* UW4 with three different promoters; CaMV 35s, rol D & PRB - lb		Increased accumalation of neutral compared with non-transformed plants. Higher shoot/root ratio of metal concentration.
Fusion protein pf ethylene forming enzyme from *P.syringae PV phaseolicola Pk2* & GUS with N + ADH promoter	Tomato	Increased ethylene product in root only GUS in roots. Dwarfed morphology.
ACCdeaminase from *E.cloaceae* UW4 with three promoters : CaMV35s, rolD & PRB - lb	Tobacco	Concentration of ethylene decreased by 70 % in transganic plants compared to non-transformed when flooded
ACC deminase from E.cloacae UW4 with three promoters CaMV35s rolD & PRB 10	Tomato	Reduction in disease symptoms but no decrease in progeres of fungal infection.
ACC deaminase from *E.cloacae* UW4 with double 35s promoter	Tobacco	GM plants had 40% higher germination rate double the biomass & were able to take up marginally high arsenate.

*Source: http://www.agbios.com, http://www.whybiotech.com

fying the amount produced under ripening or stress conditions is the goal of a wide array of transgenic strategies (Stearns and Glick, 2003). Table 5 summarizes the types of transgenic plants that have been created by insertion of genes from bacterial sources that alter the level of ethylene produced.

Another bacteria, which aroused interest in recent years, is *Photorhabdus luminescens* that dwell inside the gut of enmatophagus nematodes. It was found that these bacteria synthesize high molecular weight protein complexes toxic to insects ranging from lepidoptera, coleoptera to disctyoptera. This complex has similar action as *Bt* and can serve as good candidate to be expressed along with *Bt* in transgenic plants (Bowen *et al.*, 1998). De Wit (1992) proposed a model of expressing both the resistance gene (*R*) and avirulence gene (*Avr*) in the plant. When this R *Avr* gene cassette is put under strict pathogen–inducible promoter, resistance reaction like HR will be activated upon pathogen infection. Meanwhile Tang *et al.* (1996) and Scrofield *et al.* (1996) showed that resistance to bacterial speck disease in tomato transgenics occurs when the resistant gene to a kinase binds to *avr – pto* transgene isolated from *Pseudomonas syringae* expressing in the tomato plants. Another avirulence gene *hrm* A from *P. syringae* in transgenic tobacco confers high level of resistance against tobacco vein molting virus, tobacco etcli virus, *Phytopthora parasitica* (Shen *et al.*, 2000).

GENES FROM FUNGAL SOURCES

Improvements in genetic transformation technology have allowed the genetic modification of almost all-important food crops like rice, wheat, maize, mustard, pulses and fruits. The estimated global area of GM crops in the year 2001 was 526 million hectares (James, 2001). To identify the important genes, which need to be introduced in the plants to improve their resistance to fungal pathogens, a lot of basic work has been done in the area of host pathogen recognition (Feys and Parker, 2000). During last decade, many resistance genes whose products are involved in recognition of invading pathogens have been identified and cloned (Takken and Joosten, 2000). Many antifungal compounds have been identified (Houlb, 2001). Production of fungus resistant transgenics can basically be studied under two categories: (*i*) production of GM crops with ant fungal molecules like proteins and toxins, and (*ii*) generation of a hypersensitive response through R genes or by manipulating genes of SR pathway. In 1970, Loon and Kammen showed that a set of protein is induced in tobacco plants after TMV infection (Loon and Kammen, 1970). These proteins were described as pathogenesis related (PR) proteins. They were found to be induced by pathogens wounding, fungal cell-wall elicitors ethylene, UV light, heavy metals, etc. hypersensitive response and systemic acquired resistance (SAR). It is important to mention here that introduction of desired gene in the host plant under constitutively high expressing promoter can cause silencing of the transgene as well as its endogenous homologue leading to a high proportion of progency losing its enhanced resistance (Chareonpornmattana *et al.*, 1999).

Fungal growth is inhibited *in vitro* by cell-wall degrading enzymes, mostly chitinases from various fungi. Some of these chitinases show synergy with PR5 proteins or other membrane affecting compounds and other fungal cell-wall hydrolases (De la Cruz *et al.*, 1995, Lorito *et al.*, 1996). Plants which co-express *Trichoderma harzianum* endo-chitnase gene and a tobacco PR5 gene (Lrito, 1996) and plants over expressing *Streptomyces* chitosinase gene (El Quakfaouri

Table 6: Coat protein-mediated transgenic resistance to viruses in crops.

Crops	Viruses
Cereals	
Maize	MDMV, MCMV
Rice	RSV, RTSV
Wheat	WSMV
Fruits	
Apricot	PPV
Cantaloupe	ZYMY, WMV2, CMV
Citrus	CTV
Grape	GCMV, GFLU, ToRSV
Muskmelon	ZYMV
Papaya	PRV
Plum	PPV
Squash	SYMV, WMV2
Vegetables	
Pepper	TSWV
Tomato	ToMV, YMV, CMV
Potato	PVX, PVY, PLRV
Lettuce	LMV, TSWV
Pea	PEMV
Cucumber	CMV
Sugar beet	BNYVV
Legumes	
Peanut	TSWV
Soybean	BPMV
Bean	BPMV

Abbreviations , MCMV- Maize Chlorotic mottle virus, MDMV- Maize dwarf Mosaic Virus, RSV – Rice Stripe virus, TRSV – Rice fungro spherical virus, WSMV – Wheat streak mosaic virus
CTV – Citrus tristeza virus, GCMV – Grape vine chrome mosaic virus, GFLV – Grape vine fanleaf virus, ToRSV – Tomato ring spot virus, YMV – Yellow mosaic virus
LMV – Lettuce mosaic virus, PERMV – Pea enation mosaic virus, BNYVV – Bean necrotic yellow vein virus, BPMV – Bean pod mottle virus
Source: http://www.bspp.org.uk/mppol/1997/011 6 fuchs, Sivamani *et al.* (2002)

et al., 1995) have been produced. Chitosinase enzyme isolated from these transgenic plants was found to be as effective as native *streptomyces* chitosinase in inhibiting fungal growth *in vitro*. A fungal chitinase gene from *Rhizopus oligosporus* confers antifungal activity to transgenic tobacco (Terakawa, 1997). Different strains of *Ustilago maydis* harbours different double stranded RNA viruses encoding anti-fungal proteinaceous killer toxins. *U. maydis* strains are resistant to the toxin produced within themselves but sensitive to the killer toxins of other strains. High-level secretion of KP4 or KP6 killer toxin in transgenic tobacco plants rendered them resistance to fungal pathogens (Park *et al.*, 1996). Induced expression of sacrotoxin IA, a bactericidal peptide from *Sarcophaga peregrina* enhanced the resistance of transgenic tobacco plants to *R. solani* and *Phythium aphanidermatum* (Mitsuhara, 2000). Many resistant genes have been isolated from bacteria which confers high degree of resistance on genetically modified plant as reported earlier. On similar lines, resistance gene of plants against fungal pathogens can be used in combination with fungal *avr* (antivirulence gene) to produce fungus resistant transgenic plants. *R- Avr,* two component system is more advanced sophisticated and broad spectrum in action in the sense that it provides resistance to any pathogen which can activate the promoter of *R-Avr* cassette provided the promoter used in tightly pathogen inducible and non-leaky (Grover and Gowthaman, 2003).

One of the earliest events in incompatible plant pathogen interaction is the oxidative burst during which active oxygen sp. eg. H_2O_2 are produced (Lamb and Dixon, 1997). H_2O_2 triggers production of Phytoalexins, PR proteins and other HR related processes. H_2O_2 also has a direct inhibitory effect on microbial growth (Wu *et al.,* 1995). Glucose oxidase (GO) an enzyme occurring in some bacteria and fungi, bring about the oxiation of B-D glucose yielding gluconic acid and H_2O_2. Expressing a GO gene from *Aspergillus niger* in potato showed increased level of H_2O_2. Such genetic modification has reduced the susceptibility to *E carotonora, P. infestans* and *Verticillium dehliae* (Wu *et al.* ,1995).

GENES FROM VIRAL SOURCES

In a number of crops, introducing a sequence of the viral genome in the target crop by genetic transformation has developed transgenic resistances to an infective virus. Virus resistant transgenics have been developed in many crops by introducing either viral coat protein (CP) or replicase gene encoding sequences. Resistance obtained by using CP is conventionally called CPMR. On the other hand, replicase mediated resistance is due to an inherent plant response known as post transcriptional gene silencing (PTGS). The other pathogen-derived approaches are use of satellite RNA and defective interfering viral genomic components.

The use of viral CP as a transgene for producing virus resistant plants is one of the most spectacular successes achieved in plant biotechnology. Numerous crops have been transformed to express viral CP and have a high degree of resistance in comparison to other untransformed plants (Table 6). The first reported evidence against TMV (Powell - Abel *et al.*, 1986) in genetically modified *tobacco* expressing the TMV - *CP* gene. The resistance was manifested as delayed appearance of symptoms as well as a reduced tittle of virus in the infected transgenic plants. The resistance against TMV using TMV-*CP* in tobacco was also reported to be effective against other *tobacco viruses* whose CP was closely related to that of TMV but not effective against viruses which were distantly related to TMV (Masuta *et al.*, 1995). Transgenic potato, expressing the CP of potato virus X (PVX) also showed resistance against PVX, but mechanism was different (Hemenmay *et al.*, 1988). The stage of the viral life cycle, at which the CPMR is effective, has been shown to vary. In TMV, it is the virus disassembly and in the long distance transport stage (Wisniewski *et al.*, 1990).

Recently, considerable efforts have been made towards understanding the molecular basis of the CPMR especially in *tobacco viruses*. These studies may lead to more rational design of CP- derived transgenes. Evidences to date suggest that if the transgenically expressed CP was mutated such that there was an increase in inter subunit interaction, the transgenic plant expressed higher levels of virus resistance (Bendahmane *et al.*, 1997, Lu *et al.*, 1998). In the case of resistance to TMV, the transgenically expressed CP sub-units are believed to recoat the nascent disassembled viral RNA which leads to a decreased pool of the available viral RNA for translation (Beachy, 1999) resulting in resistance. Substantial yield increase observed in field trials of transgenic papaya and squash (Table 7) has established CPMR as the most favoured strategy to engineer resistance against many viruses. The success of CPMR has prompted the production of transgenic plants expressing multiple CP genes from more than one virus. Several important crops have been engineered for virus resistance using CPMR approach and released for commercial cultivation. These include tomato resistant to TMV, tomato mosaic virus (ToMv) and cucumber mosaic virus (CMV), cucumber resistant to CMV, squash resistant to Zucchini yellow mosaic virus ZYMV, WMV2 and CMV, potato resistant to PVX, potato virus Y (PVY) and potato leaf roll virus (PLRV)' papaya resistant to papaya ring spot virus, PRSV (more details can be found at the website http:// www.bspp.org.uk/mppoi/1997 / 0116 fuchs). Transgenic tobacco containing the *CP* gene of three viruses has developed resistance to all three of them i.e. tomato spotted wilt virus (TS WV), tomato chlorotic spot virus and groundnut ring spot virus (Prins *et al.*, 1995).

Since *CP* plays a major role in the vector transmission, CPMR confess additional advantage of resistance to vector inoculation in a majority of cases. For example, transgenic rice expressing high level of rice stripe virus *CP* gene expressed resistance to virus inoculation by plant hopper (Hayakawa *et al.*, 1992). The discussion on CPMR would not be complete without reference to the most successful story of resistance to PRSV in papaya. Papaya production in Hawaii suffered due to high incidence of PRSV in 1950s. Transgenic papaya containing *CP* gene grown from 1991 to 1993 remained virus free for 25 months and it was subsequently crossed with other popular variety resulting in a dramatic increase from 5,600 kg/ha to 112,000 kg/ha (www.plant.uoguelph.ca/safe food/gmo/ papayarep.htm). Replicase (Rep) protein-mediated resistance against a virus used in GM of tobacco against TMV containing the 54 kDa putative Rep gene was first studied by Golemboski *et al.* (1990). Gene constructs of Rep genes that have been used for resistance include full length, truncated or mutated genes. Many of the above resistance response have now been shown not to require protein synthesis and to be mediated at the RN level. This type of resistance remains confined only to a narrow spectrum of virus. To make the resistance broad based, it may be necessary to pyramid such genes from several dissimilar viral sources into the test plant genome (Dasgupta *et al.*, 2003).

Movement proteins (MP) are essential for cell-to-cell movement of plant viruses. These proteins have been shown to modify the function of plasmodesmata thereby allowing the virus particles or their nucleoprotein derivative to spread to neighbouring cells. The conferred resistance is believed to be based on the competition between wild type virus encoded MP and preformed dysfunctional MP to bind to the plasmodesmatal sites (Lapidot *et al.*, 1993). This resistance may be effective against related or unrelated virus as observed in TMV resistance in tobacco using MP derived from brome mosaic virus, suggesting functional conservation of this protein among several virus (Cooper *et al.*, 1995). However, viral movement is mediated by a set of three onverlapping genes, known as triple gene block (TGB) in potex-, Carla- and hordeiviruses. Expression of the modified central 12 kDa *TGB* gene of PVX was shown to confer MP-derived resistance in potato to potexvirus PVX and Carlavirus potato virus M and potato virus S (Seppanen *et al.*, 1997). However, resistance was onvercome when inoculated with viruses lacked a TGB, like PVY. This indicated that the resistance depended upon the interaction of the viral derived and transgene derived MPs.

Besides using the genomic components of an infectious virus, a strategy exploiting the use of satellite RNA associated with certain viruses received greater attention. Some strains of CMV encapsidate satellite RNA (sat-RNA) in addition to the tripartite messenger sense, single stranded RNA genome. The presence of sat-RNA modulates the symptoms induced by the

HV and often depresses HV accumulation in different host species. So the genetically modified tobacco plants expressing multiple or partial copies of CMV sat-RNA showed attenuated symptoms when challenged with CMV (Bahlcombe *et al.*, 1986). Some of the plants also showed delayed symptom development with the congante virus (Toush *et al.*, 1994). The sod-RNA has been tested as an effective biocontrol agent in field it trials with success (McGruvey and Kaper, 1993), with tomato showing a high level of tolerance to CMV confessed by sat-RNA. However, as sat- RNA spreads epidemically sufficient caution will have to be exercised in adopting this technology.

Table 7: Comparative performance of GM virus resistant plants

Host	Transgene	% Yield Incease
Tomato	TMV CP	40
Tomato	CMV Satellite	14
Potato	PVX + PVY CP	38
Squash	CMV + ZYMV + WMV2 CP	97
Squash	ZYMV + WMV2.CP	90
Squash	ZYMV CP	77
Papaya	PRSV CP	90

*Source : Gonsalves (1996).

ABIOTIC STRESSES IN AGRICULTURE

Rapid and continuing increase in population implies a greater demand for food. It is projected that by 2010 our food grain demand will be 246 million tons and 294 million tons by 2020 as against our current production level of 208 million tons (Kumar, 1998). The major impediments to increased crop production are drought, unfavourable climatic and soil conditions resulting in salt stress, low and high temperature stress, flooding stress, chemical stress, oxidative stress and other related stress types. There is hardly a land mass in India which is not influenced by one or the other stress factors. The average yields of any region are lower than national average, meaning that potential productivity is much higher than the average yields (Aggarwal *et al.*, 2000). The contribution due to osmotic stress is a common denominator in water stress, salt stress and low temperature stress. However, a great deal of success has been observed in the production of genetic modified crops resistant to herbicide, virus, fungus, etc. but due to complex genetic mechanisms that govern abiotic stress, a limited success only has been possible. The selective reports on abiotic stress tolerant transgenics produced so far are shown in Table 8. The selective websites that contain information on different aspects of molecular biology and biotechnology related to abiotic stress are http://www. stress-genomics.org/ heat shock html and http://www.plant stress.com.

Several groups in India are working on cellular responses trigerred by abiotic stress factors on microbial, animal and plant system. Pioneering contribution has been made by Uma Prasad and Gowrishankar (1998) in identification of several transporters including Proll (glycine betaine uptake) Prop (Proline uptake) and kdp (K-uptake related with water stress adaptation in *E. coli.* Analysis of transcriptional regulation of the genes encoding these transporters has also been reported (Reddy and Gowrishankar, 2000). Cloning of several osmoresponsive genes from a marine nitrogenfixing cyanobacterium *Anabena torulosa* using substractive RNA

Table 8: Production of Abiotic Stress tolerant GM Crops.

Gene	Protein	Source	Trans-host	Comments	Reference
cnb 1	Calcineurin B1	*S.cerevisiae*	*N. tabacum*	Ca ++ binding protein. Na Cl tolerance increased	Pardo *et al.*(1995)
gr	Glutathione reductase	*E.coli*	*N.tabacum*	Component of O_2 Scavenging system. GM Crops show 3 fold increase in photo oxidative stress	Aono *et al.* (1993)
gst /gpx	Glutathinone –S-Transferase and glutathione peroxidase	*E.coli*	*N.tabacum*	Detoxification of herbicides and toxic substances stimulation of seedling growth under chilling and salt stress conditions	Roxas *et al.* (1995)
sat	Serine acetyl transferase	*E.coli*	*N. tabacum*	Resistance to oxidative stress	Blaszcysyts *et al* . (1999)
gpd	NAD+ dependent Glyceraldehydes (3) phosphate dehydrogenase	*Pleurotus*	*S.tuberosum Sajor-caju*	Potato plants showed salt stress tolerance	Jeong *et al.* (2001)
bet A	Choline dehydrogenase	*E.coli*	*N.tabacum*	Better survival in high salt levels	Lilius *et al.* (1996)
bet	Chloine dehydrogenase	*E.coli*	*Synchococc us Sps.*	Survival of plants under salt stress showing protetive role of glycine betaine to Rubisco protein	Nomura *et al.* (1998)
bet A	Choline dehy drogenase	E.coli	*N. tabacum*	Increased stress tolerance deu to glycine betaine biosysthesis	Holmstrom *et al.*(2000)
bet B	Betaine aldehycle dehydrogenase	*E.coli*	*N.rabaccum*	Better growth in osmotic stress conditions	Holmstrom (1998)
cod A	Choline oxidase A	*Arthrobacter globiformis*	*A. thaliana*	GM crops tolerant to salt and cold	Hayashi *et al.* (1997)
cod A	Choline Oxidase A	*Arthrobacter globiformis*	*O.sativa*	Increased tolerance to salt and low temperature stress.	Sakamoto *et al.* (1998)
Cod A	Choline oxidase A	*Arthrobacter globiformis*	*Brassica juncea*	Enhanced capacity to germinate under salt stress	Prasad *et al.* (2000)
ect A *ect B* *ect C*	L-2,4 diamino Butyric acid acetyl transferase, L-2, 4 diamino Butyric acid Transaminase L-ectoine synthase	*Halomonas elongata*	*N .tabacum*	Increased tolerance to hyperosmotic stress	Nakayama *et al.* (2000)
mtl D	Mannitol –1 Phosphate dehydrogenase	*E.coli*	*N.tabacum*	Increased growth in salt stress	Tarcyznski *et al.* (1993)
mtlD	Mannitol –1 Phospate dehydrogenase	*E.coli*	*A.thaliana*	Tolerance to salt stress	Thomas *et al.* (1995)
otsA *otsB*	Trehalose-6-Phosphate synthetase	*E.coli*	*N.tabacum*	Increased biomass production under stress	Pilon-Smits *et al.* (1998)
tps1	Trehalose 6- -phosphate synthase	*S.cerevisiae*	*N.tabacum*	Trehalose accumalation and improved drought tolerance	Romero *et al.* (1997)
hal1	Protein involved in regulation of K+ transport	*S.cerevisiae*	*Lycopersicum esculentum*	Higher kind of salt tolerance and retain more K+ than controls	Gisbert *et al.* (2000)

hybridizaton has been performed (Apte and Haselkorn, 1990). However, the work on production of abiotic stress tolerant plants is still in its infancy and requires significant inputs in the identification isolation and cloning of moral abiotic stress tolerance related genes from the wide germplasm available. Different laboratories in India are engaged in this process. Majumdar's group at the Bose Institute, Kolkata is focusing a metabolic engineering of pathways leading to osmoprotectant biosynthesis under stress conditions. The bacterially expressed protein from cloned genes of rice and porteresia are being used. George Thomas group at SPIC foundation Chennai is involved in engineering *hal1* gene from yeast in eggplant and rice. At the same time, this group has also been trying to study the role of antiporters in salt tolerance from rice and an algae *Dunaliella* (Grover *et al.*, 2003).

Global Scene

The estimated global area of GM crops for 2004 is 81 million hectares or 202.5 million acres. During the nine year period 1996 to 2004, the global area of GM crops increased by almost 50 fold from 1.7 million to 81 million hectares. This high rate of adoption reflects the growing acceptance of GM crops by farmers in both industrial and developing countries (Table1) (James, 2004). Globally, transgenic soyabean occupied 33.3 million hectares in 2001 while transgenic corn in second place at 9.8 million hectares, transgenic cotton at 6.8 million hectares and canola at 2.7 million hectares. (Table 2). The unprecedented rapid adoption of transgenic crops during the first six years' period reflects the significant multiple benefits realized by large and small farmers. There is a growing body of evidence that clearly demonstrates the improved weed and insect-pest control attainable with transgenic herbicide tolerant and insect-resistant *Bt* crops that also benefit from lower input and production costs. Despite the ongoing debate on GM crops millions of large and small farmers in both, industrial and developing countries, continue to increase their plantings of GM crops because of the significant multiple benefits they offer. The recent survey indicates that area under GM crops cultivation would have exceeded 81 million hectare by 2005. List of GM crops approved for commercial use along with product name/company / country and year of manufacture has been complied from http:// www.ag bios.com (Table 9).

Table 9: GM Crops approved for commercial use.

Crop	Genetically modified trait uses	Introduced gene	Origin of gene	Product name company	Country
1	2	3	4	5	6
Argentine Canada	Gly phosphate herbicide tolerance Improved weed control	5-enolypyruvyl shikimate-3-phosphate synthase (EPS PS) gene	CP4 strain of *Agrobacterium tumefaciens*	Roundup Ready/ Monsanto Co	USA Japan
	Phosphino thricin (PPT) herbicide tolerance, specifically	PPT –acetyl transferase (PAT) gene	*Streptomyces viridochromogenes*	Liberty Link/Bayer Crops Science	Canada Japan US Austrialia

1	2	3	4	5	6
	glufosinate ammonium Improved weed control Oxynil herbicide Tolerance, including bromoxynil and ioxynit	*bxn* gene	*Klebsiella pneumoniae (sub sp. Ozaenae)*	BXN /Aventis Crop Science	Canada Japan
	Improved weed control Glufosinate Ammonium	Phosphino thricin-N-acetyltransferase (PAT) gene	*Streptomyces hygroscopicus*	Invigor/ Bayer/ Crop Science	Canada Japan USA Australia
	Herbicide Tolerance and fertility restored Improved weed control	Barnase Barstargene	*Bacillus amyloliquefaciens*		
Chicory	Glufosinate Ammonium Herbicide Tolerance and fertility restored Improved weed control and high yields	Phosphino thricin acetyl-transferase (PAT) gene Barnase gene	*Agrobacterium tumefaciens* mediated plant Tansformation *Bacillus amyloliquefaciens*	Seedlink/Bejo Zaden BV	European Union USA
Cotton	Resistance to lepidoteran pests Improved insect Protection	*Cry1 AC* *Cry2 Ab*	*B.thuringiensis kurstati*	Bollard Monsanto Co.	Japan, Asutralia, China, Mexico, South Africa, Argentina, India
	Oxynil, ioxynil, bromoxynil Tolerance Improved weed control	*bxn* gene	*Klebsiella pneumoniae Sub sp.ozaenae*	BXN / Calgene Inc.	US Japan
	Resistance to lepidopteran insects Oxynil, bromoxynil Tolernce Improved weed and pest control	*bxn* gene *Cry1 Ac* gene EPSPS gene	*Klebsiella pneumoniae* *Bacillus thuringiensis* CP4 strain of *Agrobacterium tumefaciens*	Bollgardwith BXN/ Calgene Inc Round up Ready/ Monsanto Co.	USA Japan USA Japan Argentina Australia

1	2	3	4	5	6
	Glyphospate Herbicide tolerance	Phosphinothricin Acetyle transferase		Bayer Crop Science	USA
	Phosphinothricin (PPT) herbicide tolerance, specifically glufosinate ammonium	(PAT) encoding bar gene	*Streptomyces hygroscopicus strain ATCC 21705*		
	Improved weed control				
flax linseed	Antibiotic Kanamycin resistance	Neomycin Phosphotransferase II (NPTII) gene	*E. coli*	Triffid University of Sakatchewan Crop Dev. Centre	Canada USA
Maize	Improved weed control Resistance to European Corn borer (strinianubilalis), glyphosphate herbicide tolerance Improved insect Resistance	*Cry 1 Ab* gene EPSPS gene	*Bacillus thuringiensis* *A.tumefaciens Strain CP4*	Monsanto Co. and Pioneer Hi Bred International Inc.	Canada Japan USA
	PPT herbicide tolerance, specifically glufosinate ammonium Improved weed control	Phosphinothricin Acetyl transferase gene *(pat)*	*Streptomyces hygroscopicus*	Dekalb Genetics Co / Bayer Crop Science	Canada USA Japan Argentina
	Resistance to European corn Borer (Ostrinia Nubilalis), a phosphinothericin (PPT) herbicide tolerance	*Cry1* Ab gene	*B.thuringiensis kurstaki*	Syngenta seeds Inc / ventis Crop Sicence Dekalb Genetics Co. / Mycogen/ Dupont	USA Canada Japan Argentina European Union
	Improved weed protection and insect resistance	Phosphinothricin Acetyl transferase *(pat)* gene	*Streptomyces hygroscopicus*	Monsanto Co.	

1	2	3	4	5	6
	Resistance to European corn borer (ostrinia nubilalis) Improved insect resistance	*Cry1* Ab gene	*Bacillus thuringiensis*	Bayers Crops Science	USA Japan Canada South Africa Argentina European Union Philippines
	Glufosinate ammonium Herbicide tolerance and male sterility	barnase ribonuclease gene	*Bacillus amyloliquefaciens*		Canada
	Improved weed control and male sterility	*pat* gene	*Streptomyces viridochromogenes*	Pioneer Hi-Bred International Inc.	USA
	Glufosinate ammonium Herbicide tolerance and fertility restored	barnase ribonucleasegene	*Bacillus amyloliquefaciens*	Monsanto Co.	
	Improved weed control and fertility Restores	*Cry1* Ab gene	*Streptomyces viridochromogenes*	Monsanto Co.	USA
	Resistance to European corn borer	*Cry3* Bb 1 gene	*B.thuringiensis*		Canada
			B.thuringiensis Subsp.		USA
	Improved Insect resistance		*Kumamotoensis*		
	Resistance to corn Root worm Improved insect resistance	Single nuclectide Substitution within ALS encoding gene	Microorganism	Syngenta Seeds Inc.	
	Imidazolinone Herbicide tolerance				Canada
	Improved weed control				Canada USA
Melon	Delayed ripening By introduction of gene that results in degradation of a precursor of ethylene, Delayed ripening	S-adenosyl methionine hydrolase encoding sam-K gene	*Escherichia coli* bacteriophage T.3	Agritope Inc.	

1	2	3	4	5	6
Polish Canola	Glyphosphale Herbide tolerance Improved weed control Gene encoding Glyphosphate degrading enzyme	EPSPS	*A.tumefaciens*	Monsanto Co.	Canada
Potato	Resistance to coloradeo otato Beetle, PLRV Improved insect Resistance	*Cry3* A gene	*B. thuringiensis* *(subsp* *Tenebrionis)*	New Leaf/ Monszanto	USA Canada
	Resistance to Colorado potato Beetle, PVY.	Delta-endotoxin Encoding *Cry3* A gene	*B.thuringiensis* *(subsp* *Tenebrionis)*	Monsanto Co.	USA Canada
	Improved protection from insect and potato virus Y	Coat protein (*cp*) gene)		Liberty link/ Aventis Crop Sicence	US
Rice	Phosphino thricin (PPT) herbicide tolerance, specifically glufosinate ammonium Improved weed control	*pat* gene	*Streptomyces* *hygroscopicus*		
Soyabean	Glyphosphate Herbicide tolerance Improved weed control	EPSPS gene	*Agrobacterium* *tumefaciens* *Strain CP4*	Roundup Ready/ Monsanto Co.	US Argentina Japan Canada Uruguay Mexico Brazil South Africa
	Phosphino thricin (PPT) herbicide tolerance Improved weed control	*pat* gene	*Streptomyces* *viridochromogene*	Liberty link/ Bayer Crop Science	US Canada Japan
Squash	Resistance to viral Infection, water melon mosaic virus (WMV)2, Zuchini yellow Mosaic virus (ZYMV) Improved viral resistance	Gene encoding protein coat	Cucumber Mosaic virus (CMV) Zucchini Yellos mosaic Potyvirus (ZYMV) and WMV2	Destiny, Liberatory/Upjohn, Seminis Vegetable Inc.	US Canada US Canada

1	2	3	4	5	6
Sugarbeet	PPT herbicide Tolerance, specifically Glufosinate ammonium Improved weed control	*pat* gene	*Streptomyces viridochromo genes*	Liberty link Bayer Crop Science Crop Science	US
	Glyphosphate herbicide tolerance/ Improved weed control	EPS PS gene	*Agrobacterium tumefaciens CP4*	Round up Ready Novartis Seed	
Tobacco	Oxynil herbicide Tolerance including Bromoxynil and Ioxynil Improved weed control	Nitrilase gene	*Klebsiella Ozaenae*	Societe National d Exploitation des Tobacs ef Allumettes	

GM TECHNOLOGY

The naturally evolved unique ability of *Agrobacterium tumefaciens* to precisely transfer defined DNA sequences to plant cells has been very effectively utilized in the design of a range of T plasmid based vectors. The current status of our understanding of *Agrobacterium* T-DNA transfer process has been reviewed by Gelvin (2000) and Zupan *et al.* (2000). Three genetic elements, *Agrobacterium* chromosomal virulence genes (*chv*), T- DNA delimited by a right border and a left border and Ti plasmid virulence gene (*vir*) constitute the T- DNA transfer machinery. In fact, *Agrobacterium* T-DNA transfer is now reviewed as universal based on successful transformation of yeast (Bundock *et al.* 1995) *Aspergillus* (de Groot *et al.*, 1998) and human cell (Kunik *et al.*, 2001). Selecting the transformed plant tissue is of vital importance and is based on the inclusion into the culture medium of substances that is toxic to plants. Many potential selectable markers are being used which are listed in Table 10 (Slater *et al.*, 2003). However, leaky expression of plant selection marker genes such as *nptII* and *hpt* in *Agrobacterium* contributes to bacterial overgrowth during selection and to the emergence of false positive results. In a strategy to eliminate leaky expression of marker genes in *Agrobacterium,* Mass *et al.* (1997) placed the intron 2 of the potato ST-1 gene in the *N*-terminal part of *ptII* gene. Lack of mRNA splicing in *Agrobacterium* completely eliminated leakiness in kanamycin selection. Modifications in co-cultivation conditions have also led to increase in efficiency of transformation. Co-cultivation of explants with *Agrobacterium* in the presence of acetosyringone, a *vir* gene inducer, has become a routine exercise in the transformation of recalcitrant crops such as maize, barley, rice etc. (Ishida *et al.*, 1996). Many physical conditions influence transformation efficiency. Uniform wounding of tobacco leaves, and sunflower apical meristern by particle bombardment (Bidney *et al.*, 1992), vacuum infiltration of *Arabidopsis* seedings (Rossi *et al.*, 1993) rapid proliferation of rice calli at 30°C (Toki, 1997) and co-cultivation at lower (22°C) temperature (Dillen *et al.*, 1997) led to an increase in the efficiency of *Agrobacterium* mediated transformation. Although markers are necessary for selectively propagating transformed cells and tissues but subsequent maintenance of markers in genetically modified crops is unnecessary. Elimination of markers is advocated since the antibiotic resistance genes may be transferred to pathogenic bacteria or the herbicide resistance genes may be transferred to weeds. Two excellent reviews on marker elimination have been published (Hohn *et al.*, 2001, Ebinuma *et al.*, 2001). The recombination system of the 2 micron plasmid of *Saccharomyces cerevisiea* involving FLP recombinase and FRT recombination site has been successfully deployed for marker elimination in maize cells (Lyzhik *et al.*, 1996). Here, a combination of lambda attachment site (*attP*) and napthalencacetiamide (NAM) were used for marker elimination by intrachromosomal recombination (Zubco *et al.*, 2000). A single step IPT- MAT vector incorporating the recombinase (*R*) and recombination sites (RS) of *Zygosaccharomyces rouxii* was reported for rice (Ebinume, 2001). The T-DNA constructed by the carried nptII/gus/hpt/RS/lipt 35S R/gfp/Rs sequences. *Agrobacterium* mediated transformation of scutellum derived callus was performed. Selection of transformants was based on the ability to form normal shoots on cytokinin minus medium. In 25% of the infected calli, regenerated normal shoots carried the *gus* gene without *ipt*. They showed that marker (*ipt*) free rice plants could be generated in a very short span of one month.

Table 10: Selectable marker genes used in GM crops*.

Selectable marker gene	Abbreviation	Source of gene	Selection Mechanism	Selective agent
Aminoglycoside adenyl	aad A	*Shigella flexneri*	Antibiotic Resistance	Streptomycin Spectinomycin
Bleomycin resistance	ble	*E.coli*	Antibiotic Resistance	Bleomycin
Dihydropterorate synthase	sul/dhps	*E.coli*	Antibiotic Resistance	Sulhonamides
Hygromycin phosphotransferase	hpt/aph IV	*E.coli*	Antibiotic Resistance	Hygromycin
Neomycin Phospotransferase-II	nptII/reo	*E.coli*	Antibiotic Resistance	Kanamycin Geneticin
Neomycin Phosphotransferase III	nptIII	*Streptococcus faecalis*	Antibiotic Resistance	Kanamycin Geneticin
Glyphosate oxidoreductase	gox	*Achromobacter LBAA*	Herbicide Resistance	Glyphosate
Phosphinotricin Acetyl transferase	bar/ pat	*Streptomyces hygroscopicus*	Herbicide Resistance	Bi-alophos Glufosinate

*Source : Slater *et al.* (2003)

PERSPECTIVES AND CONCLUSIONS

Molecular tools give us an opportunity to develop genotypes that carry resistance traits. The resistance needs to be protected by taking lessons from our past experiences with chemical pesticides. *Bt* has rightly emerged as a powerful tool of plant protection in agriculture in a substainable manner. Although, still not universal in its application and total in its protection,

B. thuringiensis will play a central role in protecting the crop from its major insect-pests. In combination with the other powerful biopesticidal proteins such as proteinase inhibitors, *Bt* will drastically reduce the consumption of chemical pesticides. There is also an urgent need to test various *Bt* delta-endotoxins, Vips, proteinase inhibitors, lectins etc for their toxicity to pests endemic to India. Special attention must be given to *H. armigera* which is the major pest on important crops and which can survive on more than 130 plant species. The durability of insect resistance in GM crops can only be ensured if integrated pest management (IPM) practices are followed. Development of fungus resistant plants has been possible by use of several strategies including the manipulation of resistance by expression of PR proteins, antifungal peptides and manipulation of biosynthsis of phytoalexins. However, in these cases the observed resistance was not absolute and was restricted to a limited number of fungi. For the antifungal compounds strategy to be successful in the long-term level of resistance in GM crops should be increased and isolating new genes and testing new combination of genes should broaden its range. Precise knowledge of both the signalling pathways involved and subsequent metabolic pathways that get triggered is essential. Also extra caution is required about the role of signalling gene in various other pathways which would lead to undesirable side effects in GM crops (Groves and Gowthaman, 2003).

Of the major crop plants in our country, transformation systems are available for only a few cereals, vegetables, fibres crops and oilseed varieties. A major push needs to be given for GM of pulses and legumes which incur the heaviest losses due to virus. Emerging techniques of functional genomics need to be harnessed to understand the molecular interaction between the viral pathogen the resistant/susceptible plants leading to resistance. This may result in novel insights at disease control. The dominant and virulent strains of each important and in the country need to be identified for obtaining the genes. Studies should also focus on the degree of variability and the recombination of viral genomes.

Meanwhile, tomato ripening was one of the first systems in which antisense technology was used to manipulate endogenous gene expression (Grierson and Schuch, 1993). Enhancement of nutritional value of agricultural produce is also deemed possible. A successful manipulation of Beta-carotene synthesis in the rice grains gives them a characteristic yellow/ orange colour. Rice is genetically enriched in provitamin A and described as 'Golden rice' (Potrykers, 2001). The GM crops are being developed that have an improved nutritional status and/or health value. The new generation of GM crops includes plants with obvious environmental benefits. GM crops can be used, for instance, to detect and treat heavy metal contamination or to remove toxic contaminants. GM crops can be used as factories to produce biodegradable, renewable and CO_2 neutral alternative to oil-based products etc. It is also clear that such an approach needs a close cooperation among microbiologists, molecular biologists, tissue culture specialists and agronomists. The combined efforts are sure to deliver crops in the near future, which will improve drastically the quality and quantity of crops.

REFRENCES

Aggarwal, P.K, Talukdar, K.K. & Mall, R.K. (2000) Potential yields of rice wheat system in the Indo Gangetic plains of India, Rice Wheat consortium paper series 10, New Delhi, India, RWC1GP, C1MMYT, p. 16.

Aono, M., Kubo. A, Saji, H., Tanaka K. & Kondo, N. (1993) *Plant Cell Physiol.* **34**:129-135.

Apte, S.K. and Harselkorn, R. (1990) *Plant Molecular Biol.* **15**: 723-733.

Baulcombe, D.C., Saunders, G.B, Bevan, M.W., Mayo, M.A, & Harrison, B.D. (1986) *Nature.* **321**: 446 –449.

Beachy, R.N., Philos Trans, R. (1999) *Soci. London.* **35**: 659-664.

Bendahmane, M., Fitchen, J.H., Zhang, G. & Beaching, R.N. (1997) *J. Virol.* **71**: 7942-7950.

Bidney, D., Scelonge, C., Martich, C., Burrus, M., Sims, L. & Hoffman, S. (1992) *Plant Mol. Biol.* **18**: 301-313.

Blaszcyzk, A., Brodzik, R. & Sirco, A. (1999) *Plant J.* **20**: 237-243.

Bowen, D., Rochelean, P., Blackburn, M., Andreev, O., Golubeva, E., Bhartia, R. & French Constant, R. (1998) *Sci.* **280**: 2129-2132.

Chareon pornmattana, S., Thara, K.V., Wang, L., Datta, S.K, Panbangred, W. & Muthukrishnan, S. (1999) *Theor Appl. Genet.* **98** : 371-378.

Chet, I. & Inbar, J. (1994) *App. Biochem. Biotech.* **48**: 37 –43.

Conne ll, T.D., Metzger, D.J., Lynch.J. & Fplster, J.D. (1998) *J. Bacteriol.* **180**: 5591-5600.

Cooper, B., Lapidot, M., Heick, J.A, Dodds, J.A. & Beachy, R.N. (1995) *Virology.* **206**: 307-313.

Dai, S. (2001) *Mol. Breeding* **7**: 25-33.

Dasgupta, I., Malathi,V.G, Mukerjee, S.K. (2003) *Curr. Sci.* **84 (3)**: 341-354.

De la Cruz, J., Pintor, T.J.A., Benitezs, T. & Liobell, A. (1995) *J. Bacteriaol.* **177**: 1864-1871.

De Wit, P.J. GM (1992) *Ann. Rev. Phytopath.* **30**: 391-481.

Dillen ,W., Clerq, J.D., Kapilas, J., Zambre, M., Van Montagi, M. & Angenon, G. (1997) *Plant J.* **12**:1459-1463.

El Quakfaoui, S., Potvin, C., Brazezinski, R. & Asselin, A. (1995) *Plant Cell Rep.* **15**: 222-226.

Ebinuma, H., Sugita, K., Matsunaga, E., Endo, S., Yamada, K. & Komamine, A. (2001) *Plant Cell Rep.* 383-392.

Estruch, J.J., Warren, G.W, Mullins, M.A.,Nye, G.J., Craig, J.A & Koziel, M.G. (1996) *Proc. Natl. Acad Sci, USA.* **93**: 5389-5394.

Feys, B.J. & Parker, J.E. (2000) *Trednds Genet.* **16**: 449-455.

Gelvin, S.B. (2000) *Ann. Rev. Plant Physiol Plant Mol Biology* **51**: 223-256.

Gisbert, C. (2000) *Plant Physiol.* **123**:393-402.

Glick , B.R. (1995) *Can. J. Microbiol.* **41**: 109-117.

Glick, B.R. & Pasternak, J.J. (1998) In *Molecular Biotechnology Principles and Applications of Recominant DNA*, ASM Press, Washington, p. 347.

Golemboski, D.B., Lomomosseff G.P. & Zaithin, M. (1990) *Proc. Natl. Acad. Sci., USA* **87**: 6311-6315.

Grierson, D. & Sahuch, W. (1993) *Philos. Trans. Royal Soc. London.* **342**:241-250.

Grover, A. & Gowthaman, R. (2003) *Curr. Sci.* **84(3)**: 330-340.

Grover, A., Aggarwal, P.K., Kapoor, A., Katiyar, A.S., Agarwal, M. & Chandramouli, A. (2003) *Curr. Sci.* **84**: 355-367.

Grover, A. & Pental, D. (2003) *Curr. Sci.* **84**: 310-321.

Gupta, C.P., Dubey, R.C. & Maheshwari, D.K. (2002) *Biolo.Fertil. Soils.* **35**: 399-405.

Hamilton, R.I., (1980) In *Plant Diseases: an Advanced Treatise*, Academic Press, New York, p. 279-303.

Hayakawa, T., Zhu, Y., Itoh, K., Kimura, Y., Izawa, T., Shimamoto, K. & Toryama, S. (1992) *Proc. Natl. Acad. Sci. USA* **89**: 9865-9869.

Hayashi, H., Mustardy, A., Deshnium, P., Ida, M. & Murata, N. (1997) *ibid.* **12**: 133- 142.

Hememway, C., Fang, R.X., Kaniewski, W.K., Chna, N.H. & Tumer, N.J.E. (1988) *EMBO J.* **7**:1273 –1280.

Hohn, B., Levy, A.A. & Puchta, H. (2001) *Curr. Opinion Biotech.* **12**: 139-142.

Holmstorm, K.O (1994) *Plant J.* **6**:749-758.

Holmstrom, K., Somersalo, S., Mandal, A., Palva, E.T. & Welin, B. (2000) *J. Exp. Biol.* **51**: 177- 185.

Houlb, E.B. (2001) *Nature Genet.* **2**: 516-527.

Howie, W., Joe, L., Newbigin, E., Suslow, T. & Dunsmuir, P. (1994) *Trans Res.* **3**: 90-98.

Hull, R.(2002) *Mattew's Plant Virology*, 4th Ed., Academic Press, New York.

Ishida, Y., Saito, H., Ohta, S., Hiei, Y., Komari, T. & Kumashiro, T. (1996) *Nature Bioteh.* **14**: 745-750.

James, C. (2001) Global Review of commercialized transgenic crops, International service for the acquisition of Agri-biotech Application (ISAAA) Briefs No.24, I Thaca, NY.

James, C. (2003) *Curr. Sci.* **84**: 303-310.

James. C. (2004)Global status of commercialized Biotech/GM crops, International service for the acquisition of Agri- Biotech Applications, Report No.32.
Jansens, S. (1997) *Crop Sci.* **37**: 1616-1624.
Jansens, S., Cornelissen, M., de Clercq, R., Reynaerts, A. & Peferear, M.J. (1995) *Econ. Entomol.* **88**: 1469-75.
Jeong, M.J., Park, S.C. & Byun, M.O. (2001) *Mol Cells.* **12**: 185-189.
Jouanin, L., Boade-Bootino, M., Girard, C., Morrot, G. & Giband, M. (1998) *Plant Sci.* **131**: 1-11.
Kozief, M.G. (1993) *Bio/Technol.* **11**: 194-200.
Kumar, P.A., Mandaokar. A., Sreenivasu K., Chakrabarti, S.K., Sharma, S.R., Bisaria S., Kaur, S. & Sharma R.P. (1998) *Molecular Breeding.* **4**: 33-37.
Kumar, P.A., Sharma, R.P. & Malik, V.S. (1996) *Adv. Appl. Microbiol.* **42**: 1-43.
Lamb, C. & Dixon R.A. (1997) *Annual Review Plant Physiology, Plant Molecular Biology.* **48**:251-275.
Lapidot, M., Gafny R., Ding, B., Wolf, S., Lucas, W.J. & Beachy, R. N. (1993) *Plant Journal.* **4**:959-970.
Lilius .G., Holmberg, N. & Bulow, L. (1996) *Biotechnology.* **14**: 177-180.
Loon. V.L.C. & Kamman V. A. (1970) *Virology.* **40**: 199-211.
Lorito –M, Woo S.L., D., Ambrosio M., Harman G.E., Hayes C.K., Kubicek, C.P. & Scala, F. (1996) *Mol. Plant Microbe Interact.* **9**: 206 -213
Lu. B., Stubbs .G., & Culver, J.N. (1998) *Virology.* **248**: 188-198.
Lyznik, L.A., Rao, K.V. & Hodges, T.K. (1996) *Nucleic Acids Res.* **24**: 3784-3789.
Maas, C. (1997) *Molecular Breeding.* **3**: 15-28.
Mandaokar, A., Goel. R.K., Bisaria S., Reddy, V.S., Altosaar, I., Sharma, R.P. & Kumar P.A. (2000) *Crop protection.* **19**:307-312.
Mandaokar, A.M., Chakrabarti, S.K., Kumar, P.A., & Sharma, R.P. (1998) *World J. Microbiol. Biotechnol.* **14**:599-601.
Masuta, C., Tanaka, H., Uehora K., Kuwata, S., Koiwai, A. & Noma, M. (1995) Proceeding National Academy Science, USA.**13**: 6117 –6121.
Mayak, S., Triosh, T. & Glick, B.R. (2004) *Plant Physol. Biochem.* **42**: 565-572.
Mayo, M.A. (1992) Biotechnology for Crop Improvement in Asia (Eds. Moss, J.P.) ICRISAT, Hyderabad.
Me Garvey, P.B. & Kaper, J.M. (1993) Transgenic Plants (Eds. Kung, S.D and Wu, R) Academic Press, New York p 277-296.
Mitsuhara, I. (2000) *Mol. Plant Microbe interaction.* **13**:860-868.
Monreal, J. & Reese .E.T. (1969) *Canadian J. Microbiol.* **15**: 689.696.
Muryoi, N., Sato, M., Kaneko, S., Kawahara, H., Obata, H., Mahmond, W.F., Griffith, M. & Glick, B.R. (2004) *J. Bacteriol.* **186 (17)** : 5561-5671.
Nakayama, H., Yoshida, K., One, H., Murooka, Y. & Shinmyo, A. (2000) *Plant Physiol.* **122**: 1239-1247.
Nayak, P. (1997) Proceedings National Academy of Science, USA. **94**:2111-2116.
Nomura, M., Hibino, T., Takabe, T., Sugiyama, T., Yokota, A. & Miyake, H. (1998) Cell Physiol. **39**: 425-432.
Pardo, J.M., Reddy, M.P. & Yang, S. (1998) Proceeding National Academy Sciences, USA. **95**: 9681-9683.
Park C.M., Berry. J.O & Brienn J.A. (1996) *Plant Mol. Biol.* **30**: 359-366.
Pattayanak, D. & Kumar P.A. (2000) *Proc. Indian Natt Sci. Acad.13 (66)*: 265-310.
Pattern C.L & Glick, B.R. (2002) *Appl. Enviorn. Mircobiol.* **68(8)**: 3795-3801.
Perlak F.J. (1993) *Plant Mole. Biol.* **22**: 313-321.
Perlak. F.J., Deaton, R.W., Armstrong. T.A., Fuchs, R.L., Sims, S.R., Greenplate.J.T. & Fischnoff, D.A. (1990) *Bio/Technology.* **8** : 939-943.
Pilon Smits, E.A. H. (1998) *J. Plant Physiol.* **152**: 525-532.
Potrykus, I. (2001) *Plant Physiol.* **125**: 1157-1161.
Powell- Abel P., Nelson, R.S., De B., Hoffman, N., Rogers, S.G., Fraley, R.T. & Beaching, R.N. (1986) *Science.* **236** 738 –743.
Prasad K V.S.K., Sharmila, P., Kumar P.A. & Pardha Saradhi P. (2000) *Molecular Breeding.* **6**: 489-499.
Prins. M., de Horan P., Luyten, R., Van Meller M., Van Grinsven, M.Q., & Goldbach, R. (1995) *Mol. Plant Microbat Int.* **8** :85-91.

Rao, N.G.V., Majumdar, A., Mandaokar, A., Nimbalkar, S.A. & Kumar, P.A. (1999) *Curr. Sci.* **77** : 336-337.
Reddy, M. & Gowrishankar.J. (2000) *J. Bacteriol.* **182**: 1978-1986.
Ren,Y. & West, C.A. (1992) *Plant Physiol.* **99** :1169-1178.
Romero, C., Belles, J.M., Vaya, J.L., Serrano, R. & Culianez –Mada F.A. (1997) *Plants* **201**: 293-297.
Rossi, L., Escudero, J., Hohn, B. & Tinland, B. (1993) *Plant Molecular Biology Rep.* **11**:220-229.
Roxas, V.P., Smith, R.K. Jr., Allen, E.R. & Alhar, R.P. (1997) *Nature Biotechnol.***15**: 988-991.
Sakamoto A., Murata, A. & Murata, N. (1998) *Plant Molecular Biology.* **38**: 1011-1019.
Schuler, T.H., Poppy, G.M. & Denholm, I. (1998) *Trends Biotechnol.* **16** :168-175.
Scofield S.R., Tobian, C.M., Rathjen, T.P., Chang, J.H., Lavelle, D.T., Michelmore, R.W. & Staskawicz, B.J. (1996) *Science.* **274**:2063-2065.
Seppanaen, P., Puska, R., Honkanen, J., Tyulkina, L.G., Federkin, O., Mo rozoo S.Y.U. & Atabekov, J.G. (1997) *Journal Virology.* **78**:1241 –1246.
Shen S, Li Q., He, S.Y., Barker K.R., Li, D. & Hunt, A.G. (2000) *Plant J.* **23**:205-213.
Shiwamani E., Huet, H., Shen, P., Ony, S.A., De Kochko, A., Fauquet, C. & Beaching, R.N. (2002) *Molecular Breeding.* **5**: 177-185.
Slater, A., Scott, N. & Fowler, M. (2003) In Plant Biotechnology. The genetic manipulation of plants. Ch:4 : Binary vectors for plant transformation. p79-101.
Slater, A., Scott, N.W. & fowler, M.R. (2003) In: Plant biotechnology Ch.6: The genetic manipulation of pest resistance p131-155. Oxford University Press.
Staskawicz, B.J., Ausubel, F.M., Baker, B.J., Ellis, J. G. & Jones, J.D.G. (1995) *Science.* **268** : 661-666.
Stewart, C.N., Jr. Adang, M.J., All, J.N., Boerma, H.R., Cardinean, G., Tucker, D. & Parrott, W.A. (1996) *Plant Physiology.* **112**: 121 –129.
Stewart, C.N., Jr. Adang M.J., All J.N., Raymer, P.L., Ramachandran, S. & Parrott, W.A (1996) *Plant Physiology.***112**:115-120.
Takken F.L.W. & Joosten, M.H.A.J. (2000) *European J. Plant Pathol.* **106**: 699-713.
Tang, X., Frederick, R.D., Zhou, J., Halterman, D.A., Jia, Y. & Martin, G.B. (1996) *Science.* **274**: 2060 –2062.
Tarczyuski, M.C., Jensen, R.G. & Bohvert, H.J. (1993) *Science.* **4**: 199-205.
Terrakawa T., Takaya N., Horiuchi, H., Koike, M. & Takagi, M. (1997) *Plant Cell Rep.* **16**:439-443.
Thomas, J.C., Sepahi, M., Arendall, B. & Bohvert, H.J. (1995) *Plant cell environment.* **18**:801-806.
Toki, S. (1997) *Plant Molecular Biology.***15**: 16-21.
Toush, D., Jacquemond, M. & Tepfer, M.J. (1994) *Journal General Virology.* **75**:1009-1014.
Uma Prasad, G & Gowrishanakar J. (1998) *Journal Genetics.* **97**: 1-11.
Vaidya , R.J., Shah, I.M., Vyas P.R. & Chhatpar, H.S. (2001) *World Journal Biotechnology.* **17**: 691-696.
Wisniewaski, L.A., Powell, P.A., Nelson, R.S. & Beaching, R.N. (1990) *Plant Cell.* **2**:559-567.
Wiwat, C., Siwayapahm, P. & Bhumiratana A. (1999) *Current Microbiology.* **39**: 134-140.
Wu G., Shortt, B.J., Lawerance, E.B., Levine, E.B., Fitzsimmons, K.C. & Shah, D.M. (1995) *Plant cell.* **7**:1357-1368.
Zubco, E., Scutt' C. & Meyer, P. (2000) *Nature Biotechnology.***18**: 442 –445.
Zupan, J., Muth, T.D., Draper, O. & Zambyski, P. (2000) *Plant Journal.* **23**:11-28.

14

Characterization of *Streptomyces* Causing Potato Scab Disease in Korea

Duck Hwan Park[1], Jeom Soon Kim[2], Soon Wo Kwon[3], Jang Hyun Hur[1], Jun Mo Cho[1] and Chun Keun Lim[1]

[1]*Division of Biological Environment, College of Agriculture and Life Sciences, Kangwon National University, Chuncheon 200-701K (Korea)*

[2]*Alpine Agricultural Experiment Station, Pyeongchang, Gangwon 232-950, Korea*

[3]*Division of Molecular Genetics, National Inst. of Agri. Science and Technology, Suwon 441-707 (Korea)*

From 1996 to 1999, potato-growing areas in Korea were surveyed for distribution and identification of potato scab pathogens. Survey results showed that potato scab was widely distributed in mass cultivation areas in the country, particularly in Je-ju island, the southern areas of the Chon-nam and Gyounggi provinces, and the alpine area of Gangwon province. Je-ju island was the most affected area by this disease. A total of 55 *Streptomyces* strains were isolated from potato scab lesions among which 40 strains were pathogenic on progeny tubers. Among the pathogenic strains, 19 strains were identified as *S. scabies*, 7 strains as *S. turgidiscabies*, 5 strains as *S. acidiscabies*, and 7 strains were observed as having distinct phenotypic properties. All these strains were classified into six distinct clusters based on their phenotypic characteristics, analyses of 16S rRNA genes and 16S-23S inter transcribed spacer (ITS) region, production of thaxtomin A, DNA relatedness, and presence of *nec1* and ORF*tnp* gene homologs, and then representative strains were selected for each cluster. Results showed that strains S33 and S27 had typical characteristics of *S. scabies* and *S. turgidiscabies* respectively, producing thaxtomin A and hybridizing with genes of *nec1* and ORF*tnp*. Strain S71 produced thaxtomin A and had phenotypic and phylogenetic properties similar to those of *S. acidiscabies*, except having a greater minimum growth pH (4.5), production of a melanoid pigment on tyrosine agar, and failure to hybridize with *nec1* and ORF*tnp* gene probes. In contrast, strains S63, S77 and S78 were phenotypically different from the previously described scab pathogens. Spore colours of strains S63 and S77 were yellow-white and pale orange, respectively, with rectiflexuous chains. Strain S78 had thin and compact spores unlike the typical *S. acidiscabies* (ATCC 49003). Phylogenetic analyses of representative strains S33, S27, and S71 based on 16S rRNA genes and 16S-23S ITS region sequences were most similar to *S. scabies, S. turgidiscabies*, and *S. acidiscabies*, respectively (more than 99.5% 16S rRNA sequences identity and 86.3% 16S-23S ITS region sequences identity). In contrast, strains S63, S77, and S78 showed low homology compared to those of previously described scab pathogens (less than 97.3%, 96.0%, and 96.3% 16S rRNA sequences identity, respectively and 71.2% and 75.7% 16S-23S ITS region sequences identity respectively).DNA-DNA hybridization results for all strains against strains S63, S77, and S78 as probe showed distant relationships to previously described pathogens. Pathogenicity was also determined by production of thaxtomin A and homologs of *nec1* and ORF*tnp* genes. In TLC, representative strains S33, S27, S71, S63, S77 and S78 produced a yellow band that co-migrated with the authentic thaxtomin A. However, thaxtomin A was not detected in chloroform extracts from oatmeal broth culture and slice tuber tissue of *S. luridiscabiei* (S63) and *S. puniciscabiei* (S77) by HPLC analysis. In addition, no homologs of *nec1* and ORF*tnp* genes in *S. acidiscabies* (S71), *S. luridiscabiei* (S63), *S. puniciscabiei* (S77), and *S. niveiscabiei* (S78) were detected by PCR and Southern hybridization analysis. Based on polyphasic evidences, *S. scabies* (S33), *S. turgidiscabies* (S27), *S. acidiscabies* (S71), *S. luridiscabiei* [S63^T (LMG21390^T = KACC 20252^T)], *S. puniciscabiei* [S77^T (LMG 21391^T = KACC 20253^T)] and *S. niveiscabiei* [S78^T (LMG 21392^T = KACC 20254^T)] were characterized as the major pathogens causing potato scab in Korea.

INTRODUCTION

Hundreds of soil-borne, saprophytic streptomycetes have been previously described, but relatively only a few species in the genus *Streptomyces* are known pathogens of plants (Loria *et al.*, 1997). Hence, pathogenicity is considered a rare phenotype in this genus. *Streptomyces* spp. are filamentous, Gram-positive prokaryotes that produce an array of agriculturally and clinically important secondary metabolites. Members of this genus have a complex morphology that produces spores to aid their dispersal. Spore chains are produced through fragmentation of aerial hyphae that form on the substrate mycelium. The shape of the spore chains, which may be straight, spiral or wavy, is an important taxonomic characteristic. The color of spores on mass sporulation is also critically important taxon (Goyer *et al.*, 1996; Locci, 1994; Loria *et al.*, 1997). In addition, spore chains are phenotypically and genetically complex, based on fatty acid profiles (Ndowora *et al.*, 1996; Paradis *et al.*, 1994), inhibitory reaction (Ndowora *et al.*, 1996), DNA-DNA relatedness (Healy and Lambert, 1991; Miyajima *et al.*, 1998), and ribosomal sequence analysis (Bouchek-Mechiche *et al.*, 2000; Takeuchi *et al.*, 1996).

Several pathogenic species of potato scab have been described to date. *Streptomyces scabies* was first described by Thaxter (1892) and re-proposed by Lambert and Loria (1989a) as the predominant species causing common scab of potatoes. A distinct acid-tolerant species *S. acidiscabies* induces symptoms similar to *S. scabies* (Lambert and Loria, 1989b). *S. turgidiscabies* was described in Japan (Miyajima *et al.*, 1998). *S. europaeiscabiei* and *S. stelliscabiei* were reported to cause common scab of potato, while *S. reticuliscabiei* was identified as a causal pathogen of netted scab in France (Bouchek-Mechiche *et al.*, 2000).

This report discussed the characterization of major scab pathogens found in Korea based on their distribution, taxonomy and pathogenicity.

DISTRIBUTION

The potato scab disease, which is widely distributed worldwide, is causing serious problems in potato cultivation and marketing in Korea. From 1996 to 1999, potato-growing areas in the country were surveyed for the identification and distribution of the potato scab disease. A total of 55 *Streptomyces* strains were isolated from potato scab lesions from different geographic locations. Strains were identified by pathogenicity test as previously described by Loria and

Kempter (1986) and by physiological and morphological characterization according to Lambert and Loria (1989a; 1989b) and Shirling and Gottlieb (1966). Among those strains, forty strains reproduced the same symptoms on progeny tubers. According to physiological and morphological characteristics, twenty-one strains were identified as previously described *S. scabies* (Kim and Lee, 1996). On the other hand, seven strains were identified as *S. turgidiscabies*, five strains were identified as *S. acidiscabies* (Kim *et al.*, 1998a; Kim *et al.*, 1998b), while seven strains were identified as having distinct phenotypic properties with those of others. Three among those seven distinct strains were identified as novel species such as *S. luridiscabiei*, *S. puniciscabiei*, and *S. niveiscabiei* based on phenotypic and phylogenetic analyses (Park *et al.*, 2003b). Therefore, in this study, potato scab pathogens were identified as *S. scabies*, *S. turgidiscabies*, *S. acidiscabies*, and three novel species as *S. luridiscabiei*, *S. puniciscabiei*, and *S. niveiscabiei* in Korea.

Serious problem of potato scab has been observed in mass cultivation areas, especially in Je-ju island, the southern areas of Chon-nam and Gyounggi provinces, and the alpine area of Gangwon province. Among those areas, Je-ju island was the most affected area by this disease. For the analysis of potato scab occurrence in Je-ju island, the relationship between geographical distribution and potato cultivars was investigated. Daejima, the cultivar used in potato growing areas of Je-ju island, was found highly susceptible to potato scab pathogens, whereas, cultivars Atalantic and Irish cobbler were found moderately susceptible, while, cultivars Superior and Jopung were resistant to potato scab pathogens (Fig. 1). In addition, cultivar Daejima is usually cultivated twice a year to obtain more yield. The second cultivation time is usually from fall to the next spring. During this period, most farmers prefer to keep potato in natural soil condition without harvesting as they can sell their produce at a higher price in the next spring season. However, this method may accelerate the rapid increase of pathogens in the soil and consequently, may increase the chance of its outbreak in Je-ju island. In contrast, cultivar Superior, which is mostly cultivated in Gangwon, Gyounggi, and Chon-nam provinces, is resistant to potato scab pathogens. Farmers in these areas often rotate potato with other crops, which may contribute to minimizing the outbreak of potato scab.

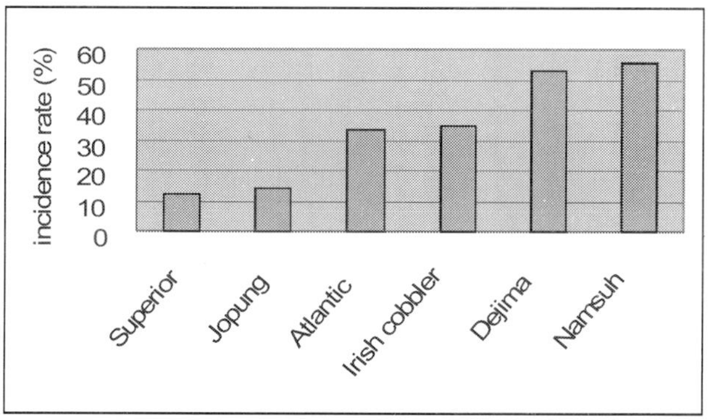

Fig. 1. Occurrence of potato scab disease on various potato cultivars (*Solanum tuberosum*) in Korea.

Hence, in the survey of potato cultivating areas in Korea, the relationship between each other crop rotation was also investigated. Results showed that bean, which is cultivated instead of potato in the next season, has the lowest disease incidence and the highest control effect to potato scab. In addition, Chinese cabbage and oat used as rotation crops instead of potato or making a fallow land, were found to be effective in controlling scab disease than continuously cultivating potato (Fig. 2).

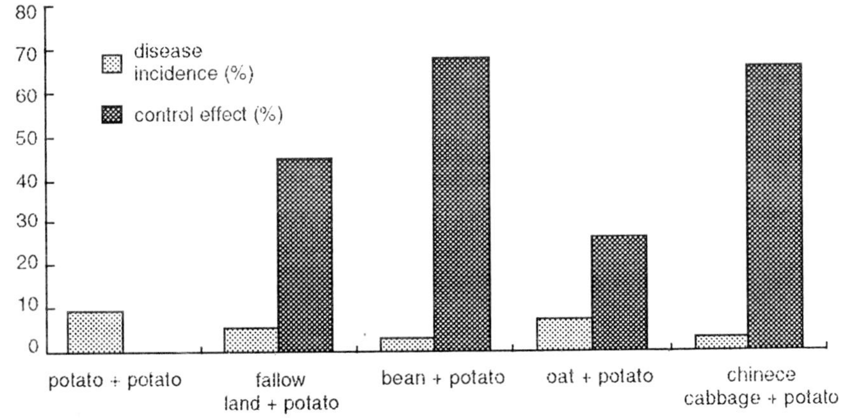

Fig. 2. Control effect of potato scab disease based on crop rotation.

Table 1. Phenotypic characteristics of *Streptomyces* causing potato scab in Korea

Characteristics	S33	S27	S71	S63	S77	S78	*S. s*[a]	*S. t*	*S. a*	*S.c*	*S.e*	*S.st*	*S.r*
Spore colour[b]	G	G	W	YW	PO	W	G	G	W	W	G	G	LG
Chain morphology[c]	S	Rf	Rf	Rf	Rf	Rf	S	Rf	Rf	Rf	S	S	Rf
Colony colour on YME[d]	TB	RY	LY	OB	GO	LY	TB	Y	Br	GB	ND	ND	ND
Melanin on tyrosine agar	+	+	+	+	+	-	+	+	-	-	+	+	-
Melanin on peptone agar	+	-	-	+	-	-	+	+	-	-	+	+	-
Utilization[e]													
L-Arabinose	+	+	+	+	+	+	+	ND	+	-	+	+	+
D-Fructose	+	+	+	+	+	+	+	ND	+	-	+	+	+
D-Glucose	+	+	+	+	+	+	+	ND	+	-	+	+	+
D-Mannitol	+	+	+	+	+	+	+	ND	+	-	+	+	+
Raffinose	+	+	-	+	+	+	+	+	-	+	+	+	+
Rhanmose	+	+	+	+	+	+	+	ND	+	-	+	+	+
Sucrose	+	+	+	+	+	+	+	ND	+	-	+	+	+
D-Xylose	+	+	+	+	+	+	+	ND	+	-	+	+	+
i-Inositol	+	+	+	+	+	+	+	ND	+	-	+	+	+
Minimum growth pH	5.0	4.5	4.5	4.5	3.5	3.5	5.0	4.5	4.0	5.0	ND	ND	ND
Growth with[f]													
5% NaCl	-	-	+	-	+	-	+	-	-	+	ND	-	-
6% NaCl	-	-	+	-	+	-	+	-	-	v	ND	-	-
7% NaCl	-	-	+	-	+	-	-	-	-	v	ND	-	-
Thallium acetate	-	-	-	+	-	-	-	-	+	ND	ND	ND	ND
Thallium acetate	-	-	-	-	-	-	-	-	-	ND	ND	ND	ND
Crystal violet	-	-	+	+	+	-	-	-	+	v	-	-	-
Phenol	-	-	-	-	+	-	-	-	+	+	ND	ND	ND
Penicillin	+	+	+	+	+	+	-	-	+	+	+	+	+
Oleandomycin	-	+	+	-	+	-	-	-	+	+	+	+	+
Oleandomycin	-	-	+	-	+	-	-	-	-	v	ND	-	-
Streptomycin	-	-	-	-	-	-	-	-	+	+	-	-	-

+, positive reaction; -, negative reaction; v, between 20 and 90% of the strains were positive; [a]Data of *S. scabies*, *S. turgidiscabies*, *S. acidiscabies*, *S. caviscabies*, *S. europaeiscabiei*, *S. stelliscabiei*, *S. reticuliscabi* from references Bouchek-Mechiche *et al*; Goyer *et al*; Lambert *et al*; Miyajima *et al*; Lambert *et al.* (2,8,15,17,26). Type strain : S.s, *S. scabies*; S.t, *S. turgidiscabies*; S.a, *S. acidiscabies*; S.c, *S. caviscabies*; S.e, *S. europaeiscabiei*; S.st, *S. stelliscabiei*; S.r, *S. reticuliscabiei*.[b]G, grey; W, white; YW, yellowish white; PO, pale orange; LG, light grey.[c]S, spiral; Rf, rectiflexuous.[d]YME, yeast malt extract; TB, tan brown; RY, red-yellow; LY, light yellow; OB, olive brown; GO, grey-orange; Y, yellow; Br, brown; GB, golden brown.[e]ND, not determined. [f]thallium acetate10μg/ml, 100μg/ml; crystal violet 0.5μg/ml; phenol, 0.1%; penicillin, 10IU/ml; oleandomycin, 25μg/ml, 100μg/ml; streptomycin, 20μg/ml.

TAXONOMY OF STREPTOMYCES

Plant pathogenic *Streptomyces* spp. have been classified by phenotypic characteristics such as colours of spores and substrate mycelium, spore chain morphology, pigment production, utilization of International *Streptomyces* Project (ISP) sugar, and resistance to inhibitory compounds (Lambert and Loria, 1989a; 1989b; Shirling and Gottlieb, 1966). In recent years, phylogenetic analyses using rRNA operon and DNA relatedness, and guanine and cytosine composition have been extensively used for identification at species level (Bouchek-Mechiche *et al.*, 2000; Kreuze *et al.*, 1999; Miyajima *et al.*, 1998; Takeuchi *et al.*, 1996). Therefore, this study classified those 40 pathogenic strains into six distinct clusters based on phenotypic characteristics of selected representative strains (Table 1). The phenotypic characteristics of a member of the first group, strain S33, were identical to those of the type strain of *S. scabies*, except that S33 failed to grow in 5% and 6% NaCl and was resistant to penicillin ($10IU^{-1}$). Analysis of 16S rRNA sequences also indicated that it is most similar to S. *scabies* strains (ATCC 49173^T, ATCC 33282, SNS-26) with sequence identities of more than 99.8%. Phylogenetic analysis also confirmed clustering of S33 with *S. scabies* strains in 100% of bootstrap replicates and, thus, was identified as *S. scabies* (Fig. 3). In addition, analysis of 16S-23S ITS region showed that strain S33 is most similar to *S. scabies* strains (ATCC 49173^T, ATCC 33282) with sequence identity of 98.7% (Fig. 4).

Strain S27, which represents the second group, was identified as *S. turgidiscabies* based on its phenotypic characteristics which were identical to those of type strain ATCC 700248 from Japan, except that S27 did not produce melanin on peptone agar and was resistant to penicillin ($10IU^{-1}$) and oleandomycin ($25ug^{-1}$). 16S rRNA sequences were found 99.5% identical to that of *S. turgidiscabies* strain (ATCC 700248^T) and 16S-23S ITS regions sequences were most similar to those of *S. turgidiscabies* with sequence identity of 94.9%.

Strain S71, which is the representative of the third group, showed similar characteristics to those of *S. acidiscabies*, except that its growth was inhibited below pH 4.5, produced melanin on tyrosine agar, and has varied reaction to several antibiotics and other inhibitory compounds. This strain was identified as most similar to *S. acidiscabies* (ATCC 49003^T), with 16S rRNA sequence identity of 99.9% and 16S-23S ITS regions sequence identity of 86.3%, and these clusters were also supported by 100% of bootstrap replicates (Figs. 3, 4).

Phenotypic and phylogenetic analyses of strains S63, S77, and S78 which were isolated from Je-ju island showed distinct characteristics from the other common pathogens of potato scab. They were characterized as novel species and their detailed descriptions are given below (Park *et al.*, 2003b).

Strain S63, which represents the fourth group, produced yellow-white spores in a rectiflexuous chain, produced melanin on tyrosine or peptone agar, and utilized all the ISP sugars. Growth of this strain was inhibited below pH 4.5 and in the presence of 5%, 6%, and 7% NaCl, thallium acetate (100μg/ml), phenol (0.1% v/v), oleandomycin (25, 100μg/ml), and streptomycin (20μg/ml). Strain S63 was close to *S. setonii* (ATCC 25497^T) and *S. griseous* subsp. *griseous* (KCTC 9080) with 99.3% and 99.0% 16S rRNA sequence identity, respectively. The sequence lengths of the 16S-23S ITS region ranged from 272 to 315bp and considerable sequence variation was observed between species (Fig 4). Strains S63 showed relatively low sequence similarity with other strains, having less than 71.2%. In terms of DNA relatedness, this strain

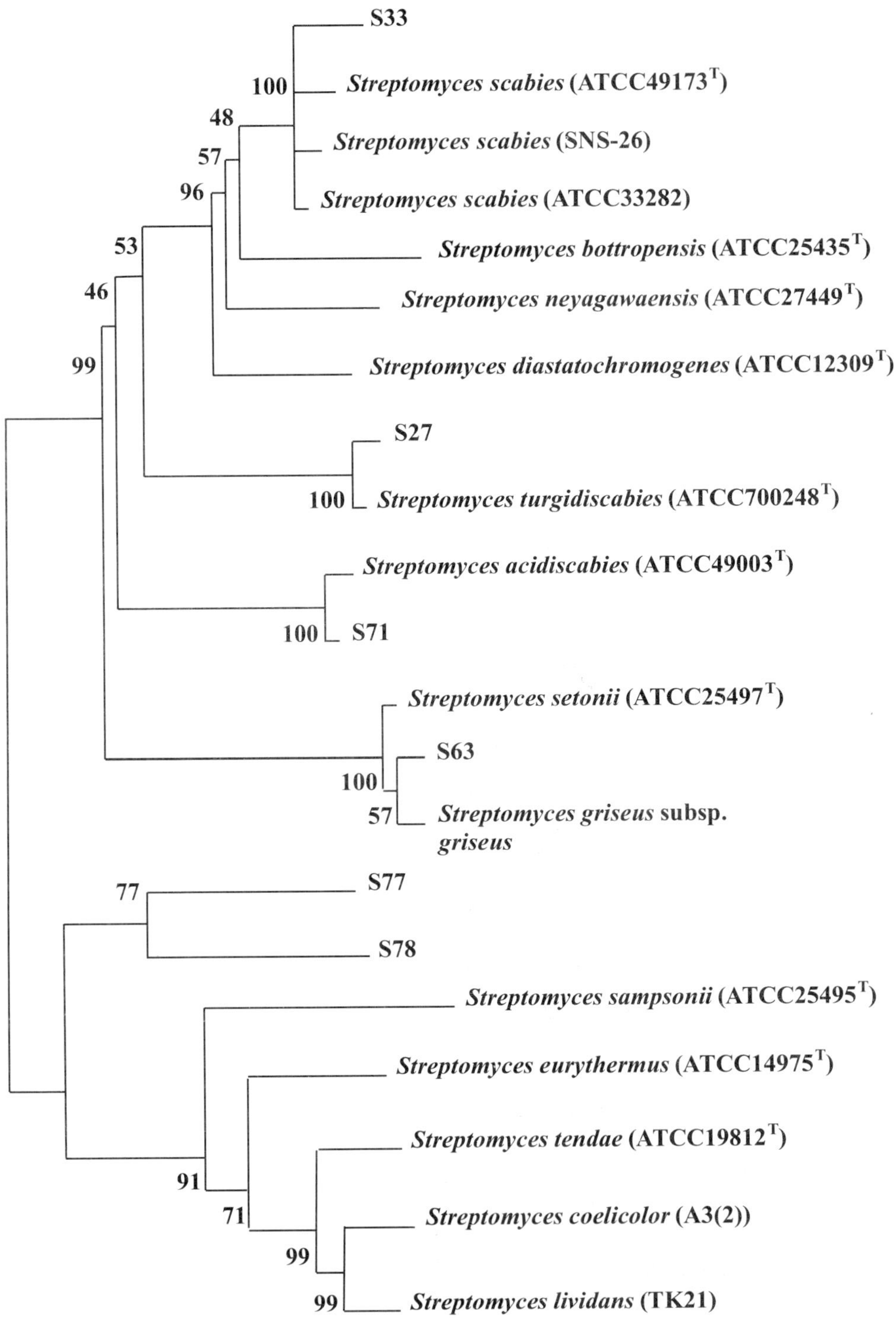

Fig. 3. Phylogenetic tree of *streptomyces* spp. based on the 16S rRNA gene sequences obtained by the neighbor-joining method.

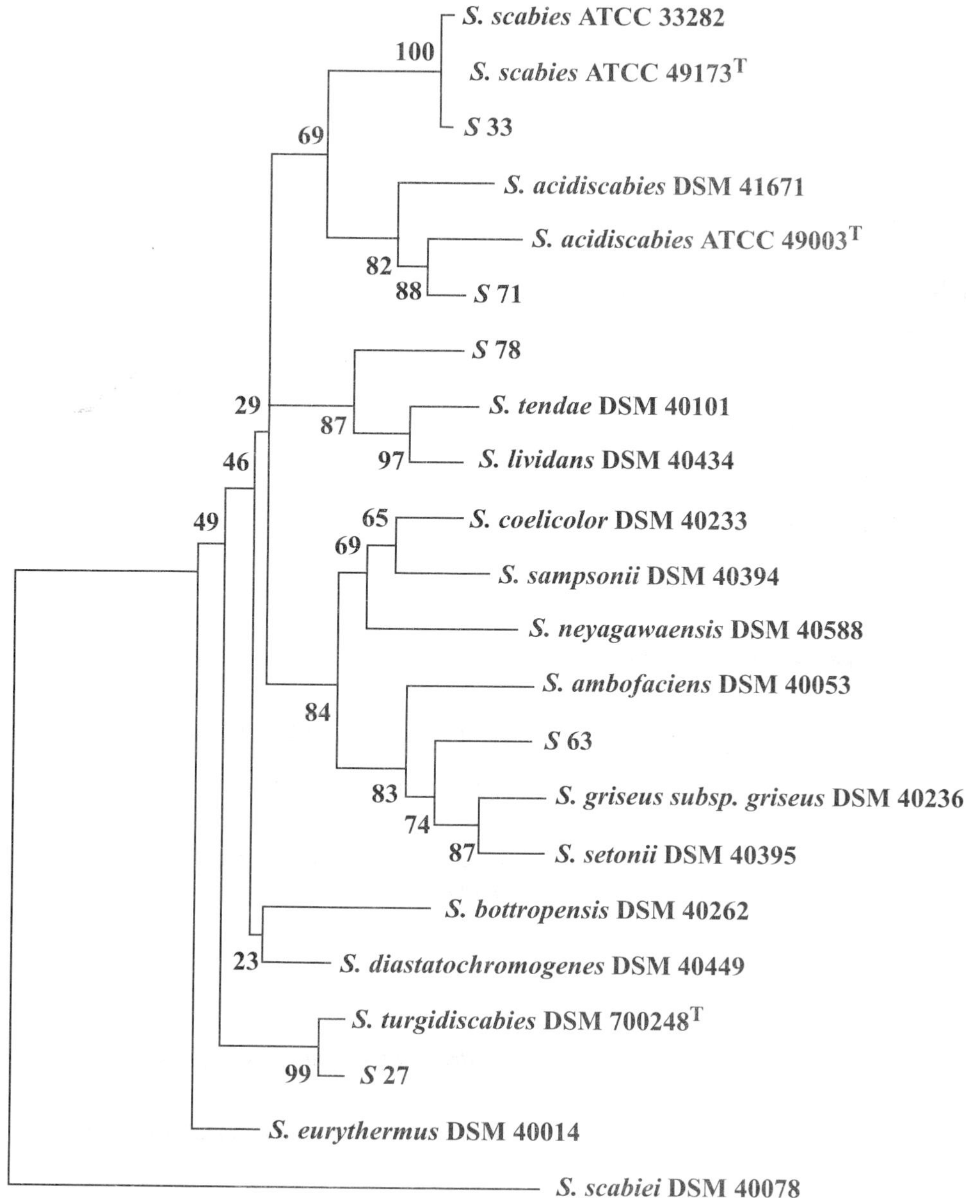

Fig. 4. Phylogenetic tree constructed from a multiple sequence alignment of the 16S-23S ITS region of *Streptomyces* spp. DNA distances were established using the neighbour-joining method (Saitou and Nei, 1987). Bootstrapping values (from 1000 performed bootstrap trials) are given at respective nodes.

was homologous by less than 16%, 16%, and 10% to *S. scabies, S. turgidiscabies*, and *S. acidiscabies*, respectively. Also, this strain showed lower homology with *S. griseus* and *S. setonii* at 10% and 7%. G+C contents was 70.3 mol%. Based on these data, this study proposed the name *S. luridiscabiei* sp. nov. for this strain.

Strain S77, included in group five, had pale orange spores in a rectiflexuous chain, produced melanin on tyrosine agar and utilized all the ISP sugars. The growth was inhibited below pH 3.5 and in the presence of thallium acetate (10, 100μg/ml) and streptomycin (20μg/ml). 16S rRNA sequences showed relatively low homology level (< 96.5%) with the other pathogenic and saprophytic strains, whereas 16S-23S ITS region was not amplified by PCR which made use of the consensus primers AM44 and AM42 in this study. DNA relatedness of this strain was less than 16%, 13%, and 11% to *S. scabies, S. turgidiscabies*, and *S. acidiscabies*, respectively. Also, this strain showed very low DNA relatedness values with other *Streptomyces.* G+C contents which was 68.3 mol% based on those data. Therefore, this study proposed the name *S. puniciscabiei* sp. nov. for this strain.

Strain S78, which represents the final group, had thin and compact spores in a rectiflexuous chain, did not produce melanin on tyrosine or peptone agar and utilized all ISP sugars. The growth was inhibited below at pH 3.5 and in the presence of 5%, 6%, and 7% NaCl, thallium acetate (10, 100μg/ml), crystal violet (0.5μg/ml), phenol (0.1%), oleandomycin (25, 100μg/ml), and streptomycin (20μg/ml). 16S rRNA sequences showed relatively low homology level (< 96.5%) with the other pathogenic and saprophytic strains and 16S-23S ITS region sequence identity also showed low homology (<75.7%). DNA relatedness of this strain was less than 21%, 17%, and 17% to *S. scabies, S. turgidiscabies*, and *S. acidiscabies*, respectively. In addition, this strain showed very low DNA relatedness values with other *Streptomyces.* G+C content was 70.1 mol%. Based on those data, this study proposed the name *S. niveiscabiei* sp. nov. for this strain.

In a nutshell, pathogens of potato scab in Korea were identified as *S. scabies, S. turgidiscabies, S. acidiscabies*, and three novel species (Park *et al.*, 2003a). *S. scabies* (S33) had grey spores in a spiral chain, *S. turgidiscabies* (S27) had grey spores in a rectiflexuous chain, *S. acidiscabies* (S71) had white spores in a rectiflexuous chain, *S. luridiscabiei* (S63) had yellow-white spores in a rectiflexuous chain, *S. puniciscabiei* (S77) had purple-red spores in a rectiflexuous, and *S. niveiscabiei* (S78) had thin and compact white spores in a rectiflexuous chain. Therefore, colour of spores and spore chain morphology of the *Streptomyces* are the main characteristic features for taxonomic position rather than other physiological characters. Moreover, molecular analyses such as 16S rRNA sequences (more than 98% similarity), 16S-23S ITS regions sequences (more than 86% similarity), pairwise hybridization of DNA (more than 70% DNA relatedness), and G+C contents (68-73%) are equally important for correct taxonomy position of any species of pathogenic *Streptomyces.*

PATHOGENICITY

Pathogenicity test for all isolates was performed according to standard Koch's postulates. Generally, pathogenicity test of strains was done several times by whole plant test and tuber slice tissue test and was repeated several times (Loria and Kempter, 1986; Loria *et al.*, 1995). To date, all the species of *Streptomyces* related to potato scab have been reported to produce thaxtomin A of the phytotoxin family. Thus, there was a need to confirm which of the representative strains of the six Korean groups produce thaxtomin A or not from the extracts of oatmeal broth culture. Results showed that representative strains S33, S27, S71, S63, S77, and S78 produced a yellow band that co-migrated with authentic thaxtomin A by TLC analysis

Table 2: Streptomyces strains used in this study, pathogenicity on whole plant (pot) and tuber slices (TB), thaxtomin A production by TLC and HPLC analysis, and hybridization to nec1 and ORFtnp genes

Strain[a]	Pathogenicity[b]		TLC[c]	HPLC[d]		Hybridization[e]	
	Pot	TB		OA	TB	*nec1*	ORF *tnp*
S. scabies							
ATCC 49173[T]	+	+	+	+	+	+	+
S33	+	+	+	+	+	+	+
ATCC33282	NA	NA	NA	NA	NA	NA	NA
SNS-26	NA	NA	NA	NA	NA	NA	NA
S. turgidiscabies							
ATCC 700248[T]	+	+	+	+	+	+	+
S27	+	+	+	+	+	+	+
S. acidiscabies							
ATCC 49003[T]	+	+	+	+	+	+	+
S71	+	+	+	+	+	-	-
S. luridiscabiei							
S63	+	+	+	-	-	-	-
S. puniciscabiei							
S77	+	+	+	-	-	-	-
S. niveiscabiei							
S78	+	+	+	+	+	-	-
S. lividans							
TK21	-	-	-	-	-	-	-

[a]T, type strain of the species; NA, not known or applicable.
[b]Pot, pot test; TB, tuber slices; +, pathogenicity on pot test and tuber slices assay; -, non-pathogenicity on both tests.[c]TLC, thin layer chromatography; +, comigrated spot similar to authentic thaxtomin A; -, no comigrated spot similar to authentic thaxtomin A.[d]HPLC, high pressure liquid chromatography; OA, oatmeal broth culture; TB, tuber slices; +, detection of thaxtomin A; -, no detection of thaxtomin A.[e]+, hybridization; -, no hybridization.

(Table 2). However, thaxtomin A was not detected from chloroform extracts of oatmeal broth culture of *S. luridiscabiei* (S63) and *S. puniciscabiei* (S77) by HPLC analysis. However, it was speculated that the production of thaxtomin A by scab pathogens would require component or signal from potato tissue. Therefore, all six Korean strains were inoculated onto tuber slices and chloroform extracts from the inoculated tuber slices were analyzed by HPLC for the detection of thaxtomin A. Thaxtomin A was not detected from *S. luridiscabiei* (S63) and *S. puniciscabiei* (S77), which clearly indicated that these novel species did not produce thaxtomin A neither in the media nor on the infected potato tissue (Table 2). Therefore, it was proven that the absence of thaxtomin A from those two strains was not due to any artifact. This is the first result showing that some pathogenic strains of *Streptomyces* do not produce thaxtomin A which is a characteristic of potato scab pathogens.

Streptomyces strains having *nec1*, necrosis gene and ORF*tnp*, necrosis gene and transposase pseudogene located 5′ to *nec1*, have also been also reported from pathogens of potato tubers (Bukhalid *et al.*, 1998; Healy *et al.*, 2000). Therefore, this study investigated the relationship between pathogenicity of six strains and the presence of homologous genes to *nec1* and ORF*tnp* genes by PCR and Southern hybridization analysis. Specifically, it was determined whether these two genes exist in all six representative strains by PCR and Southern blotting. In this study, *nec1* gene and the 5′ transposase pseudogene ORF*tnp* were successfully amplified by polymerase chain reaction (PCR) from all studied strains except for *S. acidiscabies* (S71) and

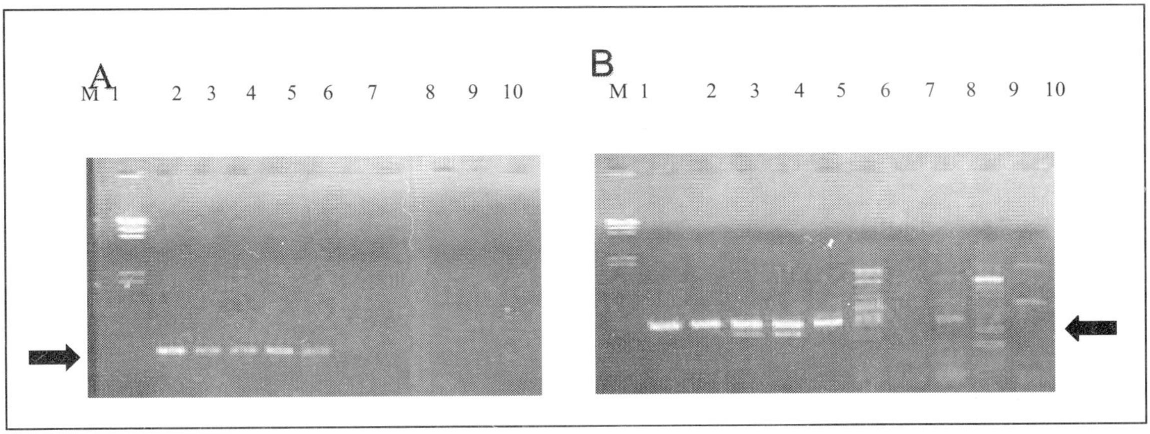

Fig. 5. Polymerase chain reaction amplification of A), nec1 (720bp); and B), ORFtnp (550bp). M: molecular maker (bacteriophage λ DNA digested with HinDIII); lanes 1: *S. scabies* ATCC 49173^{T}; 2: S33; 3: S. *turgidiscabies* ATCC 700248^{T}; 4: S27; 5: *S. acidiscabies* ATCC 49003^{T}; 6: S71; 7: S63; 8: S77; 9: S78; 10: *S. lividans* ATCC 19844. Arrows indicate amplified nec1 (A: 720bp) and ORFtnp (b: 550bp) fragments.

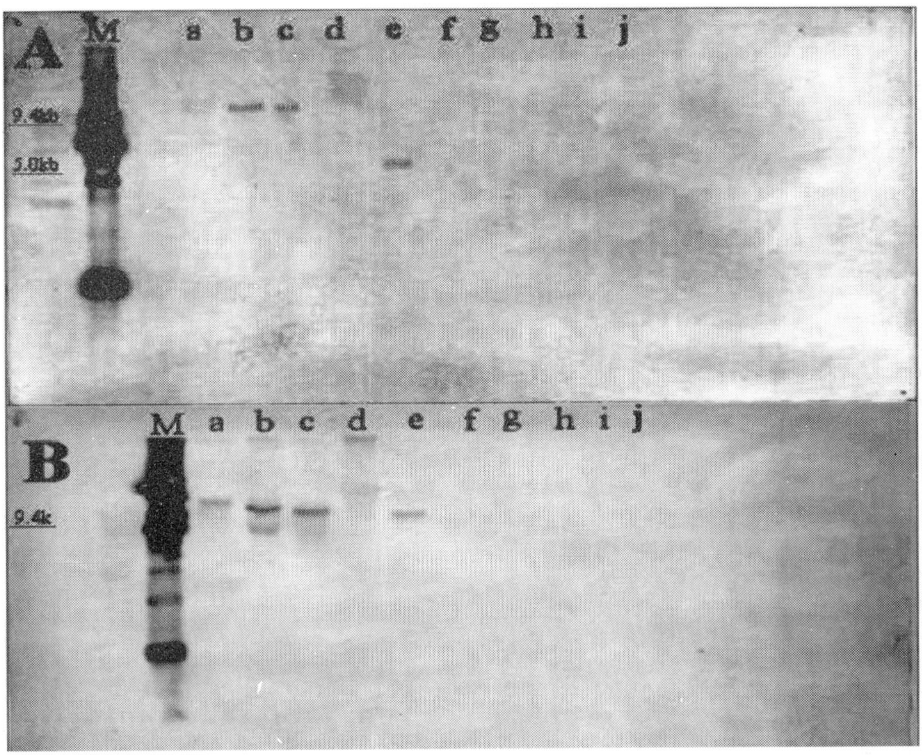

Fig. 6. Southern hybridization analysis of a BamHI digest of representative *Streptomyces* strains with (A) nec1 (720bp) and (B) ORFtnp (550bp) used as DNA probes. Lanes a and b are S. scabies ATCC 49173^{T} and S33; c and d, *S. turgidiscabies* ATCC 700248^{T} and S27; e and f, *S. acidiscabies* ATCC 49003^{T} and S71; g, S63; h, S77; i, S78; and j, *S. lividans* ATCC 19844. Sizes in base pairs are indicated for detected fragment of the nec1 and ORFtnp based on reference (Bukhalid *et al.*, 2002).

three novel Korean strains, *S. luridiscabiei* (S63), *S. puniciscabiei* (S77) and *S. niveiscabiei* (S78) where multiple or no bands were produced (Fig. 5). However, successful amplification by PCR only shows the possibility of homologous DNA sequences matched to primers, not the presence of *nec1* and ORF*tnp* homologs. Thus, to confirm the presence of *nec1* and ORF*tnp* homolog genes in Korean strains, total genomic DNA from all strains was restricted with the endonuclease *Bam*H1 and subjected to Southern hybridization analysis. The *nec1* and ORF*tnp* homologs were present in all strains tested except for *S. acidiscabies* (S71) and the three novel Korean strains, *S. luridiscabiei* (S63), *S. puniciscabiei* (S77) and *S. niveiscabiei* (S78) (Fig. 6). Therefore, this is the first report of *S. acidiscabies* strains lacking a *nec1* gene homolog. Surprisingly, the presence or absence of *nec1* and ORF*tnp* homologs was not linked to thaxtomin A biosynthesis genes in Korean strains *S. luridiscabiei* (S63) and *S. puniciscabiei* (S77). These two novel strains were also found to be morphologically, physiologically and pathologically diverse. Results of this study suggest that these novel pathogens may have a unique mechanism for pathogenicity which might have evolved separately from other known scab pathogens.

REFERENCES

Bouchek-Mechiche, K., Gardan, L., Normand, P. & Jouan, B. (2000) *Int. J. Syst. Evol. Microbiol.* **50**: 91-99.

Bukhalid, R. A., Chung, S. Y. & Loria, R. (1998) *Mol. Plant-Microbe Interact.* **11**: 960-967.

Bukhalid, R. A., Takeuchi, T., Labeda, D. & R., Loria (2002) *Appl. Environ. Microbiol.* **68**: 738-744.

Goyer, C., Faucher, E. & Beaulieu, C. (1996) *Int. J. Syst. Bacteriol.* **46**: 635-639.

Healy, F. G. & D. H. Lambert (1991) *Int. J. Syst. Bacteriol.* **41**: 479-482.

Healy, F. G., Wach, M., Krasnoff, S. B., Gibson, D. M. & Loria, R. (2000) *Mol. Microbiol.* **38**: 794-804.

Kim, J. H. & W. H., Lee (1996) *Kor. J. Plant Pathol.* **12**: 109-115

Kim, J. S., Park, D. H., Choi, Y. C., Lim, C. K., Hong, S. Y., Lee, S. D., Hahm, Y. I. & W. D. Cho (1998a) *Kor. J. Plant Pathol.* **14**: 689-692.

Kim, J. S., Park, D. H., Lim, C. K., Choi, Y. C., Hahm, Y. I. & W. D., Cho (1998b) *Kor. J. Plant Pathol.* **14**: 551-554.

Kreuze, J. F., Suomalainen, S., Paulin, L. & J. P. T., Valkonen (1999) *Phytopathol.* **89**: 462-469.

Lambert, D. H. & R., Loria (1989a) *Int. J. Syst. Bacteriol.* **39**: 387-392.

Lambert, D. H. & R., Loria (1989b) *Int. J. Syst. Bacteriol.* **39**: 393-396.

Locci, R. (1994) *Eur. J. Plant Pathol.* **100**: 179-200.

Loria, R. & B. A., Kempter (1986) *Plant Dis.* **70**: 1146-1148.

Loria, R., Bukhalid, R. A., Fry, B. A. & R. R., King (1997) *Plant Dis.* **81**: 836-846.

Loria, R., Bukhalid, R. A., Creath, R. A., Leiner, R. H., Olivier, M. & J. C. Steffens (1995) *Phytopathol.* **85**: 537-541.

Miyajima, K., Tanaka, F., Takeuchi, T. & S., Kuninaga (1998) *Int. J. Syst. Bacteriol.* **48**: 495-502.

Ndowora, T. C. R., Kinkel, L. L., Jones, R. K. & N. A., Anderson (1996) *Phytopathol* **86**: 138-143.

Paradis, E., Goyer, C., Hodge, N., Hogue, R., Stall, R. & C., Beaulieu (1994) *Int. J. Syst. Bacteriol.* **44**: 561-564.

Park, D. H., Kim, J. S., Cho, J. M., Hur, J. H. & C. K., Lim (2003a) *Plant Dis.* **87**: 1290-1296.

Park, D. H., Kim, J. S., Kwon, S. W., Wilson, C., Yu, Y. M., Hur, J. H. & C. K., Lim (2003b) *Int. J. Syst. Evol. Microbiol.* **53**: 2049-2054.

Saitou, N. & M., Nei (1987) *Mol. Biol. Evol.* **4**: 406-425.

Shirling, E. B. & D., Gottlieb (1966) *Int. J. Syst. Bacteriol.* **16**: 313-340.

Takeuchi, T., Sawada, H., Tanaka, F. & I., Matsuda (1996) *Int. J. Syst. Bacteriol.* **46**: 476-479.

Thaxter, R. (1892) *Conn. Agric. Expt. Sta. Rept.* **1891**: 155-160.

15

Secondary Metabolite-mediated Antagonism against Soil-borne Plant Pathogens

M. Senthil Kumar[1], Bhavesh Kumar[2], P. Dureja[3], and K. Annapurna[4]

[1, 4] *Division of Microbiology, Indian Agricultural Research Institute, New Delhi-110012 (India)*

[2] *Department of Botany and Microbiology, Gurukul Kangri University, Haridwar-249404 (India)*

[3] *Division of Agricultural Chemicals, Indian Agricultural Research Institute, New Delhi-110012 (India)*

Microorganisms constitute a rich source of novel compounds with diverse chemical structures and biological activities. A systematic screening of microbial metabolites using relevant and sensitive detection methods can yield novel compounds with interesting biological profile as the lead molecules for further development into drugs. One of the ecological advantages of microbial agents is that they are highly selective, infecting or killing a very narrow range of target pests. Secondary metabolites produced by a microbe are chemicals which serve as biocontrol agents. i.e chemical defense against pathogens. This article provides an overview of the research carried out across the country on use of secondary metabolites in agriculture, particularly of bacterial origin.

INTRODUCTION

The term secondary metabolite means that the organisms produce these compounds in the late log or stationary phase. Such compounds are not essential for the cell metabolism. Certain strains of microorganisms are important biological components of agricultural soils that are suppressive to diseases caused by pathogenic fungi on crop plants. The biocontrol abilities of such strains depend essentially on aggressive root colonization, induction of systemic resistance in the plant and the production of diffusible or volatile antifungal antibiotics. Evidence that these compounds are produced *in situ* is based on their chemical extraction from the rhizosphere and on the expression of antibiotic biosynthetic genes in the producer strains colonizing plant roots. Fluorescent pseudomonads inhibit phytopathogens by producing secondary metabolites with antibiotic activity eg. phenazines, pyrroles, acetylpholoroglucinols and cyanides (Deepti and Johri, 2003).

Biological control research has gained considerable attention and appears promising as a viable alternative to chemical control strategies (Rebafka *et al.*, 1993; Trigalet *et al.*, 1994). Bio-agents reduce the pathogen attack and its diseases producing activities have induced resistant in host and created an environment for unsuccessful disease development. Different microorganisms including both beneficial and deleterious one, heavily populate the rhizosphere region. Therefore, there is an opportunity to find rhizosphere-competent bacteria which may act as potential biocontrol agent. Bio-agents act against pathogens through antagonism. The activity of biocontrol is not restricted to only one of these and indeed efficient biocontrol agents affect pathogens by a combination of mechanisms (de Boer *et al.*, 2003).

A number of disease suppressive secondary metabolites have been characterized, including antibiotics, siderophores, volatile organic compounds (VOCs), lytic enzymes, bacteriocins and phytoalexins having different mechanisms.

ANTIBIOTICS

An intuitive, simple explanation of how the biological control of soil-borne pathogens could work was discussed at the 1963 international symposium entitled 'Ecology of soil-borne plant pathogens—prelude to biological control' (Baker and Snyder, 1965). The idea was that antagonistic microorganisms could compete with pathogens, particularly by producing antibiotic compounds. In soil, these antibiotics could interfere with pathogen development, for example, during spore germination and the onset of root interaction (Baker and Snyder, 1965). However, there was little evidence forty years ago that antibiotics were actually produced in soil and, if they were, they could account for biocontrol activity. The observation made by Wright in 1956 that the biocontrol fungus *Trichoderma viride* produced detectable amounts of the antifungal antibiotic gliotoxin in soil amended with straw remained unparalleld for many years and the crucial role of antibiotics in the biological control of root diseases remained speculative.

The first clear-cut experimental demonstration that a *Pseudomonas* antibiotic can suppress plant disease in an ecosystem was made by Thomashow and Weller in 1988. They identified a phenazine antibiotic phenazine-1-carboxylic acid (PCA) as biocontrol factor produced by *Pseudomonas fluorescens* 2-79. This strain, originally isolated from wheat, was found to suppress take-all disease caused by the fungal pathogen in *Gaeumannomyces graminis* var. *tritici* (Ggt) on wheat. Pyocyanine, which is derived from PCA in two enzymatic reactions also acts as a biocontrol factor in environmental isolates of *P. aeruginosa* that suppress several plant diseases of fungal origin. The phenazines which are analogous of flavin coenzymes, inhibit electron transport and are known to have various pharmacological effects on animal cells. In the presence of ferripyochelin, phenazines catalyse the formation of hydroxyl redicals, which damage lipids and other molecules (Britigan *et al.*, 1992). Interestingly, reduced phenazine-1-carboxamide can release soluble Fe^{2+} ions from insoluble $Fe^{3+}(OH)_3$ at neutral pH, which resist the possibility that phenazines might contribute to iron mobilization in soils. Pyocyanine, PCA and other phenazines are typical secondary metabolites of bacteria, in particular of pseudomonads (Hassan and Fridovich, 1980; Dieter and Keel, 2003), and are, by definition, dispensable for primary metabolism. However, under appropriate conditions they can be of benefit to the producer organism by contributing to its ecological fitness and competitiveness in natural habitats such as the rhizosphere (Mazzola *et al.*, 1992).

A large number of antibiotics have been reported from different fluorescent pseudomonads including agrocin-84, agrocin-434, 2, 4-diacetylphloroglucinol, herbicolin, oomycin, phenazine, pyoluteorin, pyrrolnitrin, pyrroles etc. They have a role to play in inhibition of pathogens (Colyer and Mount, 1984; Gutterson *et al.*, 1986; James and Gutterson *et al.*, 1986). Fluorescent *Pseudomonas* spp. have been reported to produce 2,4-diacetylfluoroglucinol (Phl) by which they provide biological control of soil borne pathogens on a wide range of crops and have a key role in the suppressiveness of some plant pathogens (Raaijmakers and Weller, 2001; Weller *et al.*, 2002; Validov *et al.*, 2005). Recently, Kang *et al.* (2004) isolated 2-Hydroxy-methyl-chromo-4-one produced by *Burkholderia* sp. MSSP, showing antifungal activities. Antibiotics from rhizobia are not new as several antibiotics have been reported such as anthranilic acid from *R. leguminosarum* (Rioux *et al.*, 1986), 4,3-Hydroxy-2,3,9-trimethoxy pterocarpan phytoalexin from *R. leguminosarum* bv *viciae*, trifolitoxin (TFX) from *R. leguminosarum* bv *trifoli* T24 (Chakraborty and Chakraborty, 1988) and rhizobitoxin from *Bradyrhizobium* spp. (Chakraborty and Purkayastha, 1983; Deshwal *et al.*, 2003).

A minor class of antifungal compounds that does not have nitrogen, 2, 4-di-acetylpholoroglucinol (Phl), has been studied extensively in the context of biological control of take-all disease of wheat. The exact mechanism of Phl action is still partially known, although it is postulated that disease suppression by this antifungal molecule is a result of interaction of specific root-associated microorganisms and the pathogen. Phl is the best-known pholoroglucinol compound in the family of related molecules that includes monoacetylpholoroglucinol and uncharacterized condensation products of Phl and monoacetylpholoroglucinol. Phl causes membrane damage to *Pythium* spp. and is particularly inhibitory to zoospores of this oomycete (de Souza *et al.*, 2003).

Another antibiotic produced by *Psuedomonas* genus is pyoluteorin, an aromatic polyketide consisting of a resorcinol ring which is derived through polyketide biosynthesis. This in turn is linked to a biochlorinated pyrrole moiety, whose biosynthesis remains unknown (Keel *et al.*, 1992; Kitten, 1998; Nowak-Thompsan *et al.*, 1999; Cuppels *et al.*, 1986). Pyoluteorin mainly inhibits the oomycetous fungi, especially *Pythium ultimum*. Seeds treated with the antibiotic show a decrease in the *Pythium* mediated damping off disease. Pyrrolnitrin has been used as an antimycotic topical antibiotic in human medicine and synthetic analogous of pyrrolnitrin has been developed for use as agricultural fungicides (Ligon *et al.*, 2000).

VOLATILE ORGANIC COMPOUNDS (VOCs)

Another typical secondary metabolite produced by certain pseudomonads, hydrogencyanide (HCN) (Castric, 1994), was recognized as a volatile antibiotic and biocontrol factor of *P. fluorescens* CHA0 by Voisard *et al.* (1989), who found that the suppression of black root-rot caused by the fungus *Thielaviopsis basicola* on tobacco depended, in part, on the production of HCN by the biocontrol strain CHA0. Discovered in 1782 by Scheele, HCN was soon identified as a potent poison, although in the early nineteenth century diluted aqueous solutions of HCN were recommended as a treatment of respiratory diseases (Dieter and Keel, 2003). The mode of action of HCN in the disease control is partially understood (Haas and Defago, 2005) but it is believed that cyanide ions derived from HCN, are the potent inhibitor of many matelo-enzymes, especially copper-containing cytochrome *c* oxidase (Blummer and Haas, 2000) which are critical in respiration of living beings. The other fungi are also inhibited by the action of HCN because HCN negative organisms, including fungal pathogens, do not resist the HCN activity (Glick *et al.*, 1998) hence are inhibited easily. Much later, HCN was found to be a potent virulence factor of *P. aeruginosa* eg. in a nematode assay (Blummer and Haas, 2000; Gallagher and Manoil, 2001; Gallagher *et al.*, 2002). *Septoria tritici* blotch and leaf rust of wheat were suppressed by cyanide producing strains of *P. putida* (Flaishman *et al.*, 1996). HCN producing fluorescent strains of *P. aeruginosa* GRC1 and *P. aeruginosa* GRC2 (Gupta *et al.*, 1999, 2001; Bhatia *et al.*, 2003; 2005) inhibited charcoal rot of peanut and sclerotial rot of *Brassica* spp. caused by *M. phaseolina* and *Sclerotiorum sclerotonium*. However, rhizobia do contain low amount of HCN in comparison to that of fluorescent pseudomonads and only 12.5% and 3% strains were found to be HCN producers by Beauchamp *et al.* (1991) and Antoun *et al.* (1998) respectively. Cyanides play a role in the suppression of root-rot in tobacco caused by *Thilaviopsis basicola* (Ahl *et al.*, 1986). Cyanide produced by rhizosphere pseudomonads has detrimental effect on plant establishment in several crops but it also suppresses root pathogens (Defago *et al.*, 1990).

In addition to HCN, the production of volatile organic compounds (VOCs) in liquid cultures has been discovered. Ryu *et al.* (2003) isolated several VOCs including 3-hydroxy-2-butanone, 2,3-butanediol, decane, tetramethyl pyrazine, undecane, decanal, dodecane, 2-undecanone, 2-tridecanone and 2-tridecanol. Of them, two VOCs viz., 2,3-butanediol and acetoin were released exclusively from two biocontrol agents (*Bacillus subtilis* GB03 and *B. amyloliquefaciens* IN937a) that trigger the induced disease resistance. Pharmacological applications of 2,3-butanediol induced resistance, while bacterial mutants blocked in the synthesis of 2,3-butanediol and acetoin were devoid in induced resistance capacities. Using transgenic and mutant lines of *Arabidopsis*, Ryu *et al.* (2003) provide evidence that the signal pathway activated by volatiles on an ethylene signalling pathway induced pathogen resistance.

SIDEROPHORES

Siderophores are low molecular weight (~ 400–1000 Da) iron-chelating secondary metabolites that bind Fe^{3+}, and transport it back to the cell and make it available for the microbial cells (Fig. 1). The secreted siderophore molecules have a very high affinity (kd = 10^{-20} to 10^{-50}) for iron and bind most of the Fe^{3+} that is available in the rhizosphere and prevent the pathogens present in immediate vicinity from proliferation because of lack of iron (O'Sullivan and O' Gara, 1992). Antagonists can prevent the proliferation of fungal phytopathogens by producing siderophores that bind most of the Fe^{3+} in the rhizosphere. The resulting lack of the iron prevents any fungal pathogen from proliferating in this immediate vicinity (Fig. 1). Kloepper *et al.* (1988) have supported this mechanism and stated that the production of siderophores that chelate and thereby scavenge the ferric iron in the rhizosphere may result in growth inhibition of other microorganisms whose affinity for iron is lower. It has been suggested that the ability to produce specific siderophores, and/or to utilize a broad spectrum of siderophores may contribute to the root colonization ability of biocontrol strains.

Siderophore production has been reported by many researchers in different groups of plant growth-promoting bacteria such as rhizobia (Arora *et al.*, 2001; Saraf and Sood, 2002 and Deshwal *et al.*, 2003), pseudomonads (Gupta *et al.*, 2002; Bhatia *et al.*, 2003), *Azotobacter*

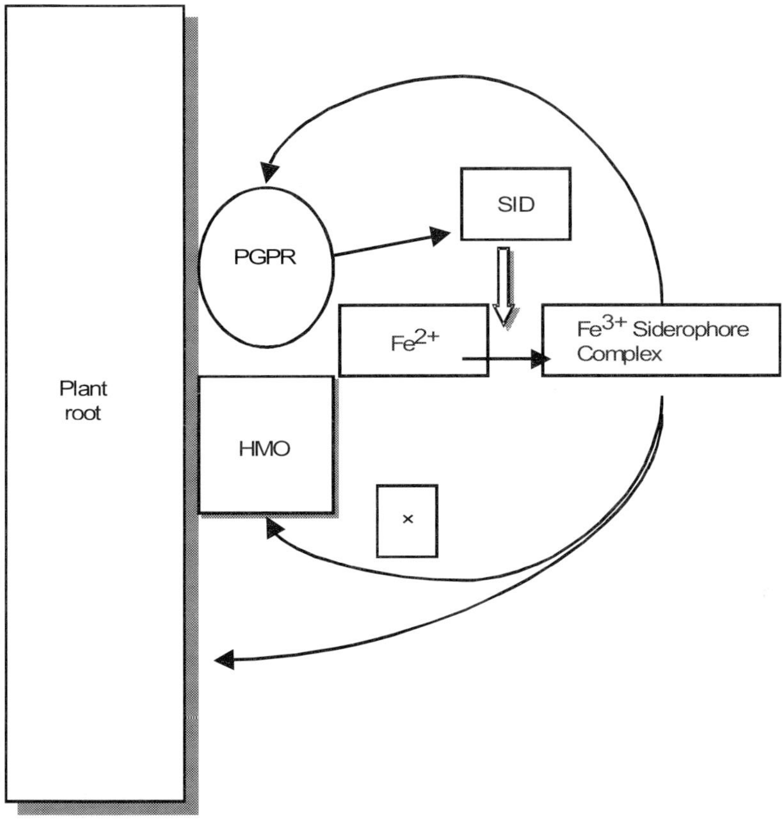

Fig. 1: Mechanism of siderophore action as biocontrol agent and nutrient to the plants. Abbreviations: PGPR, plant growth-promoting rhizobacteria; SID, siderophore; HMO, harmful microorganisms.

(Budzikiewicz *et al.*, 1992), *Azospirillum* (Shah *et al.*, 1992), *Burkholderia* sp. (Pandey *et al.*, 2005). Liles *et al.*, (2000) discovered that *Legionella pneumophila* (aquatic bacteria) produced a non-classical siderophore legiobactin and recently it was suggested that siderophore production was conserved among the members of legionellae (Starkenburg *et al.*, 2004). Various authors have described the role of siderophore in the inhibition of several fungal pathogens such as *M. phaseolina* etc. (Arora *et al.*, 2001; Deshwal *et al.*, 2002; Bhatia *et al.*, 2003; Pandey *et al.*, 2005). There are different types of siderophores such as hydroxamate, catechol, pyoverdin, etc, reported in different groups of microorganisms (Table 1).

Bano and Musarart (2003) characterized a *Pseudomonas aeruginosa* strain NJ-15 as a potential biocontrol agent inhibiting in varying degrees *F. oxysporum*, *T. harzianum*, *A. alternata* and *M. phaseolina*. The strain produced indole acetic acid (IAA) in tryptophan supplemented medium and exhibited a significant production of both the siderophore and hydrogen cyanide. However, the study did not conclusively prove if the antifungal activity was because of one or both mechanisms.

Table 1: Types of siderophores produced by different PGPR and rhizobia.

Siderophore type	*PGPR*
Catechol	*Rhizobium* RA-1
	Azospirillum lipeferum M
	Bradyrhizobium sp.
Hydroxamate	*S. meliloti*
	P. aeruginosa GRC_2
	Pseudomonas PS1 and PS2
Ferribactin	*P. fluorescens*
Pyoverdin	*P. aeruginosa*
Isopyoverdin	*P. putida*
Azotobactin	*Azotobacter vinelandii*
Aerobactin	*Aerobacter aerogenes*
Enterochelin	*E. coli*
Enterobactin	*Salmonella typhimurium*
Pyochelin	*P. aeruginosa*
Quinolobactin	*P. fluorescens*
Rhizobactin	*S. meliloti*
	R. leguminosarum
Hydroxamate	*R. leguminosarum* bv trifolii
Anthranalate	*R. leguminosarum* bv viciae
Vicibactin	*R. leguminosarum* bv viciae
Phenolate/2,3 DHBA	*Rhizobium* sp.
Rhodotorulate	*R. leguminosarum* bv viciae MNF 7104

Studies from Gnanamanickams laboratory (Velusamy and Gananamanickam, 2003) has shown the production of 2, 4-diacetylphloroglucinol by plant-associated bacteria and its role in suppression of rice bacterial blight. Out of the 637 fluorescent bacterial strains screened, 278 isolates showed antibiosis against the pathogen *Xanthomonas oryzae* pv. *Oryzae* (*Xoo*). PCR amplification of the Phl gene gave positive results only in 27 of these isolates. No PCR products were amplified from the non-producing strains. Confirmation of the sub-set of 27 positive samples was done by co-chromatography of ethyl acetate extracts of the cultures. The purified antibiotic Phl was further screened for its inhibition of *Xoo*.

After extensive screening of fluorescent pseudomonads strain PUPa3 that exhibited broad-spectrum antifungal activity towards several phytopathogenic fungi. Strain PUPa3 produced hydroxamate-type siderophore, IAA and phosphatase. It did not produce HCN. None of the genes coding for *Phz, Phl, Plt, Prn* were detected. Crude extract obtained from PUPa3 was brownish in colour which on further purification gave a greenish-yellow metabolite exhibiting broad-spectrum antifungal activity. The purified compound deformed the mycelia and inhibited spore germination of *Macrophomina phaseolina*. This greenish-yellow compound with a melting point of 246°C was characterized to be PCN (phenazine-1-carboxamide) based on spectral data.

BACTERIOCINS

Bacteriocins are proteic molecules synthesized for various lineages of Gram-positive and Gram-negative bacteria when exposed to stressful conditions. Bacteriocins have been characterized as molecules of high antimicrobial property even at low concentrations, provoking the pathogenic inhibition by antibiosis. These substances have their synthesis mediated for genetic mechanisms and develop their lethal action on the microbial cell by multiple mechanisms that can act of isolated or concomitant way culminating with microbial cell killing. This secondary metabolite class presents characteristic of stability to heat, low pH, refrigeration and freezing, and resistance to weak organic solvents, salts and enzymes. On the other hand, they are very sensitive to proteolytic enzymes action. Bacteriocins could appear as potential agents to be applied in biocontrol systems.

In the mechanism, bacteriocins are adsorbed to the receptor on the surface of the bacterial cells, eliminate electric charge of the inner membrane and stop biosynthesis of ATP. Consequently, biosynthesis of cell biomolecules viz., DNA, RNA, proteins and phospholipids is disturbed, resulting in inhibition of bacterial cells. Eventually, it may cause death of the cell. Bacteriocin-mediated biocontrol was first reported by Ryder and Jones (1990). In their report, *Agrobacteriaum radiobacter* K84 found to produce a bacteriocin i.e. agrocin 84 which controlled the crown gall disease caused by *A. tumefaciens*. The other example of bacteriocin-mediated inhibition of plant pathogens is the production of the metabolite namely, fluoricin BC-8 by an antagonistic bacteriaum *P. fluorescens* BC-8. Recently, Parret *et al.* (2005) have reported novel lectin-like bacteriocins (LlpA) of biocontrol strain *Pseudomonas fluorescens* Pf-5. Antibacterial activity of the recombinant proteins was evidenced by gel overlay assays. Analysis of the antimicrobial spectrum indicated that LlpA is able to inhibit pathogenic *P. fluorescens* strains, as well as the related mushroom pathogen *Pseudomonas tolaasii*.

PHYTOALEXINS

Phytoalexins are low molecular weight flavonoid-like antimicrobial metabolites produced in plants in response to stress or bacterial/fungal invasion. These metabolites have also been considered to be involved in mechanism of plant disease control. Accumulation of these metabolites in carnation plants has been reported with *Pseudomonas* sp. WCS417 (van Peer *et al.*, 1991). Paul *et al.* (2005) isolated a bacterial strain from soil, belonging to the genus *Bacillus* was found to be an antagonist of this disease causing fungus. The fungal attack on the grapevine acts as an elicitor to the production of phytoalexins like resveratrol. This compound was also formed when the leaves of the grape plants were inoculated with the bacteria alone, and this activity was enhanced when a mixture of the pathogen and the antagonist bacteria was applied. Since, resveratrol in wine is considered to be beneficial to human health provided moderate consumption, this bacteria can be used as a potential biological control agent as well as a biological elicitor of resveratrol. The article includes the details of the fungal parasite, its biological control and resveratrol elicitation.

OTHER METABOLITES

In addition to secondary metabolites, some other metabolites eg. lytic enzymes have also been proven their role in the biological control of soil borne plant pathohens. Recently, the hydrolytic enzymes have received considerable attention because they play a role in controlling diseases due to plant pathogens. The enzymatic digestion or deformation of cell wall components of phytopathogenic fungi occur by the enzymes chitinase and β-1,3-glucanase of several plant growth-promoting bacteria. Microorganisms capable of lysing other organisms are widespread in natural ecosystems. The hyperparasitism and lysis of propagules in soil are some logically methods of biological control of soil-borne plant pathogens (Vaidya *et al.*, 2001). Hyperparasitism occurs when a fungus exists in intimate association with another fungus which derives some nutrients while conferring no benefit in return. Hyperparasitism and lysis of propagules in soil is a logically satisfying method of biological control of soil-borne plant pathogens by microbial antagonists. The production of cell-wall degrading or lytic enzymes (Table 2) such as chitinase (EC 3.2.1.14), β-1,3-glucanase (EC 3.2.1.39), proteases (EC 3.4.21.14) and lipase (EC 3.1.1.3) (Fridlender *et al.*, 1993; Lim and Kim, 1995; Dunne *et al.*, 1997; Vaidya *et al.*, 2001; Vivekanathan *et al.*, 2004) degrade fungal cell-walls resulting into lysis of wall material leading to cell death. Chitinase and â-1,3-glucanase are the main enzymes involved in degradation of fungal cell-walls since chitin and β-1,3-glucan are the major components of most fungal cell-walls (Henis and Chet, 1975; Lim and Kim, 1995). To date, chitinase has been the most investigated lytic enzyme, and its importance in natural ecosystems has been proven by cloning its gene into non-chitinolytic bacteria.

Bacillus subtilis has been used for many years in attempts to control plant pathogens and increase plant growth (Podile *et al.*, 1985; Podile and Dube, 1988; Podile, 1995; Podile and Prakash, 1996). One of the major mechanisms involved in the bacterial antagonism is parasitism via degradation of the cell-wall. Chitin, the un-branched homopolymer of N-acetyl glucosamine in a β-1, 4 linkage is a structural component of cell-walls in most of the fungi (other than oomycetes). *B. subtilis* strain AF-1 decreased wilt incidence in pigeon pea (Harish *et al.*, 1998) which was due to the production of antifungal substances *i.e.* chitinase. A similar study by Basha and Ulaganathan (2002), showed the antagonistic effect of *Bacillus* strain BC121 against *Curvularia lunata* to be because of production of chitinase. The strain isolated from the rhizopshere of sorghum, gave a zone of inhibition of 0.5-1 cm in dual plate assay. Scanning electron microscopic observations showed a clear hyphal lysis and degradation of fungal cell-wall. The strain produced a clear halo zone on chitin agar medium (0.5% colloidal chitin). A negative mutant of this strain lacking chitinolytic activity seems to have lost its antagonistic ability. This proved that the antagonistic activity of strain BC121 was due to its chitinase production.

Table 2: Different types of lytic enzymes produced by different plant growth-promoting.

Lytic Enzymes	Antagonists	Fungal Pathogen
Chitinase & β-1,3-Glucanase	*Pseudomonas stutzeri*	*Fusarium solani*
Chitinase & β- 1,3-Glucanases	*Alcaligens xylosoxydane* *Pseudomonas cepacin*	*Rhizoctonia bataticola, Fusarium* sp. *Rhizoctonia solani, Sclerotium rolfsii, Pythium ultimum*
Chitinase	*Serratia marcescens*	*S. rolfsii*
Chitinase & β-1,3-Glucanase	*Stachybotrys elegans*	*R. solani*
Protease	*Stenotrophomonas maltophilia*	*Pythium ultimum*

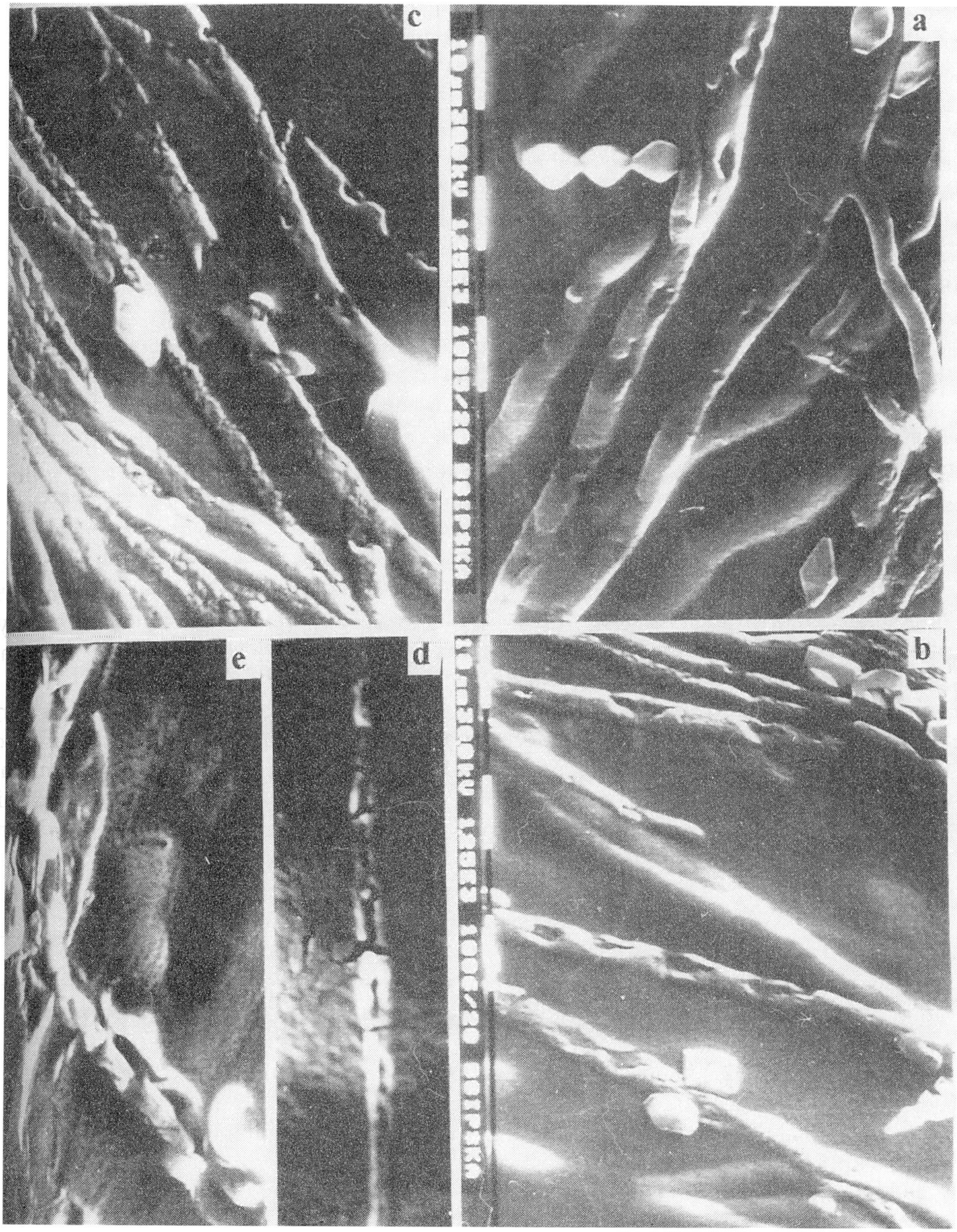

Fig. 2: Scanning electron photomicrographs of post interaction events in *Sclerotinia sclerotiorum* due to *Pseudomonas* GRC_2, (a), hyphal perforations during early stage of growth and (b) during late stage of growth; (c) hyphal lysis; (d, e) hyphal fragmentation.

In our particular observations, *P. aeruginosa* LRP7 released extracellular chitinase and β-1, 3-glucanase enzymes, which are key enzymes in the decomposition of fungal hyphal wall. The action of lytic enzymes by *P. aeruginosa* LRP7 was quite effective on *M. phaseolina* and *F. oxysporum* because their substrate chitin is involved in cell-wall composition. The cell-wall of the *F. oxysporum* composed of 47% chitin and 14% laminarin (Skujins *et al*, 1965). Therefore, it is likely that cell-wall lysis caused by concerted action of both chitinase and β-1, 3-glucanase. Microscopic abnormalities in fungal pathogens under the influence of *P. aeruginosa* LRP7 clearly showed degeneration and degradation of the hyphae of *M. phaseolina* and *F. oxysporum*. Abnormal hyphal swelling, digestion of fungal cell-wall, cytoplasm coagulation, hyphal perforation, hyphal tip degradation and halo cells formation (Fig. 2) appeared in mycelia of *F. oxysporum* were caused by *P. aeruginosa* LRP7 mediated chitinolytic enzymes. In *M. phaseolina* hyphal abnormalities such as abnormal hyphal growth (zig-zag appearance of hyphae) hyphal swelling, mycelial digestion, halo cell formation, degradation of mycelial tips and loss of structural integrity of sclerotia were also observed. Thus, it showed the hyphal wall interaction with bacterial cells rapidly lysed the hyphae, causing leakage of the fungal cytoplasm. Further, chitinase deficient mutant of LRP7 i.e. LRP7-P7 later confirmed the involvement of enzymes, because during its interaction no major hyphal deformities occurred against *M. phaseolina* and *F. oxysporum*. Thus, role of lytic enzymes in the digestion of the cell-wall component of *M. phaseolina* and *F. oxysporum* is established. Although, the degree of fungal inhibition was recorded but still it was not similar to that obtained by wild type *P. aeruginosa* LRP7. Gupta *et al.* (2001) also demonstrated the loss of sclerotial integrity, hyphal shriveling, mycelial and sclerotial deformities and hyphal lysis in *M. phaseolina* by *Pseudomonas* GRC_2.

Globally interest in antibiotics from soil bacteria and in the regulatory mechanisms governing the production of these secondary metabolites has grown during the past two decades, essentially because of their demonstrated ecological function in the suppression of plant root diseases. In India, experiments using bacteria for biocontrol of fungal pathogens have been undertaken, however without much research into understanding its mechanism of action. In contrast, much more is known about the secondary metabolites of fungal origin which have been isolated, identified and their structures formulated (Dipanker *et al.*, 2003, Shyamli *et al.*, 2003). *Trichoderma harzianum* Rifai is the most prominent hyphal fungi used as biocontrol agent and extensively exploited by plant pathologists due to its high efficacy broad-spectrum activity and ease in isolation and cultivation (Chet, 1987).

Much work in India has been concentrated on the use of bacterial cell or its extract for *in vitro* antagonistic studies (Table 3). *Rhizobium* (Balasundaram and Sarabhoy, 1988; Gupta *et al.*, 2001) and other bacteria have been used as antagonistic agent against some plant pathogenic fungi(Ramesh *et al.*, 2002;) *In vivo* biocontrol studies with the plant has been carried out using the bacterial culture as seed treatment mainly under pot conditions (Pal *et al.*, 2000; Deshwal *et al.*, 2003). Large scale field trials using the bacterial biocontrol agents have not been taken up with the exception of *Bacillus thuringiensis*.

In our laboratory, we had initiated work on isolation and identification of secondary metabolites showing antifungal activities against a number of soil-borne plant pathogenic fungi. Of the 204 bacterial isolates screened against the fungi in *in vitro*, only 12 (none is a fluorescent pseudomonad) showed inhibition of the fungi. Interestingly, none of the seven fluorescent pseudomonads from the 204 isolates screened inhibited the fungal pathogens. Since

Table 3: Commercially available bacterial inoculants for control of economic agricultural crops.

Antagonist	Product	Target pathogen	Intended crop	Country of origin
Pseudomonas cepacia	Blue circle	*Rhizoctonia solani, Pythium* sp.	Vegetable crops	Stine Microbials, Madison WI (USA)
Pseudomonas fluorescens	Dagger G	*Rhizoctonia solani, Pythium ultimum*	Cotton	Ecogen Inc, Langhorne PA (USA)
Bacillus subtilis	Epic	*Rhizoctonia solani, Fusarium* sp.	Cotton, legumes	Gustafson Inc, Dallas (USA)
Agrobacterium radiobacter	Galltrol, Norbac 84-C	*Agrobacterium tumefaciens*	Trees, shrubs	Ag Biochem, Inc. Urinda CA (USA)
Bacillus subtilis	Kodiak	*Pythium ultimum, Rhizoctonia solani*	Cotton, legumes	Gustafson, Inc. Dallas TX (USA)
Streptomyces griseovirides	Myostop	*Alternaria* sp., *Fusarium* sp.	Vegetable, ornamentals	Kermira Biotech, Helsinki, (Finland)
Pseudomonas fluorescens bv 5	Victus	*Phytopithora tolaasii*	Mushrooms	Sylvan, Kattanning, PA (USA), Millar, FC; Sylvan, PC)
Agrobacterium radiobacter	Bakuterohzu	Crown gall disease	Rose	Japan
Pseudomonas fluorescens	Bio Coat	*Fusarium* sp.	Radish	Australia
P. fluorescens	Bioshiel		Cereals, pulses and vegetables	Anu Biotech International Ltd. Faridabad (Hariyana)
Bacillus subtillis	Biotok	*Coriticum invisium, C. theae*	Tea	Tockalia Exoerimental Station, Tea Research Association, Jorhat (Assam)
P. fluorescens	Pant Biocontrol Agent - 3	Causative pathogens of soil borne diseases	Pulses and vegetables CROPS	G.B.Pant University, PantNagar (Uttaranchal)

all 12 isolates did not produce HCN and were negative for phenazine and Phl production, we are concentrating on identification of polar compounds.

CONCLUSION

'Suppressive soils' contain rhizobacteria that are able to control plant diseases that are caused by fungi or bacteria. The mechanisms responsible for this biocontrol activity include competition for nutrients, niche exclusion, induced systemic resistance (ISR) and the production of anti-fungal metabolites (AFMs). Most of the identified *Pseudomonas* biocontrol strains produce AFMs, of which phenazines, pyrrolnitrin, 2, 4-diacetylchloroglucinol (Phl) and pyoluteorin are most frequently detected classes. However, new AFMs belonging to the class of cyclic lipopeptides, such as viscosinamide (Nielsen *et al.*, 1999) and tensin (Nielsen *et al.*, 2000) have been discovered. In India, biocontrol work with bacteria has concentrated using the cells directly or the cell extract. Serious attempts at isolating these metabolites from cells have not been made. The 6th International workshop on Plant Growth Promoting Rhizobacteria held from 5-10 Oct 2003 in Calicut, India, showcased the various groups of people working in biocontrol area. It showed the large potential of microbes in controlling various plant pathogens predominant being fluorescent pseudomonads. However, we must explore other bacteria for their potential in biocontrol. Also the focus should be on isolating and identifying the secondary metabolites inhibitory against pathogens. This will have industrial implications also. Certainly bacteria are a great source of novel products which can be used as biocontrol agents. These secondary metabolites are the resource pool for developing eco-friendly mechanisms of pathogen control and enhancing crop productivity.

REFERENCES

Ahl, P., Volsard, C. & Defago, G. (1986) *J. Phytopathol* **116**: 121-134.

Antoun, H., Beauchamp, C.J., Goussard, N., Chabot, R. & Lalinde, R. (1998) *Plant Soil* **204**: 57-67.

Arora, N.K., Kang, S.C. & Maheshwari, D.K. (2001) *Curr. Sci.* **81**: 673-677.

Baker, K.F. & Snyder, W.C. (1965) In *Ecology of soil borne plant pathogens: Prelude to biological control.* University of California Press, Berkeley.

Balasundaram, V.R. & Sarabhoy, A.K. (1988) *Ind. Phytopathol.* **41**: 128-130.

Bano, M. & Musarrat, J. (2002) *Curr. Microbiol.* **43:** 182-186.

Basha, S. & Ulaganathan, K. (2002) *Curr. Sci.* **82**: 1457-1462.

Beauchamp, C. J., Dion, P., Kloepper, J.W. & Antoun, H. (1991) *Plant Soil.* **132**: 273-279.

Bhatia, S., Dubey, R.C. & Maheshwari, D.K. (2005) *Indian Phytopathol.* **58:** 17-24.

Bhatia, S., Bhaita, S., Dubey, R.C. & Maheshwari D.K. (2003) *Ind. J. Exp. Biol.* **41**: 1442-1446.

Blummer, C. & Haas, D. (2000) *Arch. Microbiol.* **173**: 170-177.

Chakraborty, U. & Chakraborty, B.N. (1988) *Can. J. Bot.* **67**: 1698-17.2.

Chakraborty, U. & Purkayastha, R.P. (1983) *Can. J. Microbiol.* **30**:285-289.

Chet, I. (1987) In *Innovative approaches to plant disease control* (Ed. Chet, I.). John Wiley & Sons, New York, pp. 137.

Colyer, P.D. & Mount, M.S. (1984) *Plant. Dis.* **68**: 703-706.

Cuppels, D.A., Howell, C.R., Stipanovic, R.D., Stossel, A. & Stothers, J.B. (1986) *Z. Naturforsch.* **41**: 532-536.

de Boer, M., Bom, P., Kindt, F., Keurentjes, J.J.B., van der Sluis, I *et al.* (2003) *Phytopathol.* **93**: 626-632.

de Souza, J.T. et al. (2003) *Phytopathol.* **93**: 966-975.

Deepti, D. & Johri, B.N. (2003) *Curr. Sci.* **85:** 1693-1703.

Defago, G., Berling, C.H., Burger, U., Hass, D., Hahr, G., Keel, C., Voisard, C., Wirthner P.H. & Wutrich, B. (1990) In *Biological control of soil-borne plant pathogens.* (Ed. Hornby, D.), CAB international, pp. 93-108

Deshwal, V.K., Pandey, P., Kang, S.C. & Maheshwari, D.K. (2003) *Ind. J. Exp. Biol.* **41**: 1160-1164.

Dipanker, P., Dureja, P. & Somayajulu, K.K. (2003) *Pesticide. Res. J.* **15**: 151-154.

Dunne, C., Moenne-Loccoz, Y., Mccarthy, J., Higgins, P., Powell, J., Dowling, D.N., and O'Gara, F. (1998) *Plant Pathol.* **47**: 299-307.

Flaishman, M.A., Eyal, Z., Zilberstein, A., Voisarsd, C. & Haas, D. (1996) *Mol. Plant-Microbe Interact.* **9**: 642-645.

Fridlender, M., Inbar, J. & Chet, I. (1993) *Siol Biol. Biochem.* **25**: 1211-1221.

Gallagher, L.A & Manoil, C. (2001) *J Bacteriol* **183**: 6207-6214.

Gallagher, L.A., Mcknight, S.L., Kuznetsova, M.S., Pesci, E.C & Manoil, C. (2002) *J. Bacteriol.* **184**: 6472-6480.

Glick, B.R., Penrose, D. M. & Li, J. (1998) *J. Theor. Biol.* **190**: 63-68.

Gupta, C.P., Dubey, R.C. & Maheshwari D. K. (2002) *Biol. Fert. Soils* **35**: 399-405.

Gupta, C.P., Dubey, R.C., Kang, S.C & Maheshwari, D.K. (2001) *Curr. Sci.* **81**: 91-94.

Gupta, C.P., Sharma, A., Dubey, R.C. & Maheshwari, D.K. (1999) *Cytobios* **99**: 183-189.

Gutterson, N.I., Layton, T.J., Ziegle, J.S. & Warren, G.J. (1986) *J. Bacteriol.* **165**: 696-703.

Haas, D. & Keel, C. (2003) *Ann. Rev. Phytopathol.* **41**: 11-53.

Harish, S., Manjula, K. & Podile, A.R. (1998) *FEMS Microbiol. Ecol.* **25**: 382-390.

Hass, D. & Defago, G. (2005) *Nat. Rev. Microbiol.* **1038**: 1129-1141.

Hassan, H.M & Fridovich, I. (1980) *J. Bacteriol.* **141**: 156-163.

Henis, Y. & Chet, I. (1975) *Adv. Appl. Microbiol.* **19**: 85-111.

James, D.W.Jr. & Guttorson, N. (1986) *App. Environ. Microbiol.* **52**: 1183-1189.

Kang, J.G., Shin, S.Y., Kim, M.J., Bajpai, V., Maheshwari, D.K. & Kang, S.C. (2004) *J. Antibiot.* **57**: 726-731.

Keel, C., Schnider, U., Maurhofer, M., Boisard, C., Laville, J., Burger, U., Wirthner, P., Hass, D. & Defago, G. (1992) *Mol. Plant Microbe Interact.* **5**: 4-13.

Kitten, T., Kinscherf, T., Mcevoy, G. & Willis, D.K. (1998) *Mol. Microbiol.* **28**: 917-929.

Kloepper, J.W., Hume, D.J., Scher, F.M., Singeleton, C., Tipping, B., Laliberte, M., Frauley, K., Kutchaw, T., Simonson, C., Lifshitzr, R., Zeleska, I. & Lee, L. (1988) *Plant. Dis.* **72**: 42-46.

Ligon, J.M. et al. (2000) *Pest Manag. Sci.* **56**: 688-695.

Liles, M.R., Averscheel, T.C., & Cianciotto, N.P. (2000) *J Bacteriol.* **182**: 749-757.

Lim, H.S., & Kim, S.D. (1995) *Curr. Microbiol.* **33**: 295-301.

Mazzola, M., Cook, R.J., Thomshaw, L.S., Weller, D.M. & Pierson L.S. (1992) *Appl. Environ. Microbiol.* **58**: 2616-2624.

Nielsen, T.H., Christophersen, C., Anthoni, U. & Sorensen, J. (1999) *J. Appl. Microbiol.* **86**: 80-90.

Nielsen,T.H., Thrane, C., Christophersen, C., Anthoni, U. & Sorensen, J. (2000) *J. Appl. Microbiol.* **89**: 992-1001.

Nowak-Thompsan, B., Chancey, N., Wing, J.S., Gould, S.J. & Loper, J.E. (1999) *J. Bacteriol.* **181**: 2166-2174.

O'Sullivan, D.J. & O'Gara, F. (1992) *Microbial. Rev.* **56**: 662-676.

Pal, K.K., Tilak, K.V.B.R., Saxena, A.K., Dey, R. & Singh, C.S. (2000) *Microbial. Res.* **155**: 233-242.

Pandey, P., Kang, S.C. & Maheshwari, D.K. (2005) *Curr. Sci.* **89**: 177-180.

Parret, A.H., Temmerman, K. & De Mot, R. (2005) *Appl. Environ. Microbiol.* **9**: 5197-207.

Podile, A.R. & Dube, H.C. (1988) *Curr. Sci.* **57**: 183-186.

Podile, A.R. & Prakash, A.P. (1996) *Can. J. Microbiol.* **42**: 533-538.

Podile, A.R. (1995) *Ind. J. Microbiol.* **35**: 199-204.

Podile, A.R., Prasad, G.S & Dube, H.C. (1985) *Curr. Sci.* **54**: 864-865.

Raaijmakers, J.M. & Weller, D.M. (2001) *Appl. Environ. Microbiol.* **67**: 2545-2554.

Ramesh Kumar, N., Arasu, T. & Gunasekaran, P. (2002) *Curr. Sci.* **82**: 1463-1466.

Rebafka, F.P., Ndungunu, B.J. & Marschner, H. (1993) *Plant Soil* **150**: 213-222.

Rioux C. R., Jordan, D. C. & Rattray, J. B. M. (1986) *Arch. Biochem.* **248**: 175-182.

Ryder, M.H. & Jones, D.A. (1990) In *Biological Control of Soil-borne Plant Pathogens* (Ed. Hornby, D.), CAB International, Wallingford, pp. 45-63

Ryu, C.M., Farag, M.A., Hu, C.H., Reddy. M.S., Wei, H.X. & Pare P.W. (2003) *PNAS* **100**: 4927-4932.

Saraf, M. & Sood, N. (2002) *J. Ind. Bot Soc.* **81**: 154-157.

Shah, H., Karkhanis, V. & Desai, A. (1992) *Curr. Microbiol.* **25**: 347-351.

Shyamli, S., Tanwar, R.S. & Dureja, P. (2003) *Pesticide Res. J.* **15:**19-22.

Skujins, J.J., Potgieter, H.J. & Alexander, M.A. (1965) *Arch. Biochem. Biophys.* **111**: 358-364.

Starkenburg, S.R., Casey, J.N. & Cianciotto, N.P. (2004) *Curr. Microbiol.* **49**: 203-207.

Thomashow, L.S. & Weller, D.M. (1988) *J. Bacteriol.* **170**: 3499-3508.

Trigalet, A., Frey, P. & Demmerry-Trigalet, D. (1994) In (Eds. Hartman, G.L. & Hayward, A.C.*) Bacterial wilt: The disease and its causative agent, Pseudomonas solanacearum,* pp. 225-234.

Van Peer, R., Niemann, G.J. & Schippers, B. (1991) *Phytopathol.* **81**: 728–734

Vaidya, R.J., Shah, P.R., Vyas, H. S. & Chhatpar, H. S. (2001) *Biotechnol.* **17**: 691-696.

Validov, S., Mavrodi, O., De La Fuente, L., Boronin, A., Weller, D. Thomasho, L. & Mavrodi, D. (2005) *FEMS Micribiol. Lett.* **242**: 249-256.

Velusamy, P. & Gnanamanickam, S.S. (2003) *Curr. Sci.* **85**: 1270-1273.

Vivekanathan, R., Ravi, M., Ramanathan, A. & Samiyappan, R. (2004) *W. J. Mictrobiol. Biotechnol.* **20**: 235-244.

Voisard, C., Keel, C., Haas, D. & Defago, G. (1989) *EMBO J.* **8**: 351-358 year?

Weller, D.M., Raaijmakers, J.M., Gardner, B.B.M. & Thomashow, L.S. (2002) *Ann. Rev. Phytopathol.* **40**: 309-348.

Yedidia, I., Shoresh, M., Kerem, Z., Benhamou, N., Kapulnik, Y. & Chet I. (2003) *Appl Environ Microbiol* ***69:*** 7343-53.

16

Biocatalysis and Biodegradation in the Context of Functional Genomics

Sung-Cheol Koh, Byung-Hyuk Kim, Jung-Hye Choi, Se-Eun Lee, Sung-Woo Shin, Sung-Kyu Shin, and Myung-Hyun Hong

Division of Civil and Environmental Systems Engineering, Korea Maritime University, Busan 606-791, (Korea)

Functional genomics aims at uncovering useful information carried on genome sequences and at using it to understand the mechanisms of biological function. Elucidating the unknown biological functions of new genes based on the genomics rationales will greatly speed up the extensive understanding of biocatalysis and biodegradation in biological world including microorganisms. DNA microarray technology generates a system for the simultaneous measurement of the expression level of thousands of genes in a single hybridization assay. Their data mining (transcriptome) strategy has two categories: differential gene expression and coordinated gene expression. Furthermore, measurement of proteins (proteome) generates information on how the transcribed sequences end up as functional characteristics within the cell, and quantitation of metabolites yields information on how the functional proteins act to produce energy and process substrates (metabolome). Various composite functional genomics databases containing genetic, enzymatic and metabolic information have been developed and, once facilitated by massive data analysis algorithms (*i.e.*, ANNs and SOM, etc.), will contribute to the understanding of the life blue print and the new discoveries and practices in biocatalysis and biodegradation that could enrich their industrial and environmental applications.

INTRODUCTION

Genomics or the complete sequencing and analysis of an organism's DNA will be transforming the way in which the biologists approach their sciences. While DNA sequence analysis is yielding important patterns, the greatest genomic richness results from assigning metabolic functions to individual genes and deriving a biological usefulness of the gene in the context of the organism and its environment (Wackett and Hershberger, 2000). Functional genomics aims at uncovering useful information carried on genome sequences and at using it to understand the mechanisms of biological function. Elucidating the unknown biological functions of new genes based on the genomics rationales will greatly speed up the extensive understanding of biocatalysis and biodegradation in biological world including microorganisms.

In this paper, we will present an overview of microbial biocatalysis and biodegradation in terms of functional genomics. The understanding of these areas will greatly facilitate the development of technologies to resolve the issues of various biotechnology-related industries including medicine, agriculture, food, and environmental biotechnology-related businesses.

FUNCTIONAL GENOMICS: CONCEPT AND ITS TECHNOLOGIES

Traditionally, the analysis of the regulation and function of genes have largely been performed by sequential studies of individual genes and proteins. In the last decade, however, a paradigm shift has been observed in which we may be able to handle many different genes in a highly parallel and rapidly serialized way: the development of DNA microarrays. These arrays consist of a highly ordered matrix of thousands of different DNA sequences that can be used to measure DNA and RNA variation in applications that include gene expression profiling, comparative genomics and genotyping (Case-Green *et al*, 1998; Lander, 1999; Brown and Botstein, 1999; Ferea and Brown, 1999). DNA microarrays generate a system for the simultaneous measurement of the expression level of thousands of genes in a single hybridization assay (Harrington *et al.*, 2000). Fluorescently labelled RNA or DNA prepared from messenger RNA is hybridized to complementary DNA on the array and then detected by laser scanning. Hybridization intensities for each DNA sequence on the array are determined using an automated process and converted to a quantitative read-out of relative gene expression levels. The data can then be further analyzed to identify expression patterns and variation that correlate with cellular development, physiology and function. A description of the typical microarray system is shown in the Fig. 1.

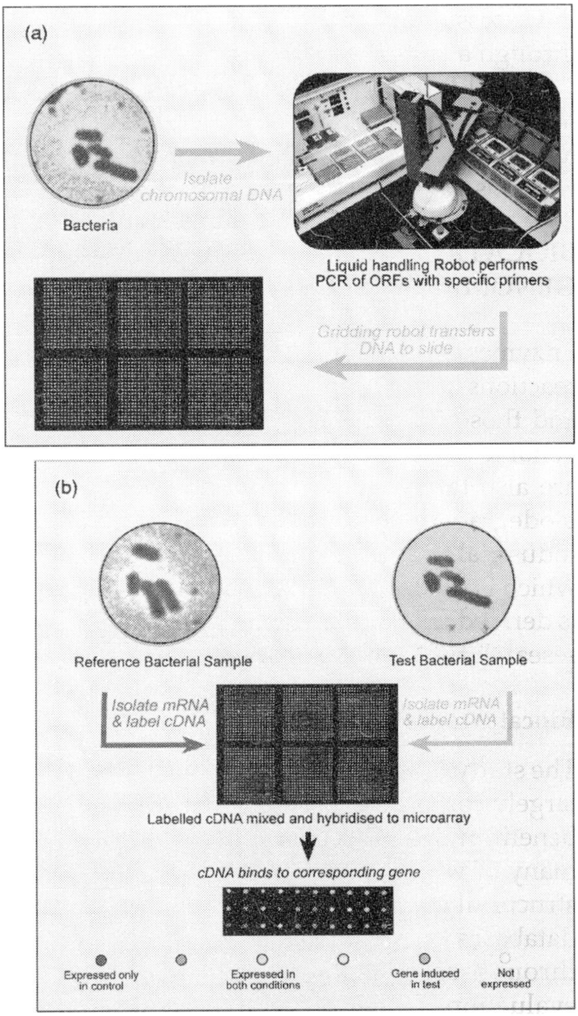

Fig. 1: Making (*a*) and using (*b*) DNA microarray for gene expression profiling (http://www. microarrays.org)

Following acquisition and processing of the fluorescent array image, there are three essential steps to meet for efficient and effective data analysis: data normalization, data filtering, and pattern identification. The methods used for data mining and interpretation are varied, ranging from simple lists of increased and decreased genes based on user-defined thresholds to the implementation of sophisticated clustering and visualization programs, such as hierarchical clustering (Eisen *et al.*, 1998) and self-organizing maps also called k-means clustering (Tamayo *et al.*, 1999; Törönen *et al.*, 1999; Tavazoie *et al.*, 1999).

The data mining strategy used depends on the experimental design and can be generally divided into two categories: differential gene expression and coordinated gene expression (Claverie, 1999). The differential gene expression approach generally consists of paired comparisons between normal/abnormal samples such as from healthy and pathological specimens or wild-type and mutant genotypes. Coordinated gene expression analysis involves the assessment of the expression levels of a large number of genes over a period of time or through a series of experimental conditions.

In addition, as a result of extensive sequencing programs the complete genomic sequence has become available for an increasing number of organisms including microorganisms. This has resulted in substantial efforts in assigning function to the identified open reading frames which is referred to as functional genomics.

BIOCATALYSIS AND BIODEGRADATION IN THE CONTEXT OF FUNCTIONAL GENOMICS

Enzymes contain the largest and most diverse group of all proteins, catalyzing various chemical reactions in the metabolism of all organisms. The completion of the human genome sequencing and those of many other organisms, including several dozens of bacteria, speeded up post-genome projects aiming at elucidating the blueprint of life from a scientific perspective. They are also targeting at uncovering new drugs and other useful materials, and at finding out biodegradation pathways of xenobiotic chemicals such as pollutants and toxins from medical, industrial and environmental aspects. All of them, however, require chemical information, which is not found in the genome, in addition to information about genes and proteins, which is derived from the genome, and chembioinformatics has been considered one of the important research fields in the post-genome age.

Biocatalysis

The study of secondary metabolites that organisms such as microbes and plants have evolved, largely for the purpose of their own survival, has historically turned out to be of immense benefit in drug discovery, providing a rich source of structurally novel bioactive molecules, many of which have become life-saving drugs (Nisbet and Moore, 1997). The exploitation of structural chemical databases comprising a wide variety of chemotypes, in conjunction with databases on target genes and proteins, will facilitate the creation of new chemical species through computational molecular modelling for pharmacological and other chemical evaluation. Development of products from novel genes and pathways typically, however, require optimal selection of the gene(s) for the commercially useful property. The selected genes are then shuffled by typically homologous recombination, resulting in a large pool of chimeric genes, which is assayed for improved properties in a variety of functional screens (Stemmer, 2002).

The studies in functional genomics will allow the increased understanding of the molecular mechanisms in each cell system. It will be then eventually possible to describe the interaction between all the components in the system, also at the quantitative level, and this will be the goal of systems biology (Nielsen and Olsson, 2002).

A new method has been introduced to improve the functional predictions based on a phylogenetic approach to functional prediction, as implemented in the Clusters of Orthologous Groups (COG) database (available at http:@www.ncbi.nlm.nih.gov/COG). This approach paves a convenient way to characterize the protein families (hence metabolic pathways) that are present or absent in any given organism (Galperin and Koonin, 1999).

Measurement of mRNA provides information on what genetic information is transcribed (the transcriptome), measurement of proteins (the proteome) yields information on how transcribed sequences end up as functional entities within the cell, and measurement of metabolites gives information on how the functional proteins act to produce energy and process materials (the metabolome) (Delneri *et al.*, 2001).

Frishman *et al.* (2001) introduced the PEDANT software system for high-throughput analysis of large biological sequence sets and the genome analysis server associated with it. The principle features of PEDANT are: (*i*) completely automatic processing of data using a wide range of bioinformatics methods, (*ii*) manual refinement of annotation, (*iii*) automatic and manual assignment of gene products to a number of functional and structural categories, (*iv*) extensive hyperlinked protein reports, and (*v*) advanced DNA and protein viewers. The main purpose of the PEDANT genome database has been to spread well-organized information on completely sequenced and unfinished genomes. In particular, the availability of structural prediction data for over 300 000 genomic proteins makes PEDANT one of the most extensive structural genomics resources available on the web (Fig. 2).

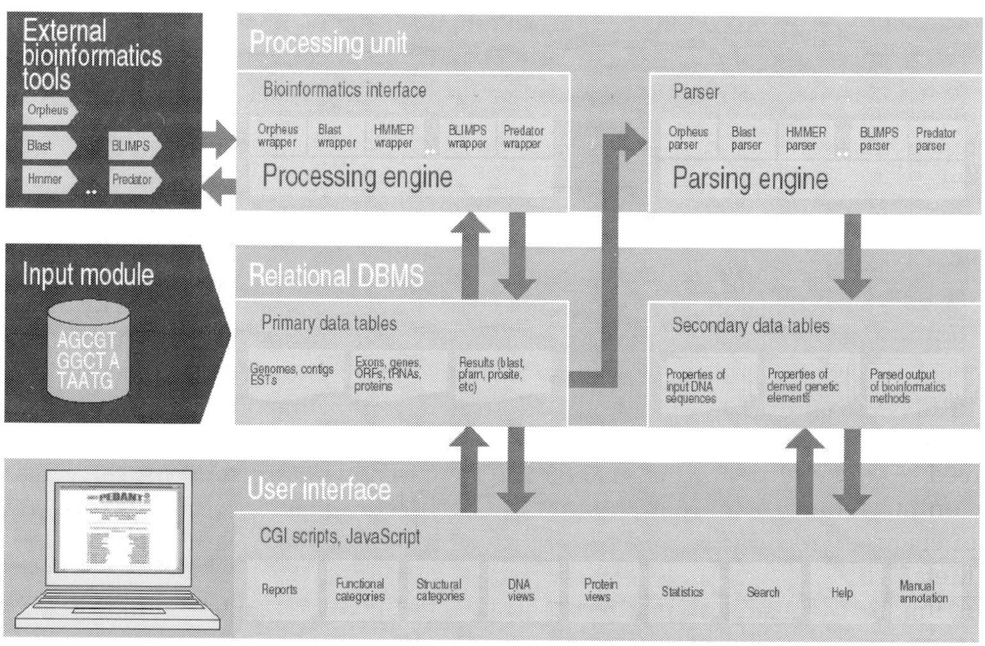

Fig. 2: The PEDANT architecture, a genome analysis server, is available at http://pedant.mips.biochem.mpg.de (Frishman *et al.*, 2001).

The LIGAND (Goto *et al.*, 2002) is a composite database composed of three sections: COMPOUND for the information about metabolites and other chemical compounds, REACTION for the collection of substrate-product relations representing metabolic and other reactions, and ENZYME for the information about enzyme molecules (Fig. 3). This has been established to fill in the gap between genomic information and chemical information, and applied to actual reconstruction of metabolic pathways in the completely sequenced organisms in the Kyoto Encyclopedia of Genes and Genomes (KEGG) (Goto *et al.*, 1998; Kanehisa, 1997; Kanehisa *et al.*, 2002). The current version of COMPOUND includes xenobiotic chemicals such as environmental pollutants and toxins, because KEGG has an agreement with UM-BBD to include biodegradation pathways of xenobiotic chemicals in KEGG/PATHWAY (Ellis *et al.*, 2001).

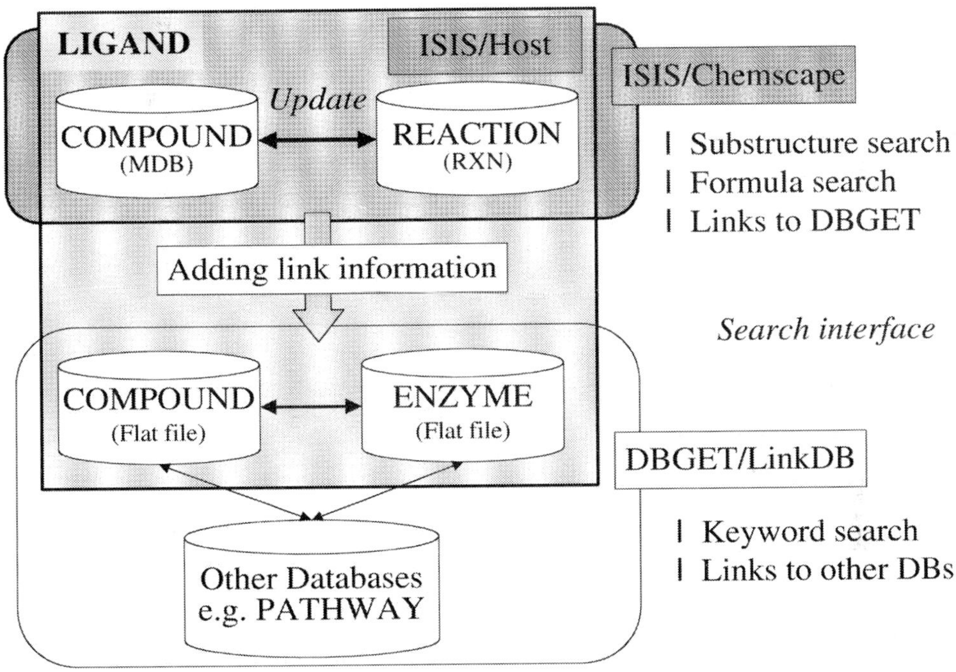

Fig. 3: Relationship between the ISIS version and the DBGET version of LIGAND (Goto *et al.*, 2002).

BRENDA (BRaunschweig Enzyme DAtabase), launched in 1987 by Dietmar Schomburg, is a comprehensive protein function database, containing enzymatic and metabolic information derived from the primary literatures (Schomburg *et al.*, 2002). Currently, the database carries data on more than 40 000 enzymes and 4460 different organisms, and includes information about enzyme–ligand relationships with numerous chemical compounds.

MetaCyc is a metabolic-pathway database that annotates 445 pathways and 1115 enzymes occurring in 158 organisms (Karp *et al.*, 2002). Applications of MetaCyc are pathway analysis of genomes, metabolic engineering and biochemistry education. The modification of a metabolic map through genetic engineering involves (*i*) inserting a new enzyme or pathway into an organism, (*ii*) replacing an existing enzyme or pathway with a substitute or (*iii*) removing an enzyme or pathway.

The University of Minnesota Biocatalysis/Biodegradation Database (UM-BBD, http://www.labmed.umn.edu/umbbd/index.html) first became available on the web in 1995 to provide information on microbial biocatalytic reactions of, and biodegradation pathways for, organic chemical compounds, especially those produced by humans (Fig. 4) (Ellis *et al.*, 2002).

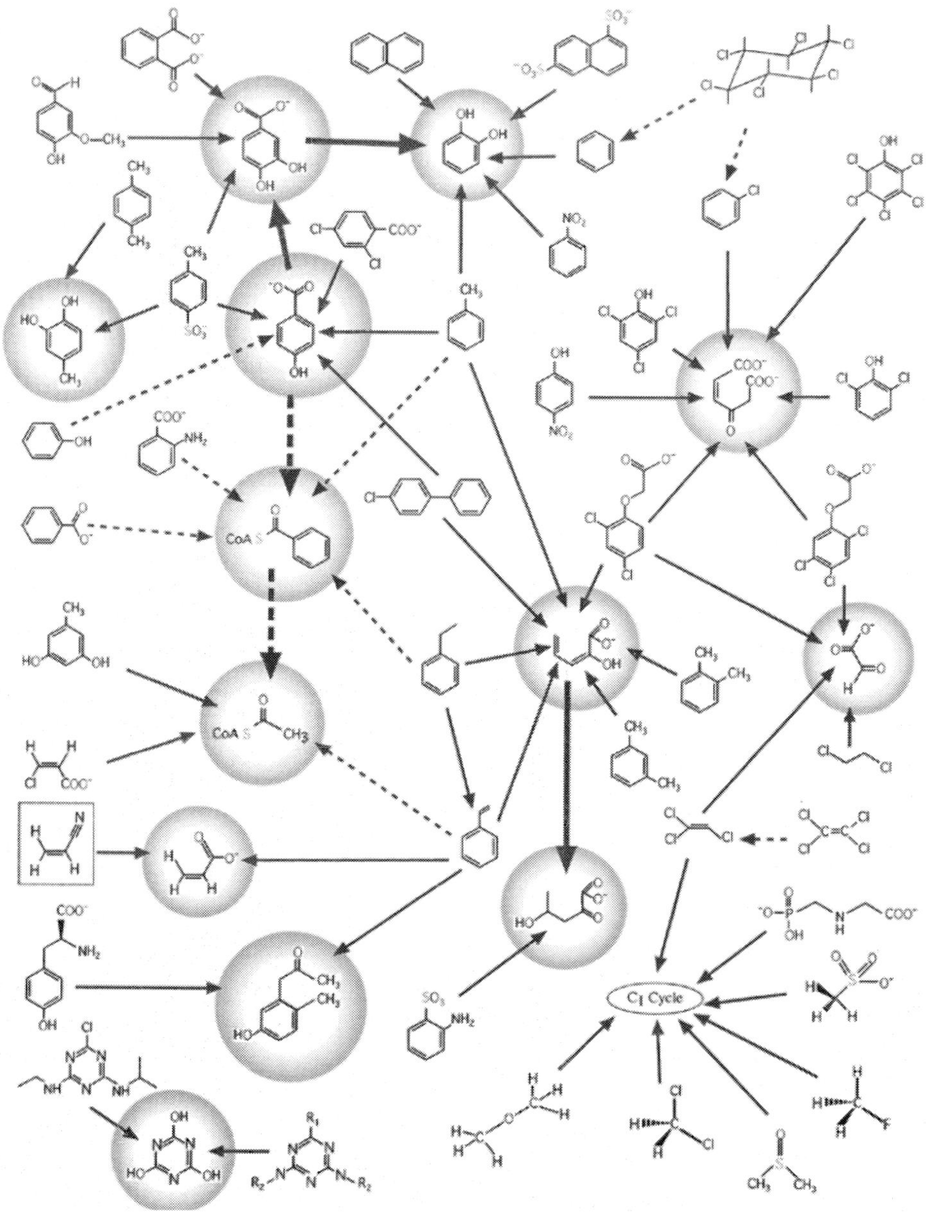

Fig. 4: Graphical overview of UM-BBD content. The circled compounds lead to intermediary metabolism (Ellis *et al.*, 2002).

The database includes the diversity of known microbial metabolic routes, organic functional groups (Ellis *et al.*, 2000) and environmental conditions under which biodegradation occurs. The database could contribute to enhance understanding of basic biochemistry, biocatalysis leading to especially chemical manufacture and biodegradation of environmental pollutants. It will be also a resource for functional genomics, since it contains information on enzymes and genes involved in specialized metabolism not found in intermediary metabolism databases, and thus can assist in assigning functions to genes homologous to such less common genes (Fig. 5) (Ellis *et al.*, 2002).

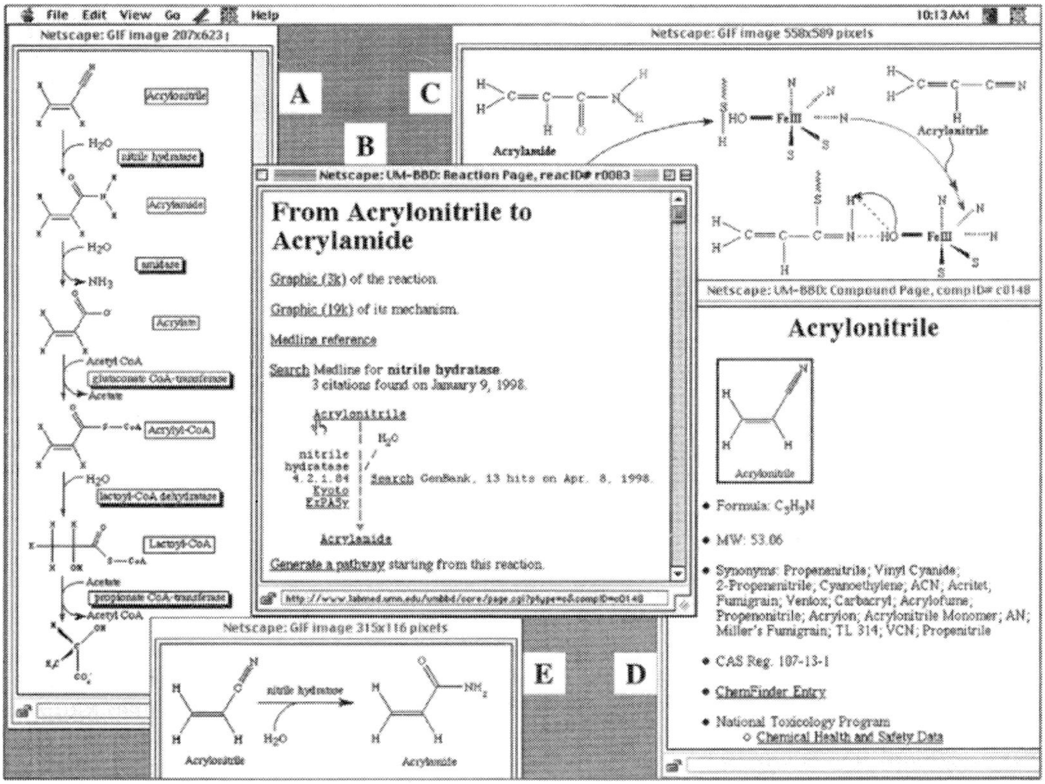

Fig. 5: Example UM-BBD pathway information. (A) Graphical pathway map for the acrylonitrile pathway; (B) reaction page for the nitrile hydrolase reaction; (C) reaction mechanism graphic for nitrile hydrolase; (D) compound page for acrylonitrile; (E) reaction graphic for nitrile hydrolase. URL, http://www.labmed.umn.edu/umbbd/acr/acr_map.html (Ellis *et al.*, 2002).

Recently, the database carries information on over 130 metabolic pathways, 800 reactions, 750 compounds and 500 enzymes. It also contains more examples of microbial metabolism of metals and metalloids; and expanded the types of information it includes to contain microbial biotransformations of, and binding interactions with many chemical elements. A guided pathway prediction system has been newly added (Ellis *et al.*, 2003). The research group has developed a system to predict microbial catabolism, using the University of Minnesota Biocatalysis/Biodegradation Database (UM-BBD, http://umbbd.ahc.umn.edu/) as a knowledge base. The present system, available on the Web (http://umbbd.ahc.umn.edu/predict/), can predict biodegradation of most of the major aliphatic and aromatic organic functional groups containing C, H, N, O, and halogens. (Hou *et al.*, 2003).

Biodegradation

Environmental applications of microarray technology has been recently reviewed and include innovative approaches in gene expression, comparative genomics and mixed community analysis (Cook and Sayler, 2003). A substantial understanding of sources of experimental and analytical error in the array experiments will facilitate the future application of array technology to environmental analysis.

Although DNA microarrays have been initially developed for medical and diagnostic applications; they are also an ideal technology to assess the sequence diversity of 16S rRNA samples from the environments. Soils are an immense source of microbial diversity, which mostly remains unexplored. Some novel methods utilizing rRNA and rDNA sequence analyses have uncovered a portion of the soil microbial diversity. An advanced step for the microbial ecological studies will be to elucidate genomic, evolutionary and functional information from microbial artificial chromosome libraries of the soil community genomes (*i.e.* the metagenome). The metagenome technology has been successfully applied for isolation of more than 15 different genes (including 12 putative agarase genes) encoding novel biocatalysts by using a combined cultivation and direct cloning strategy (Voget, 2003).

Sophisticated analyses that are based upon molecular phylogenetics, DNA microarrays, functional genomics and *in situ* activity measurements will provide a large amount of new data, potentially enhancing our understanding of the structure and function of soil microbial ecosystems and the interactions that occur within them (Phelps *et al.*, 2002; Torsvik and Øvreås, 2002).

Studies of the natural diversity and abundance of sulphate-reducing bacteria (SRB) in relationship to hydrocarbon degradation could be greatly facilitated by application of microarray technology. In order to assess the utility of the microarray format for analysis of environmental samples, oil-contaminated sediments from the coast of Kuwait were analyzed (Koizumi *et al.*, 2002). Here the DNA microarray successfully detected bacterial nucleic acids from these samples, but probes targeting specific groups of SRB did not give positive signals. The results of this study demonstrate the limitations and the potential utility of DNA microarrays for microbial community analysis.

To determine the potential of DNA array technology for assessing functional gene diversity and distribution, a prototype microarray was constructed with genes involved in nitrogen cycling: nitrite reductase (*nirS* and *nirK*) genes, ammonia mono-oxygenase (*amoA*) genes, and methane mono-oxygenase (*pmoA*) genes from pure cultures and those cloned from marine sediments (Wu *et al.*, 2001). Here hybridization signal intensity within a certain range of sequence identity and size was affected by sequence divergence and probe length, respectively. The prototype functional gene array did show differences in the apparent distribution of *nir* and *amo*A and *pmo*A gene families in sediment and soil samples. These results indicate that the glass-based microarray hybridization technique has potential as a tool to reveal functional gene composition in natural microbial communities. However, more efforts are needed to improve sensitivity and quantitation and to understand the issues of specificity. Denef *et al.* (2003) were, however, able to single copy genes coding aromatic oxygenase of an organism in a soil community degrading polychlorinated biphenyls (PCBs).

For cultivation-independent detection of sulphate-reducing prokaryotes (SRPs) an oligonucleotide microarray consisting of 132 16S rRNA gene-targeted oligonucleotide probes (18-mers) was designed in the way that it had hierarchical and parallel (identical) specificity for the detection of all known lineages of sulphate-reducing prokaryotes (SRP-PhyloChip) and subsequently evaluated with 41 suitable pure cultures of SRPs (Loy *et al.*, 2002). Consistent with previous studies, the SRP-PhyloChip showed the occurrence of *Desulfomicrobium* spp. in the tooth pockets and the presence of *Desulfonema*- and *Desulfomonile*-like SRPs (together with other SRPs) in the chemocline of the mat.

Even if whole genomic DNA-DNA hybridization technique has an advantage in microbial species determination, it is not widely used because of its laborious implementation procedure. Cho and Tiedje (2001) have developed a method based on random genome fragments and DNA microarray technology that avoid the disadvantages of whole-genome DNA-DNA hybridization. The cluster analysis of the hybridization profiles using this technique elucidated taxonomic relationships between bacterial strains tested at resolution of species to strain level, indicating that this approach is useful for the bacterial identification as well as determining the genetic distance among bacteria. To quantify target genes in biological samples using DNA microarrays, Cho and Tiedje (2002) recently have employed reference DNA to normalize variations in spot size and hybridization. This approach for designing quantitative microarrays and the equation derived from the study could provide a simple and convenient way to estimate the target gene concentration from the hybridization signal ratio.

The challenges of applying microarray hybridization for use in environmental studies are problems associated with specificity, sensitivity and quantitation. Gene expression ratios generated from spotted-glass microarray experiments possess several drawbacks including the inability to relate ratios to absolute expression levels or to compare experimental conditions not measured with the same control. Dudley *et al.* (2002) discussed future applications of method to measure RNA expression on the absolute scale of number of transcripts per cell from any organism. Recent advances in genomic analysis and stable isotope probing technology appear to resolve the relationships between structure and function in microbial communities (Wellington EM, 2003). The full value of microbial genomic hybridizations can facilitate an understanding of microheterogeneity, bacterial speciation and taxonomic relationships (Murray *et al.*, 2001). Here, they used DNA microarrays constructed with full length ORFs from *Shewanella oneidensis*, MR-1, which were hybridized with genomic DNA from nine other *Shewanella* species and *Escherichia coli* K-12.

Data Analysis

Artificial neural networks (ANNs) may provide a useful tool for recognizing patterns in complex, non-linear data sets such as those associated with predicting gene expression patterns from the DNA microarray data (Urakawa *et al.*, 2002). ANNs will be particularly advantageous over the conventional statistical methods because they can deal with the inherent variability associated with biological data. A self-organizing map (SOM) is an unsupervised neural network protocol that is particularly useful for data visualization (Kohonen, 1997). The SOM algorithm simultaneously allows a representative set of reference vectors of the training data and positions them on a regular two-dimensional grid of neurons so that it can be easily visualized (Wang, *et al.*, 2001). It has been successfully used for the analysis and organization of large data files. For example, Petri *et al.* (1999) used the SOM algorithm to analyze yeast gene expression and show that SOM is an excellent tool for the analysis and visualization of gene expression profiles. Back propagation ANNs (BPANNs) were used to model the microbial growth curves of *S. flexneri*. The developed model appeared to be reasonably accurate in simulating both training and test time-dependent growth curves as affected by temperature and pH. (Basheer and Hajmeer, 2000). The BPANNs were also applied to analyse microbial community compositions of surface and subsurface marine sediments and sediments lining burrows of marine polychaetes and hemichordates by comparing ester-linked phospholipid fatty acid (PLFA) profiles. Results of cross-validation using BPANNs and linear discriminant analysis (LDA) disclosed that BPANNs had significantly fewer incorrect classifications (2.7%) than LDA (8.4%) (Noble *et al.*, 2000).

Adaptive double self-organizing map (ADSOM), an improved algorithm of SOM, has a flexible topology and performs data partitioning and cluster visualization simultaneously without requiring a prior knowledge about the number of clusters (Ressom *et al.*, 2003). This is because it combines features of SOM with two-dimensional position vectors which serve as a visualization tool to detect the number of clusters present in the data. This has been proven by identifying number of clusters in publicly available yeast gene expression data (Ressom *et al.*, 2003). Xinglai *et al.* (2003) developed a new framework for microarray data analysis based on hidden Markov models. Here, they were able to mine biologically meaningful gene groups expressed during yeast cell cycle by clustering analysis. SOM was also able to visualize the similarity of genes (hence relationships) in a more reliable way than two conventional methods, multidimensional scaling and hierarchical clustering (Nikkilä *et al.*, 2002) (Figure 6) and to identify additional genes with similar expression patterns (Torkkola *et al.*, 2001). SOM was used to examine codon usage heterogeneity in the *E. coli* O157 genome, which contained 'O157-unique segments' (O-islands), and it was revealed that SOM is a powerful tool for characterization of horizontally transferred genes (Kanaya *et al.*, 2001).

PERSPECTIVES AND CONCLUSIONS

From the standpoints of medicine, agriculture, and food industries, metabolic engineering has been directed primarily to useful metabolite overproductions. The parallel and recent advances in the resolution and acquisition time of biological data, especially structural and functional genomics, will greatly contribute to the systemic view of biology of which the metabolic engineering may take advantage. Moreover, microorganisms play a significant role in ecosystem function and sustainability. Dissecting the structure and composition of microbial communities and their responses and adaptations to environmental perturbations such as toxic pollutants, agricultural and industrial practices will be critically important in finding ways of working out these complicated ecological and environmental issues. The functional

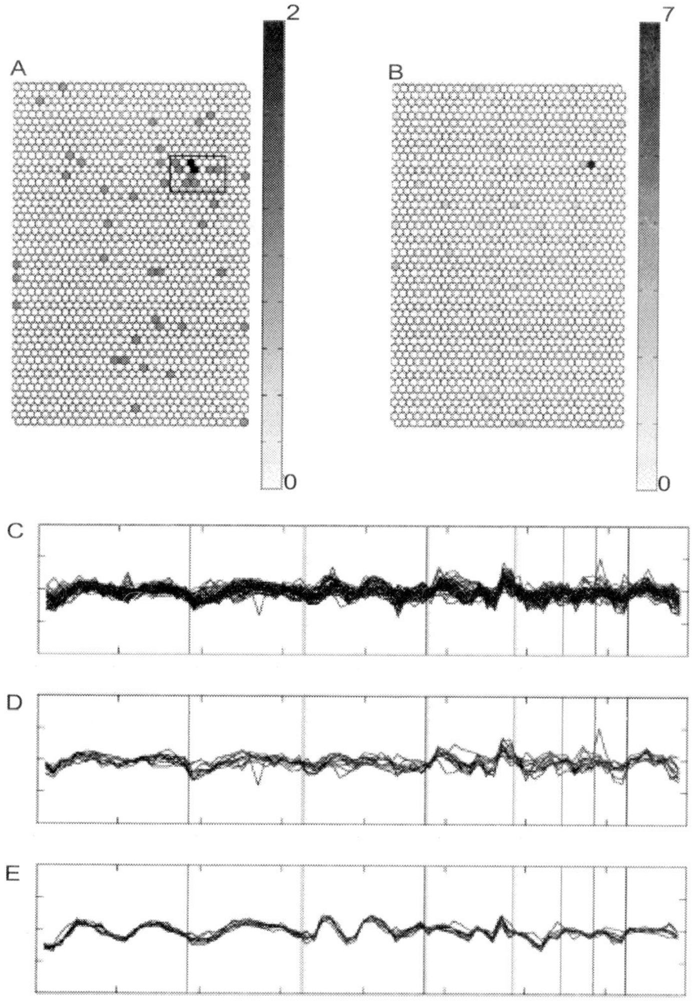

Fig. 6: Analysis of the glycosylation (A) and histone (B) genes based on their distributions on the SOM. (A) Shows the distribution and number of genes with the descriptor 'glycosylation protein' in each SOM unit. The boxed area indicates units chosen for subsequent analysis. (B) Shows the distribution and number of genes with the descriptor 'histone protein' in each SOM unit. The expression profiles of all the genes found in the boxed area of the panel A are shown in panel C. In panel D are the profiles of the genes with 'protein glycosylation' in the descriptor, and in panel E the profiles of the genes with 'histone protein' in the descriptor. (Nikkilä *et al.*, 2002).

genomics connected by large-scale enzyme databases and facilitated by massive data analysis algorithms (*i.e.* once ANNs and SOM, etc.) will greatly facilitate new discoveries and practices in biocatalysis and biodegradation.

ACKNOWLEDGEMENTS

Part of this paper was presented in 2002 International Meeting of the Federation of Korean Microbiological Societies, Osong, Chungbuk, Oct. 22-23, 2002.

REFERENCES

Basheer, I.A. & Hajmeer M. (2000) *J. Microbiol Methods.* **43:** 3-31. Brown, P.O. & Botstein D. (1999) *Suppl. Nat. Genet.* **21**: 33-37.

Case-Green, S.C., Mir. K.U., Pritchard C.E. & Southern E.M. (1998) *Curr. Opin. Chem. Biol.* vol. 404-410.

Cho, J.C. & Tiedje J.M. (2001) *Appl Envir. Microbiol.* **67**: 3677-3682.

Cho, J.C. & Tiedje J.M. (2002) *Appl. Envir. Microbiol.* **68**: 1425-1430.

Claverie, J.M. (1999) *Hum. Mol. Genet.* **8**: 1821-1832.

Cook, K.L. & Sayler G.S. (2003) *Cur. Opin. Biotechnol.* **14**: 311-320. Delneri, D., Brancia F.L. & Oliver, S.G. (2001) *Curr. Opin. Biotechnol.***12**: 87-91.

Denef, V.J., Park J., Rodrigues J.L., Tsoi T.V., Hashsham S.A. & Tiedje J.M. (2003) *Appl. Environ. Microbiol.* **5:** 933-43. Dudley, A.M., Aach, J., Martin, S.A. & Church, G.M. (2002) *Proceedf. Nat. Acad. Sci. U S A.* **99**: 7554-7559 Eisen, M.B., Spellman, P.T., Brown, P.O. & Botstein, D. (1998) *Proc. Natl. Acad. Sci. USA.* **95**: 14863-14868.

Ellis, L.B.M., Hershberger, C.D. & Wackett, L.P. (2000) *Nucleic Acids Res..* **28**: 377–379.

Ellis, L.B.M., Hershberger, C.D., Bryan, E.M. & Wackett, L.P. (2001) *Nucleic Acids Res.* **29**: 340–343.

Ellis, L.B.M., Hershberger, C.D. & Wackett, L.P. (2002) *Nucleic Acids Res.* **27**: 373-376

Ellis, L.B.M., Hou, B.K., Kang, W. & Wackett, L.P. (2003) *Nucleic Acids Res.* **31**: 262-265

Ferea, T.L. & Brown, P.O. (1999) *Curr. Opin. Genet. Dev.* **9**: 715-722.

Frishman, D., Albermann, K., Hani, J., Heumann, K., Metanomski, A., Zollner, A. & Mewes, H.W. (2001) *Bioinform.* **17**: 44-57.

Galperin, M.Y. & Koonin, E.V. (1999) *Funct. Genom. Enz. Eevol.* **106**: 159-70.

Goto, S., T. Nishioka & Kanehisa, M. (1998) *Bioinform.* **14**: 591–599.

Goto, S., Okuno, Y., Hattori, Y., Nishioka, T. & Kanehisa, M. (2002) *Nucleic Acids Res.* **30**: 402-404

Harrington, C.A., Rosenow, A. & Retief, J. (2000) *Curr. Opin. Microbiol.* **3**: 285-291

Hou, B.K., Wackett, L.P. & Ellis, L.B. (2003) *J. Chem. Inf. Comput. Sci.* **43**: 1051-7. Kanaya, S. & T. Ikemura (2001) *Gene.* **276**: 89-99.

Kanehisa, M. (1997) *Trends Genet.* **13**: 375–376.

Kanehisa, M., Goto, S., Kawashima, S. & Nakaya, A. (2002). *Nucleic Acids Res.* **30**: 42–46.

Karp, P.D., Riley, M., Paley, S.M. & Toole, A.P. (2002) *Nucleic Acids Res.* **30**: 59-61

Kohonen, T. (1997) In: *Self-organizing maps.* 2nd extended edition. Springer, Berlin.

Koizumi, Y., Kelly, J.J., Nakagawa, T., Urakawa, H., Saïd, E.F., Saleh, A.M., Fukui, M., Urushigawa, Y. & Stahl, D.A. (2002) *Appl. Envir. Microbiol.* **68**: 3215-3225.

Lander, E. (1999) *Suppl. Nat. Genet.* **21**: 3-4.

Loy, A., Lehner, A., Lee, N., Adamczyk, J., Meier, H., Ernst, J., Schleifer, K.H. & Wagner, M. (2002) *Appl. Environ. Microbiol.* **68**: 5064-5081.

Murray, A.E., Lies, D., Li, G., Nealson, K., Zhou, J. & Tiedje, J.M. (2001) *Proc. Natl. Acad. Sci. USA.* **98**: 9853–9858.

Nielsen. J. & Olsson, L. (2002) *FEMS Yeast Res.* **2**: 175-81. Nikkilä, P., Törönen, S., Kaski, J., Venna, E., Castrén & Wong, G. (2002) *Neural Networks.* **15**: 953-966.

Nisbet, L. & Moore, M. (1997) *Curr. Opin. Biotechnol.* **8**: 708-712

Noble, P.A., Almeida, J.S. & Lovell, C.R. (2000) *Appl. Environ. Microbiol.* **66**: 694-9 Øvreås, L. (2000) *Ecol. Letts.* **3**: 236-251.

Törönen, P., Kolehmainen, M., Wong, G. & Castrén, E. (1999) *FEBS Letters.* **451**: 42-146. Phelps, T.J., Palumbo, A.V. & Beliaev, A.S. (2002) *Curr. Opin. Biotechnol.* **13**: 20-24

Schomburg, I., Chang, A., Hofmann, O., Ebeling, C., Ehrentreich, F. & Schomburg, D. (2002) *Trend. Biochem. Sci.* **27**: 54-56

Stemmer, W.P. (2002) *Sci. W. J.* **2:** 130-131. Ressom, H., Wang, D. & Natarajan, P. (2003) *Neural Networks.* **16**: 633-640

Tamayo, P., Slonim, D., Mesirov, J., Zhu, Q., Kitareewan, S., Dmitrovsky, E., Lander, E.S. & Golub, T.R. (1999) *Proc. Natl. Acad. Sci. USA.* **96**: 2907-2912.

Tavazoie, S., Hughes, J.D., Campbell, M.J., Cho, R.J. & Church, G.M. (1999) *Nat Genet.* **22**: 281-285.
Torkkola, K., Gardner, R.M., Kranich, T.K. & Calvin, M. (2001) *Inform. Scis.* **139**: 79-96.
Torsvik, V. & Øvreås, L. (2002) *Curr. Opin. Microbiol.* **5**: 240-245
Törönen, P., Kolehmainen, M., Wong, G. & Castrën, E. (1999) *FEBS Lett.* **451**: 142-146.
Urakawa, H., Noble, P.A., Fantroussi, S.E., Kelly, J.J. & Stahl, D.A. (2002) *Appl. Environ. Microbiol.* **68**: 235-244.
Voget, S., Leggewie, C., Uesbeck, A., Raasch, C., Jaeger, K.E. & Streit, W.R. (2003) *Appl. Environ. Microbiol.* **69**: 6235-6242. Wackett, L.P. & Hershberger, C.D. (2001) *Amer. Soc. Microbiol.* Washington, D.C.
Wang, H-C., Badger, J., Kearney, P. & Ming, Li. (2001) *Mol. Biol. Evol.* **18**: 792-800
Wellington, E.M., Berry, A. & Krsek , M. (2003) *Curr. Opin. Microbiol.* **6**: 295-301. Wu, L., Thompson, D.K., Li, G.H., Hurt, R.A., Tiedje, J.M. & Zhou, J. (2001) *Appl. Environ. Microbiol.* **67**: 5780-5790.
Xinglai, J., Jesse, L.L. & Zhirong, S. (2003) *FEBS Letters.* **542**: 125-131.

17

Bioremediation of Agrochemicals and Heavy Metals in Soil

Javed Musarrat and Sabina Zaidi
Department of Agricultural Microbiology,
Faculty of Agricultural Sciences,
Aligarh Muslim University, Aligarh-202002 (India)

A plethora of chemical substances including agrochemicals and heavy metals are continuously being released in substantial amounts into the environment. Several of them adversely influence the ecosystem, causing toxicity to flora and fauna including humans. Conventionally, the methods employed for attenuation and/or elimination of such pollutants include the physical containment of affected material, chemical degradation and incineration of the compounds. Such methods are often not adequate, as they require the collection and concentration of the pollutants, which, in many cases, are dispersed in the environment and cannot be recalled. Thus, more appropriate and cost-effective alternatives have been explored. Biodegradation of xenobiotics offers a suitable alternative for bioremediation of many pollutants with the help of microorganisms introduced in the environment. Bioremediation is relatively inexpensive technology opted for treatment of a variety of pollutants at contaminated sites. Its applicability and potential for success depends upon the (i) biodegradability of the specific chemical pollutants, (ii) microbial community structure, and (iii) environment factors. In many cases, bioremediation relies upon the intrinsic degradative capacity of the indigenous microflora for removing pollutants without further treatment (intrinsic bioremediation). Also, the rate of metabolism of the indigenous microbial community can be increased through environmental engineering (biostimulation) and/or altering the microbial community through seeding with specialized cultures (bioaugmentation). Indeed, the most commercial applications of bioremediation predominantly employ naturally occurring microorganisms. Furthermore, the use of genetically modified microorganisms and plants for remediation of agricultural and industrial pollutants is likely to gain impetus in the near future.

INTRODUCTION

Decades of industrial growth and development of modern agricultural system have left a legacy of soil and water polluted with hazardous heavy metals and agrochemicals. Due to the inadequacy and increased cost of physical and chemical methods available, the need for eco-friendly and cheaper methods has been strongly realized. This has led to the exploitation of microbes for bioremediation purpose. This review emphasizes on the distribution and extent of environmental contaminants and the various options available for remediating both, the organic (agrochemicals) and inorganic (heavy metals) contaminants from the ecosystem. It also accentuates the current status of research in the area of biodegradation and bioremediation and presents the newer approaches available for microbial tracking in the environment.

Agrochemicals: Historical Perspective

The development and application of agrochemicals for controlling the insect-pests and weeds dates back to the ancient Romans. During 16th century, the mixture of honey and arsenic has been exploited for controlling ants. Subsequently, the moderate amounts of arsenical compounds were used as insecticides in China. Later, in the year 1705, the mercuric chloride and copper sulphate have been introduced as a wood preservative and smut spore's inhibitor, respectively. The Bordeaux mixture, developed in the year 1885, has been used to control the downy mildew of vine. In addition, several inorganic compounds *viz.* nicotine sulfate, sulphur, copper arsenite (Paris green), calcium arsenate and lead arsenate have been used for controlling insect-pests in field crops (Klassen *et al.,* 1982). Extensive applications of arsenical insecticides leading to residual toxicity have stimulated the quest for less hazardous pesticides. Therefore, the organic compounds such as tar, petroleum oils and dinitro-o-cresol have been introduced. Eventually, the compound dinitro-o-cresol replaced tar oil for control of aphid eggs, and was patented as selective herbicide against weeds in cereals crops. Gradually, with the exception of sulphur, all inorganic chemicals were almost completely substituted by synthetic organic pesticides. The era of modern organic pesticides started with the synthesis of the alkylthiocyanate insecticides as well as saliyanilide and dithiocarbamate fungicides for controlling a range of pathogenic fungi causing the scabs and rots of fruits and potato blight. Also, the 2,4-dinitro-(1'methyl-n-heptyl) phenylcrotonate or dinocap and chloranil (tetrachloro-1,4-benzoauionine) have been developed in the year 1938, as protectant fungicides against powdery mildew. Furthermore, the dichlorodiphenyltrichloroethane (DDT) has been discovered as a powerful insecticide for control of louse-borne typhus and malaria-carrying mosquitoes *(services/PBT/ddt.cfm).* Besides, certain organic compounds *viz.* azobenzene, ethylene dibromide, ethylene oxide, methyl bromide and carbon disulphide were developed as fumigants. The insecticidal properties of hexachlorocyclohexane have been recognized in the year 1942, and prompted the emergence of several insecticidal chlorinated hydrocarbon compounds such as phenothiazine, p-dichlorobenzene, naphthalene and thiodiphenylamine (Brooks,1974).

Finally, the organochlorines were replaced with the organophosphates as a powerful group of insecticides. The development of organophosphates stemmed from war-time research on nerve gases which resulted in the formulation of powerful insecticides such as schradan (octamethylpyrophosphoramide) and malathion. Another, closely related group of insecticides discovered in Switzerland in the year 1947 belongs to the carbamate esters. Simultaneously, the herbicidal activity of the phenoxyacetic acids *viz.* 2-methyl-4-chloro-phenoxyacetic acid (MCPA) and 2,4-dichlorophenoxy acetic acid (2,4-D) were discovered at Imperial Chemical Industries, England. These compounds can be easily translocated in plants, and are extensively used for the selective control of broad-leaved weeds in cereal crops. Furthermore, the bipyridinium compounds, diquat and paraquat were introduced in 1958 as fast acting herbicides, which could be readily absorbed by plants, causing desiccation of the foliage. Since then, several new formulations of chemicals including synthetic pyrethroids, systemic fungicides such as oxathiins, benzimidazoles, thiophanates and pyrimidines, certain antibiotics, triazoles, and strabulurins have been developed for controlling various types of newly emerging pests, microorganisms and weeds.

Global Distribution and Use of Agrochemicals

Agrochemicals have made an important contribution in increasing the agricultural productivity world over. Their application is regarded as the most effective and accepted means of plant protection. In the last several decades, the use of pesticides has grown steadily at about the rate of 11% per annum (Carvalho *et al.*, 1998). In United States alone, the pesticide usage in agriculture has increased to about 33-fold (*navigate.htm*). About three quarters of the total agrochemicals are reported to be used in developed countries, mainly in North America, Western Europe and Japan. Although, the amount of agrochemicals being used in developing countries is relatively small as compared with the developed nations, it is nonetheless substantial and growing rapidly.

The global distribution profile of agrochemical usage is depicted in Fig.1. In recent years, there has been a boom in the global agrochemical market, and new insecticides, herbicides and fungicides and their formulations are continuously being introduced with greater levels of activity. The data revealed that in the United States about 62 % of planted acreage is treated annually with pesticides. Similar to developed nations, some Asian countries are also the potential users of agrochemicals. Therefore, the total sale of agrochemicals in the Asian region has grown from US$ 5.0 billion in 1983 to US$ 10.3 billion in 2003 (ESCAP, 2000). Overall, the global market has been estimated to be around US$ 17.5 billion, which was likely to increase to US$ 27 billion by the year 2005.

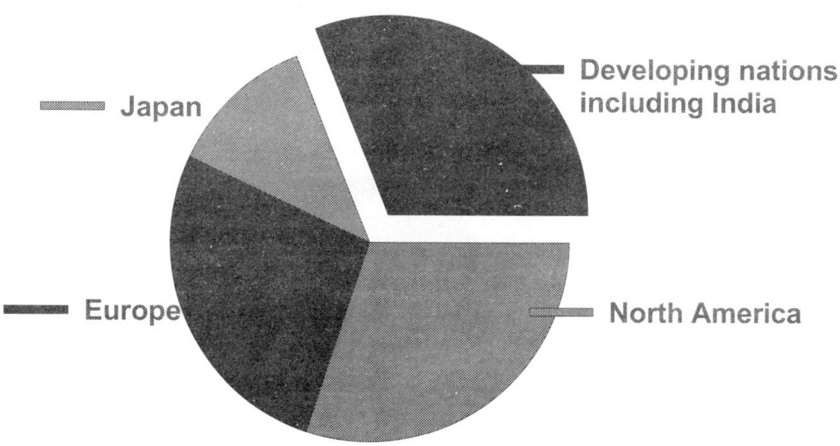

Fig.1: Global distribution of agrochemicals.

Production of Agrochemicals in India

The production of agrochemicals started in India in the year 1952 with the establishment of a production plant of hexachlorocyclohexane (BHC) near Calcutta. India, being the largest manufacturer of agrochemicals in south-east Asia, is placed on twelfth position globally (Mathur, 1999). Production of technical grade pesticides exhibited consistent growth in India, from 5,000 metric tons (MT) in 1958 to 100,000 metric tons (MT) in the last 50 years. Out of the total 145 registered pesticides, about 85 % of technical grade pesticides are being manufactured in India. An estimated demand for pesticides in terms of value is around Rs 22 billion, which is about 2 % of the total world market. The major consumption of pesticides in India is for cotton crops (45%), followed by paddy and wheat. Thus, the pattern of agrochemical usage in India is different from rest of the world. The data in Fig. 2 exhibit that in India 76 % of the pesticides are used as insecticides as compared to 44 % used globally (Mathur, 1999). However, India spends \$3/ha on agrochemicals as compared with \$24, \$255 and \$633/ha invested by Philippines, South Korea and Japan, respectively. In spite of the comparatively low usage of agrochemicals in India, the extent of contamination of produce and food products with pesticide residues in the country is alarming. About 20 % of food products reportedly contain pesticides residues above tolerance levels as compared to only 2 % globally. Of all the agrochemicals used in India, about 40 % are organochlorines and 30 % are organophosphates. Even some of the highly hazardous pesticides *viz* monocrotophos, phorate, phosphamidon, ethion, methyl parathion, dimethoate and several more are continuously and indiscriminately being used in India (*www.teriin.org/news/terivsn/issue31/pesticid.htm*).

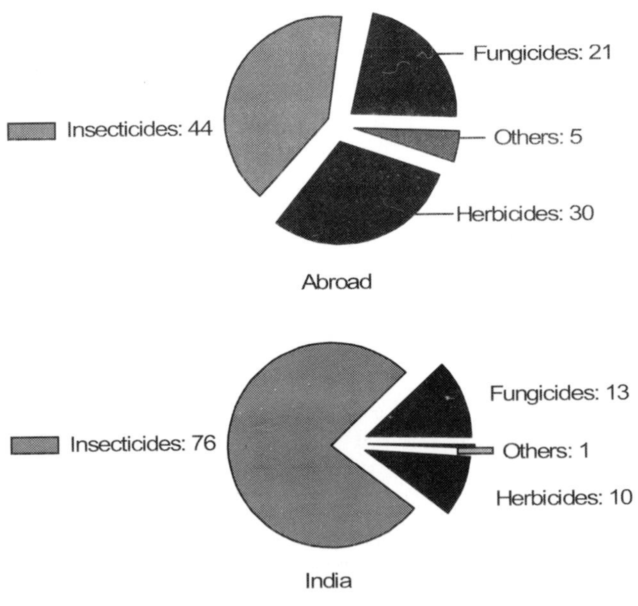

Fig. 2: Pattern of agrochemical consumption in India and abroad. The values represent the distribution of specific agrochemicals in terms of percentage.

BIOREMEDIATION: A NATURAL APPROACH FOR AGROCHEMICAL DEGRADATION

The term 'bioremediation' is defined as the management or utilization of biological community to detoxify, immobilize or mineralize contaminants in an ecosystem. The prefix *'bio'* means 'living' and *'remediation'* means 'to clean'. It is a technology that uses biological systems for degradation and transformation of diverse chemicals and metals to simpler and non-toxic forms. Bioremediation manages the physical, chemical, and biological properties of soil and optimizes the capability of the microbial community to detoxify the soil in a cost-effective manner. Its cost-effectiveness as compared to physical and chemical technologies, especially for dilute contaminants is the main driving force for its large-scale applications (Table1).

Table 1. Application and Economics of Bioremediation

Application	Physical/Chemical Treatment	Bioremediation	Cost Benefit	Ref
Chlorinated hydro-carbon-industrial site.	Excavation; cost estimated at $15 million.	Bioventing; estimated cost, $ 2 million.	Project cost savings, $ 13 million.	Rosenberg (1993)
Multiple contaminants-Superfund site Containing Benzene, Toluene, Ethyl benzene, Xylene (BTEX) and arsenic	Pump and treat capping; cost estimated at $ 50 million.	In situ biostimulation, oxygen sparging, and Bioventing plus biological immobilization of metals; estimated cost, $ 2 million.	Project cost savings, $ 48 million.	Rosenberg *et al.* (1992)
Petroleum hydrocarbon contaminated soil-urban Brownfield site.	Excavation and off-site disposal; projected cost, $3 million	Bioventing on site; estimated cost, $ 0.2 million	Project cost savings, $ 2.8 million	Riser-Robertis (1992)
TCE- contaminated soil and groundwater-industrial site.	Pump and treat; projected cost, $ 20 million.	Bioventing; projected cost, $ 2 million.	Project cost savings $ 20 million.	Stoner (1994)
Marine oil Spills	Physical washing; cost estimated at $1.1 million per kilometer of oiled shoreline.	Biostimulation through fertilizer addition; estimated cost at $ 0.005 million/km of oiled shoreline.	Project cost savings of over $ 1 million per kilometer of oiled shoreline.	Sander *et al.* 1991)
Soil contamination of petroleum hydrocarbons (Ogden UT)	Excavation and off-site disposal; projected cost, $3 million	Phytoextraction and rhizodegradation by alfalfa, poplar, juniper, fescue; estimated cost $ 0.4 million.	Project cost savings, $ 2.6 million	Lee & Banks (1993)
Soil contamination of Polycyclic aromatic hydrocarbons (Texas)	Pump and treat; projected cost, $ 20 million.	Vegetative cover rhizodegradation by mulberry; estimated cost $ 2 million.	Project cost savings $ 18 million.	Liet *et al.* (1992)
Soil & Ground water contamination of radionuclides and heavy metals (Upton, NY& Ashtabula, OH)	Pump and treat capping; cost estimated at $ 50 million.	Rhizofiltration & Phytoextraction by sunflowers and hybrid poplars; estimated cost $ 3 million	Project cost savings of over $ 47 million.	Lyngby & Brix (1989); Lewis& MacDonell (1990)

Principles of Bioremediation

Understanding of the bioremediation processes is derived from the combination of biochemical-microbiological-ecological processes (Tate, 2000), summarized as under:

(*i*) *The relationship of comparative biochemistry apply equally both in the axenic culture of microorganisms and in field soil.* A compound mineralized by one or a combination of pathways or processes in culture can be anticipated to be similarly transformed in a field soil.

(*ii*) *The microbial growth requirements are the same whether in the laboratory culture or in the field.* If a growth factor is required for a specific enzyme to function, it must be supplied in the soil system. Similarly, the microbes and enzymes are controlled by the same physical limitations (eg., pH, temperature) both in the field and lab conditions.

(*iii*) *The limitations resulting from ecological interactions including the need to accommodate the presence of other microbes as well as adaptation to the physical and chemical properties of the microsite wherein the microbe lives.*

A microbe amended to a soil ecosystem with the intent of biostimulation and bioaugmentation processes must not only possess the requisite genetic information and be capable of expressing that capability *in situ*, but it must also have the capacity to become, at least temporarily, a part of the overall soil microbial community. It must express the requisite biochemical activity in the community, and get established in the community and accommodate all means of interactions (amensalism and competition, etc) with the other microflora at least until its presence is no longer required for bioremediation. Such microbes are referred to as rhizospheric competent bacteria with intrinsic bioremediation potential.

Applicability of Bioremediation

The targets for bioremediation must be viewed not only in terms of structure but also in terms of the matrix containing the target. Thus, the applicability of bioremediation can be considered for each of the environmental states of matter (*i*) solids (soil, sediment, and sludge), (*ii*) liquids (groundwater, industrial wastewater), (*iii*) gases (industrial air emissions) and (*iv*) sub-surface environments (saturated and vadose zones). The various approaches to bioremediation are outlined in Fig. 3. The utility and limitations of this technology, and its scientific and engineering underpinning have been extensively reviewed (Alexander, 1994; Baker and Herson, 1994; Borden *et al.*, 1994; Cookson, 1995; Hinchee and Olfenbutell, 1991; Hinchee *et al.*, 1994; Levin and Gealt, 1993; Means and Hinchee, 1994; Rosenberg, 1993). The general approaches to bioremediation are the (*i*) Intrinsic bioremediation (a natural biodegradation process), (*ii*) Biostimulation (environmental modification, through nutrient application and aeration), and (*iii*) Bioaugmentation (addition of microbes). The biological community exploited for bioremediation generally consists of the native soil microflora. However, higher plants can also be manipulated to enhance toxicant removal (Phytoremediation), especially for remediation of metal contaminated soils (Alkorta and Garbism, 2001; Salt *et al.*, 1998).

Bioremediation of Organic Contaminants

Soil is an extremely dynamic and complex system, which exhibits continuous interplay between the soil biota and abiotic components. In this natural process, the soil microbial diversity is regarded as a major factor responsible for many biochemical transformations, including degradation of diverse organic compounds (Rister-Roberts, 1992; Rosenberg *et al.*, 1992; Sander *et al.*, 1991; Singer and Finnerty, 1984; Spain,1995; Stoner, 1994; Unterman, 1996). The chemical structures of organic pollutants exert profound influence on the metabolic abilities of microbes. Thus, harnessing the microbiological activities for biodegradation of recalcitrant hazardous chemicals and its implementations, forms the basis for the increased use of biotreatment systems. Predominantly, the noxious chemicals such as polychlorinated biphenyls (PCB's), trichloroethylenes (TCE's), polycyclic aromatic hydrocarbons (PAH's), pesticide residues, certain explosives and other toxic organic pollutants deposited in soils and waters are the targets. In recent years, there has been an increased interest in the ability of bacteria to degrade man made chemicals and applying these bacteria for bioremediation of polluted environments. In this review, the emphasis has been laid specifically on the biodegradation of agrochemicals and bioremediation of inorganic contaminants, particularly heavy metals.

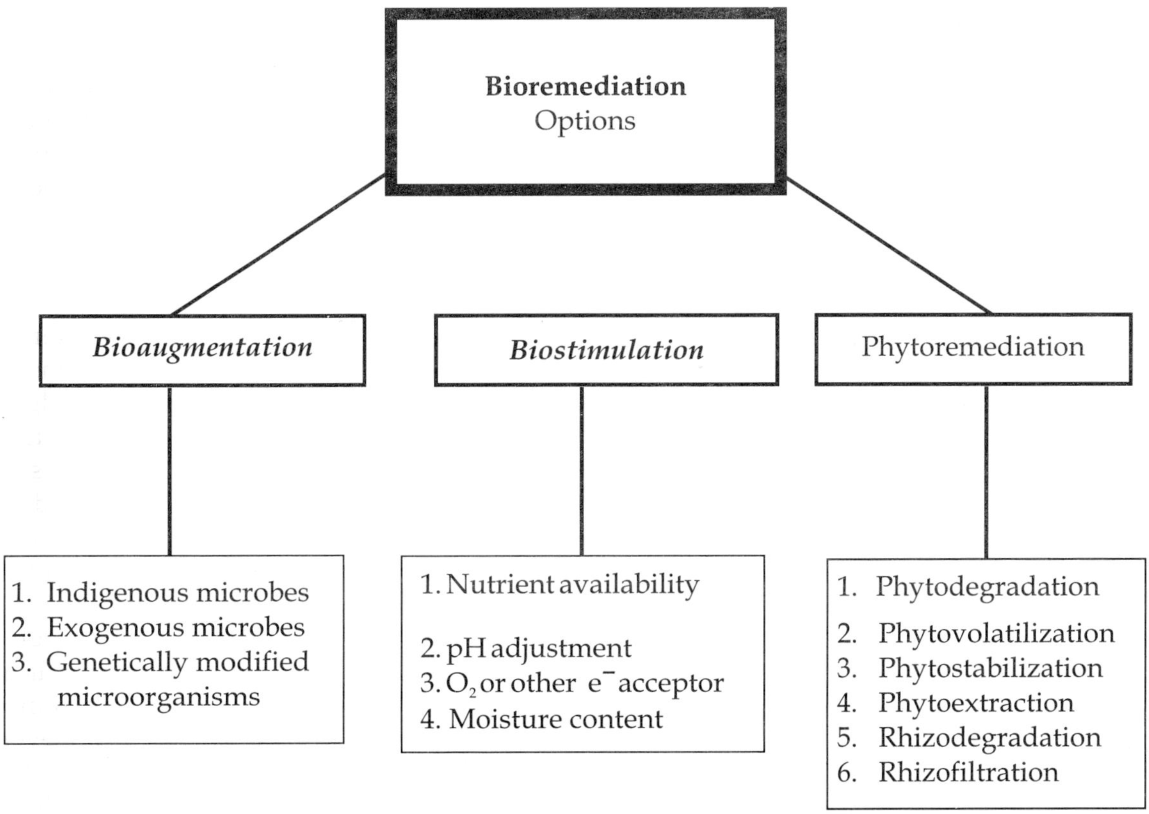

Fig. 3: Bioremediation approaches for environmental clean-up.

Microbial degradation of agrochemicals: Agrochemicals are the integral component of modern agricultural system. However, the extensive and indiscriminate use of agrochemicals for controlling plant disease causes serious imbalance in the agro-ecosystem. Although, the agricultural soil serves as an ultimate sink for agrochemicals, the prevalence of soil microflora with bioremediation potential contributes substantially in reducing the pollution load. Thus, a large number of pesticides are rapidly degraded in soil, due to efficient mechanisms of recruitment and assembly of diverse catabolic pathways in soil bacteria. Major agrochemical contaminants include: the chlorinated aromatic compounds, organophosphates and carbamates (Chaudhary and Chapalamaduga, 1991, Felsot *et al.*, 1981). The incessant exposure of soil microorganisms to sub-lethal doses of agrochemicals develops the enhanced detoxification potential, and acquisition of new traits (Donnelly *et al.,* 1993). Such microorganisms could be exploited for bioaugmentation of contaminated soil in cost-effective manner for bioremediation and soil health restoration vis-à-vis other mechanical technologies. The rhizospheric competent bacteria isolated and characterized in our laboratory have also demonstrated efficient degrading ability of selected agrochemicals viz., 2,4-D, phorate and carbofuran (Bano and Musarrat, 2004; 2003b; Musarrat *et al.*, 2000). The 2,4-dichlorophenoxyacetic acid (2,4-D) is a widely used post-emergence systemic herbicide for controlling broad-leaf vegetation (Beste, 1983). Phorate [O, O-diethyl-S-(ethylthio)methyl phospho-radiothioate] is a systemic insecticide. Carbofuran (2,3-dihydro-2, 2 dimethy-7-benzofuranoyl N-methylcarbamate), a cholinesterase inhibitor is widely used as carbamate insecticide for controlling the soil-dwelling and foliar feeding insects *viz.*, corn root worm, wire worms, boll weevils, mosquitoes, alfalfa weevils and white grubs (Trabue *et al.*, 1997; Trotter *et al.*, 1991). These agrochemicals have been specifically chosen for assessment of their degradation kinetics by indigenous rhizobacteria. Our published data clearly demonstrate the efficient degradation of these selected agrochemicals by the soil isolates (Bano and Musarrat, 2004; 2003a; Musarrat *et al.,* 2000). Substantial growth of our soil isolates NJ-10 (*Pseudomonas* sp.), NJ-15 (*Pseudomonas aeuroginosa*) and PS-1 (*Rhizobium sp.*), PS-2 (*Pseudomonas putida*) and PS-3 (*Proteus* sp.) in mineral salt medium containing 2,4-D and phorate, respectively validated the ability of these strains to mineralize the selected compounds as a carbon and energy sources (Bano and Musarrat, 2003a; Musarrat *et al.,* 2000). Moreover, the carbofuran degrading bacterial strains NJ-101 (*Pseudomonas aeuroginosa*), NJ-115 (*Pseudomonas putida*) and NJ-125 (*Pseudoxanthomonas mirabilis*) exhibiting growth in the mineral salt medium and nitrogen-free mineral salt medium supplemented with 200μgml^{-1}carbofuran suggest the utilization of carbofuran as both the carbon and nitrogen source (Bano and Musarrat, 2004). The degradation isotherms suggested a time-dependent disappearance of agrochemicals following the first-order rate kinetics (Bano and Musarrat, 2004; Bano and Musarrat, 2003a; Musarrat *et al*, 2004) as also suggested earlier (Baskaran *et al*, 1996; Szeto *et al*, 1990). Comparative analysis of our degradation data revealed the differential rate of agrochemical dissipation in soil in the order as 2,4-D > Carbofuran > Phorate with the half-lives ($T_{1/2}$) of 10, 15 and 20 days respectively. Several other degrader bacteria exhibiting pathways of agrochemical degradation such as hydrolysis, photolytic oxidation, and microbial transformation have been isolated and characterized (Racke, 1992; Szeto *et al.*, 1990). However, the information available on the microbial decomposition of majority of agrochemicals in the environment is limited.

Interestingly, our agrochemical degrading bacterial strains NJ-15 and NJ-101 exhibited resistance to rifampicin, while the strains NJ-10, NJ-15, PS-2, PS-3, NJ-101, NJ-115 exhibited resistance to nalidixic acid. Resistance to these drugs is not mediated by plasmids and it could be due to mutations (Courvalin, 1990). The resistance to rifampicin is known to be mediated by mutation in ββ sub-unit of RNA polymerase (Sippel and Hartmann, 1968). Indeed, the resistance to rifampicin is unusual among soil bacteria (Nautiyal, 1997). The chromosomal nature of the mutation affords greater stability than occurs with the plasmid-borne markers, and is non-transferable (Compeau *et al.*, 1988). The acquisition of resistance traits against environmental conditions including toxic substances helps in survival of microorganisms in intimidating ecological niches. Therefore, the use of such degrader strains for bioremediation of soils contaminated with agrochemicals is a focus of considerable attention owing to their potential for restoring or reclaiming sites in a cost-effective manner. Such bio-resource could be recommended for bioremediation of pesticides contaminated soil after thorough evaluation of their efficacy in field.

Bioremediation of inorganic contaminants: Microbes encounter metals and metalloids of diverse nature in the environment. Of particular interest are the base metals including vanadium, chromium, manganese, iron, cobalt, nickel, copper, zinc, molybdenum, silver, cadmium and lead; the precious metals gold and silver; and the metalloids arsenic, selenium, and antimony (Breirley, 1986). Microbes can detoxify metals by valence transformation, extracellular chemical precipitation or volatilization. Such microbes combat high concentrations of heavy metals by (a) inactivation of metals (b) alteration of the site of inhibition (c) impermeability of the metals and (d) other by-pass mechanisms (Belliveau and Trevors, 1989). A variety of metals can be enzymatically reduced by microbial metabolic processes that are not related to metal assimilation (Oremland *et al.*, 1994). The toxic forms of metal can be converted to more tolerable non-toxic forms by bacterial reduction. For instance, the reduction of toxic selenate and selenite to the insoluble and much less toxic elemental selenium could be exploited to enhance removal of these anions from contaminated sites (Gerrard *et al.*, 1974; Heider and Boeck, 1994; Kessi *et al.*, 1999; Stalder *et al.*, 1997). Bacterial biomass can as well bioaccumulate heavy metals both in live and dead states through intracellular accumulation and extracellular adsorption, respectively, giving an effective alternative for small scale remediation purposes (Chang and Hong, 1994). Bacteria can remediate heavy metals by a variety of mechanisms including bioaccumulation, biosorption and bioreduction.

Bioaccumulation and biosorption of metal: Microorganisms can physically remove heavy metals and radionuclides from solution through either bioaccumulation or biosorption. Bioaccumulation is the retention and concentration of a substance by an organism. In this process, the metals are transported from outside the microbial cell through cell membrane into cytoplasm, where the metal is sequestered and immobilized. However, in biosorption the positively charged metal ions are sequestered primarily through adsorption to the negative ionic groups on cell surfaces, such as capsules or slime layers. Binding sites on microbial cell surfaces usually are carboxyl residues, phosphate residues, SH groups or hydroxyl groups. The amount of metal biosorbed to the exterior of bacterial cells often exceeds the amount predicted, depending upon the charge density on the cell surface. The charged functional groups serve as nucleation sites for deposition of various metal-bearing precipitates. Many studies have demonstrated the efficiency of metal removal by microbial biomass in a range of reactor formats and under varying physical and chemical conditions (Beveridge and Koval,

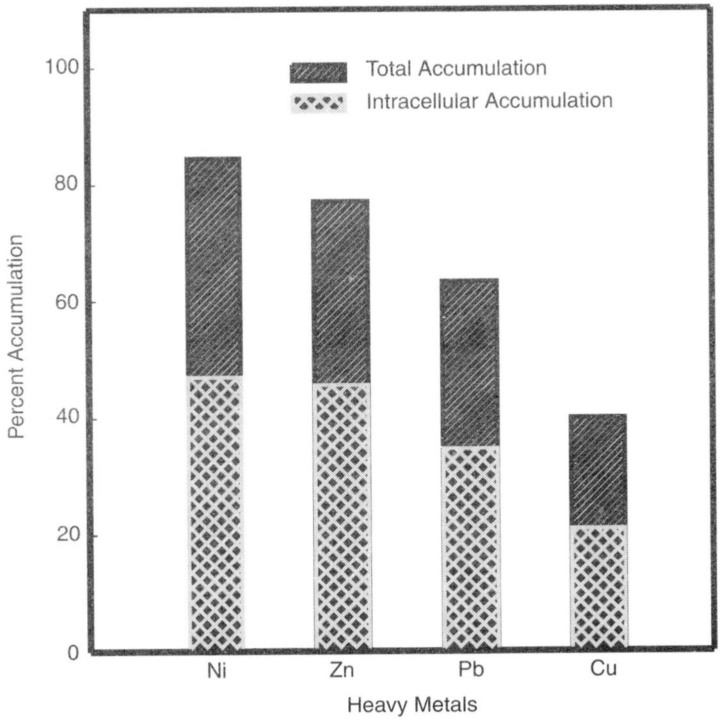

Fig. 4: Bioaccumulation of heavy metals by viable cells of isolate SJ-101 in minimal low phosphate medium at 30°C (Adopted from: Zaidi & Musarrat (2004)).

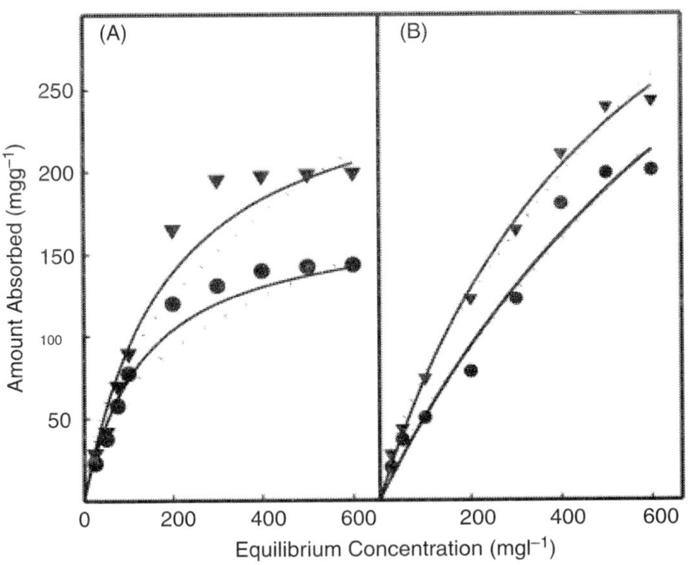

Fig.5: Adsorption isotherms of Ni in concentration range of 0 - 600 mgl^{-1} with the biomass of isolateSJ101 (?) and amberlite IR-120 (?) both at 25^{0}C (A) and 40^{0}C (B). The soli and broken lines represent the fitting of data by Langmuir and Freundlich models (Adopted from: Zaidi & Musarrat (2004)).

1981; Beveridge, 1978; Chang and Hong, 1994; Hartmeier and Berends, 1995; Holan and Volesky, 1993; Horikoshi *et al.*, 1981; Volesky, 1990b; Volesky and Chong, 1995). Certain bacterial species involved in biosorption of heavy metals are summarized in Table 2. Comparative analysis of data in Table 2 revealed that the soil isolate SJ-101 (*Bacillus* sp.) exhibit much higher capacity of intracellular nickel (Ni) accumulation. Besides, the biomass of strain SJ-101 showed significantly higher Ni adsorption *vis-à-vis* chemical adsorbent (Amberlite IR-120) (Zaidi and Musarrat, 2004). The enhanced biosorption and bioaccumulation capacity of this strain is attributed to the anionic nature of its cell surface. Thus, the strain SJ-101, due to its high Ni binding and sequestration potential, could be exploited as a valuable bioinoculant for Ni bioremediation (Figs. 4 and 5).

Table 2. Sorption Capacity of Bacteria for Various Heavy Metals.

Metals	Organisms	Sorption Capacity (mmol/g)	References
Au	*B. subtilis (cell wall)*	0.4010	Beveridge (1986)
Co	*E. coli AB 264 (envelope)*	0.0020	Beveridge & Kovel (1981)
Cr	*Streptomyces nouressi*	0.0340	Mattushika et al. (1993b)
Cu	*B. amyloliquefaciens*	0.0437	Hartmeier & Berends (1995)
	B. subtilis	2.9900	Walker *et al.* (1989)
	B. subtilis (walls)	0.5300	Beveridge (1978)
	E.coli AB 264(envelope)	0.0900	Beveridge & Kovel (1981)
Fe (III)	*E. coli AB 264(envelope)*	3.5810	Beveridge (1978)
	A. subtilis (cell wall)	3.5900	Beveridge (1986)
Hg	*Pseudomonas sp.*	0.0089	Chang & Hong (1994)
	B. amyloliquefaciens	0.0437	Hartmeier & Berends (1995)
	E. coli AB 264(envelope)	0.0640	Beveridge & Kovel (1981)
La (III)	*B. cereus*	0.0330	Muller *et al.* (1989)
	P. aeuroginosa	0.1440	Muller *et al.* (1989)
	P. aeuroginosa	0.3970	Muller *et al.* (1989)
Mn	*B. subtilis*	0.8010	Beveridge (1978)
	E. coli AB264 (envelope)	0.1400	Beveridge, & Koval (1981)
Ni	*B. subtilis (walls)*	0.1070	Beveridge (1978)
	Bacillus sp. (SJ-101)	4.1600	Zaidi & Musarrat (2004)
Pb	*B. amyloliquefaciens*	0.0437	Hartmeier & Berends (1995)
	E. coli AB264 (envelope)	0.1520	Beveridge & Koval (1981)
	B. subtilis (walls)	0.5430	Walker *et al.* (1981)
U	*B. subtilis AHU 1390*	0.2219	Horikoshi *et al.* (1981)

Bioreduction: Bioreduction is a process of biologically catalyzed redox reaction that lead to either immobilization or solubilization of metals. Microorganisms are known to reduce a wide variety of multivalent metals that pose environmental problems (Gorby *et al.*, 1998; Gurby and Lovely, 1992; Lloyd and Mackasie, 1996; Lovely, 1993; Lovely *et al.*, 1991; Nealson and Saffarini, 1994; Roden and Zachara, 1996; Weilinga *et al.*, 2001; Wildung *et al.*, 2000). Direct enzymatic reduction involves the use of oxidized forms of these contaminants as alternate electron acceptors. The oxidized forms of metals are highly soluble in aqueous media and are generally the most mobile species in aerobic ground water. However, the reduced species are highly insoluble and precipitate out from solution.

Redox reactions leading to immobilization. Direct enzymatic reduction of soluble uranium

U (VI), chromium Cr (VI) and technetium Tc (VII) to insoluble species is well-documented (Henrot, 1989; Lovely and Phillips, 1992; Sabaty *et al.*, 2001). Earlier studies demonstrated the conversion of U (VI) to U (IV) through microbial binding, which form intracellular precipitate, and can be easily mined by a process referred as 'microbial mining' (Lovely and Phillips, 1992). Metal-reducing bacteria also reduces technetium Tc (VII) to Tc (IV) (Henrot, 1989), forming oxide mineral TcO_2 and likewise tellurium (Sabaty *et al.*, 2001) to form gray deposits inside the cell matrix. Such strains could efficiently serve as 'bio-ores' for the expensive and rare earth metals. Also, the bioreduction of hexavalent chromium occurs under both the aerobic and anaerobic conditions. A number of Cr (VI) reducing microbial strains including *Oscillatoria* sp., *Arthrobacter* sp., *Agrobacter* sp., *Pseudomonas ambigua, Chlamydomonas* sp. (algae), *Chlorella vulgaris* (algae), *Zoogloea ramigera,* and anaerobic sulphate-reducing bacteria have recently been isolated from chromate-contaminated water, soil, and sediment (Lloyd *et al.*, 2002; Shriver *et al.*, 1994; Smith and Gadd, 2000; Suzuki *et al.*, 1992; VanRoy *et al.*, 1995). Metal reducing organisms also reduce certain metalloids like arsenic and selenium (Ahmann *et al.*, 1994; Kashiwa *et al.*, 2000; Oremland *et al.*, 1998). In aerated soils and aquatic environments, selenium occurs predominantly in the form of selenite and selenate oxyanions SeO_3^{-2} and SeO_4^{-2} that are freely available to living organisms. Selenate (SeO_4^{-2}) and selenite (SeO_3^{-2}) seem to be the most abundant forms of bioavailable selenium in the environment, and both can serve as electron acceptors for many microorganisms, including some phototropic bacteria. It is known that microbial reduction of selenate (Se^{6+}) into elemental selenium (Se^0) via selenite plays an important role in detoxification of soluble Se in the natural environment. Several species of bacteria are capable of reducing sub-millimolar levels of Se (VI) to Se (0). These include *Wollinella succinogenes* (Tomei *et al.*, 1992) *Desulfovibrio desulfuricans* (Tomei *et al.*, 1995), *Pseudomonas stutzeri* (Lortie *et al.*, 1992) and *Enterobacter cloacae* (Losi and Frankenberger, 1997). Also, a selenate and selenite respiring strain, isolated and characterized in our laboratory, significantly reduces the soluble selenium oxyanions to non-toxic and insoluble elemental selenium. Bacterial cells accumulating insoluble elemental selenium were easily distinguishable due to deep red pigmentation, which increases in a concentration dependent manner upto 12 mM selenium oxyanions (our unpublished observation). Thus, owing to its high capability of reducing selenate to its insoluble elemental form, the strain could be exploited as potential bioinoculant for bioremediation of selenium contaminated sites.

Redox reaction leading to solubilization: Metal-reducing bacteria also promote the mobilization of insoluble forms of some heavy metals. Solubilization of adsorbed and co-precipitated metals may occur by direct or indirect microbial processes. The solubilization of toxic heavy metals and radionuclides from co- precipitates requires at least partial solubilization of oxide minerals (Lovely and Phillips, 1992). Such dissolution reactions have been shown to release cadmium, nickel, and zinc into solution during reduction of geothite (a form of iron oxide) by an anaerobic *Clostridium* species (Lovely, 1991). It has been demonstrated that metal-reducing bacteria solubilize plutonium dioxide (PuO_2) in the presence of the synthetic chelator (Rusin et al., 1994). This process may provide means of mobilizing plutonium (Pu) from contaminated soils and sediments. Direct reduction of iron oxides provides potential for releasing a wide range of heavy metals and radionuclides that were co-precipitated and immobilized in sub-surface sediments. Presumably, the organic acids formed by the metabolic activity of microbes lowers the pH of the system so that it interferes with the electrostatic forces that hold heavy metals and radionuclides on the surface of iron or manganese oxide minerals. Displacement of cations by hydrogen ions may lead to solubilization of the surface-associated metal. In some cases, the organic metabolites also serve as chelating agents that can form soluble metal-ligand complexes. These chelating agents, such as dicarboxylic acids, phenolic compounds, ketogluconic acids and salicylic acids, have been shown to promote the dissolution of a wide range of heavy metals and radionuclides, including PuO_2, copper, uranium, thorium and nickel oxides, and can accelerate the movement of metals in soils and sediments. Since only limited metals have been tested, extensive and systematic research is needed in this area.

PHYTOREMEDIATION: A BENEFICIAL ALTERNATIVE

Phytoremediation (the term *phyto* means 'plant' and *remediate* mean to 'clean-up') is a natural process in which the plants are used to remove pollutants from soil and water. Plants can degrade or transform both the organic and metal contaminants by acting as filters or traps. It is the direct use of living green plants for *in situ* risk reduction in contaminated soil sludges, sediments and ground water. Growing or harvesting plants on a contaminated site as a remediation method is an aesthetically pleasing, solar energy driven passive technique that can be used to clean-up contaminated sites with low to moderate levels of contamination. Plants and associated microbes can degrade these pollutants or at least limit their spread in the environment. Various enzymes that are capable of metabolizing pollutants have been identified in plants. Nevertheless, the understanding of the mechanism of degradation of organic pollutants by plants is still limited *vis-à-vis* bacteria-assisted bioremediation. It is also difficult to assess the contribution of rhizospheric microbes (bacteria and fungi associated with plant roots) to the overall success of plant-assisted phytoremediation of organics. Nevertheless, the initial progress in utilizing plants for the clean-up and containment of organic pollutants warrants serious evaluation of this area of research. Indeed, an integrated approach is required to effectually clean-up or to remediate the organics from ecosystem (Raskin *et al.*, 1997). There are several ways in which plants can be used for the phytoremediation of organic contaminants *viz.*, phytoextraction, phytodegradation, rhizodegradation and phytovolatization. These techniques can be used, along with or in some cases in place of mechanical clean-up methods, to reduce the load of contaminants such as metals, pesticides, solvents, explosives, crude oil and polyaromatic hydrocarbons and landfill leachetes. Plants have also been used for metal bioremediation; they can be compared to solar driven pumps for extraction and concentration of elements from the environment (Alkorta and Garbisu, 2001; Brown *et al.*, 1994; Raskin *et al.*, 1994; Salt *et al.*, 1998). Following the harvest of pollutant-enriched plants, the weight and volume of contaminated material can be further reduced by ashing or composting. Metal enriched plants can be disposed of as hazardous material or, if economically feasible, used for metal recovery (Salt *et al.*, 1995). However, the ability to accumulate hea y metals varies significantly between species and between cultivars within species. Lately, the possibility of planting metal hyper-accumulator crops over a low-grade ore body or mineralized soil, and then harvesting and incinerating the biomass to produce commercial 'bio-ore', has been proposed (Brooks *et al.*, 1998).This process is referred to as 'Phytomining' (Anderson *et al.*, 1999).

BIOAUGMENTATION

Bioaugmentation is the enrichment or isolation of specific microorganisms from the contamination site and their subsequent addition to the target site, after sub-culturing in bulk, for enhanced *in situ* biodegradation of pollutants (Walter, 1997). This treatment process is recommended at sites where the indigenous microflora does not display sufficient activity for effective bioremediation or biostimulation. Bioaugmentation could be performed in different ways, such as using (*i*) microorganisms indigenous to the application target (*ii*) microorganisms from similar soil in same geographical location (*iii*) constructed or forced mutated microbes (Heuer *et al.*, 1995) and (*iv*) microorganisms with additional genetic material such as plasmids and viruses, with the intention of dispersing the appropriate genetic material amidst to indigenous populations. An important objective in bioaugmentation is to maximize the survival of microorganisms released into the environment. The survival of microorganisms freely released into soil is influenced by numerous factors. These include nutrient availability, moisture, physical and chemical characteristics of the environment, predation, parasitism, and competition from the indigenous microbial population. Any one or a combination of these environmental factors can adversely affect the ability of an augmenting population to carry out biodegradation at useful rates over a prolonged period (England *et al.*, 1993; Goldstein *et al.*, 1985; Lobenz, 1995; Portier *et al.*, 1988; Recorbet *et al.*, 1992). The study of specific microorganisms in environmental samples is complicated by the tremendous complexity of natural ecosystem. For instance, the soil environment has been approximated to contain between 10^9 to 10^{10} cells of more than 10^3 distinct genotypes per gram (Torsvik *et al.*, 1990). To monitor a specific genotype in these complex environments, several molecular and microbial ecology methods have been developed (Atlas, 1992; Jansen, 1995; Prosser, 1994). Many of these methods are based on tracking cells tagged with biomarkers or bioreporters. A number of sensitive and selective biomarkers/bioreporters are currently available. Since the scorable phenotypes and the relevant methods for detection vary considerably, the use of a given biomarker/bioreporters will depend on the nature of the environmental or industrial samples to be analyzed. Thus, tracking and monitoring of the bioinoculant introduced in soil is utmost important for the assessment of effective bioaugmentation.

REPORTER SYSTEMS AND THEIR SIGNIFICANCE IN TRACKING BIOINOCULANTS

Importance of a reporter system has been emphasized over the years considering antibiotic resistance as one of the strong parameters. However, increasing incidence of antibiotic resistant bacteria in the environment has raised serious concern on this issue (Armstrong *et al.*, 1981; Atlas and Bartha, 1992). Thus, the need for construction of novel reporter systems has been realized. Reporter systems consist of special genes that are either inherently encoded on extra-chromosomal DNA (plasmid) or artificially inserted into the genome of microbes. Some of these reporter genes *viz. xylE* (Prosser, 1994), *lacZY* (Winstanley, 1989), *lux* (Drahos *et al.*, 1986), and *gfp* (Meighen, 1991) are briefly discussed.

***xylE* reporter gene:** This gene codes for catechol 2,3-dioxygenase (C230). Bacterial strains tagged with this gene can be detected and identified by induced expression of the enzyme at 37 to 42°C, followed by spraying catechol over the culture plates for visual detection. The enzyme transforms catechol to a yellow coloured hydroxymuconic semialdehyde product yielding yellow bacterial colonies (Atlas and Bartha, 1992).

***lacZY reporter gene*:** The *lac* Z gene codes for the enzyme β-galactosidase that hydrolyzes lactose to D-galactose and D-glucose (Atlas and Bartha, 1992; Prosser, 1994; Winstanley, 1989). The *lac* Y gene encodes for lactose permease that transports lactose into bacterial cells. Bacterial colonies expressing *lac* ZY genes turn blue-green upon exposure to chromogenic dye X-Gal (5-chloro-4-bromo-3-indoyl-β-D-galactopyranoside). This reporter gene system served as a useful marker for tracking *Pseudomonas* and several other bacteria in the environment.

***lux reporter gene*:** Bioluminescence is a phenotypic marker based on the introduction of genes for light emission. Originally cloned from *Vibrio fischeri* and *Vibrio harveyi*, the *lux* operon includes *lux* AB coding for the luciferase enzyme, *lux* CD coding for reductase and transferase polypeptides of the fatty acid reductases, and *lux* E, which codes for synthetase (Drahos *et al.*, 1986). The light emitting reaction involves the oxidation of reduced $FMNH_2$ and a long-chain fatty aldehyde with the emission of blue-green light (Drahos *et al.*, 1986). Luminescence can be measured using a luminometer, scintillation counter, or photomultiplier. It may also be directly imaged using photographic or X-ray films and charge-coupled devices (Atlas and Bartha, 1992; Heim *et al.*, 1994). Application of *lux AB* gene as a marker has been demonstrated in tracking *Pseudomonas* sp. UG14, a phenanthrene degrader, isolated from a creosote contaminated soil sample (Drahose, 1988; Hemming and Drahose, 1984). The benefit of this technique for environmental applications is the ability to measure light in real-time without prior extraction of cells from the environment.

***gfp reporter gene*:** Another development in phenotypic markers is a fluorescent genetic tag called the green fluorescent protein (gfp) (Flemming, 1999; Meighen, 1991; Providenti *et al.*, 1995). The *gfp* genes were isolated from jellyfish *Aequorea victoria* and the gfp gene product emits green light upon excitation with blue light. The chromophore, responsible for gfp fluorescence, is p-hydroxybenzylidene-imidazole, which is completely contained within the coding sequence of protein and does not require any unusual cofactors. This reporter gene also enables *in vivo* visualization of differential gene expression and protein localization when the *gfp* genes are fused directly to the genes encoding the protein of interest (Providenti *et al.*, 1995; Winstanley, 1989). Presently, *gfp* genes have been incorporated into several bacterial genomes including *Pseudomonas* sp. UG14Gr (Dyke *et al.*, 1996) and *Moraxella* sp. strains G21 and G25 (Stearns, 1995).

Nucleic acid probes and DNA microarray for tracking bioinoculants: DNA hybridization technology is commonly used to detect the fate of introduced DNA sequences and avoid dependency on phenotypic expression of introduced genes. The technology is based on the formation of double-stranded DNA from complementary DNA sequences. A homologous complementary probe is applied to the targeted denatured nucleotide sequence bound to a filter. Probe labels may be radioactive or non-radioactive. However, the former is less preferred due to its short half-lives, hazardous nature and special waste disposal. Excess probe is washed from the filter and the presence of probe is visualized using X-ray film or by chemical colour production (Youvan, 1995). Using the appropriate probes, specific DNA sequences such as nitrogen fixation genes (nif H), sulfite reduction (dsv A/B) nodulation (nod) genes, mercury resistance (merA) genes and many other specific genes of metabolic pathways can be detected in culturable cells using colony hybridization. Lately, the application of cDNA microarray has revolutionized the process of expression profiling of environmentally important genes (Errampalli, 1999). The author has custom designed a DNA microarray for an aquifer isolate Pseudomonas stutzeri strain KC, and suggested its application for tracking this bacterium at the bioaugmented site. This bacterium is well known for transforming carbon tetrachloride to carbon dioxide (Criddle *et al.*, 1990; Musarrat and Hashshani, 2003). Besides, the DNA microarray technology has widespread applications in biological and agricultural research including measurement of gene copy number (Tatara *et al.*, 1993), gene mapping (Pinkel *et al.*, 1998), gene profiling in yeast (Chee *et al.*, 1996; Lashkari *et al.*, 1997), arabidopsis (Spellman *et al.*, 1998) and in bacteria (Errampalli, 1999; Schena *et al.*, 1995; Tao *et al.*, 1999). To date about 40 microbial genomes have been completely sequenced and at least 100 microbial genomes are currently being sequenced (Richmond *et al.*, 1999). With the increasing number of fully sequenced bacterial genomes, the practical problems in tracking of bioinoculants in the environment, filamentous microbial populations in activated sludge, biological nitrogen and phosphorus removal, strain level differentiation of critical pathogens, and bioaugmentation can be adequately addressed by using this high throughput and highly parallel technique. Advancement of research in this area has led to the rapid development and commercialization of DNA microarrays for several microbes (Schena *et al.*, 1995; Tao *et al.*, 1999).

FUTURE PROSPECTS

Development of Genetically Modified Microbes and Plants for Bioremediation

Improvement of the degradative capabilities of microbes by altering degradation genes is important for developing efficient strains and technique for bioremediation. For this purpose, bacteria with complete catabolic pathways for pollutant degradation can be constructed by natural genetic exchange between different microbes. For instance, the plasmid mediated transfer of the chromosomally encoded dehalogenase I gene from *Pseudomonas putida* PP3 to other bacteria has resulted in transconjugants with a new ability to degrade halogenated compounds (De Saizieu *et al.*, 1998). Genetic engineering techniques can create new catabolic pathways from existing pathways which degrade analogue compounds. Thus, the genetic engineering tools may be utilized to insert into genome, the genes capable of aggravating remedial performance (Newman *et al.*, 1997). This may enhance the ability of bioinoculants to metabolize many of the compounds that are of environmental concern. Similarly, the transgenic tobacco plant resistant to cadmium has been developed with stable integration of human metallothionein (MT) genes. Thus, introduction of metal binding peptides and proteins into the plants is a compelling strategy to enhance metal-tolerance and/or accumulation. These metal binding proteins or peptides should be preferentially metal specific such that only the toxic metals *viz.* Cd, Hg, Pb are sequestered and not essential trace metals like Zn. *In vitro* grown transgenic seedlings expressing the fusion protein accumulated 60-70% less Cd in their shoots than control plants (Elmayan and Tepfer, 1994). Another major breakthrough in this field has been achieved by Rugh *et al.* (Rugh, 1998) by inserting altered mercuric ion reductase gene (*mer*A) into *Arabidopsis thaliana* creating a mercury resistant transgenic plant that volatizes mercury to atmosphere. Further studies are warranted on metal uptake, translocation and storage in order to develop a clear cut understanding for metal distribution to non-edible tissue or the harvestable tissue in plant used for phytoremediation (Alkorta and Garbism, 2001).

Strategies for Effective Delivery of Bioinoculant for Bioremediation

Encapsulated microbial cells in a porous biopolymer have been tested for effectiveness of *in situ* bioremediation. The encapsulation maintains the survival and metabolic activity of microorganisms in a more controlled microenvironment within the heterogeneity of the soil environment (Cassidy *et al.*, 1996; De Borne, 1998). Use of biodegradable and non-toxic algal biopolymers such as alginate or *k*-carrageenan has been suggested to be the effective approaches for environmental applications (Cassidy *et al.*, 1997). Encapsulated microbial cells have been effectively used in bioreactors to detoxify or remediated liquid suspensions of toxic substances. In liquid medium, the encapsulated cells have demonstrated significant increase in ability to tolerate and mineralize toxic compounds (Trevors *et al.*, 1993; Van Elsas, 1992). Although, the application of microbial immobilization technology in bioremediation is still in its infancy, but promises to offer enhanced effectiveness of microbial inoculants in environmental applications. However, certain limitations including oxygen availability, morphological alterations of cells inside beads (smaller cells) and effects of contaminant availability need further assessment and exploration.

Another method of increasing the survival of introduced bioinoculants is the immobilization on a suitable matrix. Indeed, immobilization increases the performance of bioinoculant as compared to the non-immobilized cells (Briglia *et al.*, 1990). A range of matrices have been used for immobilization are: polyurethane (PUR) foam (Briglia *et al.*, 1990), granular clay or lava slag (Weisel *et al.*, 1993), loofah sponge (Ogbonna *et al.*, 1994) and biologically activated carbon (Weber and Rehm, 1993). Immobilization studies have demonstrated the higher survival and sustained activity of the bioinoculants in the environment (Weber and Rehm, 1993).

CONCLUSION

Bioremediation is increasingly being recognized as a promising technology. Traditionally, it works by supplementing the contaminated area with indigenous and/or foreign microorganisms with inherent biodegradation potential. Another common approach for bioremediation is bio-stimulation, which relies on the ability of bioinoculants for providing abundant nutrients to naturally occurring microorganisms in order to facilitate the process of degradation of local contaminants. However, the use of foreign bacteria may disrupt the existing food chain, contribute to the depletion of essential ground nutrient, and even produce naturally occurring but locally harmful by-products. Although, the bioremediation technology has enormous potential and prospects for effective environmental clean-up, the inherent risk of ecological disruption owing to possible nutrient toxicity and over-stimulation cannot be ignored. Too much of a particular nutrient can disrupt the equilibrium. Under such a stressed condition the bacteria may turn off their major metabolic functions and focus on their survival or they may die compromising with the task of bioremediation. Nevertheless, studies on indigenous rhizospheric microflora, genetically modified microbes and improvements in techniques for monitoring and tracking bioinoculants have opened the gateways for exploiting efficient microorganisms for improving the bioremediation process, as an alternative to chemical or physical remediation strategies. In addition, new technologies such as microbial encapsulation and phytoremediation have provided novel approaches to accentuate the existing bioremediation processes. Furthermore, systematic and concerted efforts are required for assessing the effects of bioremediation and phytoremediation on the food chain and natural recycling as well as for making the process more safer, economical and eco-friendly.

REFERENCES

Ahmann, D., Roberts, A.L., Krumholz, L.R. & Morel, F.M.M. (1994) *Nature.* **371:** 750-758.

Alexander, M. (1994) *Biodegradation and Bioremediation*, Academic Press, San Diego, California.

Alkorta, I. & Garbisu, C. (2001) *Biores. Technol.* **77:** 229-236.

Anderson, C. W. N., Brooks, R. R., Chiarucci, A., Lacoste, C. J., Leblanc, M., Robinson, B. H., Simcock, R. & Stewart. R. B. (1999) *J. Geochem. Explor.* **67(1-30):** 407-415.

Armstrong, J.L., Shigeno, D.S., Calomiris, J.J. & Seidler, R.J. (1981) *Appl. Environ. Microbiol.* **42:** 277-283.

Atlas, R.M. & Bartha, R. (1992) *Microbial Ecology: Fundamentals and application*, 3rd ed. The Benjamin/ Cummings Publishing Co., Redwood City, California.

Atlas, R.M. (1992) *Biodegradation* **3:** 137-146.

Baker, K. H. & Herson, D.S. (1994) *Bioremediation* McGraw-Hill, New York.

Bano, N. & Musarrat, J. (2004) *FEMS Microbiol. Lett.* **231:** 13-17.

Bano, N. & Musarrat, J. (2003) *Lett. Appl. Microbiol.* **36:** 1-5.

Bano, N. & Musarrat, J. 2003 *Curr. Microbiol.* **46:**324-328.

Baskaran, S., Bolan, N.S., Rahman, A. & Tillman, R.W. (1996) *New Zealand J. Agri. Res.* **39:** 297-310.

Belliveau, B.H. & Trevors, J.T. (1989) *Appl. Organometal Chem.* **3:** 283-294.

Beste, C.E. (1983) *Herbicide Handbook of the Weed Science Society of America*, Weed Science Society of America, 5th ed, Champaigne, IL. pp 515

Beveridge, T.J. & Koval, S.F. (1981) *Appl. Environ. Microbiol.* **42:** 325-330.

Beveridge, T.J. (1978) *Can. J. Microbiol.* **24:** 89-95.

Beveridge, T.J. (1986) *Biotechnol. Bioengg. Symp.* , New York, USA, **52:** 127-140.

Borden, R.C., Vogel, T.M., Thomas, J. M. & Ward, C.H. (1994) *Handbook of Bioremediation.* Lewis Publishers, Boca Raton, FL.

Breirley, C.L. (1986) *Immobilization of ions by biosorption.* (Eds. Eccks, H. & Hunt, S.) Ellis Horwood Ltd. Chichester, England, pp. 105-117.

Briglia, M., Nurmiaho-Lassila, E.L., Vallini, G. & Salkinoja-Salonen, M. (1990) *Biodegradation* **1:** 273-281.

Brooks, G. T. (1974) *Chlorinated Insecticides*, Vol. **1,** CRC Press, Boca Raton, FL, pp. 186-194.

Brooks, R.R., Chambers, M.F., Nicks, L.J. & Robinson, B.H. (1998) *Trends Plant Sci.* **3:** 359-362.

Brown, S.L., Chaney, R.L., Angle, J.S & Baker, A.J.M. (1994) *J. Environ. Qual.* **23:**1151-1157.

Carvalho, F.P., Nhan, D.D, Zhong, C., Tavares, T. & Klaine, S. (1998) *Results of an International Research Project, Tracking Pesticides in the Tropics* IAEA, Bulletin, pp.24-30.

Cassidy, M.B., Lee, H & Trevors, J.T. (1996) *J. Ind. Microbiol.* **16:** 79-101.

Cassidy, M.B., Lee, H & Trevors, J.T. (1997) *Appl. Environ. Microbiol.* **47:** 108-113.

Chang, J.S. & Hong, J. (1994) *Biotechnol. Bioengg.* **44:** 999-1006.

Chaudhry, G.R. & Chapalamaduga, S. (1991) *Microbial. Rev.* **55**: 59-65.

Chee, M., Yang, R., Hubbell, E., Berno, A., Huang, X.C., Stern, D., Winkler, J., Lockhart, D.J., Morris, M.S. & Fodor, S.P. (1996) *Science* **274:** 610-614.

Compeau, G., Al-Achi B. J., Platosouka, E. & Levy S.B. (1988) *Appl. Environ. Microbiol.* **54**: 2432-2438.

Cookson, J.T. Jr. (1995) *Bioremediation Engineering: Design and Application.* McGraw-Hill, New York.

Courvalin, P. (1990) *Antimicrobial Agents and Chemotherapy* **34:** 681-684.

Criddle, C.S., Dewitt, J.T., Gribic-Garlic, D. & McCarty, P.L. (1990) *Appl. Environ. Microbiol.* **56:** 3240-3246.

De Borne, F.D. (1998) *Mol. Breeding* **4:** 83-90.

De Saizieu, A., Certa, U., Warrington, J., Christopher, G., Keck, W. & Mous, J. (1998) *Nat. Biotech.* **16:** 45-48.
Donnelly, P.K., Entry, J.A. & Crawford, D.L. (1993) *Appl. Environ. Microbiol.* **59**: 2642-2647.
Drahos, D.J. (1988) *The release of genetically engineered microorganisms*, (Eds. Sussman, M., Collins, G.H., Skinner, F.A. & Stewart, D.E.) Academic Press, San Diego, pp. 181-191.
Drahos, D.J., Hemming, B.C. & McPherson, S. (1986) *Biotechnology* **4:** 439-444.
Dyke, V.M.I., Lee, H. & Trevors, J.T. (1996) *J. Chem. Biotechnol.* **65:** 115-122.
Elmayan, T & Tepfer, M. (1994) *Plant J.* **6:** *433-440.*
England, L. S., Hung, L. & Trevors, J. (1993) *Soil Biol Biochem.* **25:** 525-531.
Errampalli, D. (1999) *J. Microbiol. Methods* **35:** 187-199.
ESCAP (2000) *State of the Environment in Asia and Pacific.* United Nation, pp. 905-911.
Felsot, A., Maddox, J.V. & Bruce, W. (1981) *Bull. Environ. Contam. Toxicol.* **26**: 781-788.
Flemming, C.A. (1999) *Appl. Environ. Microbiol.* **60:** 1606-1613.
Gerrard, T.L., Telford, J.N. & Williams, H.H (1974) *J. Bacteriol.* **119:** 1057-1060.
Goldstein, R. M., Mallory, L.M. & Alexander, M. (1985) *Appl. Environ. Microbiol.* **50:** 977-983.
Gorby, Y.A., Caccavo, F.J. & Bolton, H. J. (1998) *Environ. Sci. Technol.* **32:** 244-247.
Gurby, Y.A. & Lovley, D.R. (1992) *Environ. Sci. Technol.* **26:** 205-207.
Hartmeier, W. & Berends, A (1995) *Meded Fac. Landbouwet Rijksuniv. Gent.* **60:** 2585-2600.
Heider, J. & Boeck, A. (1994) *Adv. Microbiol. Physiol.* **35:** 71-109.
Heim, R., Prasher, D.C. & Tsien, R.Y. (1994) *Proc. Natl. Acad. Sci. USA* **91:** 12501-12504.
Hemming, B.C. & Drahos, D.J. (1984) *J. Cell Biochem. Suppl.* **8b:** 252-259.
Henrot, J. (1989) *Health Phys.* **57:** 239-245.
Heuer, H., Dwyer, D.F., Timmis, K.N. & Wagner, I. (1995) *Microb. Ecol.* **29:** 203-220.
Hinchee, R.E., & Olfenbuttel, R.F. (1991) *In situ Bioreclamation: Applications and Investigations for Hydrocarbon and Contaminated Site Remediation.* Buttterworth-Heinemann, Boston.
Hinchee, R.E., Anderson, D.B., Metting, F.B. Jr. & Sayles, G.D. (1994) *Applied Biotechnology for Site Remediation.* Lewis Publishers, Boca Raton, FL
Holan, Z.R. & Volesky, B. (1993) *Biotechnol. Bioengg.* **43:**1001-1009.
Horikoshi, T., Nakajima, A. & Sakaguchi, T. (1981) *Eur. J. Appl. Microbiol. Biotechnol.***12:** 90-99.
Jansen, J.K. (1995) *Curr. Opin. Biotechnol.* **6:** 275-283.
Kashiwa, M., Nishimoto, S., Takahashi, K., Ike, M. & Fujita, M (2000) *J. Biosci. Bioengg.* **89:** 528-533.
Kessi, J., Ramuz, M., Wehrli, E., Spycher, M. & Bachofen, R (1999) *Appl. Environ. Microbiol.* **65(11):** 4734-4740.
Klassen, W., Ridgway, R.L. & Inscoe, M. (1982) *Insect Suppression with Controlled Release Pheromone Systems,* Vol. **1** (Eds. Kydoneius, A. and Beroza, M.), CRC Press, Boca Raton, FL, pp. 14-21.
Lashkari, D.A., DeRisi, J.L., McCusker, J.H., Namath, A.F., Gentile, C., Hwang, S.Y., Brown, P.O. & Davis, R.W. (1997) *Proc. Natl. Acad. Sci. USA* **94:** 13057-13062.
Levin, M.A. & Gealt, M.A. (1993) *Biotreatment of Industrial and Hazardous Waste.* McGraw Hill, New York.
Lloyd, J.R. & Mackasie, L.E. (1996) *Appl. Environ. Microbiol.* **62:** 578-582.
Lloyd, J.R., Macaskie, L.E. & Mabbett, A.N. (2002) *Biotechnol. Bioengg.* **79(4):** 389-397.
Lobenz, G. (1995) *Bioremediation Rep.* **4:** 1-2.
Lortie, L., Gould, W.D., Rajan, S., McCready, R.G.L. & Cheng, K.J (1992) *Appl. Environ. Microbiol.* **58:** 4043-4044.
Losi, M.E. & Frankenberger, Jr W.T. (1997) *Appl. Environ. Microbiol.* **63:** 3079-3084.
Lovley, D.R. & Phillips, E.J.P. (1992) *Environ. Sci. Technol.* **26:** 2228-2234.
Lovley, D.R. (1991) *Microbiol. Rev.* **55:** 259-287.
Lovley, D.R. (1993) *Annu. Rev. Microbiol.* **47:** 263-290.
Lovley, D.R.& Phillips, E.J.P. (1992) *Environ. Sci. Technol.* **26:** 2228-2234.
Lovley, D.R.., Phillips, E.J., Gorby, Y.A. & Landa, E. (1991) *Nature* **350:** 413-416.
Mathur, S.C. (1999) *Pesticide Information*, Vol XXIV (4): pp. 9-23.
Means, J. & Hinchee, R.E. (1994) *Emerging Technology for Bioremediation of Metals.* Lewis Publishers, Boca Raton, FL.

Meighen, E.A. (1991) *Microbiol Rev.* **55:** 123-142.
Mullen, M.D., Wolf, D.C., Ferris, F.G., Beveridge, T.J, Flemming, C.A. & Bailey, G.W. (1989) *Appl. Environ. Microbiol.* **55:** 3143-3149.
Musarrat, J. & Hashshani, S.A. (2003) *Terat. Carcinogen. Mutagen..* **1:** 1-12.
Musarrat, J., Bano, N. & Rao, R.A.K. (2000) *World J. Microbiol. Biotechnol.* **16**, 495-497.
Nautiyal, C.S. (1997) *Curr. Microbiol.* **35**: 52-58.
Nealson, K. & Saffarini, D. (1994) *Ann. Rev. Microbiol.* **48:** 311-343.
Newman, L.A., Strand, S.E., Choe, N., Duffy, J. & Ekuan, G. (1997) *Environ. Sci. Technol.* **31:** 1062-1067.
Ogbonna, J.C., Liu, Y.C., Liu, Y.K. & Tanaka, H. (1994) *J. Ferment. Bioeng.* **78:** 434-437.
Oremland, R.S., Bindi, A.B. & Blum, J.S. (1998) *Appl. Environ. Microbiol.* **171:** 19-30.
Oremland, R.S., Blum, J.S., Culbertson, C.W., Vesscher, P.T., Miller, L.G., Dowdle, P. & Strohmaier, F.E. (1994) *Appl. Environ. Microbiol.* **60:** 6311-6319.
Pinkel, D.R., Segraves, R., Sundar, D., Clark, S., Poole, I., Cowbel, D., Collins, C., Kuo, W.L., Chen, C., Zhai, Y., Dairkee, S.H., Ljung, B.M., Gray, J.W. & Albertson, D.G. (1998) *Nat. Genet.* **20:** 207-211.
Portier, R., Bianchini, M., Fujisaki, K., Henery, C. & McMillin, D. (1988) *Schriftenr Ver Wassser Boden Lufthyg Berlin-Dahlem.* **80:** 273-292.
Prosser, J.I. (1994) *Microbiol.* **140:** 5-17.
Providenti, M.A., Greer, C.W., Lee,H. & Trevors, J.T. (1995) *World J. Microbiol. Biotechnol.* **65:** 115-122.
Racke, K.D. (1992) *Organophosphates: Chemistry, Fate, and Effects,* (Eds. Chambers, J. E. & Levi, P.E.), Academic Press, San Diego, pp.47-73.
Raskin, I., Kumar, P.B.A.N., Dushenkov, S. & Salt, D.E. (1994) *Curr. Opin. Biotechnol.* **5:** 285-290.
Raskin, I., Smith, R.D. & Salt, D.E. (1997) *Curr. Opin. Biotechnol.* **8:** 221-226
Recorbet, G., Steinberg, C.& Faurie, G. (1992) *FEMS Microbiol. Ecol.* **101:** 251-260.
Richmond, C., Glasner, J., Mau, R., Jin, H. & Blattner, F. (1999) *Nucleic Acids Res.* **29:** 3821-3835.
Riser-Roberts, E. (1992) *Bioremediation of Petroleum Contaminated Sites.* CRC Press, Boca Raton, FL.
Roden, E.E. & Zachara, J.M. (1996) *Environ. Sci. Technol.* **30:** 1618-1628.
Rosenberg, E. (1993) *Microorganisms to Combat Pollution.* Kluwer Academic Publishers, Dordrecht, The Netherlands.
Rosenberg, E., Legmann, R., Kushmaro, A., Taube, R., Adler, E. & Ron, E.Z. (1992) *Biodegradation* **3**:337-350.
Rugh, C.L. (1998) *Nat. Biotechnol* **16:** 925-928.
Rusin, P.A., Brainard, J.R., Strietelmeier, B.A., Tait, C.D., Ekberg, S.A., Palmer, P.D., Newton, T.W. & Clark, D.L. (1994) *Environ. Sci. Technol.* **28:** 1686-1690.
Sabaty, M., Avazeri, C., Pignol, D. & Vermigold, A. (2001) *Appl. Environ. Microbiol.* **67:** 5122-5126.
Salt, D. E., Smith, R.D. & Raskin, I. (1998) *Annu. Rev. Plant Physiol. Plant Mol. Biol.* **49:** 643-668.
Salt, D. E., Blaylock, M., Kumar, P.B.A.N., Dushenkov, V., Ensley, B.D., Chet, I. & Raskin, L. (1995) *Biotechnology* **13:** 468-475.
Sander, P., Wittich, R.M., Fortnagel, P., Wilkes, H. & Franke, W. (1991) *Appl. Environ. Microbiol.* **55**: 1624-1629.
Schena, M., Shalon, D., Davis, R.W. & Brown, P.O. (1995) *Science* **270:** 467-470.
Shriver, D.F., Atkins, P.W. & Langford, C.H. (1994) *Inorganic Chemistry*, 2nd edition, Oxford, UK: Oxford University Press. pp 200-250.
Singer, M.E. & Finnerty, W.R. (1984) *Petroleum Microbiology.* (Ed. Atlas, R.M.), Macmillan, New York. pp. 1-59.
Sippel, A.E. & Hartmann, G.R. (1968) *Biochim. Biophys. Acta.* **157:** 218-219.
Smith, W.L. & Gadd, G.M. (2000) *J. Appl. Microbiol.* **88:** 983-991.
Spain, J.C. (1995) *Biodegradation of Nitro Aromatic Compounds*. Plenum Press, New York.
Spellman, P.T., Sherlock, G., Zhang, M.Q., Iyer, V.R., Anders, K., Eisen, M.B., Brown, P.O., Bostein, D. & Futcher, B. (1998) *Mol. Biol. Cell.* **9:** 3273-3297.
Stalder, V.F., Gurleyuk, V.H., Bachofen, R. & Chasteen, T.G (1997) *J. Ind. Microbiol. Biotechnol.* **19:** 98-103.
Stearns, T. (1995) *Curr. Biol.* **5:** 262-264.

Stoner, D.L. (1994) *Biotechnology for the Treatment of Hazardous Waste.* Lewis Publishers, Boca Raton, FL.

Suzuki, T., Miyata, N., Horitsu, H., Kawai,K., Takamizawa, K., Tai, Y. & Okazakai, M. (1992) *J. Bacteriol.* **174:** 5340-5345.

Szeto, S.Y., Price, P.M., Mackenzie, R.J. & Vernon, R.S. (1990) *J. Agricult. Food Chem.* **38:** 501-504.

Tao, H., Bausch, C., Richmond, C., Blattner, F. & Conway, T. (1999) *J. Bacteriol.* **181:** 6425-6440.

Tatara, G.M., Dybas, M.J. & Criddle, C.S. (1993) *Appl. Environ. Microbiol.* **59:** 2126-2131.

Tate, R.L. (2000) *Soil Microbiology*, 2nd ed., John Wiley & Sons, Inc. New York, pp 464-494.

Tomei, F.A., Barton, L.L., Lewanski, C.L. & Zocco, T.G (1992) *Can. J. Bacteriol.* **38:** 1328-1333.

Tomei, F.A., Barton, L.L., Lewanski, C.L., Zocco, T.G., Fink, N.H. & Sillerud, L.O. (1995) *J. Ind. Microbiol.* **14**:329-336.

Torsvik, V., Goksoyr, J & Doae, F.L.(1990) *Appl. Environ. Microbiol.* **56:** 782-787.

Trabue, S.L., Feng, X., Ogram A.V. & Ou. L.T (1997) *J. Environ. Sci. Health B.* **32**: 861-867.

Trevors, J.T., Van-Elass, J.D., Lee, H. & Wolters, A.C. (1993) *Appl. Microbiol. Biotechnol.* **39:** 637-643.

Trotter, D., Kent, R. & Wong, M. (1991) *Crit. Rev. Environ. Cont.* **21(2)**: 137-176.

Unterman, R. (1996) *Bioremediation: Principles and Applications.* (Eds Crawford, R.L. & Crawford, D.L.) University Press, Cambridge. pp. 209-253.

Van Elsas, J.D. (1992) *Biol. Fertil. Soils.* **14:** 14-22.

VanRoy, S., Peys, K., Dresselaers, T. & Diels, L.(1995) *Res. Microbiol.* **148:** 526-528.

Volesky, B (1990b) *Biosorption of Heavy Metals*, CRC Press, Boca Raton, FL, pp.7-44.

Volesky, B. & May-Philllips, H.A. (1995) *Appl. Microbiol. Biotechnol.* **42:** 797-806.

Volesky, B. & Chong, K.H. (1995) *Biotechnol. Bioengg.* **47:** 451-460.

Walter, M.V. (1997). Bioaugmentation, (Eds. Hurst, C.J., Knudsen, G.R., McInciney, M.J., Strtzenbeach, L.D. & Walter, M.V.) *Manual of Environmental Microbiology* ASM Press, Washington, pp. 753-757.

Weber, W. J. Jr. & Rehm, H.J. (1993) *Appl. Microbiol. Biotechnol.* **39:** 110-116.

Weilinga, B., Mizuba, M.M., Hansel, C.M. & Fendorf, S. (2001) *Environ. Sci. Technol.* **34:** 522-527.

Weisel, I., Wubker, S.M. & Rehm, H.J.(1993) *Appl. Microbiol. Biotechnol.* **61:** 448-455.

Wildung, R.E., Gorby, Y. A., Krupka, K.M., Hess, N.J., Li, S.W., Plymale, A. E., Mckinley, J.P. & Fredrickson, J.K. (2000) *Appl. Environ. Microbiol.* **66:** 2451-2460.

Winstanley, C. (1989) *Appl. Environ. Microbiol.* **55:** 771-777.

Youvan, D.C. (1995) *Science* **268:** 264-270.

Zaidi, S. & Musarrat, J. (2004) *J. Environ. Sci. Health A.* **A 39(3):** 681-691.

18

Bioremediation of Organochlorine and Organophosphate Pesticides by Using Microorganisms

Manisha Kapur, Sumeet Labana and R. K. Jain
Institute of Microbial Technology
Sector 39-A, Chandigarh-160036, (India)

Pesticides are widely applied in Indian agricultural fields as insecticides, fungicides, herbicides or rodenticides. Commonly used pesticides eg. organochlorines or organophosphates are toxic and recalcitrant and have detrimental effects on human health. Their accumulation in the environment over the years has been considered hazardous and several of them have therefore been banned by different agencies. Attempts are being made towards decontaminating the pesticide-contaminated sites. Several suitable microorganisms have been identified that can degrade these pesticides or their hydrolysed products. Using such microorganisms bioremediation strategies are being developed for treatment of a variety of organic pollutants. This article reviews the various types of pesticides, their potential use in India, environmental impacts and microbial accumulation upon them with special emphasis on organochlorine and organophosphate pesticides. It also discusses the application of microbial potential for decontamination of pesticide-contaminated environments with this group of pesticides.

INTRODUCTION

Pesticides are synthetic chemical substances that are used for preventing, destroying, repelling or mitigating living organisms regarded as pests. Since 1940s, the use of pesticides has grown steadily at about 11% a year. Pesticides and fertilizers play a central role in agriculture and contribute to an enhanced production of food. It is hard to imagine the agriculture of the 21st century without the use of these substances. For many years, our society has enjoyed the benefits of using pesticides. There are huge quantities of pesticides being used in all arenas of life starting from household pesticides to public health care and in agriculture. An assumption behind much of this use has been that pesticides will either degrade quickly or bind to soil particles until they eventually break down into harmless substances. Only within the last decade the scientists have learnt that pesticides on the land can leach through the soil and enter the groundwater below. For this reason, the impact of pesticides on the quality of surface and groundwater has become an issue of national importance. While 50% of the nation depends upon groundwater as a source of drinking water, about 95% of all households in rural areas nationwide use groundwater as their primary source of drinking water.

CLASSIFICATION OF PESTICIDES

Pesticides are variously classified into different categories based on their chemical composition or the purpose for which they have been made or based on their health effects. They could be organophosphates or organochlorines or carbamates or synthetic pyrethroids as per chemical classification. Pesticides are also divided according to the goal of their use against insects (insecticides), fungi (fungicides), weeds (herbicides), rodents (rodenticides) etc. Pesticides, especially insecticides and fungicides, are more heavily applied for tropical cash crops—such as banana, coffee, cotton and vegetables—than for crops in temperate regions. For example, the application of pesticides in banana plantations in Costa Rica attained 45 kg per hectare, whereas the comparable average application of pesticides in Japan for crops is 10.8 kg (Carvalho *Et al.*, 1998). Experts have estimated that only a minor fraction of applied pesticides—less than 0.1% reaches the target pest species, excess pesticide moves throughout the environment potentially contaminating soil, water and biota. It is critical to characterize the fate and non-target toxicity of these pesticides to confidently assess the risk associated with their use. It is now increasingly believed that many of the environmental problems that have come to light in the past 40 years are not isolated from each other at all but rather have been caused mainly by organochlorine and organophosphate group of pesticides.

Organochlorines are carbon-based substances that have one or more chlorine atoms. Universally important organochlorine pesticides (OCPs) are para, dichloro diphenyltrichloroethane (p,p′DDT), benzene hexachloride (BHC), chlordane, hepatachlor, aldrin, dieldrin, endrin, endosulfan. Pesticides like DDT, though they are not supposed to be used for agricultural purposes, find their illegal way into this use in India. The production, use and disposal of organochlorines create numerous problems. While not all organochlorines have the same impacts, the chemicals as a class include carcinogens, endocrine disruptors and substances that harm nervous, reproductive and immune systems. Almost every organochlorine studied has been linked to some environmental or human health harm.

The second category of the deadly pesticides includes the organophosphate pesticides (OPs), that are among the most widely used classes of pesticides in the world, and about 30% of total pesticides being used in India belong to this category. Annual usage amounts of OPs in the USA are in the order of 70 million pounds, with approximately 3.7 million pounds of OPs being applied in Canada (*Environment Canada*, 1996). Parathion and methyl parathion are two popular organophosphate pesticides which are widely used. These hydrolyze to form *p*-nitrophenol (PNP), which subsequently builds up in the soil. The United States Environmental Protection Agency (USEPA) has listed PNP as a priority pollutant necessitating the removal of this compound from soil. High-level exposure to organophosphates results in acetylcholine (ACh) accumulation, which interferes with muscular responses and produces serious symptoms in vital organs, eventually leading to death. Repeated or prolonged exposure to organophosphates can cause delayed cholinergic toxicity and neurotoxicity (De Frank, 1991).

The unique properties of such compounds make them environmentally persistent with a global distribution and thus they exert chronic-toxic effects on wildlife and humans.

Carbamates are ester derivatives of N-substituted carbamic acid and are a widely used class of pesticides. This group includes a number of important pesticides such as carbofuran, carbaryl, aldicarb, propoxur and methomyl. Insecticides of this group have very high mammalian toxicity. The major concern for such pesticides is from accidental spills such as the Bhopal gas disaster. Another concern is due to a decrease in the degradation efficiency and failure to control intended pests.

Synthetic pyrethroids are used in agriculture for insect control on field crops because of their properties such as short field-life and relatively low mammalian toxicity. Pyrethroids, when drained into lakes or river water, turn harmful due to changes in their chemical structure by biological processes.

PESTICIDE USAGE IN INDIA

India, unlike many other developed countries, produces and uses many deadly pesticides which have been banned years ago in other countries. Table 1 depicts the consumption levels of these various pesticides globally and in India.

Recovering from the euphoria of the Green Revolution, India is now battling the residual effects of the extensively used chemical fertilizers and pesticides in the country's soil. The injudicious and indiscriminate use of these pesticides has resulted in widespread contamination of soil, drinking water and food products. The decade from 1980 to 1990 in India alone saw the area under pesticides increased a whopping 20-fold from six million hectares to 125 million hectares. After a high annual consumption of nearly 75000 MT reached in the early 90's, interventions in the form of Integrated Pest Management (IPM) practices have only now started to show a declining trend in the use of pesticides in India. However, India's consumption of pesticides per hectare is low compared to world averages-0.5 kg/ha against Korea's 6.60 kg/ha and Japan's 12.0 kg/ha. Despite this low consumption of pesticides, the contamination of food products in the country is alarming. About 20% of Indian food products contain pesticide residues above tolerance level compared to only 2% globally. A nation-wide survey of food contaminants, conducted by the Indian Council of Medical Research (ICMR), revealed excessive levels of a cancer-causing fungal product in cereals and toxic metals in other food products of India (Anonymous, 1993). It has also detected high levels of pesticide residues in cows milk and metals like arsenic, cadmium and lead in infant canned products and in turmeric.

A study by the Uttar Pradesh state's water works department in September 2003 (Anonymous, 2003a) reported that groundwater sources in 36 districts in northern U.P. were contaminated with traces of cadmium, fluoride, arsenic, nitrates and lead. Low concentrations of arsenic can slowly build up in the body, eventually causing cancers, skin diseases and other illnesses. Prolonged exposure to arsenic can cause kidney, liver, intestinal, neurological, cardiovascular and respiratory disorders. A recent issue of *Indian Farming*, an ICMR publication, reported that the carpet industry in UP's Bhadoi had discharged effluents in open drains contaminating soil, water and plants with chromium (Anonymous, 2003b). In Jalandhar, leather industry effluents mixed with sewage water has led to higher concentrations of contaminants in soil, potato leaves and tubers. Recently, toxic pesticides in concentrations far higher than permitted by the European Economic Commission were found in 12 cold drink brands sold in and around New Delhi ((http://www.cseindia.org/html/cola-indepth/softdrinks report.pdf.).). Lab tests revealed that all samples, including those of Coke and Pepsi contained residues of four toxic pesticides and insecticides—lindane, DDT, malathion and chlorpyriphos in the soft drinks. The environmental organization blamed the high levels of pesticide residues on the chemical contamination of groundwater used as the basic raw material in the soft drink industry.

PESTICIDES, ENVIRONMENT AND HEALTH

The environmental impacts of pesticide residues and their effect on human health are matters of concern. Pesticides are bio-accumulative (build up in fatty tissues of animals) and bio-magnifying. Organisms at the top of the food chain are most adversely affected as these pollutants accumulate in maximum quantity in them through a process of bio-magnification. Usually pesticide residues can be found in air, water and soil beyond levels considered as safe. Pesticides are known for causing deaths of other organisms like rare species of eagles, peacocks, fish and other aquatic life, bees, beneficial organisms in farming etc. Extensive studies have shown that pesticides affect the soil health by killing the vital microbial fauna present in the soil. Each gram of soil may contain millions of microbes that are important for sustaining plant life. Pesticides not only kill the microbes in soil, they persist in the environment and accumulate in the food chain. Pesticides are carcinogenic, mutagenic and teratogenic and are also reported to possess estrogenic activity (Malley, 1997).

Studies have shown that pesticides can cause health problems, such as birth defects, nerve damage, cancer and other effects that might occur over a long period of time. WHO estimates one million pesticide poisoning cases and 20,000 deaths every year globally. Persistent organic pollutants (POPs) due to their lipophilic nature, have a tendency to accumulate in fatty tissue, making women most vulnerable to toxic contaminants, since women have a higher percentage of body fat, accompanied by regular changes through the reproductive cycle. Recent studies on the estrogen-like behaviour of DDTs and PCBs in humans suggest the implications of these compounds in breast cancer (Calle *et al.*, 2002).

This review will mainly focus on the organochlorine and organophosphate group of pesticides. Some key examples of these groups of pesticides have been discussed in detail in the following sections of the review.

Organochlorine Pesticides

Organochlorines (OCs) are a diverse group of unique class of pesticides because of their cyclic structure, number of chlorine atoms and low volatility. They were released into the environment in the past decades through their use as pesticides or industrial products. They have been extensively used for controlling the diseases of humans and domestic animals that are carried by insects and mites and also against insect-pests which cause great damage to the agricultural crops. Due to their efficiency as insecticides, these compounds were considered a boon to the field of agriculture and medical entomology. OCs are classified into four categories: diclorodiphenyl ethanes (eg. DDT), cyclodienes (eg.-endosulfan), chlorinated benzenes [eg. hexachlorobenzene (HCB)] and cyclohexanes [eg. hexachlorocyclohexane (HCH)].

Organochlorine pesticides are effective against a variety of insects. Hexachlorobenzene and pentachlorophenol had been primarily used as fumigants. These chemicals were introduced in the 1940s and are used rarely in the United States today because of their environmental persistence. The USEPA banned the use of many of these chemicals during the 1970s and 1980s. Although many of these chemicals are no longer produced in the United States, they continue to be used by other countries.

OC pesticides enter the environment from disposal of contaminated wastes into landfills, emissions from waste incinerators and releases from manufacturing plants that produce these chemicals. Although there are no major sources in polar regions, these pesticides have been measured in air, sea-water, ice and mammals in Antarctic (Iwata *et al.*, 1993). DDT, hexachlorocyclohexane, toxaphene and other organochlorines are found in air, snow, water, suspended sediments, birds, fish, marine flora, mammals and humans in the Arctic (Muir *et al.*, 1992). In aquatic systems, OCs are adsorbed onto sediments in water that can then bioconcentrate in marine mammals. Toxaphene is now the most abundant pesticide in arctic aquatic life (Muir *et al.*, 1992) and concentration of HCH is higher in arctic waters than in more southern waters (Iwata *et al.*, 1993, Muir *et al.*, 1992). Because these chemicals are soluble in fat, they are found in higher concentrations in fatty foods. Diets that contain fats that may be contaminated with organochlorine pesticides (eg. contaminated milk and dairy products, fish, whales) lead to increased exposure to these chemicals. Studies have been carried out on pesticidal contamination of foodstuffs from different parts of the world (Mitra *et al.*, 1999). Children can be exposed to these chemicals through breast milk and *in utero* through the placenta. In studies conducted during the past years, milk samples from Delhi and Ludhiana were found to contain p,p'DDT and p,p'TDE respectively (John *et al.*, 2001).

Microbial degradation of organochlorines: Microorganisms play important roles in metabolism of organochlorines. It has been shown that microorganisms play an important role in the degradation/detoxification of xenobiotics like OCs and the process of microbial degradation has been shown to reduce the toxicity of xenobiotics by 2 to 800 times (Munnecke *et al.*, 1982). Recently, the selection of microbes from the natural environment that can degrade organochlorines has received much attention (Lal *et al.*, 1995;, Johri *et al.*, 1996). Microbial degradation is the main process affecting the environmental persistence of pesticides (Adhya *et al.*, 1981). Removal of unwanted residues, as well as pesticide efficacy, is ultimately dependent on the presence, number and enzymatic capability of soil microorganisms. However, bioavailability and physical parameters such as pH and soil type strongly influence the rate at which degradation occurs (Adhya *et al.*, 1981).

Pesticide metabolism by soil microorganisms has been investigated extensively over last two decades. Many studies have documented the microbial degradation rates of selected compounds and whether these compounds serve as substrates for microbial growth (Alexander, 1999). Microorganisms mainly degrade OC insecticides by reductive dechlorination, dehydrochlorination, oxidation and isomerization of the parent molecule. The details of these types of reactions have been described and discussed in the following sections.

Reductive dechlorination: Reductive dechlorination of OCs is an important microbial reaction. The reaction proceeds by replacing a chlorine atom on a non-aromatic carbon with a hydrogen atom. A classic example of reductive dechlorination is the conversion of DDT to DDD. This reaction has been found to occur in several species of bacteria such as *Pseudomonas,*

Bacillus, Micrococcus, Arthrobacter ((Patil *et al.*, 1970), *Clostridium* (Cohisa *et al.*, 1980), Yeast (Kallman and Andrews, 1963), soil actinomycetes (Chacko *et al.*, 1966) and in fungi like *Trichoderma viride* (Matsumura and Boush, 1968), *Mucor alternans* (Anderson *et al.*, 1970). White rot fungi have also been found to be capable of mineralizing DDT by reductive dechlorination. Three strains of *P. chrysosporium* plus *Pleurotus ostreatus, Phellinus weirii* and *Polyporus versicolor* were able to mineralize 5.3-13.5% of added ^{14}C-radiolabelled DDT, diclofol, and methoxychlor over 30 days under ligninolytic growth conditions (Bunpus and Aust, 1987).

Degradation of insecticide lindane to gamma-3, 4, 5, 6-tetrachlorocyclohexane (gamma-TeCCH) by *Clostridium rectum* (Ohisa and Yamaguchi, 1970) and *P. putida* (Mutsumura *et al.*, 1976) also proceeds via reductive dechlorination. Bacteria that are able to use lindane anaerobically produce γ-TeCCH (Ohisa and Yamaguchi, 1970; Ohisa *et al.*, 1980). *Clostridium* and several species of Bacillaceae and Enterobacteriaceae are able to degrade lindane anaerobically (Jagnow *et al.*, 1977). Aerobically grown facultative anaerobes are also able to dechlorinate lindane during aerobic incubation with glucose, pyruvate or formate as substrates.

Dehydrochlorination: This reaction involves the simultaneous removal of hydrogen and chlorine from organochlorine insecticides. Typically the reaction takes place between the saturated chlorinated carbon and the adjacent hydrogen on the neighbouring carbon. Two classical examples of such reaction are:

(*i*) Formation of DDE from DDT. This mechanism is commonly observed in several marine species of algae, diatoms and phytoplankton such as *Cylindrotheca closterium* (Kenega, 1974), *Skeletonema costatum, Cyclotella nana, Isochrysis galbana* ((Rice and Sikka, 1973), *Nitzschia* sp. (Miyazaki and Thorsteinson, 1972).

(*ii*) The second classical reaction of dehydrochlorination is the formation of gamma-PCCH from lindane. Cultures have been reported to exhibit this mechanism of dehydrochlorination such as *Pseudomonas* (Benezet and Matsumura, 1973), *E. coli* (Vonk and Quirjns, 1979), *Citrobacter freundii* (Jagnow *et al.*, 1977), *Chlorella vulgaris* and *Chlamydomonas reinhardii* ((Sweeney, 1969).

Oxidation: The oxidation reactions are important reactions of OC degradation in higher organisms but are less common during microbial degradation. This is probably because of the lack of mixed function oxidase system in microorganisms (Matsumura, 1975). Examples of oxidation reaction are the formation of hepatachlor epoxide from hepatachlor (Lichtenstein and Schulz, 1960) and the formation of kelthane from DDT (Matsumura and Bousch, 1968).

Isomerization: Microbial isomerization reactions involve the conversion of γ-BHC to a-BHC (Beneze*t* and Matsumura, 1973), dieldrin to photodieldrin (Matsumura, 1975) and d-ketoendrin to endrin (Mendel *et al.*, 1967). In each of these cases, the isomers differ from the parent compound due to the difference between the position of the chlorine atom in the benzene ring.

Biological isomerization is a rare event. The biological interconversion of lindane to other isomers of HCH and formation of other metabolites have been reported in plants, animals and in microbes (Benezel and Mutsumura, 1973; Deo *et al.*, 1994; Macholy and Kujawa, 1985; Srinathi *et al.*, 1985). A *Pputida* strain capable of producing α-HCH from γ-HCH in laboratory conditions has been isolated (Benezel and Mutsumura, 1973). This reaction was catalyzed by nicotinamide adenine dinucleotide (NAD) and to a lesser extent by flavin adenine dinucleotide (FAD). The high levels of α-HCH and β-HCH in the environment may be possibly due to the isomerization of γ-HCH.

DDT

DDT is one of the most extensively studied insecticides. Several microorganisms are known to degrade DDT [1, 1-bis(4-chlorophenyl-2, 2, 2-trichlorophenyl) ethane] anaerobically (Wedemeyer, 1966). The primary metabolic mechanism is the reductive dechlorination of DDT to DDD. A number of studies on DDT metabolism in microorganisms have shown that under anaerobic conditions, reductive dechlorination of DDT predominates, whereas under aerobic conditions, dehydrochlorination is the dominant reaction.

Table 1: Consumption levels of various pesticides globally and in India.

Types of pesticides	Consumption percentage (%)	
	Global	India
Insecticides	~29	~63
Herbicides	~44	~14
Fungicides	~21	~21
Others	~6	~2

Cell-free extracts of *Aerobacter aerogenes, Pseudomonas fluorescens, E. coli* and *Klebsiella pneumoniae* dechlorinated p-p′-DDT to 1, 1-dichloro-2, 2-bis(4-chlorophenyl) ethane (DDE) exclusively under anaerobic conditions. When oxygen was provided, PCPA was formed implicating bacterial ring cleavage (Pfaender and Alexander, 1972). *A. aerogenes* was incubated with DDT in parallel experiments conducted aerobically and anaerobically (Mendel *et al.*, 1969; Wedemeyer, 1967). In these experiments, it was shown that DDT was degraded to DDD under both conditions, but anaerobic conditions favored high yield of DDD. However, Johnson *et al.* (1967) using pure cultures of microorganisms found that no species, including *A. aerogenes* produced DDD under aerobic conditions, but most produced DDD under anaerobic conditions. Several species of bacteria, algae and fungi have been isolated that are capable of metabolizing DDT. Some of the organisms are listed in Table 2. The various stages of aerobic and anaerobic degradation pathway of DDT are described in Fig. 1. The new aerobic catabolic pathway of DDT degradation provides the possibility of cloning catabolic genes responsible for the degradation of DDT, enabling us to understand the capacity and evolution of microbes involved in the concomitant degradation of PCBs, DDT and related chloroaromatic compounds (Lal *et al.*, 1995).

Table 2: Microorganisms involved in the metabolism of DDT.

Microorganisms	Metabolite	Reference
Pseudomonas sp.	DDD	Patil *et al.*(1970)
Bacillus sp.	DDD	Patil *et al.*(1970)
Micrococcus sp.	DDD	Patil *et al.*(1970)
Arthrobacter sp.	DDD	Patil *et al.*(1970)
Klebsiella pneumoniae	DDD	Yang *et al.* (2003)
Aerobacter aerogenes	DDD	Yang *et al.* (2003)
Streptomyces aureofaciens	DDD	Chacko *et al.* (1966)
Streptomyces annamoneus	DDD	Chacko *et al.* (1966)
Cylindrotheca closterium	DDE	Kenaga *et al.* (1974)
Skeletonema constatum	DDE	Rice & Sikka (1973)

Cyclotella nana	DDE	Rice & Sikka (1973)
Isochrysis galbana	DDE	Rice & Sikka (1973)
Nitzschia sp.	DDE	Miyazaki & Thorsteinson (1972)
Trichoderma viridae	DDD, DDE, DDNS, DDA	Matsumura & Boush (1968)
Mucor alternans	Solvent soluble and water soluble metabolites	Anderson *et al.* (1970)

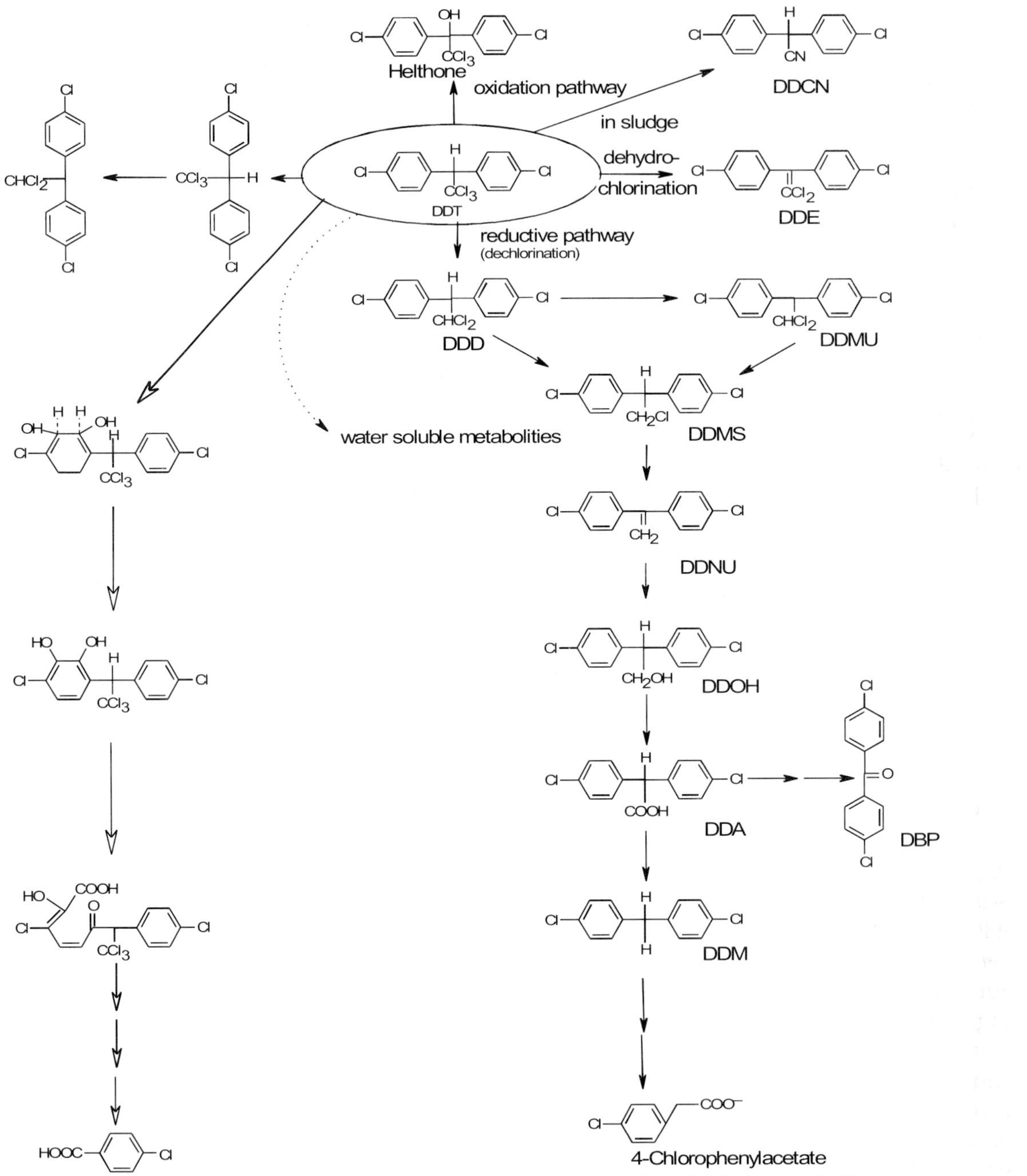

Fig. 1: Microbial Degradation Pathway of DDT (Adapted from Siddiquie *et al.*, 2003).

Hexachlorocyclohexane (HCH)

Hexachlorocyclohexane (HCH), popularly known as benzene hexachloride (BHC) is an organochlorine pesticide with theoretically eight possible isomeric forms. Out of these, four (alpha, beta, gamma and delta) forms predominate in the technical product. The gamma isomer, commonly known as lindane, is the only isomer with insecticidal activity. The other isomers are used either as fungicides or to synthesize other chemicals. Technical grade HCH contains all four isomers but mostly the alpha isomer. These four isomers of HCH are generally stable to light, high temperature, hot water and acid. However, they can be dechlorinated in the presence of alkali (Karnath and Deo, 1997). These isomers are very persistent in the environment and resistant to microbial degradation (Agnihotri *et al.*, 1994). Because of their high lipid solubility, bioaccumulation of HCH residues in aquatic and terrestrial animals is high (Lal *et al.*, 1986).

The rapid disappearance of γ-BHC from soil and aquatic environments has been attributed to its susceptibility to degradation by microorganisms (MacRae *et al.*, 1967; Yule *et al.*, 1967). Like DDT metabolism, microbial degradation of γ-BHC has been observed to occur more rapidly under anaerobic conditions (Kohnen *et al.*, 197564). Lindane is degraded aerobically as well as anaerobically. The general pathway for the microbial degradation of lindane has been shown in Figure 2. The major metabolites of γ-BHC in microorganisms are γ-PCCH and γ-TCCH. The former is formed by dehydrochlorination of γ-BHC and the latter is formed by dechlorination of γ-BHC. Some of the microorganisms involved in γ-BHC degradation have been shown in Table 3.

Table 3: Microorganisms involved in the metabolism of γ-BHC.

Microorganisms	Metabolite	Reference
Clostridium sp.	γ-TCCH	MacRae *et al.* (1967)
Pseudomonas putida	γ-PCCH, γ-TCCH, CO_2	Benezet & Matsumura (1973)
E.coli	*γ-PCCH, α-BHC*	*Vonk & Quirjns (1979)*
Cirtobacter freundii	γ-PCCH, γ-TCCH	Jagnow *et al.* (1977)
Chlorella vulgaris	*γ-PCCH*	*Sweeney (1969)*
Chlamydomonas reinhardii	*γ-PCCH*	*Sweeney (1969)*

Only four aerobes have been reported to degrade γ-HCH. *P. putida* ATCC 17484 was reported to degrade γ-HCH through dechlorination (Mutsumura *et al.*, 1976). *Pseudomonas* NO_{62} degrades γ-HCH to γ-PCCH, γ-TCCH and tetrachlorobenzene (Tu, 1976). A strain of *Pseudomonas paucimobilis* SS86, isolated from an upland experimental field was found to degrade γ-HCH through dehydrochlorination (Senoo and Weda, 1989; Wada *et al.*, 1989). Another *Pseudomonas* sp. strain isolated from the rhizosphere of an HCH treated sugarcane plant, was found to degrade α and γ-HCH and also β-HCH under aerobic conditions (Sahu *et al.*, 1993). Nagata *et al.* (1993) identified γ-PCCH, 1, 2, 4-TCB, 2, 5-DCP, 2, 5-DDOL and 2, 5-DCHQ as metabolites of HCH degradation by *P. paucimobilis* UT26 (*Sphingomonas paucimobilis*). UT26 is a unique microorganism that utilizes γ-HCH as its sole source of carbon and energy under aerobic conditions (Imani *et al.*, 1989). Johri *et al.* (1996) and Imani *et al.* (1991) initially characterized the degradative pathway of lindane in *Sphingomonas* UT26. In UT26, lindane is converted by two steps of dehydrochlorination to a chemically unstable intermediate compound

1, 4-TCDN, which is then metabolized to 2, 5-DDOL by two hydrolytic dehalogenation steps via another chemically unstable intermediate 2, 4, 5-DNOL (Nagata *et al.*, 1994).

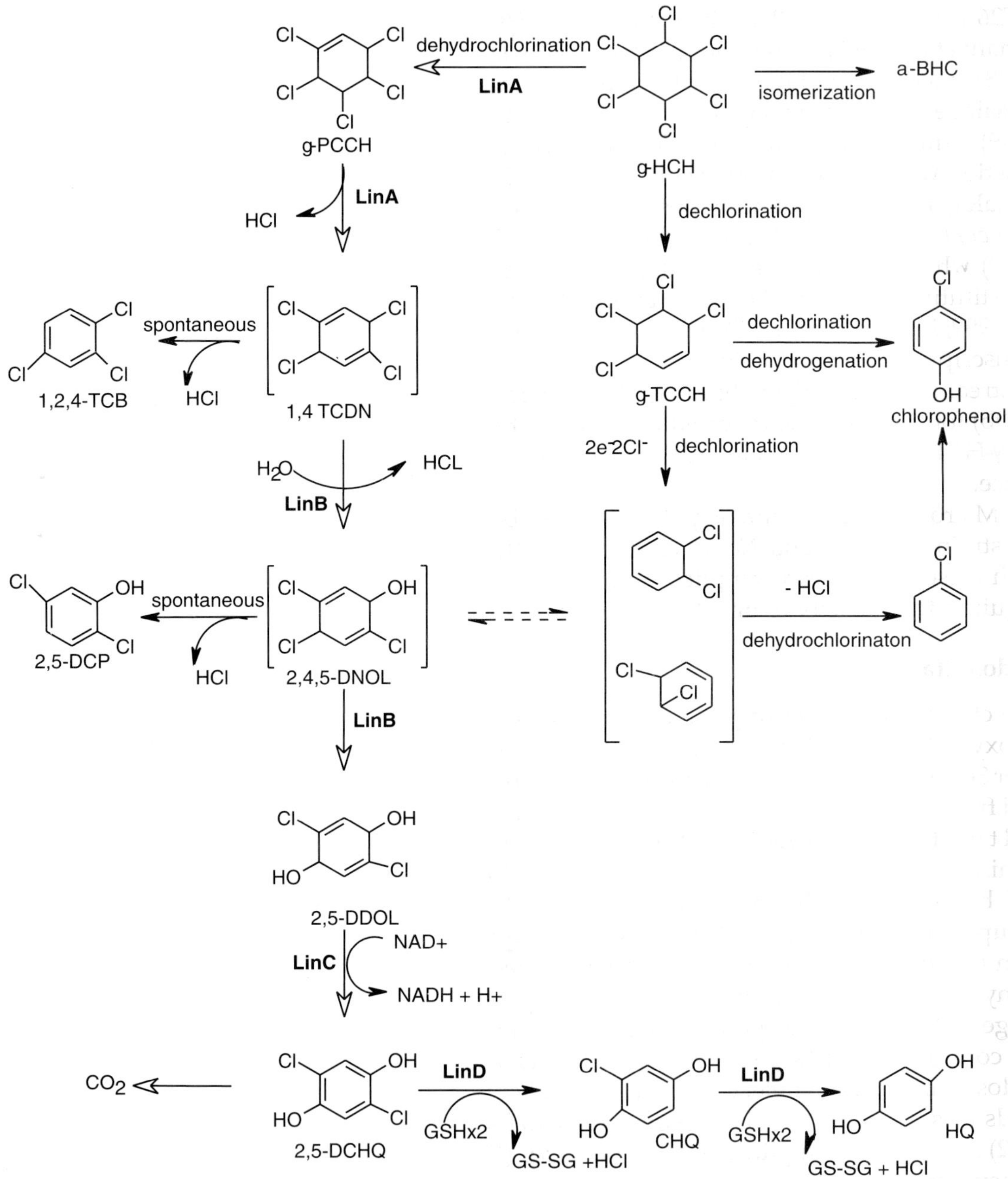

Fig. 2: Microbial Degradation Pathway of g-HCH (Adapted from Nagata *et al.*, 1999).

Five structural genes (*linA, B, C, D* and *E*) and one regulatory gene (*linR*) are involved in the degradation of γ-HCH in *S. paucimobilis* UT26. Four genes *linA, linB, linC* and *linD* have been cloned, the products of which are involved sequentially in the degradation of g-HCH in UT26 (Imani *et al.*, 1991; Miyauchi *et al.*, 1998; Napata *et al.*, 1993; Nagata *et al.*, 1994). *linA* (Imani *et al.*, 1991), *linB* (Napata *Et al.*, 1993), *linC* (Nagata *et al.*, 1994), *lin D* (Miyauchi *et al.*, 1998) encode for dehydrochlorinase, halidohydrolase, dehydrogenase and reductive dehalogenase respectively. Fifth gene *linE* encodes ring cleavage oxygenase (Miyauchi *et al.*, 1999). Among them, *linA* encodes an enzyme that catalyzes a unique dehydrochlorination reaction whose mechanism is still unknown. Dehalogenation is a key step in the degradation of halogenated compounds. These genes, in contrast to the previously studied catabolic genes (Lal *et al.*, 1995), do not exist in the form of a cluster except for *linD* and *linE* (Miyauchi *et al.*, 1999) which form an operon. All lin genes are nearly 1Kb or less in size. *Lin A, linB, linC* are constitutively expressed (Nagata *et al.*, 1999) whereas *linD* and *linE* are inducible (Miyauchi *et al.*, 1999; Nagata *et al.*, 1999). The expression of *linD* and line genes is regulated by a *LysR* type transcriptional regulator (*linR*) (Miyauchi *et al.*, 2002) which is encoded by a gene located upstream of *linE* (Miyauchi *et al.*, 1999; Nagata *et al.*, 1999). In a recent study, Nagata *et al.* (199b) for the first time reported two different types of dehalogenases, *LinA* and *LinB* that act on γ-HCH or its metabolite in *S. paucimobilis* and were found to localize in the periplasmic space.

Microbial degradation of g-HCH has also been reported in cyanobacteria such as *Anabaena* sp. strain PCC7120 and *Nostoc ellipsossorum* (Kurits and Wolk, 1995). It was seen that when both these strains were supplied with the fcb ABC operon from *Arthrobacter globiformis*, they acquired the capacity to dechlorinate 4-CB (Tsoi *et al.*, 1991).

Endosulfan

The chlorinated cyclic sulphite diester endosulfan (Thiodane, bicyclo-[2.2.1]-2-heptane-5, 6-bisoxymethylene sulphite) is a broad spectrum insecticide that has been used extensively for over 30 years on a variety of crops such as cotton, paddy, sorghum, oilseeds, coffee, vegetables and fruit crops. Endosulfan is often classified as a cyclodiene and has the same primary action and target as other cyclodienes (Casida, 1993). However, it has chemical and physical properties significantly different from other cyclodiene insecticides that affect both its environmental and biological fate. In particular, endosulfan has a relatively reactive cyclic sulphite diester group (Van Woerden, 1963) and, as a consequence, its environmental persistence is lower than that of many other insecticides. Compared to other insecticides, it has lower toxicity to many species of beneficial insects, mites and spiders (Goebel *et al.*, 1982). Because of its abundant usage and potential transport, endosulfan contamination is frequently found in the environment at a considerable distance from the point of its original application (Milles and Pfeuffer, 1997). Endosulfan has been detected in the atmosphere, soils, sediments, surface and rainwater and foodstuffs (USD, 1990). It is extremely toxic to fish and aquatic invertebrates (Sunderam *Et al.*, 1992) and has been implicated in genotoxicity (Chaudhari *et al.*, 1999, neurotoxicity (Paul and Balasubramanian, 1997).

Detoxification of pesticides through biological means is receiving increasing attention as an alternative to existing methods, such as incineration and landfills. Microorganisms have increasingly been investigated as a source of xenobiotic-degrading enzymes. Several studies have been carried out in recent years that have reported the isolation of bacterial co-culture (Awasthi *et al.*, 1997; Awasthi *et al.*, 2003) and mixed cultures (Miles and Moy, 1979; Sutherband *et al.*, 2000, 2002a) capable of degrading endosulfan. Mukherjee and Gopal (Mukharjee and Gopal, 1994) reported the degradation of β-endosulfan by *Aspergillus niger*. Other fungi that are also able to degrade endosulfan are *Trichoderma harzianum* (Katayana and Matsmurs, 1993), *Phanerochaete chysosporium* (Kullman and Matsumura, 1996) and *Mucor thermohyalospora* MTCC1384 (Shetty *et al.*, 2000). In a recent report, Siddique *et al.* (2003) reported the isolation of *Fusarium ventricosum* and *Pandoraea* sp., which degraded about 90% and 83% of 100 ppm endosulfan respectively in 15 days using it as carbon and energy source. Other bacteria, isolated by this group earlier could degrade about 70% endosulfan. Some other reports of microbial species are also there that are able to degrade endosulfan (Kwon *et al.*, 2002; Lee *et al.*, 2003).

The commercial endosulfan insecticide is a mixture of two diastereoisomers, α-endosulfan and β-endosulfan in a ratio of 7:3. *Mycobacterium* sp. strain ESD metabolizes the isomers to endosulfan sulphate and endosulfan monoaldehyde (Sutherland *et al.*, 2000). *Mycobacterium* sp. strain ESD has been isolated by providing endosulfan as the sole source of sulphur in successive enrichment cultures initially inoculated with endosulfan contaminated soil. The formation of both endosulfan sulphate and endosulfan monoaldehyde by *Mycobacterium* sp. strain ESD occurs as a response to sulphur starvation and does not occur in the presence of relatively low levels of inorganic sulphate or sulphite. A cosmid library of this train ESD DNA was constructed in *Mycobacterium-E. coli* shuttle vector and screened for activity in *Mycobacterium smegmatis*, a species that does not degrade endosulfan (Sutherland *et al.*, 2002b). By using this method, a single cosmid that conferred sulphur dependent endosulfan degrading activity on the host strain was prepared. On this basis, an open reading frame (esd) was identified within this promoter that when expressed behind a constitutive promoter in a mycobacterial expression vector, conferred sulphite and sulphate independent β-endosulfan degradation activity on the host stain.

Organophosphate Pesticides

Organophosphates comprise a large family of compounds that have appeared in the environment in recent evolutionary times. These chemicals are applied to crops, buildings, ornamental plants and lawns. Agricultural uses include field applications on corn, cotton, canola, alfalfa and nuts. Exterminators use organophosphate pesticides in residential and commercial structures, and certain pest control products for cats and dogs also contain organophosphorus compounds. People are commonly exposed to organophosphate pesticides through eating fresh and processed vegetables, contacting pesticide-contaminated surfaces, breathing air near pesticide applications (both indoors and outdoors), and drinking pesticide-contaminated water. The multiple use and ubiquitous nature of these pesticides result in widespread exposure to these chemicals for most people. Symptoms of exposure include nausea, headaches, twitching, trembling, excessive salivation and tearing, inability to breathe because of paralysis of the diaphragm, convulsions, and at higher doses, death. The organophosphate pesticides are highly toxic to freshwater fish, aquatic life, marine organisms and birds. The USEPA considered that the general agricultural uses of chlorpyriphos posed a serious hazard to wildlife and honeybees and in 1988 concluded that the use of diazinon in turf and golf course treatment posed a widespread and continuous hazard to birds because of the bird die off events linked to diazinon use. Residential uses of chlorpyriphos and diazinon have been banned by the USEPA. In a study in India, it was found that through breast milk, infants consumed 4.1 times more malathion than the average daily intake levels recommended by the World Health Organization. A correlation analysis between mothers' age and the content of the chemicals accumulated in breast milk indicated a substantial degree of correlation for malathion (Sanghi *et al.*, 2002). In an analysis of pesticide residues in soft drinks in India, chlorpyriphos was present in 100% of the samples and in 75% of the samples, chlorpyriphos residues exceeded the recommended limit by 3.9 to 7.6 times (C.F.T.R.I, 2003).

Organophosphate pesticides are sometimes called phosphotriesters as they contain three phosphoester linkages in which a central phosphorus atom is linked either to an oxygen or a sulphur; in addition they have R1 and R2 groups that are either ethyl or methyl in structure, and a leaving group R which is specific to the individual organophosphate. A general structure of organophosphate pesticides is given below:

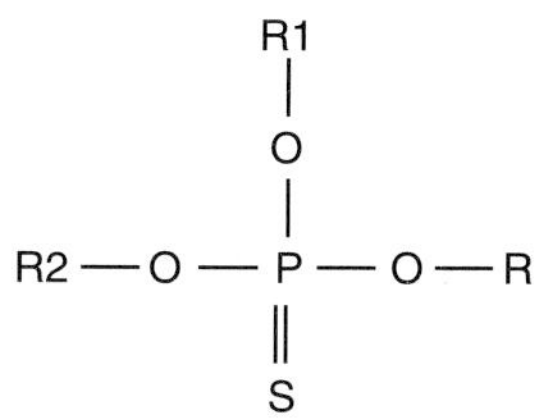

In spite of the useful properties of organophosphates such as low persistence and high potency at low doses, their widespread and injudicious usage have caused environmental and health problems.

***Microbial degradation of organophosphates*:** Organophosphate pesticides are toxic to higher organisms, being powerful inhibitors of the enzyme acetylcholinesterase (De Frank, 1991) but provide an abundant source of nutrition for bacteria. The relatively low persistence of organophosphate pesticides is attributed to their susceptibility to hydrolysis by soil microbial enzymes. There is a great deal of interest in bacteria capable of metabolizing these compounds as well as the enzymes involved in their hydrolysis because of their role in determining the fate of pesticides in the environment and their potential for environment friendly *in situ* detoxification/bioremediation (Mulbry and Karns, 1989). Studies on microbial degradation are useful in the development of strategies for the detoxification of insecticides by microorganisms. The principle reactions involved in the breakdown of organophosphates are hydrolysis, oxidation, alkylation and dealkylation (Singh et al., 1999). Microbial degradation through hydrolysis of P-O-alkyl and P-O-aryl bonds as shown below is considered to be the most significant step in the degradation of organophosphate compounds.

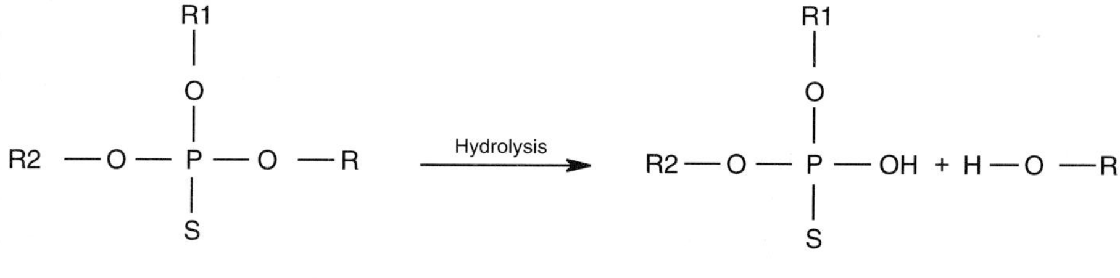

Parathion

Parathion (*O,O*-diethyl-*O*-4-nitrophenyl phosphorothioate) is the most well studied organophosphate pesticide in terms of its metabolism and is, therefore, discussed in detail here. Although it is relatively less persistent, this pesticide and its metabolites, particularly *p*-nitrophenol (PNP) have caused environmental problems (Barik and Sethunathan, 1978). It is well established that microbial metabolism is the major means of parathion degradation in soil. The major degradation products have been identified as PNP and aminoparathion (Miyanoto *et al.*, 1966; Munnecke, 1976). Table 1 shows the various degradation products formed by different bacteria. The primary degradative pathway is initial hydrolysis to yield PNP and diethylthiophosphoric acid (DETP). The secondary pathway involves the formation of paraoxon (diethyl *p*-nitrophenyl phosphate) that subsequently hydrolyses to form PNP and DETP. The third degradative pathway involves reduction of parathion under low oxygen conditions to yield *p*-aminoparathion followed by hydrolysis to *p*-aminophenol and DETP (Montgomery *et al.*, 1997). Two *Pseudomonas* strains isolated from soil and sewage utilized parathion as source of carbon and hydrolyzed it to PNP (Rosenberg and Alexander, 1979). A *Flavobacterium* sp. strain ATCC 27551 isolated from flooded soil hydrolyzed both diethyl (parathion and diazonin) and dimethyl (methyl parathion and fenitrothion) phosphorothioates, while another organism *Pseudomonas* sp. ATCC strain 29353, isolated from the same soil, hydrolyzed only diethyl phosphorothioates (Adhya *et al.*, 1981). Chaudhry *et al.* (1988) also isolated two mixed bacterial cultures that utilized both methyl parathion and parathion as sole source of carbon. A *Pseudomonas* sp. isolated from these mixed cultures degraded parathion to PNP only, while mixed cultures degraded the pesticides completely. These as well as other studies established that parathion is generally first hydrolyzed to PNP and DETP which are then further metabolized (Barik *et al.*, 1979; Munnecke *et al.*, 1982). There are a number of bacterial species which are capable of degrading PNP, the toxic metabolite of parathion, and it has been discussed in detail later. The second metabolite of parathion, DETP was found to be metabolized by enrichment culture obtained from cattle dip solution containing coumaphos (Shetten, 1988). These cultures mineralized DETP to sulphate and phosphate while utilizing ethyl moieties as a carbon and energy source. Cook *et al.* (1980) isolated *P. acidovorans* from sewage sludge that utilized DETP as a sole sulphur source. These studies suggest that hydrolysis of parathion and subsequent degradation of PNP and DETP by soil microorganisms may be responsible for rapid removal of these chemicals from the environment.

The degradation pathway of parathion is shown in Fig. 3. The hydrolysis of parathion to DETP and PNP reduces the toxicity of parathion by more than 100-fold (Serdar *et al.*, 1989). The enzyme responsible for catalyzing this hydrolytic step is organophosphate hydrolase (OPH E.C. 8.1.3.1). This enzyme has broad substrate specificity and catalyzes the hydrolysis of several other organophosphate pesticides such as methyl parathion, paraoxon, dursban, coumaphos, cyanophos and diazonin and even some organophosphate chemical warfare agents such as sarin and soman (Richins *et al.*, 1997). The OPH enzyme is encoded by the *opd* gene. This gene has been isolated from several bacteria but the most widely characterized parathion degraders are *Flavobacterium* sp. ATCC 27551 isolated from a rice field in the Philippines (Sethunathan and Yoshida, 1973) and *Pseudomonas diminuta* MG isolated from Texas from an enrichment culture (Serdar *et al.*, 1982) which possess high levels of this constitutively expressed enzyme. The *opd* gene has been cloned and sequenced from both *P. diminuta* MG (Serdar *et al.*, 1989) and *Flavobacterium* ATCC 27551 (Mulbery and Karns, 1989) and the genes were found to be identical although the plasmids carrying them were very different (Mulbery *et al.*, 1987; Serdar and Gibson, 1985). The *opd* gene has been isolated from taxonomically different bacteria and from geographically different regions of the world (Horne *et al.*, 2002; Mulbery *et al.*, 1986; Serdar and Gibson, 1985); however, the fact that it has a similar DNA sequence suggests that the genes may be part of a mobile genetic element or transposon (Mulbery *et al.*, 1987). The widespread distribution of *opd* gene has been suggested to be due to its lateral transfer by a combination of transposition and plasmid transfer (Siddavattam *et al.*, 2003, 139).

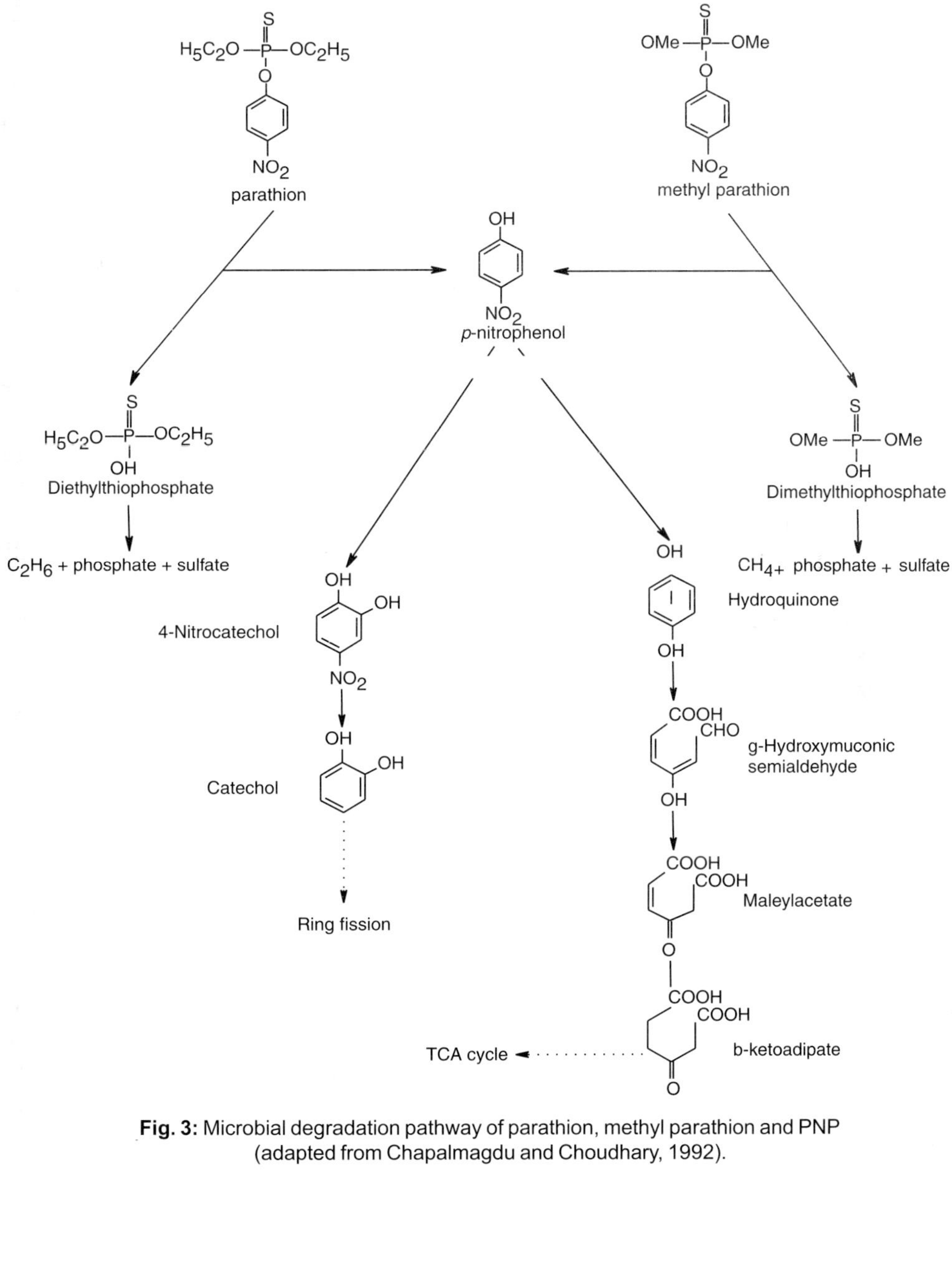

Fig. 3: Microbial degradation pathway of parathion, methyl parathion and PNP (adapted from Chapalmagdu and Choudhary, 1992).

Considering the potential use of phosphotriesterases for the bioremediation of pesticides, substantial efforts have been made to understand the nature of the mechanism and substrate-binding determinants of OPH. The crystal structure of OPH has been elucidated (Benning *et al.*, 2001) and it is reported to be a homodimeric protein with a monomeric molecular weight of 35 kDa (Desoueli *et al.*, 1999; Mac Daniel *et al.*, 1988). It belongs to a superfamily of metal-containing amidohydrolases (Holm and Sander, 1997). OPH acts on a broad range of substrates, some of which are turned over at near diffusion rates whereas others are processed at more modest rates (Caldwell *et al.*, 1991, Scanlan and Reid, 1995). Structures of OPH have been obtained with the inhibitors diethyl 4-methylbenzylphosphonate, diisopropyl methyl phosphonate and triethyl phosphate (Vanhooke *et al.*, 1996). These inhibitors are substrate analogues and bind in the active site of the enzyme giving indications of how the protein binds substrate. Pockets to accommodate the phosphate substituents have also been identified. Structural and kinetic studies carried out with this enzyme have led researchers to propose a plausible enzyme mechanism; it involves a single in-line displacement attack (S_N2) (Lewis *et al.*, 1988) with the metal-bridging hydroxide ion acting as the nucleophile (Benning *et al.*, 2000).

A naturally occurring variant of OPH enzyme designated opdA capable of degrading broader range of organophosphates was recently isolated from an *Agrobacterium radiobacter* strain (Horne *et al.*, 2002) which was very similar in sequence to the OPH from *P. diminuta* MG and *Flavobacterium* sp. ATCC 27551. The sequences of opdA and OPH were found to be similar with 90% identity at the amino-acids level (Horne *et al.*, 2002) and the sequence differences are thought to be responsible for the variation between the substrate specificities of OPH and opdA. The latter has higher k_{cat} values for substrates with shorter side chains and can hydrolyze fenthion and phos*met* for which OPH has no activity (Horne *et al.*, 2002). Mulbry and Karns (1989) purified three distinct hydrolases from Gram-negative soil bacteria which had similar affinities for parathion, but differed in their sizes, their affinities for the structurally related organophosphates parathion and EPN (*O*-ethyl-*O*-4-nitrophenyl phenylphosphonothioate), their cellular locations and their response to divalent cations and dithiothreitol. Siddavattam *et al.* (2003) recently described a novel gene in *Flavobacterium* sp. ATCC 27551 linked to *opd* that is involved in the degradation of PNP and is likely to act in concert with *opd* in the degradation of parathion.

Methyl Parathion

Methyl parathion (*O*, *O*-dimethyl *O*-(*p*-nitrophenyl) phosphorothioate) is widely used throughout the world and its residues are regularly detected in a range of fruits and vegetables. It is a highly toxic pesticide classified by WHO in the class of extremely hazardous pesticides and its high acute toxicity has prompted several countries like USA, Indonesia, Sri Lanka and Tanzania to ban its use, and several others like Colombia, Korea, China and Japan to restrict its use. However, it is regularly misused in many developing countries. Parathion and methyl parathion have similar chemical structures so it seems reasonable that the enzymes degrading these two pesticides may have homology with each other. Indeed, a methyl parathion degrading *Pseudomonas* sp. possessed DNA homologous to the *opd* gene encoding OPH (Chaudhary *et al.*, 1988). However, Zhongli *et al.* (2001) reported a novel gene designated *mpd* from another methyl parathion degrading *Pseudomonas* sp. strain M6 that is completely different from *opd* genes. *Pseudomonas putida*, isolated from a rice field soil by Lakshmi and Lalithakumari (1994), utilized methyl parathion as the sole source of carbon and/or phosphorus and could also utilize the intermediates of methyl parathion degradation i.e. PNP, hydroquinone and 1,2,4 benzenetriol (the breakdown products of PNP). The bacteria produced the enzyme organophosphorus acid anhydrase which hydrolyzed methyl parathion to PNP (Lakshmi and Lalithakumari, 1994). A *Pseudomonas* sp. strain 50541 isolated from a pesticide waste disposal site was found to oxidize methyl parathion rapidly and completely (Chapalamagdu and Chaudhary, 1992). A *Flavobacterium balustinum* strain has also been isolated from agricultural soils in Andhra Pradesh, India that could utilize methyl parathion as sole source of carbon. A large indigenous plasmid involved in the degradation of OP pesticides was identified in this soil isolate (Somara and Siddavattam, 1995). The plasmid borne *opd* gene of this organism was cloned and sequenced and the sequence was found to be 98% similar to previously reported *opd* genes (Somara *et al.*, 2002). The major degradation product of methyl parathion is PNP as in case of parathion (Fig. 3).

p-Nitrophenol

Since the major metabolite of both parathion and methyl parathion is PNP, which has been listed as a priority pollutant by USEPA (Keith and Telliard, 1979), its fate in the environment is of potential concern and much work has been done on this compound. PNP can be further utilized as a source of carbon and energy by several bacterial species (Hanne *et al.*, 1993; Kadiyala and Spain 1998; Leung *et al.*, 1997; Nishino and Sapin, 1993; Spain *et al.*, 1979). There are two pathways by which PNP is degraded (Fig. 3). In one pathway, which was elucidated in a *Moraxella* sp. by Spain and Gibson (Spain and Gibson, 1991), PNP is degraded via formation of hydroquinone followed by ring fission to form gg-hydroxymuconic semialdehyde which is further converted to maleylacetate. The other pathway which is more common in Gram-negative bacteria, involves formation of 4-nitrocatechol followed by production of 1, 2, 4 benzenetriol which is then cleaved to maleylacetate (Jain *et al.*, 1994). Maleylacetate enters the TCA cycle via formation of bb-ketoadipate and is thus mineralized to CO_2 and H_2O.

In our laboratory, we have isolated three organisms *Arthrobacter protophormiae* RKJ100, *Burkholderia cepacia* and *Ralstonia* sp. SJ98 which can utilize PNP as the sole source of carbon and energy. We have elucidated the complete pathway of PNP degradation in strain RKJ100 which occurs via formation of benzoquinone, hydroquinone, maleylacetate and β-ketoadipate (Chauhan *et al.*, 2000). This pathway has been included in the University of Minnesota Biocatalysis/ Biodegradation Database. A plasmid of ~65 kb was found to be responsible for harboring the genes for PNP degradation in this strain (Chauhan *et al.*, 2000). Kinetic studies on biodegradation of PNP by these strains have also been carried out (Bhushan *et al.*, 200017). Resting cells induced with PNP were used for studying the degradation kinetics for strains RKJ100 and SJ98. The apparent Vmax value of PNP degradation by strain RKJ100 was 0.28 and that for strain SJ98 was 0.23 mM as determined from Michaelis-Menten curves drawn between PNP concentration and rate of PNP degradation, while the maximum rates of PNP degradation (Vmax values) were found to be 7.81 and 3.38 mmM PNP degraded/min/mg dry biomass respectively, as determined from Lineweaver-Burk plots (Bhushan *et al.*, 2000b). The intermediates of PNP degradation, benzoquinone and hydroquinone, were found to inhibit PNP degradation in strain RKJ100 noncompetitively and competitively, respectively. In strain SJ98, these intermediates inhibited PNP degradation in an uncompetitive manner. Strain SJ98 was also found to be capable of using 3-methyl 4 nitrophenol, the breakdown product of fenitrothion, as the sole source of carbon and energy (Bhushan *et al.*, 2000a). Studies to determine the degradation pathway of this compound revealed catechol to be the major intermediate of the degradation pathway. Studies to understand the degradation pathway of PNP at the molecular level are currently being carried out.

GENETIC ENGINEERING STUDIES

The slow growth and low culture yields of native OPH producing strains makes them uneconomical for practical use. Therefore, efforts have been made by several researchers to improve the applicability of OPH for pesticide bioremediation. Introduction of recombinant plasmids containing *opd* gene into *E. coli* under the control of its own promoter resulted in poor expression of the gene. However, when cloned downstream to an *E. coli* promoter sequence (such as the lac promoter), greatly increased expression of the *opd* gene was achieved (Mulbry and Karns, 1989; Serdar and Gibson, 1985; Serdar *et al.*, 1989).

In a study by Richins *et al.* (1997), the rate of hydrolysis of paraoxon and parathion was increased seven times by expressing OPHs on the cell surface of *E. coli* as a Lpp-OMP-OPH fusion. However, these recombinant strains were relatively unstable and hence a new plasmid was constructed to express OPH on the cell surface under control of a tightly regulated *tac* promoter. Production of active OPH onto the cell surface was highly host specific and a high rate of parathion degradation was observed from strains JM105 and XL1-Blue (Kanewa *et al.*, 1998) . Shimazu *et al.* (Shimazu *et al.*, 2001) were able to genetically engineer a *Moraxella* sp. with OPH expressed on its surface which was capable of simultaneous degradation of parathion, methyl parathion, paraoxon as well as PNP. Walker and Weasling (Walker and Keasling, 2002) engineered *P. putida* KT2442 with the genes responsible for parathion hydrolysis and mineralization of PNP enabling this genetically engineered strain to use parathion as sole source of carbon and energy. Parathion formed a separate dense, non-aqueous liquid phase but was still bioavailable. Site directed mutagenesis has been applied to the various residues involved in the active sites, resulting in OPH mutants with improved catalytic characteristics against paraoxon (Lai *et al.*, 1996). In another study, directed evolution of OPH using a surface-display OPH library was used for screening of OPH variants with enhanced activity towards poorly degraded substrates. This approach resulted in a variant with methyl parathion hydrolyzing activity that was 25-fold higher than that of the wild type (Cho *et al.*, 2002). Directed evolution was also used to evolve OPH into a series of mutant enzymes with catalytic properties similar to those of opdA. The mutants had improved catalytic efficiency for methyl-parathion, methyl-paraoxon and coumaphos-o-analogue (Yang *et al.*, 2003). A consortium comprising two bacteria that were genetically engineered to each carry out part of the pathway for parathion degradation (*E. coli* carrying gene for parathion hydrolase and *P. putida* KT2440 containing a PNP-inducible plasmid borne operon encoding the genes for PNP mineralization) effectively worked together to break down parathion and prevent accumulation of PNP (Gilbert *et al.*, 2003).

Other Organophosphate Pesticides

Coumaphos (*O*, *O*-diethyl-*O*-(3-chloro-4-methyl-2-oxo-2H-1-benzo-pyran-7-yl) phosphorothioate) has a half life of about 300 days in soil and water (Kearny *et al.*, 1986). Three bacteria were isolated from enrichment cultures developed using cattle-dip vat solutions as inoculum (Shelton and Somich, 1988). These isolates B1, B2 and B3 hydrolyzed coumaphos to chlorferon and DETP. Chlorferon was further metabolized by B1 and B2 to α-chloro-β methyl-2, 3, 4-trigydroxy-trans-cinnamic acid. Parathion hydrolase from *Flavobacterium* sp. 27551 was also found to hydrolyze coumaphos, yielding chlorferon and DETP (Kearmy *et al.*, 1986). A fenitrothion degrading bacterium *Burkholderia* sp. strain NF100 was isolated from fenitrothion treated soil. The degradation involved initial hydrolysis of organophosphate bond to form 3-methyl-4-nitrophenol which was further metabolized to methylhydroquinone. The fenitrothion degradation ability was encoded on two plasmids pNF1 and pNF2 (Hayatsu *et al.*, 2000). Miyamoto *et al.* (Miyanoto *et al.*, 1966) studied the degradation of fenitrothion by *Bacillus subtilis*. The major metabolite was aminofenitrothion and minor metabolites were dimethyl thiophosphoric acid and desmethyl fenitrothion. Attempts to isolate chlorpyriphos degrading bacteria from chlorpyriphos treated soils have generally been unsuccessful (Racke *et al.*, 1990; Mallick, 1999) which has been attributed to the toxic effect of 3,5,6-trichloro-2-pyridinol (TCP), the main degradation product of chlorpyriphos on soil microorganisms (Racke *et al.*, 1990). Chlorpyriphos can nevertheless be degraded by bacteria capable of degrading other organophosphate pesticides such as *Flavobacterium* sp. ATCC 27551 and *Arthrobacter* sp. (Misra *et al.*, 1992; Sethunathan and Yoshida, 1973). However, Singh *et al.* (Singh *et al.*, 2003) were recently able to isolate a bacterium capable of using chlorpyriphos as the sole carbon source. Not much is known about the microbial degradation of diazinon (*O*, *O*-diethyl-O-2-isopropyl-4-methyl-6-pyrimidye phosphorothioate). Although *Flavobacterium* sp. ATCC 27551 was isolated originally as diazinon degrading bacterium, it was well characterized only with respect to parathion hydrolysis. Two *Pseudomonas* sp. isolated from sewage sludge degraded diazinon, producing DETP in the culture medium (Rosenberg and Alexander, 1979). A *Pseudomonas* sp. that hydrolyzed several organophosphates including parathion, methyl parathion, dursban, paraoxon, aminoparathion and other methoxy-or ethoxy- substituted organophosphates, was found to hydrolyze diazinon as well (Barik and Munnecke, 1982). In another study, Barik *et al.* (Barik *et al.*, 1982) found that cell-free extracts from two strains of *Arthrobacter* sp. SB3 and SB4, hydrolyzed diazinon and the Km values of the enzyme for the two strains were 1.3 and 2 ìmol/mg protein/min respectively. It had broad pH and temperature optima of 6-9 and 25-36°C respectively.

APPLICATION OF MICROORGANISMS FOR THE BIOREMEDIATION OF ORGANOCHLORINE AND ORGANOPHOSPHATE PESTICIDES

Bioremediation is a technology that utilizes the metabolic potential of microorganisms to clean up contaminated environments. Over billions of years, microorganisms have evolved an extensive range of enzymes, pathways and control mechanisms in order to degrade an array of compounds. In addition, microbes have the unique biochemical asset of being able to use molecular oxygen to catalyze the oxidation of numerous natural and man-made chemicals thereby initiating reaction sequences that enter the central pathway of metabolism such as Kreb's cycle or the fatty acid spiral. Microbial degradation critically affects the fate and behaviour of pesticides in soil. Ideally, bioremediation results in the complete mineralization of contaminants to O_2 and CO_2 without the build-up of intermediates. Bioremediation processes can be broadly categorized into two types: *ex situ* and *in situ*. *Ex situ* bioremediation technologies include bioreactors, biofilters, land farming and some composting methods. *In situ* bioremediation technologies include bioaugmentation, bioventing, biosparging, biostimulation, liquid delivery systems and some composting methods. *In situ* treatments tend to be more attractive because they require less equipment, generally cost less and are more environment friendly. One important characteristic of bioremediation is that it is carried out in non-sterile open environments that contain a variety of organisms. Of these, bacteria, such as those capable of degrading pollutants, usually have central roles in bioremediation, whereas other organisms (eg. fungi and grazing protozoa) also affect the process. A deeper understanding of the microbial ecology of contaminated sites is therefore necessary to further improve bioremediation processes. The natural environment harbors a wide range of microorganisms that have crucial roles in bioremediation.

An innovative technology developed by GRACE bioremediation technologies, Ontario, Canada by the name of 'DARAMEND' for the bioremediation of soils and contaminated with a variety of organic pollutants. DARAMEND has been demonstrated to be an effective treatment approach for soil containing organochlorine pesticides and herbicides and chlorinated phenols (such as PCP) and benzenes (eg.- HCB). This technology has been successfully implemented at industrial sites in the US and Canada and has proven effective for solid phase *in situ* or *ex-situ* bioremediation. Using this technology, lindane concentrations in the soil were reduced by 95% during 251 days (15 cycles) of treatment. Analysis for other isomers of HCH revealed 92% removal of all isomers during the same time period. Two highly successful field scale projects for the treatment of soil containing toxaphene and DDT were successfully completed in southeastern USA. In the first, toxaphene and DDT were reduced by 98% and 82% respectively, while in the second treatment toxaphene and DDT concentrations were reduced by 96% and 89%.

Bioremediation using white-rot fungi is a promising technology that is currently being researched extensively. This technology can be used in an *ex situ* or *in situ* manner. The white-rot fungi are physiological rather than taxonomic grouping, comprising those fungi that are of extensively degrading lignin (a heterogenous polyphenolic polymer) within lignocellulosic substrates. Most studies focus on the ability of *Phanerochaete chrysosporium* to degrade persistent compounds, but *Phanerochaete sordida*, *Pleuotus ostreatus*, *Phellinus weirii*, and *Polyporus versicolor* have also been successful in laboratory studies (Safferman *et al.*, 1995). These fungi are capable of degrading chlordane, lindane and DDT, which makes them useful for the remediation of pesticide-contaminated sites (Alexander, 1999). The white rot fungus, *Phanerochaete chrysosporium* is capable of degrading DDT and also several chlorinated organic compounds (Morgan *et al.*, 1991; Paszezynski and Crawford, 1995). These fungi are also involved in the degradation of γ-HCH. *P. chrysosporium* cultured under ligninolytic conditions has been reported to partially mineralize lindane in liquid cultures and in corn-cob amended silt loam soil (Kennedy *et al.*, 1990). Fungal lignin degrading system (LDS) involved in degradation mainly consists of lignin peroxidase (LIP), manganese peroxidase (MnP) and laccase. The white rot fungi have been shown to produce different combinations of these extracellular enzymes which are involved in various mechanisms of transformation of pesticides. White-rot fungi could be used to inoculate a composing operation. However, large quantities of the fungus are required to remediate a site due to the very slow nature of compound degradation (Alexander, 1999).

Parathion hydrolases encoded by genes related to the *opd* gene of *Flavobacterium* sp. ATCC 27551 have a number of characteristics which make them attractive for use in pesticide waste disposal strategies. All are constitutively expressed in their native hosts, have broad pH and temperature optima and have relatively broad substrate ranges for structurally similar organophosphate pesticides (Brown, 1980; Munnecke, 1976). The utility of these enzymes to facilitate bioremediation has been demonstrated in several studies. Both, native and recombinant OPH have been used for the detoxification of organophosphate pesticides. The enzyme has been immobilized on various supports such as nylon membrane, tubing, powder (Caldwell and Raushel, 1991), porous glass and silica beads (Munnecke, 1979) and used in enzyme reactors for the detoxification of organophosphates. OPH was expressed on the surface of *E. coli* cells and immobilized on a non-woven, polypropylene fabric for use in detoxifying contaminated wastewaters (Mulchandani *et al.*, 1999). In batch operations, the immobilized cells degraded 100% of 0.8 mM paraoxon; in addition other organophosphates such as diazinon, coumaphos, and methylparathion were also hydrolyzed efficiently. Cells of a recombinant *E. coli* strain containing the *opd* gene immobilized within a macroporous gel matrix poly (vinyl alcohol) cryogel were successfully applied for bioremediation of hazardous waste containing high concentrations of coumaphos (Kim *et al.*, 2002).

Flavobacterium sp. ATCC 27551 was used in a pilot scale study for detoxification of waste containing high concentrations of coumaphos (Kaerns *et al.*, 1987). In another study, indigenous organisms were used in biofilter columns for detoxification of coumaphos bringing down its level from approximately 1500 mg/l to 0.1 mg/l in 7-10 days (Mulbry *et al.*, 1996). In a soil microcosm study, inoculation of *Flavobacterium* sp. ATCC 27551 and *Arthrobacter* sp. into soil containing chlorpyrifos in flooded and non-flooded conditions resulted in disappearance of the pesticide in 28 days (Mallick *et al.*, 1999). An acclimated culture of parathion-degrading bacteria was used to study the feasibility for bioremediation purposes. Its inoculation into parathion-contaminated soil resulted in complete degradation in soil containing 1250 kg/ hectare within 35 days. However, the culture was unable to effectively degrade parathion higher than this concentration (Barles *et al.*, 1979).

CONCLUSION

The use of pesticides is indispensable to modern agriculture. However, their indiscriminate and rampant use, particularly in developing countries has led to serious concern over their impact on our health and environment as pesticide residues are being detected in our foods, soil, water and air with alarming regularity and frequency. In order to fight this menace, researchers have been trying to explore ways and means of effectively disposing of these toxic compounds. Bioremediation methods for destruction of xenobiotics such as pesticides are being increasingly explored and adopted in preference to physico-chemical methods because of their low cost, higher efficacy and environment friendliness. It is now well established that most pesticides are susceptible to microbial degradation. Microorganisms can, therefore, be exploited for detoxification and safe disposal of these deadly but necessary chemicals. Isolation of suitable microorganisms and biochemical characterization of the pathways and enzymes involved in degradation is a prerequisite for their use in bioremediation strategies. For organochlorines, the principal microbial reactions are reductive dechlorination, dehydrochlorination, oxidation and isomerization, while for organophosphates they are hydrolysis, reduction, dealkylation and oxidation, although not all reactions occur with any one insecticide. The challenge now lies in exploiting the existing degradative routes as well as to develop novel catabolic capabilities for rapid and safe bioremediation of pesticide containing sites.

Recently, novel detoxification strategies using genetically engineered organisms with enhanced degradative capabilities have been studied. For example, the use of genetically engineered *E. coli* with surface expressed organophosphorus hydrolase, the enzyme responsible for drastically reducing the toxicity of organophosphate pesticides, for more efficient degradation of organophosphates. The use of immobilized enzymes for pesticide hydrolysis has also shown promising potential for pesticide detoxification. To enhance the overall degradation of less efficient substrates, enzymes with improved hydrolytic activity have also been developed with the help of techniques such as directed evolution. Hence, the development of biochemical and genetic techniques has expanded the potential for experimental advances in biodegradation so that it has become practical to isolate and modify useful genes from diverse microbial strains. The future holds great promise for continuing advances in basic research and for successful application of this research for solving the environmental problems posed by these chemicals.

Table 4: Microorganisms involved in the metabolism of parathion.

Microorganism	Metabolite	Reference
Bacillus subtilis	Aminoparathion	90
B. subtilis, B. megaterium, B. cereus, B. macerans, P. aeruginosa, P. ovalis, P. aurofaciens, Alcaligenes viscolactis, E. coli, Serratia plumuthica	*Aminoparathion*	42
Serratia plumuthica		
Rhizobium japonicum, Rhizobium meliloti	Aminoparathion (major) PNP (minor)	85
Flavobacterium sp.	PNP	133
Bacillus sp.	PNP	138
Mixed micrbial clutures	PNP, Diethylthiophosphate, paraoxon and aminoparathion	107
Pseudomonas stutzeri	Diethylthiophosphate and PNP	32

A¯KNOWLEDGEMENTS

The help of Mr. Gunjan Pandey, Ms. Debarati Paul, Mr. Narinder Sharma and Ms. Anuradha Ghosh is gratefully acknowledged. The financial assistance provided by the Council of Scientific and Industrial Research (CSIR) and Department of Biotechnology (DBT), India is also acknowledged.

REFERENCES

Adhya, T.K., Barik, S. & Sethunathan, N. (1981) *J. Appl. Bacteriol.* **50**: 167-169.

Agnihotri, N.P., Gajbhiye, V.T., Kumar, M. & Mahapatra, S.P. (1994) *Environ. Mon. Assesm.* **30**: 105-112.

Aislabie, J. & Lloyd-Jones, G. (1995) *Aust. J. Soil Res.* **33**: 925-942.

Alexander, M. (1999) *Biodegradation and Bioremediation,* 2nd ed., Academic Press, New York , pp. 26-45.

Anderson, J.P.E., Lichtenstein, E.P. & Wittingham, W.F. (1970) *J. Econ. Entomol.* **63**: 1559-1599.

Anomymous (1993) *Pesticide News*, September, **No. 15**: 15.

Awasthi, N., Manickam, N. & Kumar, A. (1997) *Bull. Environ. Contam. Toxicol.* **59**: 928-934.

Awasthi, N., Singh, A.K., Jain, R.K., Khangarot, B.S. & Kumar, A. (2003) *Appl. Microbiol. Biotechnol.* **62**: 279-283.

Barik, S. & Munnecke, D.M. (1982) *Bull. Environ. Contam. Toxicol.* **29**: 235-239.

Barik, S. & Sethunathan, N. (1978) *J. Environ. Qual.* **7**: 364-368.

Barik, S., Munnecke, D.M. & Fletcher, J.S. (1982) *Biotechnol. Lett.* **4**: 795-780.

Barik, S., Wahid, P.A., Ramakrishna, C. & Sethynathan, N. (1979) *J. Agric. Food Chem.* **27**: 1391-1392.

Barles, R.W., Daughton, C.G. & Hsieh, D.P.H. (1979) *Arch. Environ. Contam. Toxicol.* **8**: 647-660.

Benezet, H.J. & Matsumura, F.(1973) *Nature.* **243**: 480-481.

Benning, M.M., Hong, S.B., Raushel, F.M. & Holden, H.M. (2000) *J. Biol. Chem.* **275**: 30556-30560.

Benning, M.M., Shim, H., Raushel, F.M. & Holden, H.M. (2001) *Biochem.* **40**: 2712-2722.

Bhushan, B., Chauhan, A., Samanta, S.K. & Jain, R.K. (2000) *Biochem. Biophy. Res. Commun.* **274**: 626-630.

Bhushan, B., Samanta, S.K., Chauhan, A.K., Chakraborti, A.K. & Jain, R.K. (2000) *Biochem. Biophy. Res. Commun.* **275**: 129-133.

Brown, K.A. (1980) *Soil Biol. Biochem.* **12**: 105-112.

Bumpus, J.A. & Aust, S.D. (1987) *Appl. Environ. Microbiol.* **53**: 2000-2008.

Caldwell, S.R. & Raushel, F.M. (1991) *Biotechnol. Bioeng.* **37**: 103-109.

Caldwell, S.R., Newcomb, J.R., Schlecht, K.A. & Raushel, R.M. (1991) *Biochem.* **30**: 7438-7444.

Calle, E.E., Frumkin, H., Henley, S.J., Savitz, D.A. & Thun, J.M. (2002) *CA. Cancer J. Clin.* **52**: 301-309.

Carvalho, F.P., Nhan, D.D., Zhong, C., Tavares, T. & Klaine S. (1998) *IAEA bulletin.* **40**: 24-30.

Casida, J.E. (1993) *Arch. Insect. Biochem. Physiol.* **22**: 13.

Chacko, C.I., Lockwood, J.L. & Zabik, M.J. (1966) *Science.* **154:** 893-895.

Chapalamagdu, S. & Chaudhry, G.R. (1992) *Crit. Rev. Biotechnol.* **12**: 357-389.

Chaudhry, G.R., Ali, A.N. & Wheeler, W.B. (1988) *Appl. Environ. Microbiol.* **54**: 288-293.

Chaudhuri, K., Selvaraj, S. & Pal, A.K. (1999) *Mutat. Res.* **439**: 63-67.

Chauhan, A., Chakraborti, A.K. & Jain, R.K. (2000) *Biochem. Biophy. Res. Commun.* **270:** 733-740.

Cho, C.M.H, Mulchandani, A. & Chen, W. (2002) *Appl. Environ. Microbiol.* **68**: 2026-2030.

Cook, A.M., Daughton, C.G. & Alexander, M. (1980) *Appl. Environ. Microbiol.* **39**: 463-468.

Daughton, C.G.& Hsieh, D.P.H. (1977) *Bull. Environ. Contam. Toxicol.* **18**: 48-56.

De Frank, J.J. (1991) *Applications of Enzyme Biotechnology* (Ed. Kelly, J.W. & Baldwin, T.O.), Plenum Press, New York, pp. 165-180.

Deo, P.G., Karanth, N.G. & Karanth, N.G.K. (1994) *CRC Crit. Rev. Microbiol.* **20**: 57-78.

DiSoudi, B., Gumsley, J.K., Lai, K. & Wild, J.R. (1999) *Biochem.* **38**: 2866-2872.

Environment Canada (1996) *Canadian water quality guidelines*, Environment Canada, Water Quality Branch, Inland Waters Directorate, Ottawa, Ont., Canada.

Environment News Network (2003) September 12.

Gilbert, E.E., Walker, A.W. & Keasling, J.D. (2003) *Appl. Microbiol. Biotechnol.* **61**: 77-81.

Goebel, H., Gorbach, S., Knauf, W., Rimpau, R.H. & Huttenbach, H. (1982) *Residue Rev.* **83**: 40-41.

Hanne, L.F., Kirk, L.L., Appel, S.M., Narayan, A.D. & Bains, K.K. (1993) *Appl. Environ. Microbiol.* **59**: 3505-3508.

Hayatsu, M., Hirano, M. & Tokuda, S. (2000) *Appl. Environ. Microbiol.* **66**: 1737-1740.

Hirakoso, S. 1968. *World Health Organisation, WHO/VBC.* **68**: 1-4.

Holm, L. & Sander, C. (1997) *Proteins, Struct. Funct. Genet.* **28**: 72-82.

Horne, I., Sutherland, T.D., Harcourt, R.L., Russell, R.J. & Oakeshott, J.G. (2002) *Appl. Environ. Microbiol.* **68**: 3371-3376.

Imani, R., Nagata, Y., Fukuda, M., Takagi, M., & Yano, K. (1991) *J. Bacteriol.* **173**: 6811-6819.

Imani, R., Nagata, Y., Senoo, K., Wada, H., Fukuda, M., Takagi, M. & Yano, K. (1989) *Agric. Biol. Chem.* **53**: 2015-2017.

Iwata, H., Tanabe, S., Sakal, N. & Tatusakawa, R. (1993) *Environ. Sci. Technol.* **27**: 1080-1098.

Jagnow, G., Haider, K. & Ellwardt, B. (1977) *Arch. Microbiol.* **115:** 285-292.

Jain, R.K., Dreisbach, J.H. & Spain, J.C. (1994) *Appl. Environ. Microbiol.* **60**: 3030-3032.

John, P.J., Bakore. N. & Bhatnagar, P. (2001) *Environ. Intl.* **26**: 231-236.

Johnson, B.T., Goodman, R.N. & Goldberg, H.S. (1967) *Science.* **157**: 560-561.

Johri, A.K., Dua, M., Tuteja, D., Saxena, R., Saxena, D.M. & Lal, R. (1996) *FEMS Microbiol. Rev.* **12**: 69-84.

Kadiyala, V. & Spain, J.C. (1998) *Appl. Environ. Microbiol.* **64**: 2479-2484.

Kallman, B.J. & Andrews, A.K. (1963) *Science.* **141**: 1050-1051.

Kaneva, I., Mulchandani, A. & Chen, W. (1998) *Biotech. Prog.* **14**: 275-278.

Karnath, N.G.K. & Deo, P.G. (1997) *Pesticides, Crop protection and Environment* (Ed. Parmar, V.S.), Oxford and IBH, New Delhi, pp. 248-257.

Karns, J.S., Muldoon, M.T., Mulbry, W.W., Derbyshire, M.K. & Kearney, P.C. (1987) *Biotechnology in Agricultural Chemistry. ACS Symposium Series 334* (Eds. LeBarson, H.M., Mumma, R.O., Honeycutt, R.C. & Duesing, J.H.) American Chemical Society, Washington ,D.C., pp. 156-170.

Katayama, A. & Matsumura, F. (1993). *Environ. Toxicol. Chem.* **12:** 1059-1065.

Kearny, P.C., Karns, J.S., Muldoon, M.T. & Ruth, J.M. (1986) *J. Agric. Food Chem.* **3**: 702-706.

Keith, L.H. & Telliard, W.A. (1979) *Environ. Sci. Technol.* **13:** 416-423.

Kenaga. E.E. (1974) *Environmental dynamics of pesticides* (Eds. Hague, R. & Freed V.H.), Plenum Publ. Corp., New York, pp. 217-273.

Kennedy, D.W., Aust, S.D. & Bumpus, J.A. (1990) *Appl. Environ. Microbiol.* **56**: 2347-2353.

Kim, J.W., Rainina, E.I., Mulbry, W.W., Engler, C.R. & Wild, J.R. (2002) *Biotechnol. Prog.* **18:** 429-436.

Kohnen, R., Haider, K. & Jagnow, G. (1975) *Environ. Qual. Saf.* **3**: 222-225.

Kullman, S.W. & Matsumura, F. (1996) *Appl. Environ. Microbiol.* **62**: 593-600.

Kuritz, T. & Wolk, C.P. (1995) *Appl. Environ. Microbiol.* **61**: 234-236.

Kwon, G.S., Kim, J.E., Kim, T.K., Sohn, H.Y., Koh, S.C., Shin, K.S. & Kim, D.G. (2002) *FEMS Microbiol. Lett.* **215**: 255-259.

Lai, K.H., Grimsley, J.K., Kuhlmann, B.D., Scapozza, L., Harvey, S.P., DeFrank, J.J., Kolakowski, J.E. & Wild, J.R. (1996) *Chimia* **50**: 430-431.

Lakshmi, R.N. & Lalithakumari, D. (1994) *Can. J. Microbiol.* **40**: 1000-1006.

Lal, R., Lal, S. & Shivaji, S. (1986) *Crit. Rev. Biotechnol.* **3**: 1-15.

Lal, R., Lal, S., Dhanraj, P.S. & Saxena, D.M. (1995) *Adv. Microbiol.* **17**: 55-95

Lee, S.E., Kim, J.S., Kennedy, I.R., Park, J.W., Kwon, G.S., Koh, S.C. & Kim, J.E. (2003) *J. Agric. Food Chem.* **51**: 1336-1340.

Leung, K.T., Tresse, O., Errampalli, D., Lee, H. & Trevors, J.T. (1997) *FEMS Microbiol. Lett.* **155**: 107-114.

Lewis, V.E., Donarski, W.J., Wild, J.R. & Raushel, F.M. (1988) *Biochem.* **27**: 1591-1597.

Lichtenstein, E.P. & Schulz, K.R. (1960) *J. Econ. Entomol.* **53**: 192-197.

Macholz, R.M. & Kujawa, M. (1985) *Res. Rev.* **94**: 119-137.

MacRae, I.C., Raghu, K. & Castro, T.F. (1967) *J. Agric. Food Chem.* **15**: 911-914.

Malley, O.M. (1997) *Lancet.* **349**: 1161 – 1166.

Mallick, B.K., Banerji, A., Shakil, N.A. & Sethunathan, N.N. (1999) *Bull. Environ. Contam. Toxicol.* **62**: 48-55.

Matsumura, F. & Boush, G.M. (1968) *J. Econ. Entomol.* **61**: 610-612.

Matsumura, F. (1975) *Toxicology of Insecticides,* Plenum Press, N.Y., pp. 325-354.

Matsumura, F., Benezet, H.J. & Patil, K.C. (1976) *J. Pestic. Sci.* **1**: 3-8.

McDaniel, C.S., Harper, L.L. & Wild, J.R. (1988) *J. Bacteriol.* **170**: 2306-2311.

Mendel, J.L., Klein, A.K., Chen, J.T & Walton, M.S. (1967) *J. Asoc. Annals Chem.* **50:** 897-903.

Mick, K.L. & Dahm, P.A. (1970) *J. Econ. Entomol.* **63**: 1151-1159.

Miles, J.R. & Moy, P. (1979) *Bull. Environ. Contam. Toxicol.* **23**: 13-19.
Milles, C.J. & Pfeuffer, A. (1997) *Arch. Environ. Contam. Toxicol.* **6**: 853-858.
Misra, D., Bhuyan, S., Adhya, T.K. & Sethunathan, N. (1992) *Soil Biol. Biochem.* **24**: 1035-1042.
Mitra, N.G., Upadhyaya, A., Sachidanand, B. & Agrawal, G.D. (1999) *Pestology.* **23**: 36-40.
Miyamoto, J., Kitagawa, K. & Sato, Y. (1966) *Jpn. J. Exp. Med.* **36**: 211-225.
Miyauchi, K., Adachi, Y., Nagata, Y. & Takagi, M. (1999) *J. Bacteriol.* **181**: 6712-6719.
Miyauchi, K., Lee, H.S., Fukuda, M., Takagi, M. & Nagata, Y. (2002) *Appl. Environ. Microbiol.* **68**: 1803-1807.
Miyauchi, K., Suh, S., Nagata, Y. & Takagi, M. (1998) *J. Bacteriol.* **180**: 1354-1359.
Miyazaki, S. & Thorsteinson, A.J. (1972) *Bull. Environ. Contam. Toxicol.* **8**: 81-83.
Montgomery, J. H. (1997) *Agrochemicals Desk Reference.* 2nd ed. (Ed. Montgomery, J.H.), Lewis Publishers, Boca Raton, Florida , pp. 325-331.
Morgan, P., Lewis, S.T. & Watkinson, R.J. (1991) *Appl. Microbiol. Biotechnol.* **34**: 693-696.
Muir, D.C.G., Wageman, R., Hagrave, B.T., Thomas, D.T., Peakall, D.B. & Norstrom, R.J. (1992) *Sci. Total Environ.* **122**: 75-134.
Mukherjee, I. & Gopal, M. (1994) *Toxicol. Environ. Chem.* **46**: 217-221.
Mulbry, W.W. & Karns, J.S. (1989) *Appl. Environ. Microbiol.* **55**: 289-293.
Mulbry, W.W. & Karns, J.S. (1989) *J. Bacteriol.* **171**: 6740-6746.
Mulbry, W.W., DelValle, P.L. & Karns, J.S. (1996) *Pest. Sci.* **48**-149-155.
Mulbry, W.W., Karns, J.S., Kearney, P.C., Nelson, J.O. & Wild, J.R. (1986) *Appl. Environ. Microbiol.* **51**: 926-930.
Mulbry, W.W., Kearney, P.C., Nelson, J.O. & Karns, J.S. (1987) *Plasmid* **18**: 173-177.
Mulchandani, A., Kaneva, I. & Chen, W. (1999) *Biotechnol. Bioeng.* **63**: 216-223.
Munnecke, D.M. (1976) *Appl. Environ. Microbiol.* **32**: 7-13.
Munnecke, D.M. (1979) *Biotechnol. Bioeng.* **21**: 2247-2261.
Munnecke, D.M., Johnson, L.M., Talbot, H.W. & Barik, S. (1982) *Biodegradation and Detoxification of Environmental Pollutants,* (Ed. Chakraborty, A.M.), CRC Press, Boca Raton, Florida, pp. 1-32.
Nagata, Y., Futamura, A., Miyauchi, K. & Takagi, M. (1999) *J. Bacteriol.* **181**: 5409-5413.
Nagata, Y., Miyauchi, K. & Takagi, M. (1999) *J. Ind. Microbiol. Biotechnol.* **23**: 380-390.
Nagata, Y., Nariya, T., Ohtomo, R., Fukuda, M., Yano, K. & Takagi, M. (1993) *J.Bacteriol.* **175**: 6403-6410.
Nagata, Y., Ohtomo, R., Miyauchi, K., Fukuda, M., Yano, K. & Takagi, M. (1994) *J. Bacteriol.* **176**: 3117-3125.
Nishino, S.F. & Spain, J.C. (1993) *Environ. Sci. Technol.* **27**: 489-494.
Ohisa, M. & Yamaguchi, M. (1978) *Agric. Biol. Chem.* **42**: 1819-1823.
Ohisa, M., Yamaguchi, M. & Kurihara, M. (1980) *Arch. Microbiol.* **125**: 221-225.
Paszezynski, A. & Crawford, R.L. (1995) *Biotechnol. Prog.* **11**: 368-379.
Patil, K.C., Matsumura, F. & Boush, G.M. (1970) *Appl. Microbiol.* **19**: 879-881.
Paul, V. & Balasubramaniam, E. (1997) *Environ. Toxicol. Pharmacol.* **3**: 151-157.
Pfaender, F.K. & Alexander, M. (1972) *Environ. Sci. Technol.* **6**: 629-632.
Racke, K.D., Laskowski, D.A. & Schultz, M.R. (1990) *J. Agric. Food Chem.* **38**: 1430-1436.
Report of analysis of pesticide residues in soft drink samples (2003) Central Food Laboratory, Central Food Technological Research Institute, Mysore, India. August, 14.
Rice, C.P. & Sikka H.C. (1973) *J. Agric. Food Chem.* **21**: 148-152.
Richins, R.D., Kaneva, I., Mulchandani, A., & Chen W. (1997) *Nat. Biotech.* **15:** 984-987.
Rosenberg, A. & Alexander, M. (1979) *Appl. Environ. Microbiol.* **37**: 886-891.
Safferman, S.I., Lamar, R.T., Vonderhaar, S., Neogy, R., Haught, R.C. & Krishnan, E.R. (1995) *Toxicol. Environ. Chem.* **50**: 237-251.
Sahu, S.K., Patnaik, K.K., Bhuyan, S. & Sethunathan, N. (1993) *Soil Biol. Biochem.* **25**: 387-390.
Sanghi, R., Pillai, M.K.K., Jayalekshmi, T.R. & Nair, A. (2003) *Hum. Exp. Toxicol.* **22**: 73-76.
Scanlan, C.S. & Reid, R.C. (1995) *Chem. Biol.* **2**: 71-75.
Senoo, K. & Wada, H. (1989) *Soil. Sci. Plant Nutr.* **35**: 79-87.
Serdar, C.M. & Gibson, D.T. (1985) *Biotechnol.* **3**: 567-571.
Serdar, C.M., Gibson, D.T., Munnecke, D.M. & Lancaster, J.H. (1982) *Appl. Environ. Microbiol.* **44**: 246-249.

Serdar, C.M., Murdock, D.C. & Rohde, M.E. (1989) *Biotechnol.* **7**: 1151-1155.

Sethunathan, N. & Yoshida, T. (1973) *Can. J. Microbiol.* **19**: 873-875.

Shelton, D.R. & Somich, C.J. (1988) *Appl. Environ. Microbiol.* **54:** 2566-2568.

Shelton, D.R. (1988) *Appl. Environ. Microbiol.* **54**: 2572-257.

Shetty, P.K., Mitra, J., Murthy, N.B.K., Namitha, K.K., Savitha, K.N. & Raghu, K. (2000) *Curr. Sci.* **79**: 1381-1383.

Shimazu, M., Mulchandani, A. & Chen, W. (2001) *Biotechnol. Bioeng.* **76**: 318-324.

Siddaramappa, R., Rajaram, K.P. & Sethunathan, N. (1973) *Appl. Microbiol.* **26**: 846-849.

Siddavattam, D., Khajamohiddin, S., Manavathi, B., Pakala, S.B. & Merrick, M. (2003) *Appl. Environ. Microbiol.* **69**: 2533-2539.

Siddique, T., Okeke, B.C., Arshad, M. & Frankenberger, W.T. Jr. (2003) *J. Agric. Food Chem.* **51**: 8015-8019.

Singh, B.K., Kuhad, R.C., Singh, A., Lal, R. & Tripathi, K.K. (1999) *Crit. Rev. Biotechnol.* **19**: 197-225.

Singh, B.K., Walker, A., Morgan, J.A.W. & Wright, D.J. (2003) *Appl. Environ. Microbiol.* **69**: 5198-5206.

Somara, S. & Siddavattam, D. (1995) *Biochem. Mol. Biol. Int.* **32**: 672-676.

Somara, S., Manavathi, B., Tebbe, C.C. & Siddavattam, D. (2002) *Ind. J. Exp. Biol.* **40**: 774-779.

Spain, J.C., Wyss, O. & Gibson, D.T. (1979) *Biochem. Biophy. Res. Commun.* **88**: 634-641.

Spain. J.C. & Gibson, D.T. (1991) *Appl. Environ. Microbiol.* **57**: 812-819.

Srimathi, M.S., Karanth, N.G.K. & Majumder, S.K. (1985) *Pest. Manag.* **5**: 21-25.

Sunderam, R.I.M., Cheng, D.M.H. & Thompson, G.B. (1992) *Environ. Toxicol. Chem.* **11**: 1469-1476.

Sutherland, T.D., Horne, I., Harcourt, R.L., Russell, R.J. & Oakeshott, J.G. (2002) *J. Appl. Microbiol.* **93**: 380-389.

Sutherland, T.D., Horne, I., Lacey, M.J., Harcourt, R.L., Russell, R.J. & Oakeshott, J.G. (2000) *Appl. Environ. Microbiol.* **66**: 2822-2828.

Sutherland, T.D., Weir, K.M., Lancey, M.J., Horne, I., Russelll, R.J. & Oakeshott, J.G. (2002) *J. Appl. Microbiol.* **92**: 541-548.

Sweeney, R.A. (1969) *Proc. Conf. Great. Lakes Res.* **12**: 98-102.

Times of India (2003) April 3.

Tsoi, T.V., Zaitsev, G.M., Plotnikova, E.G., Kosheleva, I.A. & Boronin, A.M. (1991) *FEMS Microbiol. Lett.* **81**: 165-168.

Tu, C.M. (1976) *Arch. Microbiol.* **108**: 259-263.

United States Department of Health and Human Services (1990) *Agency for toxic substances and disease registry, Atlanta, GA.*

Van Woerden, H.F. (1963) *Chem. Rev.* **63**: 557-571.

Vanhooke, J.L., Benning, M.M., Raushel, F.M. & Holden, H.M. (1996) *Biochem.* **35**: 6020-6025.

Vonk, J.W. & Quirjns, J.K. (1979) *Pest. Biochem. Physiol.* **12**: 68-74.

Wada, H., Senoo, K. & Takai, Y. (1989) *Soil Sci. Plant Nutr.* **35**: 71-73.

Walker, A.W. & Keasling, J.D. (2002) *Biotechnol. Bioeng.* **78**: 715-721.

Wedemeyer, G. (1966) *Science.* **152**: 647-649.

Wedemeyer, G. (1967) *Appl. Microbiol.* **15**: 569-574.

Yang, H., Carr, P.D., McLoughlin, S.Y., Liu, J.W., Horne, I., Qiu, X., Jeffries, C.M.J., Russell, R.J., Oakeshott, J.G. & Ollis, D.L. (2003) *Prot. Eng.* **16**: 135-145.

Yule, W.N., Chiba, M. & Morley, H.V. (1967) *J. Agric. Food Chem.* **15:** 1000-1004.

Zhongli, C., Shunpeg, L. & Guoping, F. (2001) *Appl. Environ. Microbiol.* **67**: 4922-4925.

19

Fungal Biosorbents: A Benign Alternative for Removal and Recovery of Heavy Metals from Wastewater

R.K.Saxena, Purnima Anand, Saurabh Saran, Jasmine Isar

Department of Microbiology, University of Delhi South Campus, Benito Juarez Road, New Delhi 110021, (India)

Biosorption is a property of certain types of microbial biomass to bind and concentrate heavy metals from even very dilute aqueous solutions. This phenomenon has gained interest because it offers the advantage of producing negligible secondary pollutants as compared to the conventional physico-chemical treatments. Here, the microbial biomass functions as an ion exchanger by virtue of various reactive groups available on the cell surface. The process can be made economical by procuring natural bulk biomass or spent biomass from various fermentation industries. Moreover, the biosorptive metal loaded by the biomass can easily be desorbed by the non-destructive means and the biomass thus, can be reused for subsequent cycles of metal removal and recovery.

In this endeavour, fungal biomass is a suitable system as it can be grown using media derived from agricultural/ industrial waste or other cheap materials. They can be engineered physiologically and genetically to provide good metal biosorbents. The new biosorbents can be manipulated for better efficiencies and multiple reuses to increase their economic attractiveness. This review emphasizes the fungal biosorption as a feasible alternative to existing treatment methods for removal and recovery of heavy metals from polluted sites.

INTRODUCTION

The discharge of heavy metals into the water bodies is grossly contaminating these ecosystems and has become a matter of concern throughout the world over the past few decades. The majority of toxic metal pollutants are waste products of industrial and metallurgical processes. Their concentrations have to be reduced to meet the new regulatory norms of the government. Among all the heavy metals, copper, chromium and zinc are the major pollutants which, when ingested beyond the permissible limit, causes health hazards to the biota of aquatic ecosystems.

The conventional process of heavy metal removal from industrial wastewater involves chemical precipitation of metals usually by lime followed by settling of the metal precipitates in a pond and / or clarifier. The major shortcomings of conventional treatment include low efficiency at low concentration of heavy metals and expensive handling and safe disposal of toxic sludges. Electrodialysis is another method. In this process, the ionic components (heavy metals) are separated through the use of semi-permeable ion-selective membranes. Application of an electrical potential between the two electrodes causes a migration of cations and anions towards respective electrodes. Because of the alternative spacing of cation and anion permeable membranes, cells of concentrated and dilute salts are formed. The disadvantage is the formation of metal hydroxides which clog them. Ultra filtration is yet another method. These are pressure driven membrane operations that use porous membranes for the removal of heavy metals. The main disadvantage of this process is the generation of sludge. Adsorption on the ion exchange or activated carbon is a recognized method for the removal of heavy metals from wastewater. However, high costs of them have limited their use in the real electroplating wastewater treatment.

New technologies are constantly being sought for removing heavy metals and toxic substances from wastewaters so that they can be safely discharged. The variety of materials tested as metal adsorbents includes coal, crushed coconut shell, peat, bark and straw. The search for new technologies involving the removal of toxic metals from wastewaters has directed attention to bisorption, based on metal binding capacities of various biological materials.

Biosoption is defined as a process in which microorganisms (dead and alive) or their derivatives are employed for sequestration of heavy metals from an aqueous environment. The transfer of metal ions from aqueous to solid biosorbent phase can be due to passive, facilitated or active transport. The uptake of heavy metals by biomass can take place by an active mode (dependent on the metabolic activity) known as bioaccumulation or by passive mode (sorption and /or complexation) termed as biosorption (Muraleedharan *et al.*, 1991).

Importance of Fungal Biosorbents

A wide variety of biological materials have been exploited for their metal adsorption capacity (Volesky, 1986; Chatterjee *et al.*, 1996; Senthil *et al.*, 2000; Redad *et al.*, 2002). Fungi, yeast and bacteria as by-products of industrial fermentation processes, while macro-algae due to bulk availability of their biomasses from natural water bodies, have attracted most of the attention. (Darnell *et al.*, 1986; Davis *et al.*, 2000; Arica *et al.*, 2001; Say *et al.*, 2001).

Out of these groups, filamentous fungi have shown great potentials. Their biomass contains relatively high percentage of cell wall material (Rosenberger,1975) which show excellent metal binding capacity. They may also exhibit intracellular metal accumulation (Gadd, 1990; Paknikar *et al.*, 1993). Many of them show high tolerance towards heavy metals. They grow in wider pH range. They are used in various industries. Most of the fungal biomass produced in these industries is disposed of either by land filling or incineration. Many authors have advocated the use of this biomass as potential biosorbent (Muzzarelli *et al.*, 1980; White and Gadd, 1990; Luef *et al.*, 1991). Zajic and Chiu (1972) were one of the first to investigate fungal biomass for practical applications of biosorption. They discovered that heat deactivated biomass of

Penicillium sp. could biosorb more uranium. Since then, many fungi, both unicellular and filamentous, have been tested for their biosorption capacity.

Table 1: Fungi for ability to bind different metals.

Metal	Fungi	References
Cadmium	*Aspergillus niger*	Kurek *et al.* (1982); Kapoor and Viraraghvan (1998); Akhtar *et al.* (1996)
	Candida tropicalis	Mattuschka *et al.* (1993)
	Mucor rouxii	Mullen *et al.* (1992)
	Penicillium chrysogenum	Kurek *et al.* (1992); Niu *et al.* 1993; Paknikar *et al.* (1993); Holan and Volesky (1995)
	Rhizopus arrhizus	Tobin *et al.* (1984); Fourest and Roux (1992)
	Rhizopus nigricans	Holan and Volesky (1995)
	Saccharomyces cerevisiae	Gadd and Mowll (1983); Ben Omar *et al.* (1997); Grafl and Schwantes (1983)
	Trichoderma viride	Sanna *et al.* (1997)
	Polyporous versicolor	Zajic and Chiu (1972)
	Phanerochaetes chrysporium	Zajic and Chiu (1972); Dey *et al.* (1997); Say *et al.* (2001)
	Trichoderma reesei	Chatterjee *et al.* (1997); Kurek *et al.* (1982)
Chromium	*Candida tropicalis*	Mattuschka *et al.* (1993)
	Penicillium chrysogenum	Paknikar *et al.* (1993)
	Rhizopus arrhizus	Tobin *et al.* (1984); Sag *et al.* (1982)
	Aspergillus niger	Kovacevic *et al.* (2000)
	Mucor mehei	Tobin and Roux (1998)
Cobalt	*Aspergillus niger*	Akhtar *et al.* (1996)
	Saccharomyces cerevisiae	Brady and Duncan (1993)
	Neurospora crassa	Rama Rao *et al.* (1997)
	Trichoderma viride	Sanna *et al.* (1997)
Copper	*Aspergillus niger*	Townsley *et al.* (1986); Akhtar et al. (1996), Kovacevic *et al.* (2000)
	Aspergillus terreus	Gulati *et al.* (2002)
	Candida tropicalis	Mc Gahren (1984)
	Candida utilis	Ross and Townsley (1986)
	Cunninghamella blakesleeana (cell wall)	Venkateswerlu and Stotzky (1989)
	Mucor rouxii	Mullen *et al.* (1992)
	Penicillium chrysogenum	Niu *et al.* (1993) Paknikar *et al.* (1993)
	Penicillium italicum	DeRome and Gadd (1987)
	Penicillium spinulosum	Townsley *et al.* (1986)
	Rhizopus arrhizus	DeRome and Gadd (1987); Tobin *et al.* (1984); Gadd (1988)
	Phanerochaetes chrysporium	Singh and Yu (1998); Say *et al.* (2001)
	Rhizopus arrhizus	Sag *et al.* (1998)
	Trichoderma viride	Sanna *et al.* (1997)
	Saccharomyces cerevisiae	Mattuschka *et al.* (1993); Brady and Duncan (1993)
	Trichoderma viride	Townsley *et al.* (1986)
	Auricularia polytricha	Galli *et al.* (2003)
Gold	*Aspergillus niger*	Kuyucak and Volesky (1988), Gee and Dudeney (1988); Gomes and Linardi (1996)
	Rhizopus arrhizus	Kuyucak and Volesky (1988)

Lead	*Aspergillus versicolor*	Cabuk *et al.* (2005)
	Aspergillus niger	Jianlong *et al.* (2001)
	Absidia orchidis	Holan and Volesky (1995)
	Penicillium sp.	Siegel *et al.* (1986)
	Penicillium	Niu *et al.* (1993)
	chrysogenum	Paknikar *et al.* (1993); Holan, Z.R. & Volesky, B. 1995
	Rhizopus arrhizus	Tobin *et al.* (1984)
	Phanerochaetes chrysporium	Yetis *et al.* (1998)
	Say *et al.* (2001)	
	Rhizopus nigricans	Holan and Volesky (1995)
	Saccharomyces cerevisiae	Grafl and Schwantes (1983)
	Aspergillus niger	Kapoor and Viraraghvan (1998)
	Rhizopus oligosporous	Akhtar *et al.* (1996)
	Trichoderma viride	Sanna *et al.* (1997)
Manganese	*Rhizopus arrhizus*	Tobin *et al.* (1984)
Mercury	*Rhizopus arrhizus*	Tobin *et al.* (1984)
	Saccharomyces cerevisiae	Grafl and Schwantes (1983)
	Saccharomyces cerevisiae	Ben Omar *et al.* (1997)
Nickel	*Candida* sp.	Mattuschka *et al.* (1993) Kambe – Honjoh *et al.* (1997)
	Rhizopus arrhizus	Fourest and Roux (1992)
	Rhizopus nigricans	Holan and Volesky (1995)
	Absidia orchidis	Holan and Volesky (1995)
	Penicillium chrysogenum	Bakkaloglu *et al.* (1998)
	Aspergillus niger	Mullen *et al.* (1992); Kovacevic *et al.* (2000)
Silver	*Rhizopus arrhizus*	Tobin *et al.* (1984)
	Saccharomyces cerevisiae	Brady and Duncan (1994)
	Mucor rouxii	Mullen *et al.* (1992)
	Aspergillus niger	Gomes and Linardi (1996)
Thorium	*Aspergillus niger*	Gadd (1988); Tsezos and Voleksy (1981)
	Aspergillus terreus	Tsezos and Voleksy (1981)
	Penicillium chrysogenum	Tsezos and Voleksy (1981)
	Rhizopus arrhizus	Tsezos and Voleksy (1981)
	Gadd (1988)	
	Saccharomyces cerevisiae	Gadd (1988)
Uranium	*Aspergillus niger*	Horikoshi *et al.* (1981), Tsezos and Voleksy (1981); Yakubu and Dudeney (1986)
	Aspergillus fumigatus	Bhainsa and D'Souza (1999)
	Aspergillus terreus	Tsezos and Voleksy (1981)
	Candida albicans	Horikoshi *et al.* (1981)
	Chaetomium sitophila	Nakajima and Sakaguchi (1986)
	Neurospora sitophila	Nakajima and Sakaguchi (1986)
	Penicillium sp.	Zajic and Chiu (1972)
	Penicillium chrysogenum	Tsezos and Voleksy (1981) Nakajima and Sakaguchi (1986)
	Penicillium janthinellum	Horikoshi *et al.* (1981)
	Penicillium lilacinum	Horikoshi *et al.* (1981)
	Rhizopus arrhizus	Tsezos *et al.* (1997); Tobin *et al.* (1984)
	Rhizopus oryzae	Horikoshi *et al.* (1981)
	Saccharomyces cerevisiae	Horikoshi *et al.* (1981) Strandberg *et al.* (1991)
	Trichoderma reesei	Horikoshi *et al.* (1981)
Iron	*Rhizopus arrhizus*	Sag *et al.* (1998)
	Trichoderma viride	Sanna *et al.* (1997)
Zinc	*Aspergillus niger*	Luef *et al.* (1991)

Kovacevic *et al.* (2000)	
Penicillium chrysogenum	Luef *et al.* (1991)
Penicillium spinulosum	Ross and Townsley (1986)
Saccharomyc cerevisiae	Grafl and Schwantes (1983)
Trichoderma viride	Sanna *et al.* (1997)
Rhizopus arrhizus	Tobin *et al.* (1984)
	Preetha and Viruthagiri (2005)

FACTORS AFFECTING BIOSORPTION

Temperature

Temperature does not seem to influence the biosorption. This is supported by Venkateswerlu and Stotzky (2000), who reported that cobalt binding was not affected by increase in temperature in *Cunninghamella blakesleeana*. Similarly, Luef *et al.* (1991) also reported temperature independent biosorption of zinc between 15-30°C by *Aspergillus niger*. In the same direction, Bengtsson *et al.* (1995) reported that the uptake of uranium by *Talaromyces emersonii* CBS 814 70 was found to remain unchanged between 15 and 60°C.

However, in some cases, an increase in temperature induces disturbances in membrane thus exposing more binding sites which in turn may elevate percentage uptake. Somers as early as in 1963 reported that the copper binding capacity of spores of *Neurospora crassa* increases with increasing temperature. Uptake of uranium by *Saccharomyces cerevisiae* and *Penicillium* sp. also increased when the temperature was raised from 30 to 50°C (Shumate *et al.*, 1978; Galun *et al.*, 1984). Venkateswerlu and Stotzky (2000) also reported increase in copper binding with increase in temperature. On the other hand, there are reports where increase in temperature has resulted in decrease in biosorption. Ag^{2+} sorption by *Saccharomyces cerevisiae* decreased from 185 μ mol Ag/g weight at 55°C (Singleton and Simmons, 1996). Bai and Abraham (2001) have reported decrease in Cr removal by *Rhizopus nigricans* at 50°C. They attributed this decrease to the possible damage caused to the binding sites in the biomass.

Hydrogen Ion Concentration (pH)

Hydrogen ion concentration (pH) seems to be the most important parameter in the biosorptive process as it affects the solution chemistry of the metals, the activity of the functional groups in the biomass and the competition of metallic ions (Friis and Myers-Keith, 1986; Galun *et al.*, 1997) generally at low pH, mobility of metal increases. At pH near or above neutrality, insoluble oxides, hydroxides and carbonates are formed (Volesky, 1990a,b). The binding sites are also known to be affected by pH. At low pH, +ve sites predominate on cell walls, whereas at higher pH, -ve sites predominate. With increase in external pH, increased binding has been observed for a variety of microorganisms. Low external pH generally decreases the rate and extent of fungal biosorption of Cu^{2+}, Cd^{2+}, Zn^{2+}, Mn^{2+} and Co^{2+} (Fuhrman and Rothstein, 1968; Gadd and Mowll, 1983; Gadd and White, 1993).

Strandberg *et al.* (1981) showed that initial rate of uranium uptake by yeast increased above pH 2.5. According to them, with decrease in pH, H^+ ions increased which compete with metal ions for binding sites thus decreasing the uptake. This phenomenon of competition between H^+ and metal ion occurs where binding takes place by an ion exchange process. With increase in pH, predominant cation UO^{2+} is replaced by hydrolyzed products such as $UO^2\ (OH)^+$, $(UO_2)_2(OH)^{2+}$ and $(UO_2)_3\ (OH_5)^{+4}$ and complexes are also formed with carbonates (Strandberg *et al.*, 1981; Tsezos and Volesky, 1981). Tsezos and Volesky (1981) also showed that thorium uptake occurs more at high pH. They documented that at pH<2, solubility of thorium is high with Th^{4+} being the main species. Furthermore, with increase in pH, various hydrolysis products appear, particularly, Th $(OH_2)\ ^{2+}$, which are taken up by the cell more efficiently.

Similarly, maximum radium uptake by *Penicillium chrysogenum* occurs between pH 7.0 – 10.0. Tobin *et al.* (1984) reported uranyl uptake to be 0.82 mM/g at pH 3.5 which reduced to 0.30 mM/g at pH 2.5. Galun *et al.* (1978) reported that uptake and binding of Ni^{2+}, Zn^{2+}, Cd^{2+} and Pb^{2+} by *Penicillium digitatum* increased with pH. They found that at low pH, H^+ and metal ions compete for the same site.

Increase of biosorption with increase in pH was also reported by Luef *et al.* (1991) in alkali treated *Aspergillus niger* for zinc. In case of yeast, efficiency to bind metal decreased from 63 – 97% to 30% with decrease in pH from 5.6 – 1.2 (Larsson *et al.*, 1992). Paknikar *et al.* (1993) also showed the maximum binding of Cu^{2+}, Cd^{2+} and Zn^{2+} between pH 4 and 6 for *Penicillium chrysogenum.*

Suhasini *et al.* (1999) reported that maximum removal of Co^{2+} (190 mg/g) by *Rhizopus* spp. based biosorbent PFB was obtained at pH 7.0. Similarly, increase in metal sorption by *Aspergillus oryzae* has been reported by Vianna *et al.* (2000).On the other hand, Niu *et al.* (1993) reported increase in Pb^{2+} biosorption by *Penicillium chrysogenum* till pH 5, which was followed by decrease at higher pH. According to them, H^+ competes with Pb^{2+} for binding site at low pH and decrease at high pH was may be due to deterioration of biomass. However, Huang *et al.* (1990) discussed that degree of ionization increases with increase in pH, which results in more tight binding of metal to active surfaces, which in turn, breaks salt linkages and thus complexes of Pb^{2+} with some soluble ligands are released from the cells.

Contrary to these reports, biosorption of MoO_4^{2+} and VO_3 decreases to zero with increase in pH from 4.5 to 5.5 in *Rhizopus arrhizus* (1984). Biosorption of chromium and lead were shown to be maximum at pH 2.0 in *Penicillium chrysogenum* (Paknikar *et al.*, 1993). They attributed this to coordination bond formation which is not pH dependent. Bai and Abraham (2001) also observed maximum sorption of Cr (V1) by *Rhizopus nigricans* at pH 2.0. They attributed this to predominance of anionic species of Cr at this low pH, which could easily interact with positively charged cell wall.

However, binding of uranyl ions is not pH dependent in the case of *Pseudomonas aeruginosa* (Strandberg *et al.*, 1981) and *Penicillium digitatum* (Galun *et al.*, 1987). It has also been found that pH effect varies with metal in the same organism suggesting that different interactions or functional groups are involved in biosorption for different metals (Venkateswerlu and Stotzky, 1989). They showed optimum pH for copper uptake to be 5.5 for *Cunnighamella blakesleeana.* However, no optimum pH was observed for binding of Co^{2+}, which increases linearly up to pH 8.0. Kovacevic *et al.* (2000) reported better biosorption by *A.niger* for copper, zinc and nickel at pH ranging from 4- 6.On the contrary pH value from 3 –7 had no influence on sorption of chromium.

Incubation Time

The time of contact of metal and microbial biomass has been shown to be very important.

Generally, biosorption is a very rapid process, completing within few hours (Aksu and Kutsal, 1990). After some time of contact, equilibrium is established, when the biomass becomes saturated and no further increase in biosorption is observed.

The binding of copper or cobalt was shown to be time dependent in *Cunnighamella blakesleeana* by Venkateswerlu and Stotzky (1989). They reported that the binding of Cu^{2+} reached of maximum of 0.62 and 0.86 µm Cu^{2+}/mg biomass after 2hrs for control, and cell wall grown in the presence of cobalt was shown to be 1.07 µm Cu^{2+}/mg and still, no saturation was observed. In the case of cobalt binding, even though less metal was bound, the pattern was similar (Venkateswerlu and Stotzky, 1989).

In another time course study, Khalid *et al.* (1993) reported that uranium sorption took place within the first four hours of contact and then equilibrium was established, after which no more sorption occurred. According to them the first phase is due to rapid physical sorption and the second one is due to structural changes and surface transformation. They reasoned that uranium attains a constant adsorption value after four hours due to the hexavalent uranium species, which are already attached upto saturation and are involved in strong bond formation between the adsorbent and adsorbate.

Brady and Duncan (Brady and Duncan, 1993) also showed that a fast initial phase of copper accumulation lasted few minutes in *Saccharomyces cerevisiae.* It involved binding of metal to cell wall. The slower second phase involved internalization of metal. Though at higher biomass concentration, the second phase was not observed. Maximum removal of uranium by *S. cerevisiae* was observed during 60-100 minutes of exposure (Omar *et al.*, 1991). Similarly, Pb^{2+} sorption by *Rhizopus oligosporus* was very rapid during the initial stages of sorption processes (20 minutes) but reached equilibrium only after 14h of exposure (Ariff *et al.*, 1999). Bai and Abrahain (2001) observed rapid adsorption of Cr ions during initial 30 minutes of the sorbent contact in case of *Rhizopus nigricans* biomass. Adsorption equilibrium was reached in less than 4 hours in *Aspergillus niger* (Jianlong, 2002). Say *et al.* (2001) reported more than 60% adsorption of Cd^{2+}, Cu^{2+} and Pb^{2+} in *Phanerochaete chysosporium* in 24 hrs.

Concentration of Metal

Initial metal concentration is an important factor determining the biosorption capacity, which generally increases with increasing concentration. It reaches a plateau at a high metal concentration, which depends upon availability of the number of binding sites in biomass.

Venkateswerlu and Stotzky, (1989) reported that in *Cunninghamella blakesleeana,* binding of copper and cobalt to cell wall samples progressively increased with increase in concentration of the metals. Brady and Duncan (1993) also showed similar observations for copper, cobalt and cadmium in *Saccharomyces cerevisiae.* Maximum specific Pb^{2+} sorption by *Rhizopus oligosporus* increased with an increase in initial lead concentration upto 200 mg/L beyond which the system attained equilibrium (Ariff *et al.*, 1999).

However, Townsley *et al.* (2002) showed that with increasing concentration of copper, removal efficiency fell rapidly in *Aspergillus niger.* On the other hand same group (Townsley *et al.*, 2002) found that non-growing mycelium of *Trichoderma viride* exhibited energy – independent Cu uptake when packed in the column of molochite. In this, Cu^{2+} removal up to 75% was not dependent on the concentration. Bai and Abraham (2001) observed a decrease in percentage adsorption of Cr ions by *Rhizopus nigricans* with an increase in the initial metal concentration from 50 to 400 mg/L. However, specific biosorption increased with increase in metal concentration.

Concentration of Biomass

The capacity of biosorption of metal by fungal biomass is strongly dependent on its concentration. During this process, increase in the biomass concentration may increase the total removal of the metal. However, specific uptake of biomass is reduced (Volesky, 1990a,b) at a given equilibrium concentration of metal. Fungal cells adsorb more metal ions at low pH as compared to high cell densities. Itoh *et al.* (1975) found this is true for Hg^{2+}, Ag^{+}, Cd^{2+}, $A1^{2+}$, Ni^{2+}, Cu^{2+} and Pb^{2+} in *Saccharomyces cerevisiae.* De Rome and Gadd (1987) reported similar trend for Cu^{2+} for several fungi. *Saccharomyces cerevisiae* and *Aspergillus niger* also showed more thorium adsorption at low cell densities.

Similar results of decrease in uptake capacity with increase in biomass concentration were also obtained for *Aspergillus niger, Penicillium chrysogenum* and *Claviceps paspali* (Luef *et al.*, 1991; Brady and Duncan, 1993). They suggested that this could be due to electrostatic interactions, interference between binding sites and reduced exposure of potential metal binding sites at high cell densities.

Effect of the Presence of Other Ions

Industrial and mine effluents are likely to contain several different metals and metal binding components, which may interfere with the recovery of the metals (Volesky, 1990a,b). However, generally the biosorption studies are carried out on metal solutions containing a single metal. Though, such studies are important (Stratton, 1987) but conclusions drawn from these may not be valid when application of this technology is considered (Singleton and Simmons, 1996). Therefore, it is necessary to study the effect of various cations and anions on biosorption of test metal.

Cation: Uptake of copper by spores of *Neurospora crassa* was found to be reduced in the presence of K^{+}, Mg^{2+} and Ca^{2+} (Somers, 1963). Cu^{2+} binding to *Penicillium spinulosum* was reduced by Mg^{2+}, Mn^{2+}, Co^{2+} and Zn^{2+} (Townsley *et al.*, 1986). Brady and Duncan (1993) showed that alkali metals like sodium had less effect on the level of copper accumulation in *Saccharomyces cerevisiae.* The percentage inhibition by cations was in the order of $Fe^{3+}>AI^{3+}>Ca^{2+}>Na^{+}>V^{5+}>Mn^{2+}>Mo^{4+}>Li^{+}$ (Kapoor and Viraraghvan, 1998).

In *Cunninghamella blakesleeana* binding of cobalt was inhibited by copper but not vice-versa. (Venkateswerlu and Stotzky, 1989). They also found that Cd^{2+} inhibited binding of cobalt but copper binding was unaffected. These observations suggested that there are two binding sites, one exclusively for copper and the other common to both, but with a high affinity for cobalt.

However, lower competitive biosorption capacities for Cd^{2+}, Pb^{2+} and Cu^{2+} were observed in *Phaenerochaete chrysosporium* as compared to those under non-competitive conditions; the biomass followed $Cu^{2+}>Pb^{2+}>Cd^{2+}$ as affinity series (Say *et al.*, 2001). Tsekova and Petrov (1981) showed that metal ions accumulation from solution decreased slightly for Co^{2+} and Fe^{2+}, but considerable for Cu^{2+} ions. It indicates that competitive inhibition of Cu^{2+} uptake occurs by other cations in the solution. They suggested that accumulation of Co^{2+} modifies the conformation of cell wall polymers and thereby masking Cu^{2+} binding sites.

In *Penicillium chrysogenum*, uptake of Pb^{2+} was not affected by Cu^{2+} and As^{2+}. However, Zn^{2+} drastically interfered with the absorption of Pb^{2+} (Niu *et al.*, 1993).

Uranium uptake by *Penicillium digitatum* was suppressed in the presence of Fe^{3+}. On the other hand, Ni^{2+}, Cu^{2+}, Zn^{2+}, Cd^{2+} and Pb^{2+} had very little effect on uranium uptake. However, uptake of uranium was inhibited by Fe^{3+}, Zn^{2+}, Cu^{2+}, Mn^{2+}, Mg^{2+}, Ca^{2+} and Ba^{2+} in *Rhizopus arrhizus* (Tsezos and Volesky, 1982, Galun *et al.*, 1984). The *Rhizopus arrhizus* biomass is unusual, as it does not adsorb appreciable amount of alkali ions like Na+. This is industrially significant because in many industrial applications, unwanted alkali metal ions are present and they may compete for uptake with desired metal ions.

In both above-mentioned cases, uranium did not inhibit iron uptake in *Rhizopus arrhizus* (Tsezos and Volesky, 1982) and *Penicillium digitatum*. Nakajima and Sakaguchi (89) studied the interaction of nine metals in solution viz. Mn^{2+}, Co^{2+}, Ni^{2+}, Cu^{2+}, Zn^{2+}, Cd^{2+}, Hg^{2+}, Pb^{2+} and $UO_2{}^{2+}$ for various fungi, yeast, bacteria and actinomycetes. They showed that out of all metals tested, Uo^{2+}, Hg^{2+}, Pb^{2+} and Cu^{2+} ions were more readily accumulated. They also showed that the accumulation of cobalt occurred to a large extent when present alone but when it was present along with other metals, its accumulation was poor.

Anions: Complexation of metal ions by inorganic or organic ligands in solution can affect metal biosorption as compared to ligand-free systems (Tobin *et al.*, 1987). They reported that at low metal concentration and high anion concentration, nearly all the metal ions would be complexed. The resulting metal - ligand complexes may be non-adsorbing, weakly adsorbing or strongly adsorbing. In their study, they found inhibition of metal biosorption by sulphate, chloride, phosphate and carbonate in *Rhizopus arrhizus*. Zhou and Kiff (1991) also reported inhibition of copper biosorption by sulphate in *Rhizopus arrhizus*. They suggested that direct competition for the copper ions by both fungal cell and anions was responsible for reduced uptake.

Anions such as phosphate, carbonate and thiosulphate are also known to precipitate metals (Gadd and Griffiths, 1980; Beyeramann, 1991), which reduce their availability thus causing reduction in biosorption. Volesky (1990a) reported reduced biosorption of uranium by *Rhizopus arrhizus* and *Penicillium chrysogenum* from sea-water due to the presence of carbonate ions. Zn^{2+} biosorption by *Rhizopus arrhizus* was inhibited in the presence of ligands in the series EDTA>SO4 $^{2-}$> $C1^{-}$ (Zhou, 1999).

However, according to Singleton and Simmons (1996), $SO_4{}^{2-}$ and $NO_3{}^{2-}$ did not have significant effect on biosorption. They reported that $SO_4{}^{2-}$ did not significantly inhibit silver biosorption by *Saccharomyces cerevisiae*. This account of effects of different physical and chemical factors shows that they vary with type of metal, organism and environmental conditions.

MECHANISM OF BIOSORPTION

The mechanisms of metal uptake have been found to be complex as many factors seem to be involved in this phenomenon (Luef *et al.*, 1991). They may be classified according to various criteria and are not fully understood. According to the dependence on the cell's metabolism, uptake mechanisms can be divided into metabolism dependent and (Akhtar *et al.*, 1996) non-metabolism dependent. Metabolism dependent process is often associated with an active defense system of microorganism, which reacts in the presence of toxic metal. However, metabolism-independent association of metal species to fungal cell walls may include ion-exchange, absorption, complexation, precipitation and crystallization (Mullen *et al.*, 1992, Stuetz *et al.*, 1993).

Precipitation

Precipitation may either be dependent on the cellular metabolism or independent of it. Ercole *et al.* (1994) suggested that in case of precipitation, the metal uptake may take place both in solution or on cell surface as in the presence of toxic compounds; microbes produce compounds that favour the precipitation process. It may be dependent on cell's metabolism, if metal removal from solution is associated with active defense system of the microorganism. Precipitation may not be dependent on the cell's metabolism, if it occurs after a chemical interaction between the metal and cell surface. Cell walls of microbial biomass mainly binding groups such as carboxyl, sulphate, phosphate and amino groups. The type of biosorption i.e., non- metabolism dependent, is relatively rapid (Kuyucak and Volesky, 1988).

Physical Adsorption

Physical adsorption takes place with the help of VanderWaal's forces. Kuyucak and Volesky (1988) hypothesized that uranium, cadmium, zinc biosorption by dead biomass of algae, fungi and yeast takes place through electrostatic interactions between the metal ions in solutions and cell walls of microbial cells. Electrostatic interactions have also been demonstrated to be responsible for copper biosorption by bacterium *Zoogloea ramigera* and alga *Chlorella vulgaris* (Aksu *et al.*, 1992).

Ion Exchange

Cell walls of microorganisms contain polysaccharides and bivalent metal ions exchange with the counter ions of the polysaccharides. Kuyucak and Volesky (1988) reported that the alginates of marine algae occur as salt of K^{+}, Ca^{+}, Na^{+} and Mg^{+}. These ions can exchange with counter ions such as Co^{2+}, Cu^{2+}, Cd^{2+} and Zn^{2+} resulting in the biosorptive uptake of heavy metal. The successful technology of biosorption of copper by fungi *Ganoderma lucidum* developed by Muraleedharan and Venkobacher, 1990.

Electron microscopic studies are confirmed with various techniques *viz* X-ray energy dispersion analysis (EDAX), EPR (Electronic Paramagnetic Resonance), FTIR (Fourier-Transform Infrared).

Biosorption Equilibrium Models : Assessment of Sorption Performance

The adsorption models are useful means for describing the degree of biosorption as a function of equilibrium metal ion concentration at constant pH and temperature conditions. These models are mathematical representations of biosorption equilibrium and by providing metal uptake capacities for various fungal strains, serve as a means of comparing different species in terms of metal removal. The two isotherms, which are widely used, are Langmuir (Adamson 1976) and Freundlich (1926) isotherms.

Table 2: An account of the analysis of mechanism of biosorption by various species.

Organism	Metal	Technique	Analysis
Penicillium chrysogenum (Niu *et al*., 1993)	Pb	TEM	a. Electron dense area in central part of cell wall b. Energy levels of Pb – complex (14 – 1 – 146 ev) are within the Pb – biopolymer complex energy range.
Penicillium digitatum (Galun *et al.* 1987)	U	EDAXU	Localized on outer surface following short exposure while longer exposure lead to intracellular depositions.
Ganoderma lucidium (Muraleedharan *et al*., 1991)		EPR	' O' dominating atom binding.
Rhizopus oligosporus	Pb	TEM, EDAX	a. TEM studies revealed depositions of Pb only on the surface of cell wall. b. EDAX revealed disappearance of peaks corresponding to Mg^{2+}, S^{2+}, K^{+} and appearance of Pb^{2+} along with existing P^{+}.
Aspergillus niger *Penicillium chrysogenum* *Mucor miehei* (Guibal *et al.* 1995)	U	FTIR Infra red	Uranyl binds to amine and sites. UO vibration band appears at a wave number which varies according to pH and nature of the metal ion species in solution.
Rhizopus spp (Suhasini *et al.* 1999)	CO	EDAX	a. EDAX revealed replacement of Based biosorbent EPR Ca^{2+} by CO^{2+} b. EPR showed the involvement of free organic radical.
Rhizopus digosporus (Aloysius *et al.* 1999)	Cd	TEM, EDAX	a. TEM revealed electron dense area on the middle section of cell wall following exposure to Cd. b. EDAX spectrum shows Conspicuous P and Cd indicating precipitation of insoluble metallic Cd as $PO4^{3-}$ precipitate.

Langmuir isotherm: The general formula of the Langmuir model equation (Garnham, 1997) is :-

$$\frac{Ce}{qe} = \frac{1}{k_L} + \frac{(a_L)}{(k_L)} Ce$$

qe - uptake of metal at equilibrium (mg/g)

Ce - concentration of metal at species in solution at equilibrium (mg/l)

a_L and k_L - constant

Qo - maximum metal uptake = $\frac{(k_L)}{(a_L)}$

Assumptions of Langmuir model:-

(i) The surface consists of adsorption sites,

(ii) All adsorbed species interact only with a site and not with each other,

(iii) Adsorption is limited to a monolayer,

(iv) Adsorption energy of all the sites is identical and independent of the presence of adsorbed species on neighboring sites.

Freundlich isotherm: The general formula of the Freundlich isotherm is

$$\ln q_e = \ln K + 1/n \ln C_e$$

K - measure of adsorbent capacity

1/n - intensity of adsorption

The Freundlich model is empirical in nature and was developed for heterogeneous surfaces (Kapoor and Viraraghvan, 1998). Freundlich isotherm is more preferred for aqueous solutions while the Langmuir isotherm is preferred more for gaseous solutions (Volesky, 1990a,b). An isotherm that is steep at the origin at low residual concentrations of the sorbate is highly desirable because it indicates a high affinity of the sorbent of the given sorbed species. Such a biosorbent would be performing well at very low concentrations of the sorbed species in the solution.

Zhou (1999) compared the Zn adsorptive capacity of five different fungi *viz Rhizopus arrhizus, Mucor racemosus, Mycotypha africana, Aspergillus* and *Schizosaccharomyces pombe* by both models. *Rhizopus arrhizus* had highest k and q_{max} value of 11.055 u mol 1-n h/g and 213u mol/g respectively.

Korenevskii *et al.* (1999) evaluated the biosorption efficiencies of micromycete cultures using Freundlich and Langmuir constants. They obtained low Ks value, which indicate high affinity of the fungal biomasses for Ag cations. Bai and Abraham (2001) observed that equilibrium data of Cr sorption by *Rhizopus nigricans* fitted well into linearized Freundlich isotherm models. Say *et al.* (2001) have reported good fit of Langmuir model to Cd, Pb and Cu sorption by filamentous fungus *Phaenerochaete chrysosporium.*

Following models are also used to describe biosorption equilibria—

(i) Brunaurer - Emmett- Teller (BET) model: The BET isotherm has the form:

$$\frac{C_e}{C_s - C_e} = \frac{1}{BQ^\circ} + \frac{(B-1)}{(BQ^\circ)} \frac{(C_e)}{(C_s)}$$

Cs - saturation concentration of metal ion

Q° - amount adsorbed per unit weight of biomass for monolayer biosorption

B - constant relating to the energy of interaction with the surface

The BET isotherm describes the multiplayer adsorption at the adsorbent surface and assumes that the Langmuir isotherm applies to each layer (Weber, 1972).

(i) **Scatchard Model:** Scatchard (1949) developed a model to describe the attraction of proteins for small molecules and ions. The interaction of metal ions with the cell surface binding sites can be described by the following equation.

$$M + X = MX$$

$$K = \frac{(MX)}{(M)(X)}$$

K - association constant

M - metal ion

X - binding site on biomass surface

MX - metal ion biosorbed on biomass

The plot of the amount of bound metal in the biomass divided by equilibrium metal ion concentration verses the amount of bound metal ion gives a typical Scatchard plot.

The equation is represented as follows:

$MX = K (X_0 - MX)$

Me - equilibrium metal ion concentration

X_o - concentration of potential binding sites

Applications of mathematical models (Schiewer and Volesky, 2000)

(*i*) It gives a quantitative description of the process.

(*ii*) It helps in optimizing the operating conditions.

(*iii*) It helps in minimizing the number of experimental runs and to pin-point the results of spot-check experiments.

(*iv*) It helps in designing of reactor and optimization of operational conditions at industrial scale.

Multi-metal Biosorption Modelling System

Chaong and Volesky (1995) described a multi – component Langmuir isotherm model in which 1:1 stochiometry was assumed between metal ions and binding sites where all metals make use of the same sites and compete for them. The authors have used 3-dimensional sorption isotherm surfaces using the software MATLAB 4.0 to evaluate the two-metal sorption performance of the biosorbent. Schiewer and Volesky (2000) have discussed models involving competition among metal cations, pH effects and effects of ionic strength and electrostatic attractions.

EFFECT OF PRETREATMENT OF THE BIOSORPTION OF HEAVY METALS

Biosorption capacity of biomass is affected by its composition and many other physico-chemical factors. Therefore, any chemical, physical or biological process which can change the chemical composition of the biomass will certainly have an effect on the biosorption potential. With this view, many investigations have been carried out to increase the biosorptive capacity of the biomass by treating the biomass before keeping in contact with the metal solution.

The pretreatment can be at two stages:

a. Growing of biomass under various nutritional conditions

b. Physical and chemical changes in already grown biomass.

Effect of nutrients

The type of the medium in which the fungus is grown strongly affects its biosorptive capacity. Shumate *et al.* (1987) showed that *Saccharomyces cerevisiae* grown on a synthetic medium was found to sequester 2 - 5 times more uranium than those of same culture grown on rich organic medium. Pre-growing mycelia of *Neurospora crassa* in the presence of high concentration of trace metal, copper, preconditioned the cell wall for increased biosorbent capacity (Subramanium *et al.*, 1983). Sears (1986) showed that the chemical composition of *Rhizopus javanicus* biosorbent biomass was affected by the divalent cation content of the growth medium. Venkateswerlu and Stotzky (1989) also showed that cell walls isolated from *Cunninghamella blakesleeana* grown in the presence of Cu^{2+} or Co^{2+} had higher binding capacity than control cell walls. They found high amount of chitosan along with phosphate ion in cobalt grown cell walls (Tsezos and Volesky, 1982).

Thorium biosorption by *Rhizopus arrhizus* was greater when fungus was grown on simple than on complex medium (Gadd and White, 1989). *Aspergillus niger* obtained from citric acid fermentation plant showed best biosorption capacity. Citric acid fermentation is carried out in presence of high amount of potassium hexacyano ferrate, which may affect the composition of cell wall. (Luef *et al.*, 1991). Similarly Cu^{2+} uptake by *Aspergillus niger* was very less when biomass was grown in malt medium (4-5%) as compared to Duff medium (45%) (Bosecker, 1993).

Physical pretreatment: Metal affinity to the biomass can be manipulated by pre-treating the biomass with detergents, heat etc. may increase the amount of the metal sorbed. The treatment may expose metal binding sites either by configurational changes or by removing masking groups (Galun *et al.*, 1987). McHale and McHale (1994) documented that any pretreatment which affects cell wall character will in turn, affect the nature of uptake. The interior of the cell also contains many metal binding components, so the treatments which permeabilize the cells such as grinding (Gadd and White, 1989) detergents or chemicals (Strandberg *et al.*, 1981) increase the uptake of metals.

Detergents

Ross and Townsley (1986) observed that a detergent treated mycelium of *Penicillium spinulosum* performed better in the sorption of copper. Gadd *et al.* (1988) showed improvement in the biosorption efficiency of the biomass of *Sacchromyces cerevisiae* and *Rhizopus arrhizus* by detergent treatment.

Heat

Whistle and Daniel (1985) reported that the heat treatment could cause a loss of amino – functional groups on the fungal surface through the non- enzymatic browning reaction. Amino-functional groups in the polysaccharides contribute to the binding of heavy metals (Loaec, 1997). However, Galun *et al.* (1987) reported that *Penicillium* biomass pretreatment at 100 C° for 5 min increased the biosorption of lead, cadmium, nickel and zinc and increase was attributed to the exposure of latent binding sites after pretreatment. Siegel *et al.* (1986) showed that heat treatment of *Penicillium spinulosum* biomass resulted in an initial increase in the uptake rate, but had no overall effect on maximum biosorptive capacity. Galun *et al.* (1987) reported that preheating the biomass at 100° C for 5 minutes enhanced uptake capacity for Ni^{2+}, Zn^{2+}, Cd^{2+} and Cu^{2+}. Panchandikar and Das (1993) pre-incubated the biomass at 85°°C for 2 hours. Heat killed biomass showed higher capacity of Zn^{2+} uptake. However, Yan and Viraraghvan (2000) reported that the biomass capacity of autoclaved *Mucor rouxii* decreased as compared to live fungus.

Chemical treatment

Mahadevan and Tatum (1965) as early as in 1965 had stated that alkali treatment of the cell removes most of the cell wall materials containing COO^- and PO_4^- groups. If these groups are important for biosorption, then it is questioned that how alkali treatment improves the biosorption capacity. In this respect, Muzzarelli *et al.* (1980) interpreted that the alkali treatment of the mycelia improved their capacity to chelate various metal ions. According to them, removal of the amorphous polysaccharides from the cell wall by alkali treatment simply generates more accessible space within β-glucan-chitin skeleton, hence allowing more ions to precipitate. They reported that biosorbent ability of *Aspergillus niger* was solely due to the presence of chitosan. In a study by Galun *et al.* (1987) , NaOH treated *Penicillium digitatum* also showed enhancemenmt of cadmium, nickel and zinc biosorption. Removal of surface impurities, rupture of cell-membrane and exposure of available binding sites for metal bio-adsorption after pretreatment may be the reason for the increase in metal uptake. Brierly *et al.* (1985) and Muraleedharan and Venkobachar (1990) showed that alkali treatment of biomass may destroy autolytic enzymes that cause putrefaction of biomass and remove lipids and proteins that mask reactive sites. The cell wall of *Mucor rouxii* was ruptured by NaOH treatment. Luef *et al.* (1991) also reported that the NaOH treated biomass of *Aspergillus niger* with NaOH has high zinc biosorption efficiency. Pretreatment of *Aspergillus niger, Rhizopus oryzae* and *Mucor rouxii* with 4M NaOH at 121°C enhanced sorption of Cu^{2+}, Cd^{2+}, Ni^{2+} and Zn^{2+} (Baik *et al.*, 2002).

Acid pretreatment of *Mucor rouxii* significantly decreased the bioadsorption of heavy metals (Yan and Viraraghvan, 2000), which is in agreement with the observation of Kapoor and Viraraghvan (1998), in case of *A. niger*. This is attributed to the binding of H^+ ions to the biomass after acid treatment. The polymeric structure of biomass surface exhibits a negative charge due to the ionization of organic and inorganic groups (Huges and Poole, 1989). Bux and Kansan (1994) suggested that the higher the biomass electronegativity, the greater the attraction and adsorption of heavy metal cations. However, the H^+ ions on the acidic pretreated *M. rouxii* biomass change the biomass electronegativity, resulting in a reduction in bio-adsorption capacity.

On the other hand, there are certain findings where it is reported that the acid pretreatment can strongly enhance the adsorption by *Aspergillus oryzae* mycelia. However, in case of *A. oryzae* live biomass after acid treatment was directly used in bioadsorption of heavy metals instead of being autoclaved and dried (Huang and Huang, 1996). They also reported that pre-treatment of *A. oryzae* by $HClO_4$ increased the bioadsorption of lead, cadmium and nickel, but it was not the case for the species of *Rhizopus oryzae*.

Recently, Cabuk *et al.* (2005) subjected the biomass of *Aspergillus versicolor* and *Metarrhizium anisopliae* var *anisopliae* to physical treatments such as heat and autoclaving and chemical treatments. According to them, dimethyl sulphoxide, hydrogen peroxide and gluteraldehyde increased biosorption of Pb^{2+} in comparison with the *A. versicolor* live biomass. *Metarrhizium anisopliae* var *anisopliae* biomass pretreated with hydrogen peroxide, gluteraldehyde, commercial laundry detergent, dimethyl sulphoxide and formaldehyde significantly improved biosorption of Pb^{2+} in comparison with live biomass.

Immobilization

Studies with fungal biomass have shown effective metal removal for a range of metallic ions (Galun *et al.*, 1987). Fungal species employed included *Penicillium* and *Aspergillus* and metal removal capability was shown to persist in mycelium following heat and alcohol treatment (Galun *et al.*, 1987; Kiff and Little, 1986). Yakubu and Dudeney (1986) suggested that *Aspergillus* pellets may be directly used as metal biosorbents. Luef *et al.* (1991) used NaOH treated *Aspergillus niger* biomass in a column without immobilization and considerable biosorption of zinc was achieved. Native pellets of *Penicillium* sp. were also used as biosorbent for Cu^{2+}, Ag^{2+} and Pb^{2+} (Pumpel and Schinner, 1993) without immobilization. However, pellets were found to break up with continued use and fragments gave rise of flow restriction in continuous flow contact vessels. The freely suspended microbial biomass include small, heterogeneous particle of low mechanical strength. Sometimes, it may also make separation of biomass and effluent difficult. For these reasons, immobilized biomass particles appear to have more potential in packed or fluidized bed reactors.

According to Gadd (1992) immobilized system has many advantages—

(a) Particle size can be controlled.
(b) Biomass can be easily separated from the effluent and regenerated.
(c) High biomass loading and minimal clogging under continuous flow are achieved (Macaskie and Dean, 1989; Tsezos, 1990).

Various matrix used for immobilization of fungi in metal biosorption studies are listed in table 3.

Table 3: Fungi and their immobilized support in absorption of metals.

Matrix	Organism	Metal	Reference
Lyophilized biomass	*Aspergillus niger* *Penicillium chrysogenum* *Mucor miehei*	U	Guibal *et al.* (1995)
Reticulated foam	*Aspergillus oryzae*	Cd	Kiff and Little1(1986)
Calcium sodium alginate	*Rhizopus arrhizus* *Aureobasidium pullulans*	U Cd	Tsezos (1990) Park *et al.* (1999)
Polyvyl /alcohol drying	*Rhizopus arrhizus*	Fe (CN)	Aksu *et al.* (1999)
Polyurethane foam	*Rhizopus oligosporus* *Aspergillus niger* *Rhizopus arrhizus*	Cd Cu	Aloysius *et al.* (1999) Tsekova and Ilieva (2001) Lewis and Kiff (1998)
Polysulfone	*Mucor rouxii*	Pb, Cd, Ni, Zn	Yan and Viraraghvan (2000)
Polyacrylamide	*Saccharomyces cerevisie*	Cu^{2+}, Co^{2+}, Zo^{2+}, Zn^{2+}, Ni^{2+}	Wilhelmi and Duncan (1995)

Brierley (1990) treated fungal biomass with xylene and then crosslinked it using formaldehyde-resorcinol or formaldehyde urea solution or polyvinyl-lactate. Combining natural pellet formation with chemical crosslinking techniques may prove worthy for stability. Ashkenazy *et al.* (1997) reported benzaldehyde in acetunitrile to be a good fixative, which retained initial Pb biosorption potential by *Saccharomyces cerevisiae.* It also makes microbial biomass resistant to microbial spoilage up to 1 year at 4°C.

Different applications are available for biomass immobilization. The main techniques that are available for the application of biosorption are based on (*i*) Adsorption on inert support: Support materials are introduced prior to sterilization and inoculation. This technique has been used by Zhou and Kiff (1991) for the immobilization of *Rhizopus arrhizus*. Bai and Abraham (2003) immobilized *Rhizopus nigricans* on polyurethane foam cubes and coconut fibers. (*ii*) Entrapment in polymeric matrices: The polymers used are calcium alginate, polyacrylamide, polysulphone, polyethylenimine etc. (Babu, 1993; Gulay Bayramoglu, 2002; Nikajima and Sakaguchi, 1993; Wong and Kwok, 1992; Bai and Abhraham, 2003).

Immobilization study has involved the development of process technology for the biosorption of heavy metals from process effluents or contaminated water. The process development has been based on the use of fungal mycelium immobilized on biomass support particles, with no organic substrate applied during metal removal. This process has potential application both in metal recovery and in the removal of highly toxic metal contaminants at trace levels.

Desorption (Elution) and Recharging of the Biomass

In order to have economical bioremediative process, it is necessary to have regeneration process i.e. recovery of metal that has been adsorbed by the biomass and potential of reusing the biomass in multiple desorption-adsorption cycles. For this purpose, it is desirable to desorb the sorbed metals and to regenerate the biosorbent material for another cycle of application. It is desirable that the desorption process should yield the metal in a concentrated form. It should restore the biosorbent to close to the original condition for effective reuse with undiminished metal uptake and no physical changes or damage to the biosorbent.

The process usually involves recharging the systems by washing in acid-base or exposure to chelating agents followed by chemical recovery of the metal (Gadd and White, 1989). The type of solution would depend on the extent of binding of the deposited metal. Sometimes, the chemical which is used for desorption can increase the uptake capacity for the next cycle (Gadd and Griffiths, 1980).

Dilute solutions of mineral acids like hydrochloric acid, sulphuric acid, acetic acid, nitric acid can be used for metal desorption from the biomass (Luef *et al.*, 1991; Pagnanelli *et al.*, 2002; Bai and Abhraham, 2003). Khalid *et al.* (1993) used 1M NH_4HCO_3 as desorbing solution in case of uranium loaded biomass of *Trichoderma harzianum*. 100% U was recovered and column efficiency remained almost same upto five cycles. Uranium-loaded biomass of *Aspergillus niger* (Yakubu and Dudeney, 1986) in compartmentalized, counter current, fludized bed column treated with 0.1M HNO_3. 0.1M HNO_3 was also used to effectively elute biosorbed Pb from *Aspergillus niger* (Jianlong *et al.*, 2001). Yan and Viraraghavan (2000) also reported comparable readsorption capacities of polysulphone immobilized *Mucor rouxii* biomass following desorption with 0.05N HNO_3. Huang and Huang (1996) reported enhanced Cu sorption by acid washed *Aspergillus oryzae*. They have thus recommended acid wash for pretreatment and regeneration of biomass. However, for chromium where optimum pH for biosorption was below 2, was not efficiently removed with 1M HC1. This differential desorption by mild acid treatment could lead to selective desorption between chromium and other metals to same or higher extent (Vianna *et al.*, 2000). Brady and Duncan (1994) also used EDTA for desorption of Cu^{2+} from *Saccharomyces cerevisiae*.

The desorption procedure affects the economy of the total process which in turn depends on where the metal is accumulated. Surface bound metal might be removed with mild chemical washing, whereas intracellular metal can be released only by destructive treatments like incineration or concentrated acid thereby preventing any reuse of that microbial biomass for further cycles of metal biosorption. This kind of destructive recovery is cost effective in those conditions where recovery of precious, strategic or rare metal is the main objective or where biomass can be produced or obtained at a very low cost (Brierly *et al.*, 1993).

If the biosorption process were to be used as an alternative to the wastewater treatment scheme, the regeneration of the biosorbent may be crucially important for keeping the process cost down and in opening the possibility of recovering the metals extracted from the liquid phase.

CONCLUSIONS

Biosorption is a poised alternative to conventional systems for removal of toxic metals from aquatic systems. In this direction, fungal biomass has proved to serve as a basis for metal biosorption and as a very competitive means for the detoxification of metal bearing industrial effluents. In nutshell, biosorption is an attractive, environmentally sound technology that makes use of cost effective sorbents and offers the advantage of producing negligible secondary pollutants as compared to the conventional physico-chemical treatments. The biosorptive metal loaded by the biomass can easily be desorbed by the non-destructive means and the biomass thus, can be reused for subsequent cycles of metal removal and recovery. Elucidation of mechanisms of metal biosorption is essential for successful exploitation of the phenomenon and the regeneration of biosorbent materials in multiple reuse cycles. Biosorption isotherm curves can be used in the evaluation of metal uptake efficiency of different biosorbents. However, the development of biosorption requires further investigation in the direction of modelling, of regeneration of biosorbents and of testing immobilized raw biomass with industrial effluents. New methodologies that aim at mathematical modelling of biosorbents systems and their effective optimization, should be developed.

ACKNOWLEDGEMENTS

Authors wish to acknowledge with thanks the help provided by Anoop, Gautam, Pritesh, Rekha and Kakoli for constructive suggestions and critically evaluating the manuscript.

REFERENCES

Adamson, A.W. (1976) In: *Physical chemistry of surfaces*, John Wiley & Sons, New York. pp 1-10.

Akhtar, N., Sastry, S. & Mohan, M. (1996) *Biometals.* **9**: 21-28.

Aksu, Z. & Kutsal, T. (1990) *Environ. Technol.* **11**: 979-987.

Aksu, Z., Calik, A., Dursun, A.Y. & Dermican, Z. (1999) *Process Biotech.* **34**: 483-491.

Aksu, Z., Sag, Y. & Kutsal, T. (1992) *Environ. Technol.* **13**: 579-586.

Aloysius, R., Karim, M.I.A. & Ariff, A.B. (1999) *W. J. Microbiol. Biotechnol.* **15**: 571-578.

Arica, M.Y., Kacar, Y. & Gene, O. (2001) *Biores. Technol.* **80**: 121-129.

Ariff, A. B., Mel, M., Hasan, M.A. & Karim, M.I.A. (1999) *W. J. Microbiol. Biotechnol.* **15**: 291-298.

Ashkenazy, R., Gottleib, L & Yannai, S. (1997) *Biotechnol. Bioeng.* **55**: 1-10.

Babu, G. R. Y. (1993) In: *Biohydrometallurgical Technologies.* TMS Publications Wyoming, USA.

Bai, RS. & Abhraham, T.E. (2001) *Biores. Technol.***79**: 73-81.

Bai, RS. & Abhraham, T.E. (2003) *Biores. Technol.* **87**: 17-26.

Baik, W.Y., Bae, J.H., Cho, K.M. & Hartmeier, W. (2002) *Biores. Technol.* **81**: 167-170.

Bakkaloglu, I., Butter, T.J., Evinson, L.M., Holland, F. S. & Hancock, I.C. (1998) *Water Sci. Technol.* **38**: 269-277.

Ben Omar, N., Larbi, M. M., Arias P.J.M. & Gonzalez, M.M.T. (1997) *Chemosphere.* **35**: 2277-2283.

Bengtsson,L., Johanson , B., Hackett, T., Mc. Halle, L. & Mc Halle, A. (1995) *Appl. Microbiol. Biotechno.* **42**: 807 –811.

Beyeramann, D. (1991) In: Metals and their compounds in the environment: Occurrence, analysis and biological relevance. Weinheim, VCH, New York, PP 491- 509.

Bhainsa, K.C. & D'Souza, F.S. (1999) *Biotech. Tech.* **13**: 695-699.

Bosecker, K. (1993) In: *Biohydrometallurgical Technologies,* TMS Publications Wyoming, USA. pp 55-64.

Brady, D. & Duncan, J.R. (1994) *Appl Microbiol Biotechnol* **41**: 149-154.
Brady, D. & Duncan, J.R. (1993) In: *Biohydrometallurgical Technologies* . Vol 2, TMS publication. Wyoming, USA
Brierley, C. L. (1990) Geomicrobiol J. 8: 201- 223.
Brierly, C. L., Kelly, D.P., Seal, K.J. & Best, D.J. (1985) In: *Biotechnology.* Oxford Blackwell, pp 163- 212.
Bux, F. & Kansan, H.C. (1994) *Wat SA.* **20**: 73 –76.
Cabuk, A., Semra, I, Cansu, F. & Calistskan, F. (2005) *Turk J Biol* **29** : 23 – 28.
Chaong, K. H & Volesky, B. (1995) *Biootechnol. Bioeng.* **47**: 451- 460.
Chatterjee, S., Asthana, R.K., Tripathi, A.K. & Singh, S.P. (1996) *Process Biochem.* **31**: 457-462.
Darnell, D.W., Greene, B., Henzl, M.T., Hosea, M., McPherson, R., Sneddon, J & Alexander, M.D. (1986) *Environ. Sci. technol.***20**: 206-208.
Davis, T.A., Volesky , V & Vieira, R.H.S.F. (2000) *Wat Res* **34(17)**: 2470 -4278
DeRome, L & Gadd, G.M. (1987) *Appl. Microbio. In Biotechnol.* **26**: 84- 90.
Dey, S., Rao, P.R.N., Bhattacharyya, B.C. & Bandyopadhyay, M. (1997) *Bioprocess Eng.* **12**: 273-277.
Ercole, C., Veglio, F., Toro, L., Ficara, G. (1994) In: *Mineral Bioprocessing* –II, Snowboard , Utah.
Fourest, E. & Roux, J.C. (1992) *Appl. Microbiol. Biotechnol.* **37**: 399-403.
Freundlich, H. (1926) Colloid and capillary chemistry. Methuen, London.
Friis, N. & Myers –Keith, P. (1986) *Biotechnol. Bioeng.* **28**: 21- 28.
Fuhrman, G.F. & Rothstein, A. (1968) *Biochem. Biophysc. Acta.* **163**: 325- 330.
Gadd, G.M. & Mowll, J.L. (1983) *FEMS Microbiol. Lett.* **16**: 45- 48.
Gadd, G.M (1978) *Microbiol. Ecol.***4**: 303-317.
Gadd, G.M. (1986) In: *Microbes in extreme environment.* Academic Press, London pp 83- 110.
Gadd, G.M. (1988) In: *Biotechnology a comprehensive treatise.* Vol 6b, VCH, Verlagsgesellschaft, Weinhien. pp 401- 433.
Gadd, G.M. & White, C. (1989) *Biotechnol. Bioengg.* **33**: 592- 597.
Gadd, G.M (1990) In: *Microbial mineral recovery.* Mc- Graw- Hill, New York, pp 249 – 275.
Gadd, G.M (1992) *FEMS Microbiol. Lett.* **100**: 197- 204.
Gadd, G.M & Griffiths, A. J. (1980) *Trans. BrMycol. Soc.* **75(1):** 91- 96.
Gadd, G.M. & White, C. (1993) *Trends Biotechnol.* **11**: 353- 359.
Galli, E , Mario, F.D., Rapana, P., Lorenzoni, P. & Angelina, R. (2003) *Lett. Appl. Microbiol.* **37**: 133-137.
Galun, M., Keller, P., Malki, D., Feldstein, H., Galun, E., Siegel, S. M. & Siegel, B. Z. (1984) *Water Air Soil Pollut.* **21**: 411-414.
Galun, M., Siegel, S. M., Cannon, M.I., Siegel, B. Z. & Galun, E. (1987) *Environ. Pollut.* 43: 209-218.
Garnham, G.W. (1997) In: *Biosorbents for metal ions*, Taylor & Francin Ltd. London pp. 11-38.
Gee, A.R. & Dudeney, A.W.L. (1988) In: *Biohydrometallurgy,* Sci. Technol. Letts. London, pp 437-451.
Gomes, N.C.M. & Linardi, V.R. (1996) *Rev. Microbiol.* **27**: 218-222.
Grafl, H. J. & Schwantes, H.O. (1983) *Zentrabl. Backtenol. Microbiol.* **177**: 57-74.
Guibal, E., Roulph, C & Cloirec, P.L. (1995) *Environ. Sci. Technol.* **29**: 2496- 2503.
Gulati, R., Saxena, R.K. & Gupta, R. (2002) *W. J. Microbiol. Biotechnol.* **17**: 1-5.
Gulay Bayramoglu (2002) *Microchem. J.* **72**: 63- 76.
Holan, Z.R. & Volesky, B. (1995) *Appl. Biochem. Biotechnol.* **53**: 133-146.
Horikoshi, T., Makajima , A. & Sakaguchi, T. (1981) *Eur. J. Appl. Microbiol. Biotechnol.* **12**: 90- 96.
Huang, C. & Huang, C.P. (1996) *Wat Res.* **30(9):** 1985-1990.
Huang, C.C., Huang, C.P. & Morechart, A.L. (1990) *Water Res.* **24**: 433- 439.
Huges, M.N. & Poole, R.K. (1989) In: *Metals and microorganisms*, Chapman and Hall, New York.
Itoh, M., Yuasa, M. & Kobayashi, T. (1975) *Plant Cell Physiol.* **16**: 1167-1169.
Jianlong, W., Xinmin, Z., Decai, D. & Ding, Z. (2001) *J. Biotecnol.* **87**: 273- 277.
Jianlong, W. (2002) *Process Biochem.* **37**: 847-850.
Kambe – Honjoh, H., Sugarawa, A., Yoda, K., Kitamoto, K. & Yamasaki, M. (1997) *Appl. Microbiol. Biotechnol.* **48**: 373-378.
Kapoor,A. &Viraraghvan, T. (1998) *Bioresour. Technol.* **63**: 109 – 113.

Khalid, A. M., Ashfaq, S. R., Bhatti, T. M., Anwar, M. A., Shamsi , A. M. & Akhtar , K . (1993) In: *Biohydrometallurgical Technologies.* Vol III" TMS Publication Wyoming USA. pp 229-307.
Kiff, R.J. & Little, D.R. (1986) In: *Immobilization of ions by biosorption.* Ellis Horwood. pp 71- 80
Korenevskii, A.A., Khamidova, K.H., Avakyan, Z.A. & Karavaiko, G.I. (1999) *Microbiology.* **68(2):** 139-145.
Kovacevic, Z. F., Sipos. L. & Briski, F. (2000) *Food Technol. Biotechnol.* **38(3)**: 211- 216.
Kurek, E., Czaban. J. & Bollag, J. M. (1982) *Appl. Environ. Microbiol.* **43(5)**: 1011-1015.
Kuyucak, N. & Volesky, B. (1988) *Biotech. Letts.* **10(2)**: 137-142.
Larsson, S., Gunnerusson, E. & lindstoem, E.B. (1992) *Biocecovery.* **2**: 107-119.
Lewis, D. & Kiff, R.J. (1998) *Environ. Tech. Letts.* **9**: 991- 998.
Loaec, M. (1997) *Water Res.* **31(5)**: 1171- 1179.
Luef, E., Prey, T. & Kubicek, C.P. (1991) *Appl. Microbiol. Biotechnol.* **34**: 688- 692.
Macaskie, L. E. & Dean, A. C. R. (1989) In: *Biosorption of heavy metals.* CRC Press Boca Raton. pp 199-248
Magykrosy, A., Laidlaw, R.D., Kilaas,C. & Echer, C. (2002) *J. Appl. Microbiol. Biotechnol.* **59**: 382-388.
Mahadevan, P.R. & Tatum, E.L. (1965) *J. Bacteriol.* **90**: 1073- 1081.
Mattuschka, B., Junghans, K. & Straube, G. (1993) In: *Biohydrometallurgical Technologies* (vol. III) TSM Publication, Wyoming, USA, pp 125- 132.
Mc Gahren, W.I. (1984) *Proc. Biochem.* **19**: 88- 90.
Mc Hale, A.P. & Mc hale, S. (1994) *Biotech. Adv.* **12**: 647-652.
Mullen, M. D., Wolf, D.C., Beveridge, T.J. & Baile, G.W. (1992) *Soil Biol. Biochem.* **24**: 129-135.
Mullen, M. D., Wolf, D.C., Beveridge, T.J. & Bailey, G. W. (1992) *Soil Biol. Biochem* **24**: 129- 135.
Muraleedharan, T.R. & Venkobacher, C. (1990) *Biotechnol. Bioeng.* **35**: 320-325.
Muraleedharan, T.R., Iyengar, L. & Venkobacher, C. (1991) *Curr. Sci.* **61(6)**: 379-385.
Muzzarelli, R. A. A., Tanfani, A.F. & Scarpini, G. (1980) *Biotechnol. Bioeng.* **22**: 885- 896.
Nakajima, A & Sakaguchi, T. (1986) *Appl, Microbiol.Biotechnol,* **24**: 59- 64.
Nikajima, A. & Sakaguchi, T. (1993) *Appl, Microbiol, Biotechnol,* **38**: 574-578.
Niu, H., Xu , X. S., Wang, J. H. & Volesky, B. (1993) *Biotechnol, Bioeng,* **42**: 785.
Omar, N.B., Merrun, M.L., Gonzalez- Munoz, M.T. & Arias, J.M. (1996) *J. Appl. Bacteriol.* **81**: 283- 287.
Pagnanelli, F., Toro, L. & Veglio, F. (2002) *Waste Manag* **22**: 901-907.
Paknikar, K.M., Palnitkar, U.S. & Puranik, P.R. (1993) In: *Biohydrometallurgical Technologies* Vol III" TMS Publication Wyoming USA. pp 229-236.
Panchandikar, V.V. & Das, R.P. (1993) *W. J. Microbiol. Biotechnol.* **9**: 280-281.
Park, J.K., Jin, Y.B. & Change, H.N. (1999) *Biotechnol. Bioeng.* **63**: 116-121.
Preetha, B & Viruthagiri, T. (2005) *African. J. Biotechnol.* **4(6)**: 506- 508.
Pumpel, T. & Schinner, F. (1993) *FEMS Microbiol. Rev.* **11**: 159-164.
Rama Rao, V.S.K.V., Wilson, C.H. & Mohan, P.M. (1997) *Biometals* **10**: 147-156.
Redad , Z., Grernte, C., Andres, Y. & LeCloirec, P . (2002) *Environ. Sci. Technol.* **36**: 2067- 2073.
Rosenberger, (1975) The filamentous fungi. Vol 2 Edward Arnold, London, pp 328- 343.
Ross, I.S. & Townsley, C.C. (1986) In: *Immobilization of ions by biosorption* IRL Press Chichester. pp 49-58.
Sag, Y., Acikel, U., Akzu, Z. & Kutsal, T. (1998) *Turkish J. Eng. Environ. Sci.* **22**: 145-154.
Sanna , G., Maddau, L., Franceschini, A. & Melis, P. (1997) *Micol It* **26**: 63-72.
Say, R., Denizli, A, A. & Arica, M.Y. (2001) *Biores. Technol.* **76**: 67-70.
Scatchard, G. (1949) *Ann. NY. Acad. Sci.* **51**: 660 –672.
Schiewer, S. & Volesky, B. (2000) In: *Remediation*, Kluwer Academic Publishers, Dordrecht, The Netherlands 139- 169.
Sears, M.E. (1986) Propagation and characterization of *Rhizopus* biosorbents. Ph.D Thesis. Mc Gill University. Montreal.
Senthil, K.S., Bharathi , S., Nityanandhi, D. & Subburam, V. (2000) *Bioresour. Technol.* **75**: 163-165.
Shumate S. E. I., Strandberg G. W. & Parrott J. (1978) *Biotechnol. Bioeng. Symp.* **8**: 13 –20.
Shumate, S.E., Strandberg, G.W. & Parrott, J. R. (1987) *Biotechnol. Bioeng.* **24**: 385-401
Siegel, S.M., Galum, M., Keller, P., Siegel, B.Z. & Galun, E. (1986) Fungal Biosorption: Acomparative study of metal uptake by *Penicillium* and *Cladosporium.* Symp. Illinois Institute of Technology and Italian national Research Council Chicago Illinosis.ppIII-77 to III-94.

Singh, C. & Yu, J. (1998) *Water Res.* **32:** 2746-2752.
Singleton, I. & Simmons, P. (1996) *J. Chem. Tech. Biotechnol.* **65**: 21-28.
Somers, E. (1963) *Ann. Appl. Biol.* **51**: 425-437.
Strandberg, G.W., Shumate, S.E. & Parrott, J.R. (1981) *Appl. Environ. Microbiol.* **41**: 237-245.
Stratton, G.W. (1987) In: *Environmental Toxicology,* Elsevier, pp 85-94.
Stuetz, R.M., Madawick, J.C. & Gee, A.R. (1993) In: *Biohydrometallurgical Technologies*, TMS Publications Wyoming, USA. pp 85-94.
Subramanium, C., Venkateswerlu, G. & Rao, S. L. N. (1983) *Appl. Environ. Microbiol.* **46**: 585- 5
Suhasini, I. P., Sriram, G., Asolekar, S. R. & Suresh, G. K. (1999) *Proc. Biochem.* **34**: 239- 247.
Tobin, J. M. & Roux, J. C. (1998) *Water Res.* **32**: 1407-1416.
Tobin, J. M., Copper, D.G. & Neufeld, R.J. (1984) *Appl. Environ. Microbiol.* **47**: 821-824.
Tobin, J. M., Copper, D. G. & Neufed, RJ. (1987) *Biotechnol. Bioeng.* **31**: 282- 286.
Townsley, C. C., Ross, I. S. & Atkins, A. S. (1986) In: *Immobilization of ions by Biosorption,* Ellis Horwood. pp. 159-170.
Tsezos, M. (1983) *Biotechnol. Bioeng.* **25**: 2025-2040.
Tsekova, K. & Ilieva, S. (2001) *Appl. Microbiol. Biotechnol.* **42**: 636-637.
Tsekova, K. & Petrov, G. (2002) *Verlag der Zeitschrift for Naturforschung Tubinegen* **57(7-8):** 629-633.
Tsezos, M. & Voleksy, B. (1981) *Biotechnol. Bioeng.* **23**: 583- 604.
Tsezos, M., Geoorgousis, Z. & Remoudaki, E. (1997) *Biotechnol. Bioeng.* **55**: 16-27.
Tsezos, M. (1990) In: *Microbial Mineral Recovery* Mc Graw-Hill, New York. pp 325-339.
Tsezos, M. & Volesky, B. (1982) *Biotechnol. Bioeng.* **24**: 955-969.
Venkateswerlu, G. & Stotzky, G. (1989) *Appl. Microbiol. Biotechnol* **31**: 619-625.
Vianna, L.N.L., Andrde, M. C. & Nicoli, R. (2000) *W. J. Microbiol. Biotechnol.* **16**: 437- 440.
Volesky, B. (1986) *Biotechnol. Bioeng.* **16**: 121-125.
Volesky, B. (1990a) In: *Biosorption of heavy metals,* CRC Press, Boca Raton. pp 140-172.
Volesky, B. (1990b) In: *Biosorption of heavy metals,* CRC Press, Boca Raton Florida. USA pp 396.
Weber, W.J. (1972) In: *Physicochemical processes for water quality control,* John Wiley and Sons, New York, pp 199- 260.
Whistle, R. & Daniel, T.R. (1985) *Food Chem.* 96- 105.
White, C. & Gadd, G. M. (1990) *J. Chem. Technol. Biotecnol.* **49**: 331-343.
Wilhelmi, B.S. & Duncan, J.K. (1995) *Biotechnol. Lett.* **17**: 1007-1012.
Wong, P.K. & Kwok, S.C. (1992) *Biotechnol. Letts.* **14(7)**: 629- 634.
Yakubu, N.A. & Dudeney, A.W.L. (1986) In: *Immobilization of ions by Biosorption,* Ellis Horwood. pp. 183-200.
Yan, G. & Viraraghvan, T. (2000) *Water SA.* **26**: 119- 123.
Yetis, U., Ozcengiz, G., Delik, F.B., Ergem, N. & Dolek, A. (1998) *Water Sci. Technol.* **38**: 323-333.
Zajic, J.E. & Chiu, Y.S. (1972) *Dev. Ind. Microbiol.* **13**: 19.
Zhou, J.L. (1999) *Appl. Microbiol. Biotechnol.* **51**: 686 – 693.
Zhou, J.L., Kiff, R.J. (1991) *J. Chem. Technol. Biotechnol.* **52**: 317.

20

Biological and Antimicrobial Properties of Coffee

Daljit S Arora and Gurinder Jeet Kaur
Microbial Technology Lab., Department of Microbiology
Guru Nanak Dev University, Amritsar-143005 (India)

Conventional therapeutic methods are now being recognised all over the world as an alternative line of treatment and finding healing power in plants dates back to prehistory. A number of plants and their products are used for their biological and antimicrobial properties. Tea and coffee are the two commonly consumed beverages worldwide. Coffee affects the human physiology in a variety of ways such as nervous system stimulant, a diuretic, improves motor activity and cognitive performance, and reduces the risk of Parkinson's disease by 50-80%. It imparts a protective effect on liver cirrhosis and lowers the risk of gastrointestinal disorders though at times it upsets the digestive system. Coffee does not contribute to cardiovascular disease (CVD) although researchers differ in their opinion. Coffee has a potential to be used as an antimicrobial agent to combat the ever-increasing menace of bacterial resistance posed due to extensive use of antibiotics. In an effort to describe this upcoming antimicrobial agent the present paper deals with its various aspects—history, chemistry, processing and biological as well as antimicrobial effects of coffee.

INTRODUCTION

The conventional therapeutic methods of the Far East (particularly China, India and Korea) are now finding increasing acceptance all over the world as an alternative line of treatment. A scientific discipline called Ethnobotany/Ethnopharmacology, which utilizes the impressive array of knowledge gathered by native people about the plants and animal products used to maintain health, further motivates the interest in alternative medicine (Heinrich, 2000). The traditional Indian medicine has a 5000-year old history and has been used by millions whereas modern medicine is merely 150 years old. The single most important therapeutic discovery in the history of medicine i.e. the antibiotics, have clearly revolutionized our ability to cure diseases caused by infectious microorganisms and prevent death. However, extensive use of antibiotics has led to ever increasing crisis of bacterial resistance, which is causing a worldwide concern (Kunnin, 1993; Austin *et al.*, 1999) and thus emphasizes the need to look for alternative medicine.

Finding healing power in plants is an ancient idea. People all over the globe have long been using infusions of hundreds, if not thousands, of indigenous plants dating back to prehistory. The renewed interest has been generated to test the sensitivity of various pathogenic microorganisms (bacteria, virus, and fungi) towards such plants or alternative resources (Klink, 1997). Plants produce many secondary metabolites that constitute an important source of pesticides, microbicides and pharmaceutical drugs (Muanza *et al.*, 1994; Baruah *et al.*, 1996; Cowan, 1999). A large number of plants known for their medicinal value have been screened for their antibacterial activity (Rozas *et al.*, 1992; Matthews and Haas, 1993; Izzo *et al.*, 1995). Plant extracts that inhibit pathogenic microorganisms without harming the host may have potential as therapeutic agents. The volatile components of several plant extracts are reported to inhibit spore germination and growth of many fungi (Dubey, 1991; Heisey and Gorham, 1992; Arora and Ohlan, 1997).

The most commonly consumed non-alcoholic beverage all over the world is tea, closely followed by coffee. Tea enjoys a unique place in our life and unlike alcoholic beverages, it has acquired a privileged place and popularity amongst all the sections of people without any bane on its back. It is no exaggeration to say that it is an all-purpose, all time drink for all people. Coffee drinking is as popular in south as tea is in northern India.

The history of the origin of coffee remains a mystery as the evergreen bush grows wild in Africa but its cultivation is generally accredited to Arabia. The name coffee might have been derived from the Arabic 'qahwah' or alternatively may have arisen due to the connection with the province Kaffa, in Ethiopia. The history of coffee dates back to 600-700 A.D., to the Ethiopian legend of Kaldi who observed his goats feeling energetic after having eaten the berries of a particular tree, thus leading to the discovery of coffee. Soon, this new discovery reached the Arabian Peninsula and the odyssey continued from there.

Before going into any studies on biological or antimicrobial effects of coffee, it is better to have some basic information about the plant and its processing. It has been claimed that coffee is second to oil in terms of world commodity trading and it provides employment to approximately 20 million people.

COFFEE PLANT AND ITS PROCESSING

The coffee plant is a small tree belonging to the genus *Coffea* of Rubiaceae family. It is a perennial plant and grows in tropical and sub-tropical regions. There are about 25 wild species but only 3 or 4 are cultivated. The most commonly grown commercial crops of coffee are: *Coffea arabica*, believed to have originated in Ethiopia; *Coffea robusta*, a native of Congo and *Coffea liberica* from West Africa. Coffee needs a warm and humid climate and it usually takes about three years for the tree to begin to produce berries and another two years for it to come to full fruitation. It is successfully grown in many parts of southern India, Madras, Coorg, Mysore, Travancore and Cochin (Chopra *et al.*, 1958; Singh *et al.*, 1983). The white flowered blossom is beautiful but lasts only a few days. It is replaced by coffee cherries which ripen rapidly from green to yellow and finally to a bright deep red. In some countries cherries are allowed to dry and shrivel on the trees and then shaken off onto sheets of canvas. But the best grades of coffee are those in which cherries are handpicked when they have reached just the right stage of mellow fruitfulness. The cherries after being picked up from the plants undergo a series of processing steps before coffee is commercially made available in the market.

There are two ways of extracting beans from the cherries: Dry/natural process and wet/washed process. In dry/natural process, the cherries after preliminary winnowing and sifting are washed, drained and spread in thin layers on the ground and exposed to sun for three to four weeks. The cherries are turned several times a day to ensure even drying. After drying these are repeatedly passed through the milling machines to remove the dried up husk, parchment and silver skin. In wet process, the cherries are taken to the floatation tank where the ripe cherries float and the debris including the green cherries sink. The pulp is removed from them by pulping machines. Residual part of the pulp sticking to the seeds is loosened during controlled fermentation of cherries over a period of 12-24 hours. Seeds are thoroughly washed with excessive water and then dried in the sun. Ultimately the coffee beans are roasted at 240°C or higher for 3-15 minutes. During roasting seeds swell, their colour changes from green to brown as the sugar gets caramelized and results in development of certain typical aroma (Holscher *et al.*, 1992) and the water-soluble constituents increase to 35% by this process. Grinding of the roasted beans is the last step in the preparation of coffee powder. The degree of grinding affects the final brew. Finer the grind, greater the extraction but there is some loss of aroma during grinding. Grinding and brewing techniques influence caffeine content in coffee.

GRADING AND AROMA OF COFFEE

Marketed coffee powders comprise of different blends of coffee. The grain size of ground coffee is an important factor for its flavour and aroma. In general, marketed coffee powder is ground to three sizes—fine, medium and coarse. Coarse ground coffee retains its flavour and aroma for a longer period suitable for preparing coffee by percolation. Good liquor is obtained by using a proportion of 90% fine and 10% coarse coffee (Meena and Sethi, 1994). The aroma of coffee is responsible for all flavour attributes other than the mouth feel—sweet, salt, bitter and sour taste that are perceived by the tongue. However, instant coffee lacks most of the aromatic volatile compounds causing a dramatic decrease in the overall flavour. Coffee aroma is very complex, involving more than 800 volatile compounds with a wide range of functional groups. In recent years, various studies have focussed on the most potent odorous constituents responsible for coffee aroma (Blank *et al.*, 1992). Twenty-nine volatile compounds have been found mainly responsible for the aroma of roasted and ground coffee. Out of these 13 are important contributors to the aroma of roasted coffee as revealed by Aroma Extraction Dilution Analysis (AEDA) (Serrazin *et al.*, 2000). Compositional changes occur in the brewed coffee as a function of brewing time and an optimum brew time is of 7-9 minutes (Lee *et al.*, 1992). Aroma of roasted coffee is changed during aging, which is further influenced by storage temperature, packaging, light, oxygen, dampness and several other factors. Changes in pH can lead to ionization of functional groups (eg. phenolic hydroxy groups) and this can alter the flavour of the product.

CHEMICAL CONSTITUENTS OF COFFEE

Coffee seeds contain 1-2% caffeine, about 0.25% trigonelline, 3-5% tannin (polyphenols), about 15% glucose and dextrin, 10-13% of fatty oils consisting chiefly of olein and palmitin and

10-13% of proteins. The main ingredients of brewed or instant coffee are carbohydrate fraction present mostly in polymer form, followed by alkaloids, polyphenols (tannins), proteins, organic acids, lipids and a small percentage of glycosides and vitamins (Strobel, 1984).

A range of carbohydrates, including polysaccharides and the low molecular weight sugars (mono-, di- and trisaccharides) are found in green coffee and sucrose is the major free sugar present. Carbohydrates fraction of green coffee beans predominantly comprises of polysaccharides as compared to monosaccharides eg. mannan, galactan, galactomannan, arabinogalactan.

The major nitrogenous components may be described in terms of three main groups of compounds: alkaloids, trigonelline together with nicotinic acid and amino acids and proteins. Caffeine (1, 3, 7-trimethylxanthine) is perhaps the best-known and controversial alkaloid found in coffee and it is present at about 1-2% on a dry weight basis in *Coffea arabica*. The highest concentration of caffeine is found in *Coffea canephora*. Caffeine is thought to be the main constituent of coffee responsible for its various adverse effects on health (Mazzafera *et al.*, 1991). The amount of caffeine per cup of regular bean coffee is twice as that of regular instant coffee. Caffeine content in a cup of coffee is as follows—percolated coffee 110 mg, instant (regular) 53 mg and decaffeinated 2 mg. Caffeine content varies in different parts and varieties of coffee. Coffee beans have maximum caffeine content compared to other plant parts and beans of robusta selections have almost double the quantity of caffeine compared to beans of arabica selection.

Trigonelline has received considerable attention as a nitrogen-containing component of coffee due to its reported anti-tumour activity. It is present at about 1% on a dry weight basis but is thermally unstable and hence can lead to other nitrogenous materials upon roasting such as pyridines and pyrroles.

Polyphenols, primarily tannins, are ligands that may form soluble or insoluble complexes with many polymers such as proteins and polysaccharides. Tannic activities of different compounds come from the caffeic acid moiety in their molecules, not from structural features of usual tannins (hydrolysable and condensed tannins) (Scalbert, 1991).

The main mineral components of coffee include potassium, magnesium, sodium, phosphorous as well as the trace components containing calcium, iron, manganese, copper and zinc. Mineral contents in coffee beans and coffee depend upon the sort of coffee and the kind of percolation used to prepare it (Strobel, 1984). However, vitamin content in coffee infusion is quite poor, with three and a half cups of coffee a man would obtain about a fourth of the minimum daily requirement of vitamins (Wellman, 1961).

TYPES OF COMMERCIAL COFFEE

All over India mainly two types of coffee are used commercially: Fine ground coffee and Instant coffee. Preparation of ground coffee has already been described under the title 'coffee plant and its processing'. However, soluble coffee, popularly known and marketed as instant coffee is a dry coffee extract and ninety percent of the coffee consumed these days in the world is instant coffee. It is prepared by extracting the soluble portion from ground coffee by percolation with hot water, evaporating the extract to a syrup and drying it in a vacuum dryer more expensive than spray drying, but it has the benefit of preserving the aroma in a better way in comparison to spray drying process. The dried product is then aromatized, mainly to restore some high and pleasing aroma notes that have been lost during various processing steps. Instant coffee has a poor aroma but is easy to prepare with consistent taste. Artificial flavours are added to improve its aroma (Strobel, 1984).

Decaffeinated Coffee

Recognition of various adverse physiological effects of caffeine in coffee prompted the demand for decaffeinated coffee, which is produced by solvent extraction or more recently, by supercritical fluid extraction with CO_2 before roasting and grinding (Katz, 1984). By this process, though 97% caffeine can be removed, there is the danger that some harmful residues of the solvent may linger in the brew. An alternative source of decaffeinated coffee would be the use of *Coffea* species that contain much lower levels of caffeine than *Coffea arabica* or *C. canephora*. A number of such species are: *C. salvatrix*, *C. eugenioides* and *C. bengalensis* etc. but they either produce few fruits; or beans yield a poor quality beverage and are, therefore, intrinsically unsuitable for commercial purposes. Under these circumstances, genetic engineering to produce transgenic caffeine-deficient *C. arabica* plants may be a more practical long-term proposition than a breeding program (Ashihara and Krozier, 1999). In some parts of the world particularly in the West, decaffeinated coffee is also consumed.

BIOLOGICAL EFFECTS OF COFFEE

Tea and coffee have been consumed socially and habitually since ancient times for their therapeutic value and most of these effects are attributed to their caffeine content. Although there is no great difference in the amount of caffeine present in two beverages, yet, tea is considered soothing while coffee is agitating; something that keeps people awake at night. It indicates that constituents of tea and coffee other than caffeine also influence the human body. Pharmacological action of caffeine and allied alkaloids affects nearly every system of the body.

Effect on Nervous System

Caffeine, a central nervous system stimulant, primarily affects higher centres. In CNS the site of action of caffeine is on adenosine receptors to modify chemical synaptic transmission. Administration of 125-250 mg of caffeine produces a sense of well being, alertness, beats boredom and allays fatigue, thinking becomes clearer even when dullness has tended to prevail after a sustained intellectual effort, improves performance and motor activity increases (Tripathi, 1991). Chronically depressed patients often use caffeine drinks as self-medication. Ingestion of caffeine has no effect on initial mood or working memory but it improves encoding of new information and counteracts the fatigue (Smith *et al.*, 1999). It has been useful in migraine and tension wherein cerebral vessels are constricted thereby counteracting the painful effect, thus giving the desired relief.

Parkinson's disease is an incurable and debilitating disease and one of the characteristics of Parkinson's patients is the reduced amount of dopamine in certain areas of brain. Regular coffee consumption over time may reduce the risk of Parkinson's disease by 50-80%. The coffee element that appears most likely to mitigate this effect is caffeine, which shows neuroprotective effect in slowing down the progression of Parkinson's disease by its ability to block the adenosine A2A receptor. It is supported strongly by various studies that more one consumes coffee, the less likely the risk of developing Parkinson's disease (Jimenz-Jimenz *et al.*, 1992; Hellenbrand *et al.*, 1996; Fall *et al.*, 1999; Benedetti *et al.*, 2000; Chen *et al.*, 2001). However, large quantities especially of coffee may lead to sleeplessness, restlessness and sometimes may precipitate convulsions (epilepsy) and an attack of schizophrenia in otherwise well controlled patients.

Effect on Gastro-intestinal System

Coffee has been shown to have profound effect on the gastrointestinal system. Regular and decaffeinated coffee increases stomach acid. Avoiding these substances should therefore aid in the healing of gastritis. Coffee consumption stimulates gall bladder contraction and an increase in distal colon motility is seen within 4 minutes of drinking coffee (Brown *et al.*, 1990) by acting on colonic opiate receptors and confirms the common belief that coffee consumption promotes desire to defecate. It stimulates secretion from stomach and small intestine but does not affect gastric emptying of a liquid meal or small bowel transit time (Boekema *et al.*, 2000). Coffee consumption may sometimes lead to pain in stomach and discomfort (Bolin *et al.*, 2000).

Gallstones are lumps composed mainly of cholesterol and can be caused by fatty diets. When gallstones are lodged in the duct leading from the gall bladder, they can trigger vomiting and cause pain in the abdomen and between the shoulder blades. The condition can also lead to inflammation of the gall bladder, the sac under the liver where bile is stored and concentrated. Drinking 2-3 cups of caffeinated coffee a day reduced the risk of developing gallstones by 40% (Leitzman *et al.*, 1999). Several mechanisms may be responsible for this result; the caffeine in coffee prevents cholesterol from crystallizing, cuts fat storage by increasing energy expenditure, reduces the fluid absorption that precedes gallstone formation and increases bile flow through the gall bladder. This study adds to a growing body of research indicating that coffee may have specific protective effects for the gastrointestinal area. Another study does not support a protective effect of coffee consumption on gall bladder disease, although coffee may decrease the risk of symptomatic gallstones in women (Ruhl & Everhart, 2000).

Coffee has a strong protective effect against cirrhosis of liver and drinking 3-4 cups of coffee a day is associated with an 80% reduction in risk of liver cirrhosis compared with those who do not drink coffee at all. Cirrhosis is a chronic disease, damaging the liver's tissue, making it unable to remove waste products from the blood. The formation of scar tissue leads to increased blood pressure in the veins between the intestine and spleen, cuasing bleeding and clotting and the possibility of severe, life threatening hemorrhage (Klatsky and Armstrong, 1992; Nakanishi *et al.*, 2000).

Effect on Cardiovascular System

Cardiovascular disease (CVD) or coronary heart disease (CHD) is associated with increased oxidative stress and dysfunction of the endothelium (cells lining the heart, blood, lymphatic vessels and various other cavities). Some antioxidants are known to reverse the endothelial dysfunction (Duffy *et al.*, 2001). Results are conflicting regarding coffee consumption and its association with blood pressure and coronary heart disease. Many studies reveal that coffee consumption increases blood pressure (Burke *et al.*, 1992; Ha *et al.*, 1999) and pulse rate (Smith *et al.*, 1999) which is attributed to one of its components, caffeine, that's why caffeine intake is avoided among hypertensive that are at greater risk of increased blood pressure (Nurminen *et al.*, 1999). Despite previous controversy on the subject, most researchers now conclude that regular coffee and caffeine use has little or no effect on blood pressure. First time caffeine use may produce immediate, minimal changes in blood pressure, which are transient (Rosmarin, 1989). No change in blood pressure appears to occur in regular users of caffeine. Even restricting caffeine does not reduce blood pressure in people with mild hypertension and moreover, caffeine does not contribute to hypertension (Wilson *et al.*, 1989; Bak and Grobbee, 1991). Caffeine dilates systematic blood vessels including coronaries and, thus, is beneficial in angina and hypertension.

Coffee intake strongly predicts coronary heart diseases and sometimes may lead to coronary death (Tverdal *et al.*, 1990) but results are still conflicting (Woodward & Tunstall-Pedoe, 1999). However, this effect is attributed to cholesterol raising effect of coffee (Thelle, 1991). Drinking coffee prepared by drip machines and percolators has been found to possess little or no effect on cholesterol while using boiled unfiltered coffee has been found to exert a worsening effect on serum lipids (Jee *et al.*, 2001). Thus, the differences in the brewing processes utilized in various countries explain the basis of conflicting observations. While researchers differed in their conclusions over time, an evidence in 1999 strongly indicates that consumption of coffee and caffeine does not contribute to CVD, neither caffeinated nor decaffeinated coffee was associated with the risk of myocardial infarction (death of heart muscle) or stroke (death of brain tissue) even for those drinking more than four cups of coffee a day (Sesso *et al.*, 1999).

Heart arrhythmias, the irregularities of the heartbeat, are not affected by coffee and caffeine consumption for the vast majority of individuals. Ingestion of 500 mg of caffeine (five to six 8 ounce cups of coffee) a day does not increase the frequency or severity of cardiac arrhythmias or ventricular tachycardia (increased heart rate) in healthy people or those with CVD (Meyers, 1991). Caffeine and theophylline increase force of cardiac contraction by elevating the tissue cAMP contraction (Wingard *et al.*, 1991).

Effect on Reproductive System

Caffeine crosses the placenta and achieves blood and tissue concentrations in the foetus that are similar to maternal concentrations. Precipitation of diabetes mellitus in children of heavy coffee consuming mothers has been demonstrated (Tuomilehto *et al.*, 1990). For this reason pregnant women should limit their intake of caffeine consumption to less than 300 mg/day that does not affect the health of the unborn child or mother. Caffeine is passed in the breast milk of lactating women. However some of the studies have failed to establish any direct link between coffee consumption during pregnancy and congenital defects in children (McDonald *et al.*, 1992). Spontaneous abortion is not directly associated with coffee consumption but the risk of spontaneous abortion is increased by a factor of 1.017 for each cup consumed per day, (Armstrong *et al.*, 1992) while another study has not shown any association between caffeine consumption, its metabolism and increased risk of spontaneous abortion during the first trimester of pregnancy (Fenster *et al.*, 1991; 1997). However, a threshold level of caffeine intake that presents a risk to pregnancy cannot be determined. Consumption of four cups of instant coffee/six cups of tea per day on an average is equivalent to 300 mg of caffeine that is considered to be within safe limits (Barone and Roberts, 1996).

The results of a study carried out to determine whether coffee consumption among men and women affect fertility, suggest decrease in fecundability among women who are coffee drinkers and this effect may be attributed to constituents of coffee other than caffeine (Curtis *et al.*, 1997). Caffeine is not considered to be among the risk factors associated with delayed conception or decreased fertility (Hatch and Bracken, 1993; Feld, 1994).

Effect on Respiratory System

The association between coffee consumption and asthma is contradictory. Some studies showed negative relation between asthma and coffee consumption whereas others have listed coffee bean handling as a cause of occupational asthma (Hunninghake, 1991) and this effect may be due to some pathogenic microorganisms, insects etc. inhabiting the surface of coffee beans. Caffeine improves airways function modestly in people with asthma for up to four hours after consumption. Theophylline has been used as a bronchial smooth muscle relaxant in the treatment of asthma and bronchitis. The broncho-dilator effect of caffeine is comparable to that of theophylline and the effect manifestation is quicker with caffeine.

Effect on Excretory System

Caffeine and other methylxanthines are mild diuretics and are helpful in treatment of oedema. The diuretic action of caffeine may be due to an increase in renal blood flow leading to an increased glomerular filtration rate (GFR), or due to a decreased reabsorption of sodium in the renal tubules. The diuretic effect of caffeine is dependent on the amount consumed and duration of intake. Urinary excretion of Na and K is elevated by 80±62 mmol and 14±12 mmol by consuming 6 cups of coffee containing 642 mg of caffeine per day (Neuhaeuser-Berthold *et al.*, 1997; Stookey, 1999; Grandjean *et al.*, 2000). Coffee consumption may be associated with lower levels of serum uric acid concentrations (Kiyohara *et al.*, 1999).

A kidney stone is a hard mass that forms when calcium oxalate or other chemicals in the urine form crystals that stick together. Both, caffeinated and decaffeinated coffee have been shown to help in the preventing the formation of kidney stones by increasing the flow of urine and decreasing its concentration (Curhan *et al.*, 1996; 1998). At the same time, positive results for decaffeinated coffee indicate that the beneficial effects of coffee may come from a factor other than caffeine.

Effect on Athletic Performance

Caffeine enhances contractile power of skeletal muscles thus athletic performance may be improved by its administration (Wellman, 1961). It has been found to be helpful in swimming, cycling and tennis. Habitual coffee drinkers perform better in a variety of cognitive performance tests compared with non-coffee drinkers (MacIntosh and Wright, 1995).

While it has long been thought that the use of caffeine during exercise might exacerbate dehydration, a recent study has demonstrated that caffeine in drinks taken before moderate exercise did not compromise hydration of the body (Wemple *et al.*, 1997). Ingestion of caffeine, an adenosine receptor antagonist, when taken before exercise has minimal effects on the metabolism and circulation in working muscles. In contrast, caffeine has been found to possess distinct effects on the sympathetic nervous system, vascular resistance, adipose tissue lipolysis; and lactate, potassium and glucose metabolism in non-exercising tissues (Graham *et al.*, 2000). A positive association between coffee drinking and its therapeutic effects has led to its use in treatment of rheumatoid arthritis patients (Targonska-Stepniak, 1999).

Coffee and Diabetes Mellitus

Ingestion of coffee or injection of caffeine is associated with elevated plasma catecholamines and free fatty acids as well as decreased insulin sensitivity, all of which might be expected to increase the incidence of diabetes. However, in recent studies coffee has been found to possess a graded inverse association with the risk of type 2 diabetes mellitus, though the reasons for this risk reduction associated with coffee remain unclear (Saremi *et al.*, 2003; Tuomilehto *et al.*, 2004).

Effect on Cancer

Coffee consumption is not associated with risk of stomach cancer, rather substantial coffee consumption is associated with a lower risk of colorectal cancer in the general population (Ekbom, 1999; Wong-Ho *et al.*, 1999). Four or more cups of coffee per day reduce the risk of colorectal cancer by 24% than among those who rarely or never drink coffee. The most likely explanation for lower risk of colorectal cancer among heavy coffee consumers is the enhanced colonic activity induced by coffee. Approximately 50-90% of women experience symptoms of fibrocystic breast disease (FBD), which include cyclic breast pain and tenderness (Love *et al.*, 1982). No association has been found between caffeine intake and tumours, FBD and breast cancer (Graham *et al.*, 1992; Folsom *et al.*, 1993).

ANTIMICROBIAL ACTIVITY OF COFFEE

As evident from the ongoing discussion the use of coffee has been well documented in literature for its effects on various biological functions. *In vitro* antibacterial activity of coffee extract for the first time was demonstrated against various bacteria causing diarrhoeal diseases. Gram-positive bacteria were found to be more sensitive than Gram-negative bacteria (Toda *et al.*, 1989). In another study carried out to see the inhibitory effect of coffee extract on the growth of various other bacteria (causing diarrhoeal diseases) *viz.*, *Staphylococcus aureus*, *Vibrio cholera* and *Salmonella typhi* were sensitive whereas *Escherichia coli*, *Shigella flexneri* and *Salmonella paratyphi* 'B' were resistant (Shetty *et al.*, 1994). Coffee extract has also been shown to inhibit the haemolysin activity of bacterial toxins i.e. *Staphylococcus aureus* alpha toxin and *Vibrio parahaemolyticus*, thermostable direct haemolysin (V_pTDH) when used at a concentration of 25% (w/v) (Okubo *et al.*, 1989). Results obtained after examining the effect of coffee extract (*Coffea robusta*) on the growth of Gram-positive bacteria (*Staphylococcus aureus*, *Bacillus cereus*, *Lactobacillus bulgaricus*, *Streptococcus lactis*, *Streptococcus faecalis*) and Gram-negative bacteria (*Escherichia coli*, *Salmonella typhi*, *Pseudomonas aeruginosa*) suggest that coffee extract inhibited the growth of all Gram-positive and Gram-negative bacteria tested; relative growth decreased at higher concentration of coffee extract and *L. bulgaricus* was the most sensitive organism. Coffee extract has also been shown to possess anti-adhesive property, thereby reducing the harmful effects of *Streptococcus mutans*, a causative agent of caries in humans (Daglia *et al.*, 2002). Based on these results coffee extract is used as a preservative in food or beverage products. Addition of coffee extract to pasteurized milk (incubated at 26-28°C for 72 hours), at a concentration of 2,4 and 6% reduced relative growth of bacteria in milk (Srikandi-Fardiaz, 1995). *Helicobacter pylori*, a major cause of type B gastritis, is strongly associated with duodenal and gastric ulcer disease and a positive dose-response relation has been found between coffee consumption and active *H. pylori* infection (Brenner *et al.*, 1997).

In vitro studies on some natural beverages (black tea, green tea, coffee and cocoa) as botanical pesticides against two important phytopathogenic bacteria—*Erwinia amylovora* (cause of fire blight in apple, pear, quince etc.) and *Curtobacterium flaccumfacensis* subsp. *poinsettiae* (cause of stem canker and leaf spot in poinsettia—*Euphobia pulcherrima*) revealed a negative response of coffee at concentrations 0.875, 1.75 and 3.5 %; however their higher concentration may yield positive results (Arora & Bhardwaj, 1997; Nas, 2004).

Coffee (*Coffea arabica*) has also been found to possess antifungal activity against *Aspergillus niger, A. oryzae, A. fumigatus, Rhizopus oligosporus, R. oryzae, Mucor pumilus* and anti-yeast activity against *Saccharomyces cerevisiae* and *Candida utilis* (Srikandi-Fardiaz, 1995). Two samples of commercial instant coffee, Bru and Nestle Select, when tested on ten wood rotting fungi (*Chaetomium globosum, Daedalea flavida, Gleophyllum trabeum, Sporotrichum pulverulentum, Daldinia concentrica, Phanerochaete chrysosporium, Phlebia radiata, Irpex flavus, Polyporus palustris* and *P. sanguineus*) inhibited their growth to variable level and the Nestle Select was found to be better than Bru Instant coffee. *D. flavida, S. pulverulentum, P. chrysosporium, I. flavus* and *P. radiata* suffered the maximum growth inhibition whereas the growth of *G. trabeum* and *C. globosum* was inhibited upto 50%; rest of the fungi were moderately inhibited by coffee samples (Arora and Ohlan, 1997).

Water-soluble polyphenols, tannins, constitute 5% in coffee and are responsible for its astringent activity. Tannins inhibit many fungi, yeast, bacteria and viruses and increasing attention is being paid to the use of tannins as antimicrobial agent eg. in wood preservation and prevention of dental caries. Tannins may deter microorganisms, either because of their effective inhibitory action or by protecting essential tissues such as wood against decay. Antimicrobial properties of tannins have been reported for bacteria (*Streptococcus mutans, Streptococcus sobrinus, Staphylococcus* spp., *Pseudomonas fluorescens and Escherichia coli*); fungi (*Fomes annosus, Crinipellis perniciosa, Aspergills niger, Botrytis cinerea, Chaetomium cupreum, Coriolus versicolor, Gloeophyllum trabeum, Penicillium* spp. and *Trichoderma viridae*) and yeast (*Saccharomyces cerevisiae*) (Scalbert, 1991). Bactericidal catechins which are nothing but tannins primarily act on and damage bacterial membranes (Ikigai *et al.*, 1993). These tannins in pure form or as gallotannic acid denature or precipitate proteins thereby altering their surface structures. The higher resistance of Gram-negative bacteria to bactericidal catechins can be explained to some extent by the presence of negatively charged lipopolysaccharides. Tannins are also found to exhibit some antiviral properties (Shetty *et al.*, 1994).

Degree of roasting besides affecting aroma and flavour of coffee, also influences its antibacterial activity. The two chemical markers, good indicators of degree of roasting and hence bacterial activity. The two chemical markers, good indicators of degree of roasting and hence antibacterial activity, identified by using HPLC method were: the sum of trigonelline and nicotinic acid contents and the 5-caffeoylquinic-acid/caffeine ratio. An antibacterial component isolated from roasted coffee possesses antibacterial activity against *Staphylococcus aureus* ATCC 25923 and *Streptococcus mutans* 9102 (Daglia *et al.*, 1994; 1998).

Over the past 20 years, more than 150 peptide antibiotics from non-microbial sources have been discovered. These antimicrobial peptides derived from the higher forms of organisms (plants, insects, and animals) are usually cationic, amphipathic and in open chain forms (secondary structure). Four recently discovered macrocyclic peptides with similar sequences and sizes from coffee plants of the Rubiaceae family can be included as antimicrobial peptides. The four macrocycles (Kalata, Circulin A (Cir A), Circulin B (Cir B) and Cyclopsychotride) tested against nine strains of microbes (Gram-negative bacteria-*Escherichia coli* ATCC 25922, *Pseudomonas aeruginosa* ATCC 27853, *Klebsiella oxytoca* ATCC 49131 and *Proteus vulgaris* ATCC 49132; Gram-positive bacteria- *Staphylococcus aureus* 29213 and *Micrococcus luteus* ATCC 49732; fungi- *Candida albicans* ATCC 37092 *Candida kefyr* ATCC 37095 and *Candida tropicalis* ATCC 37097) have shown significant activity against Gram-positive and negative bacteria, moderate activity against two fungi strains and no activity against *C. albicans*. Kalata and CirA are particularly potent against *S. aureus* whereas CirB and Cpt are active against both Gram-positive and Gram-negative bacteria. Despite their cyclic structures and cystine-knot disulphide motif, they are amenable to chemical synthesis through the orthogonal cyclization method. Furthermore, chemical synthesis provides a useful approach to confirm their structures. These macrocyclic peptides are different from the known, open-chained plant Defensins that are larger in size. Thus they represent a class of structural motifs for antimicrobial peptides with rigid, compact and macrocyclic structures. These structural features may be useful in understanding the antimicrobial mechanisms of peptide antibiotics and may lead to design of peptide mimetics (Tam *et al.*, 1999).

Caffeine, a purine alkaloid, is first degraded to paraxanthine (1,7-dimethylxanthine), theobromine or theophylline in humans and other animals. Since the first evidence of caffeine's action on the central nervous system, an increasing number of studies have been carried out relating caffeine with several physiological effects. Anti-mutagenic components in coffee including caffeine inhibit the cancer causing effect of various microorganisms (Giovannucci, 1998). Although high concentrations are required for bactericidal action, yet caffeine is regarded as toxic for bacteria. However, some microorganisms have the ability to grow in the presence of caffeine and survival would be related to their capacity to degrade the alkaloid. Although fungi growing on caffeine have also been isolated but most of the studies have been done with bacteria isolated from soil, mainly those belonging to the Pseudomonads group, with particular attention to *Pseudomonas putida*. Bacterial strains used for caffeine degradation studies have been obtained through enrichment—soil as well as culture medium enrichment to induce the appearance of mutants (Sarkar & Chakraborty, 1991). *P. putida* strain was isolated by enriching the medium with 2-5% of caffeine as sole source of carbon and nitrogen (Middlehoven & Bakker, 1982; Gluck & Lingens 1987). However, using a different approach by collecting soil samples taken from under the coffee plants, *S. marcescens* strain; several strains of *P. putida* and other bacteria were isolated and a strain of *P. putida* showed an impressive ability to grow in high concentrations of caffeine—25 g/l in liquid medium and at 50 g/l in solid medium

(Mazaffera *et al.*, 1994; Yamaoka-Yano & Mazzafera, 1998). The direct isolation of mutants from the soil without any enrichment is a strong indication that because of competition for organic nutrients, bacteria growing in soil under coffee plants have developed mechanisms to degrade the caffeine released by the plants (leaves, fruits and litter) or in other words, there was natural enrichment.

At present the output of coffee amounts to about several million tons and almost the same amount of extraction residue of coffee is formed which is disposed of as waste. Results obtained from *in vitro* and green house trials showed that coffee extract of waste possesses antibacterial activity against 3 pathogens (*Pseudomonas syringae* pv. *pisi* race 1 and 2) and *P. syringae* pv. *phaseolicoli* (race 1) causing plant disease but the effects varied depending upon the strain and the test method (Alstorm, 1992). Apparently no study has been carried out on antibacterial substances contained in coffee residue except one 1, 3, 7-trimethylxanthine or caffeine. An antimicrobial substance 3′-4′-Dihydroxyacetophenone (3′4′-DHAP) is contained in coffee residue and can be used as a preservative agent not only in foods but also in cosmetics and drugs (Nishina *et al.*, 1994).

Coffee husk and pulp are the by-products obtained during coffee processing. Coffee husk is obtained when harvested coffee is processed by the dry method and coffee pulp is produced by the wet method. It is estimated that millions of tons of coffee pulp/husk produced every year is disposed of and inappropriate disposal of this product, in rivers for instance, may cause serious environmental pollution (Porres *et al.*, 1993). Coffee husk or pulp can be used for feeding farm animals but its use is limited, mainly because of its not being easily digestible by cattle due to high content of caffeine and tannins in it. Tannins, known to confer astringency to foodstuffs and complex proteins, affect food digestibility; caffeine presents a somewhat bitter taste, thereby makes the food less palatable and interferes with nutrient availability and absorption in the gastrointestinal tract (Clifford & Ramirezmenenzes, 1991; Aregheore 1998). It further demands that caffeine concentration that affects palatability of coffee husk/pulp must be reduced. Many microorganisms have been reported to degrade caffeine and this potential must be utilized to decaffeinate coffee husk and pulp, so as to increase their digestibility and commercial value.

Use of Microorganisms to Decaffeinate Coffee husk and Pulp

Coffee fruits are a rich source of microorganisms and 248 fungal cultures have been isolated from coffee plants and soil collected from plantation areas (Boccas *et al.*, 1994). Approximately 272 strains of filamentous fungi from soil, fruits and leaves using a culture media containing coffee extract, coffee extract plus sucrose, and coffee pulp extract have been isolated, whereby *Aspergillus* and *Penicillium* possessed the highest ability to degrade caffeine (Roussos *et al.*, 1995). More than 20 bacterial strains collected from soil under the coffee plants showed predominance of *Pseudomonas* spp., which were also the most efficient caffeine degraders (Yamaoka-Yano & Mazzafera, 1998). A total of 754 isolates of bacteria, yeast and fungi were obtained during the maturation and natural processing (sun-dried) of coffee fruits where bacteria (fermentative, cellulolytic and pectinolytic) were the most predominant microorganisms (Silva *et al.*, 2000). This abundance and diversity of microbial species certainly explain why Porres *et al.* (1993) ensiled fresh coffee pulp with or without addition of sugarcane molasses and observed reduction of caffeine from 13-63% and also indicated the presence of

80% in coffee husk, thus improving its nutritional value (Yamaoka-Yano and Mazzafera 1998). Despite this promising potential, little work has been done with microorganisms specially or specifically selected to degrade caffeine. In regard to tannins, which are also responsible for the low palatability of coffee husk/pulp, information is even scarcer. *Lactobacillus plantarum* has been found to degrade tannin upto 90% in coffee pulp. Therefore, for each animal type or category, the use of coffee pulp needs to be further investigated. Adaptation of micro-scale laboratory experiments to larger scale could show the viability of this alternative.

Despite all these studies, the fact remains that the consumption of caffeine at an average intake of 300mg/day has no adverse effects in the vast majority of the adult population and is considered well within the safe levels. Caffeine does not accumulate in the body over a course of time and is normally excreted within several hours of consumption. The rate of caffeine elimination varies between individuals and this may be as a result of genetic factors affecting the enzyme involved in the metabolism or due to certain lifestyle factors eg. smoking, alcohol consumption etc. A number of different factors affect individual tolerance to caffeine including the amount digested, frequency of caffeine consumption and individual metabolism. It is widely recognized that gradual tolerance develops with prolonged caffeine use. Although the studies mainly focused on caffeine intake from coffee but a possible influence of other constituents of coffee cannot be excluded.

In the fight against diseases, wherein microorganisms are trying to establish their superiority by developing resistance to commonly used antimicrobial agents, there is a need to continuously look for newer antimicrobials to combat the problem effectively. No doubt, tea, the commonly consumed beverage, has been in use as a household remedy for various ailments which is further supported by number of reports regarding its biological as well as antimicrobial effects (Toda *et al.*, 1989; Arora, 1998; Arora and Mahajan, 2000; Mckay and Blumberg, 2002). Similarly, coffee another non-alcoholic beverage has been shown to possess many properties that affects almost every system of the body. There is a need to optimize biological activities associated with coffee and further enhance its antimicrobial potential.

REFERENCES

Alstorm, S. (1992) *J. Phytopathol.* **136**: 329-334

Aregheore, E.M. (1998) *Vet. Human Toxicol.* **40**: 35-39.

Armstrong, B.G., McDonald, A.D. & Sloan, M. (1992) *Am. J. Pub. Health.* **82**: 85-87.

Arora, D.S. (1998). *Antibiot. Chemother.* **2**: 4-5.

Arora, D.S. & Bhardwaj, S.K. (1997) *Geobios.* **24**: 127-131.

Arora, D.S. & Mahajan, V. (2000) *Microbes: Agriculture, Industry and Environment.* (Eds. Maheshwari, D.K., Dubey, R.C., Prasad, G. and Navneet), Bishen Singh Mahendra Pal Singh Publishers, Dehradun, India, pp. 159-163

Arora, D.S. & Ohlan, D. (1997) *J. Basic Microbiol.* **37**: 159-165.

Ashihara, H. & Crozier, A. (1999) *J. Agri. Food Chem.* **47:** 3425-3431.

Austin, D.J., Kristinsson, K.G. & Anderson, R.M. (1999). *Proc. Natl. Acad. Sci.* USA. **96:** 1152-1156.

Bak, A.A.A. & Grobbee, D. (1991) *Am. J. Clin. Nutr.* **53:** 971-974.

Barone, J.J. & Roberts, H.R. (1996) *Food Chem. Toxic.* **34:** 119-129.

Baruah, P., Sharma, R.K., Singh, R.S. & Ghosh, A.C. (1996) *J. Essential. Oil Res.* **8:** 411-412.

Benedetti, M.D., Bower, J.H., Maraganore, D.M., McDonnell, S.K., Peterson, B.J., Ahlskog, J.E., Schaid, D.J. & Rocca, W.A. (2000) *Neurol.* **55:** 1350-1358.

Blank, I., Sen, A. & Grosch, W. (1992) .*Z Lebensem-Unters-Forsch.* **195:** 239-245.

Boccas, F., Roussos, S., Gutierrez, M., Serrano, L. & Viniegra, G.G. (1994) *J. Food Sci. Technol.* **31:** 22-26.

Boekema, P.J., Lo, B., Samsom, M., Akkermans, L.M.A. & Smout, A.J.P.M. (2000) *Eur. J. Clin. Investigation.* **32:** 129-134.

Bolin, T.D., Korman, M.G., Hansky, J. & Stanton, R. (2000) *J. Gastroenterol Hepatol.* **15:** 35-39.

Brenner, H., Rothenbacher, D., Bode, G. & Adler, G. (1997) *BMJ.* **315:** 1489-1492.

Brown, S.R., Cann, P.A. & Reed, N.W. (1990) *Gut.* **31:** 450-453.

Burke, B., Belini, L.J., German, R., Grosskopf, S., Ritchie, J., Pudden, I.B. & Rogers, P. (1992) *J. Clin. Epidemiol.* **45:** 1061-1070.

Chen, J.F., Xu, K., Petzer, J.P., Staal, R., Xu, Y.H., Beilstein, M., Sonsalla, P.K., Castagnoli, K., Castagnoli Jr., N & Schwarzschild, M.A. (2001) *J. Neurosci.* **21:** RC **143:** 1-6.

Chopra, R.N., Chopra, I.C., Handa, K.L. & Kapur, L.D. (1958) *Indigenous Drugs of India*. 2nd edition. U.N. Dhur & Sons Pvt. Ltd., Kolkata, pp. 79-84.

Clifford, M.N. & Ramirezmenenzes, J.R. (1991) *Food Chem.* **40:** 35-42.

Cowan, M.M. (1999). *Clin. Microbiol. Rev.* **12:** 564-582.

Curhan, G.C., Willet, W.C., Rimm, E.B., Spiegelman, D. & Stampfer, M.J. (1996) *Am. J. Epidemiol.* **143:** 240-247.

Curhan, G.C., Willet, W.C., Speizer, F.E. & Stampfer, M.J. (1998) *Ann. Intern. Med.* **128:** 534-540.

Curtis, K.M., Savitz, D.A. & Arbuckel, T.E. (1997) *Am. J. Epidemiol.* **146:** 32-41.

Daglia, M., Cuzzoni, M.T. & Dacarro, C. (1994) *J. Agri. Food Chem.* **42:** 2273-2277.

Daglia, M., Papetti, A., Dacarro, C. & Gazzani, G. (1998). *J. Pharm. Biomed. Anal.* **18**: 219-225.

Daglia, M., Tarsi, R., Papetti, A., Grisoli, P., Dacarro, C., Pruzzo, C. & Gazzani G. (2002) *J. Agri. Food Chem.* **50**: 1225-1229.

Dubey, R.C. (1991) *Indian Phytopathol.* **44**: 241-244.

Duffy, S.J., Keaney Jr. J.F., Holbrook, M., Gokce, N., Swerdloff, P.L., Frei, B. & Vita, J.A. (2001) *Circulation.* **104**: 151-156.

Ekbom, A. (1999) *Gut.* **44**: 597.

Fall, P.A., Fredrikson, M., Axelson, O. & Granerus, A.K. (1999) *Mov. Disorders.* **14**: 28-37.

Feld, B.K. (1994) *Environ. Nutr.* **17**: 1-6.

Fenster, L., Eskanazi, B., Windham, G.C. & Swan, S.H. (1991) *Epidemiol.* **2**: 168-174.

Fenster, L., Hubbard, A.E., Swan, S.H., Windham, G.C., Waller, K., Hiatt, R.A. & Benowitz, N. (1997) *Epidemiol.* **8**: 515-523.

Folsom, A.R., McKenzie, D.R., Bisgard, K.M., Kushi, L.H. & Sellers, T.A. (1993) *Am. J. Epidemiol.* **138**: 380-383.

Giovannucci, E. (1998). *Am J Epidemiol.* **147**: 1043-1052.

Gluck, M. & Lingens, F. (1987). *Appl. Microbiol. Biotechnol.* **25**: 3343-3340.

Graham, T.E., Helge, J.W., MacLean, D.A., Kiens, B. & Richter, E.A. (2000) *J. Physiol.* **529**: 837-847.

Graham, S., Zielezny, M., Marshall, J., Priore, R., Freudenheim, J., Brasure, J., Haughey, B., Nasca, P. & Zbed, M. (1992). *Am. J. Epidemiol.* **136**: 1327-1337.

Grandjean, A.C., Reimers, K.J., Bannick, K.E. & Haven, M.C. (2000) *J. Am. Coll. Nutr.* **19**: 591-600.

Ha, J.S., He, J., Whelton, P.K., Suh, I.I. & Klag, M.J. (1999) *Hypertension.* **33**: 647-652.

Hatch, E. & Bracken, M. (1993) *Am. J. Epimediol.* **138**: 1082-1091.

Heinrich, M. (2000) *Phytother Res.* **14:** 479-488.

Heisey, R.M. & Gorham, B.K. (1992) *Lett. Appl. Microbiol.* **14:**136-139.

Hellenbrand, W., Boeing, H., Robra, B.P., Seilder, A., Vieregge, P., Nischan, P., Joerg, J., Oertel, W.H., Schneider, E. & Ulm, G. (1996) *Neurol.* **47:** 644-650.

Holscher, W., Vitzthum, O.G. & Steinhart, H. (1992) *J. Agri. Food Chem.* **40:** 655-658.

Hunninghake, G.W. (1991) In: *Harrison's principles of Internal Medicine* (Eds. Wilson, J.D., Braunwald, E., Isselbacher, K.J., Petersdorf, R.G., Martin, J.B., Fauci, A.S. & Root, R.K.) 12th Ed. McGraw Hill Inc, New York, pp. 1053-1055.

Ikigai, H., Nakae, T., Hara, Y. & Shimamura, T. (1993) *Biochem. Biophy. Acta.* **1147:** 132-136.

Izzo, A.A., diCarlo, G., deFusco, R., Borelli, F., Capasso F., Fasulo M.P. & Autore, G. (1995) *Phytother Res.* **9:** 281-286.

Jee, S.H., He, J., Appel, L.J., Whelton, P.K., Suh, I.I. & Klag, M.J. (2001) *Am. J. Epidemiol.* **153:** 353-362.

Jimenez-Jimenez, F.J., Mateo, D. & Gimenez-Roldan, S. (1992) *Mov. Disorders.* **7:** 339-344.

Katz, S.M. (1984) *Coffee and Health* (Eds. MacMohan, B. and Sugimora, T.), Banbury Reports Series, Cold Spring, Harber, **17:** 11-20

Kiyohara, C., Kona, S., Honjo, S., Todoroki, I., Sakurai, Y., Nishiwaki, M., Hamada, H., Nishikawa, H., Koga, H., Ogawa, S. & Nakagawa, K. (1999) *British J. Nutr.* **82:** 125-130.

Klatsky, A.L. & & Armstrong, M.A. (1992) *Am. J. Epimediol.* **136:** 1248-1257.

Klink, B. (1997) *Drug Top.* **141:** 99-100.

Kunnin, C.M. (1993) *Ann. Intern. Med.* **118:** 557-561.

Lee, T., Kempthorne, A.R. & Hardy, J.K. (1992) *J. Food Sci.* **57:** 1417-1419.

Leitzman, M.F., Willet, W.C., Rimm, E.B., Stampfer, M.J., Spiegelman, D., Colditz, G.A. & Giovannucci, E. (1999) *JAMA.* **281:** 2106-2112.

Love, S.M., Gelman, R.S. & Silen, W. (1982) N. Engl. J. Med. **307:** 1010-1014.

MacIntosh, B.R. & Wright, B.M. (1995) Can. J Appl. Physiol. **20:** 168-177.

Matthews, P.D. & Haas G.J. (1993) *J. Food Protect.* **56:** 66-68.

Mazzafera, P., Crozier, A. & Magalhaes, C. (1991) *Phytochem.* **30:** 3913-3916.

Mazzafera, P., Olsson, O. & Sandberg, G. (1994) *Microb. Ecol.* **31:** 199-207.

McDonald, A.D., Armstrong, B.G. & Sloan, M. (1992) *Am. J. Public Health.* **82:** 91-93.

McKay, D.L. & Blumberg, J.B. (2002) *J. Am. Coll. Nutr.* **21:** 1-13.

Meena, M.R. & Sethi, V. (1994) *J. Food Sci. Technol.* **31:** 68-70.

Menon, B. (1991) *Femina.* April 23-May **7:** 23-24.

Meyers, M.G. (1991) *Ann. Intern. Med.* **114:** 147-150.

Middelhoven, W.J. & Bakker, C.M. (1982) *European J. Appl. Microbiol. Biotechnol.* **15:** 214-217.

Muanza, D.N., Kim, B.W., Euler, K.L. & Williams, L. (1994) *Int. J. Pharmacogonosy.* **32:** 337-345.

Nakanishi, N., Nakamura, K., Suzuki, K. & Tatara, K. (2000) *Individual Health.* **38:** 99-102.

Nas, M.N. (2004) *Turk J. Agri. For.* **28:** 57-61.

Neuhaeuser-Berthold, M., Beine, S., Verwied, S.C. & Luehrmann, P.M. (1997) *Ann. Nutr. Metabol.* **41:** 29-36.

Nishina, A., Kajishima, F., Matsunga, M., Tezuka, H., Inatomi, H. & Osawa, T. (1994) *Biosci. Biotech. Biochem.* **58:** 293-296.

Nurminen, M.L., Nittymen, L., Korpela, R. & Vappatalo, H. (1999) *Eur. J. Clin. Nutr.* **53:** 831-839.

Okubo, S., Ikigai, H., Toda, M. & Shimamura, T. (1989) *Lett. Appl. Microbiol.* **9:** 65-66.

Porres, C., Alvarez, D. & Calzada, J. (1993) *Biotehnol. Adv.* **11:** 519-523.

Rosmarin, P.C. (1989) *Progress Cardiovascular Dis.* **32:** 239-245.

Roussos, S., Aquiahuatl, M.A., Trejo-Hernandez, M.R., Perraud, I.G., Favela, E., Ramakrishna, M., Rambault, M. & Viniegra-Gonzalez, G. (1995) *Appl. Microbiol. Biotechnol.* **42:** 756-762.

Rozas, A., Hernandez, L., Pereda-Miranda, R. & Mata, R. (1992). *J. Ethnopharmacol.* **35:** 275.

Ruhl, C.E. & Everhart, J.E. (2000) *Am. .J Epidemiol.* **152:** 1034-1038.

Saremi, A., Reid, M.T. & Knowler, W.C. (2003) *Diabetes Care.* **26:** 2211-2212.

Sarkar, P.K. & Chakraborty, R. (1991) *Indian J. Expt. Biol.* **29:** 679-680.

Scalbert, A. (1991) *Phytochem.* **30:** 3875-3883.

Serrazin, C., LeQuere, J.L., Gretsch, C. & Liardon, R. (2000) *Food Chem.* **70:** 99-106.

Sesso, H.D., Gaziano, J.M., Buring, J.E. & Hennekens, C.H. (1999) *Am. J. Epidemiol.* **149:** 162-167.

Shetty, M., Subbannayya, K. & Shivananda, P.G. (1994) *J. Communicable Dis.* **26:** 147-150.

Silva, C.F., Scwan, R.F., Dias, E.S. & Wheals, A.E. (2000) *Int. J. Food Microbiol.* **60:** 251-260.

Singh,U., Wadhwani, A.M. & Johri, B.M. (1983) *Dictionary of Economic Plants in India.* 2nd Ed. ICAR, New Delhi, pp. 39- 55.

Smith, A.P., Clark, R. & Gallagher, J. (1999) *Physiol. Behaviour.* **67**: 9-17.

Srikandi-Fardiaz. (1995) *ASEAN Food J.* **10:** 103-106.

Stookey, J.D. (1999) *Eur. J. Epidemiol.* **15:** 181-188.

Strobel, R.G.K. (1984). *Coffee and Health* (Eds. MacMohan, B. & Sugimora, T.). Banbury Reports Series, Cold Spring Harber, **17**: 21-40

Tam, J.P., Lu, Y.A., Yang, J.L. & Chiu, K.W. (1999) *Biochem.* **96:** 8913-8918.

Targonska-Stepniak, B. (1999) *Rheumatologia.* **37:** 40-52.

Thelle, D.S. (1991) *BMJ.* **302:** 804.

Toda, M., Okubo, S., Hiyoshi R. & Shimamura, T. (1989) *Lett. Appl. Microbiol.* **8:** 123-125.

Tripathi, K.D. (1991) *Essentials of Medical Pharmacology.* 2nd Ed. Jaypee Brothers, pp. 443-446.

Tuomilehto, J., Hu, G., Bidel, S., Lindstorm, J. & Jousilathi, P. (2004) *JAMA* **291:** 1213-1219.

Tuomilehto, J., Tuomilehto-Wolf, E., Virtala, E. & Laporte, R. (1990) *BMJ.* **300:** 642.

Tverdal, A., Stensvold, I., Solvoll, K., Fross, O.P., Lund-Larsen, P. & Bjartveit, K. (1990) *BMJ.* **300:** 566-569.

Wellman, F.L. (1961) *Coffee* (Ed. Polunin, N.), Interscience Publishers Inc. New York.

Wemple, R.D., Lamb, D.R. & McKeever, K.H. (1997) *Int. J. Sports Med.* **18:** 40-46.

Wilson, P.W., Garrison, R.J., Kannel, W.B., McGee, D.L. & Castelli, W.P. (1989) *Arch. Intern. Med.* **149:** 1169-1172.

Wingard, L.B., Brody, T.M., Larner, J. & Schwartz, A. (1991) *Human Pharmacology: Molecular-to-Clinical* (Ed. Kimkit), Wolf Publishing Ltd., London, pp. 203, 459.

Wong-Ho, C., Swanson, C.A., Lissowska, J., Groves, F.D., Sobin, L.H., Nasierowska-Guttmejer, A., Radziszewski, J., Regula, J., Hsing, A.W., Jagannatha, S., Zatonski, W. & Blot, W.J. (1999) *Int. J. Cancer.* **81:** 871-876.

Woodward, M. & Tunstall-Pedoe, H. (1999) *J. Epidemiol. Community Health.* **53:** 481-487.

Yamaoka-Yano, D.M. & Mazzafera, P. (1998) *Allelopathy J.* **5:** 23-24.

Index